STONE AGE INSTITUTE PUBLICATION SERIES
NUMBER 1

THE OLDOWAN:
Case Studies Into the Earliest Stone Age

Edited by Nicholas Toth and Kathy Schick

Stone Age Institute Press · www.stoneageinstitute.org
1392 W. Dittemore Road · Gosport, IN 47433

COVER PHOTOS

Front, clockwise from upper left:

1) Excavation at Ain Hanech, Algeria (courtesy of Mohamed Sahnouni).

2) Kanzi, a bonobo ('pygmy chimpanzee') flakes a chopper-core by hard-hammer percussion (courtesy Great Ape Trust).

3) Experimental Oldowan flaking (Kathy Schick and Nicholas Toth).

4) Scanning electron micrograph of prehistoric cut-marks from a stone tool on a mammal limb shaft fragment (Kathy Schick and Nicholas Toth).

5) Kinesiological data from Oldowan flaking (courtesy of Jesus Dapena).

6) Positron emission tomography of brain activity during Oldowan flaking (courtesy of Dietrich Stout).

7) Experimental processing of elephant carcass with Oldowan flakes (the animal died of natural causes). (Kathy Schick and Nicholas Toth).

8) Reconstructed cranium of Australopithecus garhi. (A. garhi, BOU-VP-12/130, Bouri, cranial parts, cranium reconstruction; original housed in National Museum of Ethiopia, Addis Ababa. ©1999 David L. Brill).

9) A 2.6 million-year-old trachyte bifacial chopper from site EG 10, Gona, Ethiopia (courtesy of Sileshi Semaw).

Back:
Photographs of the Stone Age Institute. Aerial photograph courtesy of Bill Oliver.

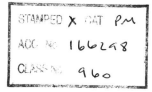
Published by the Stone Age Institute.
ISBN-10: 0-9792-2760-7
ISBN-13: 978-0-9792-2760-8
Copyright © 2006, Stone Age Institute Press.

To our Berkeley professors

J. Desmond Clark
Richard L. Hay
F. Clark Howell
Glynn Ll. Isaac

"Example is the school of mankind, and they will learn at no other."

-- Edmund Burke, 1796

CONTRIBUTORS

William Anderst is in the department of Orthopedic Surgery at the University of Pittsburg.

Jesus Dapena is a professor of Kinesiology at Indiana University, Bloomington.

Fernando Diez-Martín is in the department of Prehistory at the University of the Basque Country in Spain.

Manuel Domínguez-Rodrigo is a professor of Prehistory at the Complutense University in Madrid, Spain. He is director of the Peninj palaeoanthropological project in Tanzania.

William Fields is an anthropologist and scientist at the Great Ape Trust in Des Moines, Iowa.

Kevin Hunt is a professor of Anthropology at Indiana University, Bloomington. He is director of the Toro-Semliki chimpanzee research project in Uganda.

Travis Pickering is a professor of Anthropology at the University of Wisconsin, Madison. He is director of the Swartkrans palaeoanthropological research project in South Africa.

Mohamed Sahnouni is a senior research scientist at the Stone Age Institute in Bloomington, Indiana. He is director of the Ain Hanech palaeoanthropological research project in Algeria.

Sue Savage-Rumbaugh is a cognitive psychologist and a lead scientist at the Great Ape Trust in Des Moines, Iowa.

Kathy Schick is a professor of Anthropology and of Cognitive Science and co-director of the Center for Research into the Anthropological Foundations of Technology (CRAFT) at Indiana University, Bloomington. She also is co-director the Stone Age Institute in Bloomington, Indiana.

Sileshi Semaw is a senior research scientist at the Stone Age Institute in Bloomington, Indiana. He is director of the Gona palaeoanthropological research project in Ethiopia.

Dietrich Stout is a lecturer at the Institute of Archaeology, University College, London.

Nicholas Toth is a professor of Anthropology and of Cognitive Science, and co-director of the Center for research into the Anthropological Foundations of Technology (CRAFT) at Indiana University. He also is co-director of the Stone Age Institute in Bloomington, Indiana.

ACKNOWLEDGEMENTS

We would like to first of all thank the following people for their support and encouragement for this publication series: Alan Almquist, Jean and Ray Auel, Margaret Barrier, Alice Corning, Henry Corning, Marty Fuller, Gordon and Ann Getty, Jon and Bonnie Henricks, Tony Hess, Clark Howell, Fritz Maytag, Ken Maytag, Carole Travis-Henikoff, and Kay Woods. Secondly, we would also like to thank the following individuals for their input into this volume: David Brill, Beth Montgomery, Amy Sutkowski, David Stewart, and Tim White. Finally, we would like to thank all of the contributors to this book for their diligence and perseverance.

THE OLDOWAN:
CASE STUDIES INTO THE EARLIEST STONE AGE
EDITED BY
NICHOLAS TOTH AND KATHY SCHICK
STONE AGE INSTITUTE PRESS, 2006

CONTENTS

Dedication iii
Contributors v
Table of Contents vii
Preface ix

PART I: THE ARCHAEOLOGICAL EVIDENCE 1

Chapter 1: An Overview of the Oldowan Industrial Complex: The sites and the nature of their evidence
Kathy Schick and Nicholas Toth 3

Chapter 2: The Oldest Stone Artifacts from Gona (2.6-2.5 Ma), Afar, Ethiopia: Implications for understanding the earliest stages of stone knapping
Sileshi Semaw 43

Chapter 3: The North African Early Stone Age and the Sites at Ain Hanech, Algeria
Mohamed Sahnouni 77

Chapter 4: The Acquisition and Use of Large Mammal Carcasses by Oldowan Hominins in Eastern and Southern Africa: A Selected Review and Assessment
Travis Rayne Pickering and Manuel Domínguez-Rodrigo 113

Chapter 5: After the African Oldowan: The Earliest Technologies of Europe
Fernando Diez-Martín 129

PART II: ACTUALISTIC APPROACHES TO THE EARLIEST STONE AGE 153

Chapter 6: A Comparative Study of the Stone Tool-making Skills *Pan, Australopithecus*, and *Homo sapiens*
Nicholas Toth, Kathy Schick, and Sileshi Semaw 155

Chapter 7: Rules and Tools: Beyond Anthropomorphism
Sue Savage-Rumbaugh and William Mintz Fields 223

Chapter 8: Sex Differences in Chimpanzee Foraging Behavior and Tool Use: Implications for the Oldowan
Kevin Hunt...243

Chapter 9: Oldowan Toolmaking and Hominin Brain Evolution: Theory and research using positron emission tomography (PET)
Dietrich Stout 267

Chapter 10: Knapping Skill of the Earliest Stone Toolmakers: Insights from the study of modern human novices
Dietrich Stout and Sileshi Semaw 307

Chapter 11: Comparing the Neural Foundations of Oldowan and Acheulean Toolmaking: A pilot study using positron emission tomography (PET)
Dietrich Stout, Nicholas Toth, and Kathy Schick..321

Chapter 12: The Biomechanics of the Arm Swing in Oldowan Stone Flaking
Jesus Dapena, William J. Anderst, and Nicholas P. Toth 333

PREFACE

This volume is the first in a series being produced by the Stone Age Institute. Built in 2003, this Institute is dedicated to the study of the Palaeolithic, essentially the archaeology of human origins, and to the dissemination of information gleaned from such studies. In terms of time, the vast majority (well over 99 percent) of the known archaeological record, and thus of human technological development, took place in the Stone Age. Furthermore, for more than 90 percent of the duration of the Stone Age, all known hominins were anatomically pre-modern (archaic) in form. Thus, the evolution of technology and the evolution of the human lineage are intricately intertwined for most of our prehistory.

The record of human technological development begins with the appearance of the earliest stone artifacts approximately 2.6 million years ago and continues with a primary emphasis on use of stone for tools for most of the ensuing time. The transition to other tools – e.g., widespread use of ceramics, use of metals, development of writing, concerted harnessing of various forms of energy (animals, mechanical devices, machines, electronics, nuclear, etc.) – is relatively recent across the planet, emerging only since the end of the last Ice Age, or within the last 10,000 years or so (somewhat earlier or later in different places depending on local resources, geography, ecology, and contacts between populations). This technological transition coincides in time and is dynamically linked with a gradual but widespread shift from a hunter-gatherer existence to a new means of subsistence, one based on controlling the reproduction of certain plants and animals. This shift to an agricultural subsistence in many parts of the world supported tremendous human population growth and spurred the development of larger, more complex societies through time – culminating in the world we live in today.

But the modern human world is a remarkably recent phenomenon. If the last 2.6 million years comprises at least 130,000 generations of tool-making ancestors, the bulk of these – c. 129,500 (over 99.6 percent) of ancestral human generations – lived a Stone Age existence, and only the last 500 or so generations have lived in the last 10,000 years since the transition began from the Stone Age to our modern world. The remarkable technological, economic, and social complexity of humans today is in reality a thin veneer that overlies an incredibly deep evolutionary foundation forged over the course of millions of years in the Stone Age. This is the reason that we have focused on the Palaeolithic and dedicated a research institute to the study of human technological and biological development in an evolutionary context.

During the past several decades, a significant amount of information and data has been collected in our quest for understanding the archaeology of human origins. New prehistoric archaeological sites have been discovered and excavated; detailed analysis of stone artifacts and faunal remains have been undertaken; and new hominin fossils have been discovered that are broadly contemporaneous with these archaeological occurrences. Radiometric and other dating techniques are refining our chronological framework, and new developments in isotopic analysis and even prehistoric DNA extraction and sequencing are helping us understand the ecological, behavioral, and phylogenetic patterns in prehistory.

A wide range of actualistic studies, including studies of taphonomy and zooarchaeology, ethnoarchaeology, geoarchaeology and site formation processes, and experimental archaeological investigations making and using stone tools, have given us much better means of understanding and interpreting the prehistoric past. These new advances have corroborated some earlier assumptions about human evolution, behavior and adaptation (e.g., earliest origins in Africa, adaptation toward meat-eating), but have also challenged other conventional notions of the Early Stone Age (e.g., early importance of hunting, food sharing, and home base models).

This volume is the first in a Stone Age Institute

Press series dedicated to disseminating new information about human evolutionary and Stone Age studies. The purpose of this volume is to present new approaches to the archaeology of human origins and provide an update to our current state of knowledge. This inaugural volume is an especially fitting commencement of this publication series, as it embodies the multidisciplinary approach central to the Stone Age Institute and so much of its research. It comprises a set of studies focusing on the earliest archaeological evidence preserved in the prehistoric record, called the Oldowan. Oldowan industries consist of early stone artifacts produced by tool-making ancestors starting about 2.6 million years ago in Africa and which ultimately document the spread of ancestral forms out of Africa into Eurasia. As these ancestors are so remote from us in time, physical form, brain size, cognitive abilities, and lifeways, it is especially challenging to understand the significance of early stone tools with regard to their manufacture, their use, and their overall role in the lives and adaptation of the protohumans that made them.

To approach such questions about the early toolmakers and the role of technology in human evolution and adaptation, we first require good evidence–the early archaeological sites constituting our database on this important time period in human evolution, and we then need relevant comparative information–critical tools to aid us in interpreting this evidence. This volume is divided into two such components, the first presenting a close and critical look at early sites and their evidence, including early sites in Africa and the record of the spread of tool-making ancestors into Europe and Asia. The second half of this volume presents a range of experimental and actualistic studies critical to interpreting this evidence and evaluating its relevance to human biological and cognitive evolution. These range from comparative studies of ape technologies in the field and the laboratory, to experimental studies of novice tool-makers, to PET (positron emission tomography) studies of brain activation and kinesiological studies of upper limb motion involved in stone tool-making.

In this Stone Age Institute Press publication series, we plan to produce volumes focusing on a range of Palaeolithic projects and topics as well as produce proceedings of international symposia organized by the Stone Age Institute. Our second volume will deal with studies in African taphonomy, or burial and preservation of organic remains, in honor of South African naturalist and palaeoanthropologist C.K. Brain. Our vision for the Stone Age Institute Press is to provide a vehicle for producing important works concerning Palaeolithic research in high-quality, well-illustrated volumes but in an expedient, cost-effective manner. It is partially a response to the ever-growing price of books produced by profit-making academic and university presses. We hope that this publication series will fill a niche in the profession for data-rich, affordable volumes communicating the results of critical field and laboratory research in paleoanthropology. It is also hoped that this publication series will facilitate extensive dissemination of this important information on human evolution and adaptation to a broader academic and general audience.

The inauguration of this Stone Age Institute Press series follows the recent establishment of the Stone Age Institute in Bloomington, Indiana. The Institute has been established as an autonomous, nonprofit organization–a federally-approved 501(c)(3)–dedicated to scientific research and education and funded through tax-exempt donations from individuals, as well as grants from foundations and corporations. The Stone Age Institute is separate from, but has very close ties to Indiana University, Bloomington, as the Institute hires its senior researchers, rotating postdoctoral associates, and support staff through this major research university.

The Stone Age Institute at the present time comprises a 14,000 sq. ft. facility with a library/great room containing the collection of books and articles of J. Desmond Clark, ten research offices, a room for preparation of publications, three laboratories, storage facilities, and an outdoor experimental archaeology area. It has been constructed on a 30-acre tract of land just north of Bloomington, Indiana, with most of the land constituting a green-space nature preserve. For more information on the Stone Age Institute, including our mission, scientific researchers, field and laboratory projects, sponsored talks, conferences, symposia, and publication series, please visit our web site at www.stoneageinstitute.org.

The Stone Age Institute's goal is to establish a critical foundation for Stone Age studies in the United States, and to our knowledge it is the only independent center specifically dedicated to Stone Age research. It has been founded to give a safe and stable haven for Stone Age studies, providing support and facilities to a cadre of full-time researchers as well as visiting scholars, postdoctoral fellows, and students. The Stone Age Institute thus provides a secure base for vital research into our Stone Age foundations, and the Stone Age Institute Press is a vital component in the dissemination of results of this research.

Nicholas Toth
Kathy Schick
Co-Directors, Stone Age Institute

PART I:
THE ARCHAEOLOGICAL EVIDENCE

CHAPTER 1

AN OVERVIEW OF THE OLDOWAN INDUSTRIAL COMPLEX: THE SITES AND THE NATURE OF THEIR EVIDENCE

KATHY SCHICK AND NICHOLAS TOTH

This chapter will present an overview of the Oldowan Industrial Complex (hereafter referred to as the Oldowan), discussing its definition, its chronological and geographic context, the nature of the Oldowan archaeological record, contemporaneous hominins, key issues, and recent trends in research over the past few decades. This introduction will provide a context and foundation for the subsequent chapters which present case studies into aspects of the Oldowan. We also hope that this chapter will serve as a reference for scholars interested in the Early Stone Age of Africa.

DEFINING THE OLDOWAN

The Oldowan is a term used for the earliest archaeological traces in Africa. The term is also sometimes used for the earliest stone age sites in Eurasia. The Oldowan is characterized by simple flaked and battered artifact forms that clearly show patterned conchoidal fracture, produced by high-impact percussion, that is unlike any found in the natural (non-hominin) world. These artifacts include battered hammerstones and simple core forms, often made on water-worn cobbles but sometimes on more angular chunks of rock. These materials herald a new chapter in the human evolutionary record and mark a significant departure from the rest of the primate world: the onset of a technology-based adaptation in which synthetic tools supplemented the biological repertoire of these creatures. In the genus *Homo*, profound changes in size of jaws and teeth, brains, body size and proportions, and geographical range began after this adaptation commenced.

Classification of Oldowan Industries

The term "Oldowan" was first used by Louis Leakey in 1936 to describe materials at Olduvai Gorge (formerly known as "Oldoway Gorge") predating Acheulean handaxe and cleaver industries (Leakey, 1936). Previous to this, Leakey had used the term: "pre-Chellean" to refer to these artifact assemblages. Leakey had also suggested using the term "Developed Kafuan" (the Kafuan now widely believed to be naturally broken pebbles or geofacts), but the prehistorian E.J. Wayland convinced him that the early Olduvai materials were distinct from the Kafuan and warranted a new name. Leakey thought the term Oldowan would probably be dropped over time (Gowlett, 1990), but in fact for the past 70 years it has usually stood as the generic term for pre-Acheulean lithic industries in Africa as well as for simple stone artifact assemblages contemporary with the Acheulean and lacking the handaxe and cleaver elements.

Mary Leakey's seminal work was published as *Olduvai Gorge Volume 3: Excavations in Beds I and II, 1960-1963* (Leakey, 1971). In Beds I and II (ca. 1.9-1.3 mya) Mary Leakey divided the prehistoric archaeological sites at Olduvai into several different industries:

1. Oldowan: beginning near the base of Bed I, ca. 1.85 mya), are assemblages characterized by "choppers, polyhedrons, discoids, scrapers, occasional subspheroids and burins, together with hammerstones, utilized cobbles and light-duty utilized flakes" (Leakey, 1971, p. 1).

2. Developed Oldowan A: Beginning in at the base of Middle Bed II (ca. 1.65 mya) "Oldowan tool forms persist, but there is a marked increase in spheroids and subspheroids and in the number and variety of light-duty tools" (Leakey, 1971, p.2). No bifaces (picks, handaxes and cleavers) were associated with these assemblages. Quartz/quartzite comes into increasing use throughout Bed II times, and lavas become proportionally less prevalent in most artifact assemblages. Chert was also available in the Olduvai basin during this time

3. Developed Oldowan B: This industry, found in Middle and Upper Bed II, is similar to Developed Oldowan A, but with more light-duty tools and some bifaces (usually small and poorly-made).

   [Beginning in Upper Bed IV, ca. 1.1 mya and outside the scope of this survey, Mary Leakey described a Developed Oldowan C, consisting of even higher percentages of light-duty scrapers, as well as higher numbers of *outils écaillés*, laterally trimmed flakes, pitted anvils and punches. There are also very low numbers of choppers as well as some small, crude handaxes, but no cleavers].

4. Early Acheulean: Beginning in upper Middle Bed II (ca. 1.5 mya), bifaces (handaxes, cleavers, picks) become prevalent at some sites in Olduvai Gorge, and these are designated as Acheulean. This tradition continued in Bed III, Bed IV, and the Masek Beds to around 0.4 mya.

During the 1960's and 1970's, other Plio-Pleistocene sites in East and South Africa were discovered (e.g. Gona, Melka Kunture, Omo, East Turkana or Koobi Fora, Chesowanja, Sterkfontein, Swartkrans) that did not have handaxe/cleaver/pick elements in their lithic assemblages. These assemblages were often assigned to the Oldowan or Developed Oldowan, although sometimes investigators have, at least for a time, applied more regional names to similar industries (e.g., "Shungura Facies" for the early Omo sites, Chavaillon, 1976). Although the cultural-historical pattern of lithic industries at Olduvai are not necessarily repeated at other depositional sequences in Africa, the Olduvai data was important in showing the emergence of new technological elements through time within a long stratigraphic sequence.

Although we now know that the Oldowan first appears in the prehistoric record around 2.6 mya, it is more difficult to define when it ends, as even some Holocene sites around the world (e.g. Tasmania) associated with anatomically modern humans exhibit very simple Oldowan-like stone technologies. Mary Leakey believed the Developed Oldowan persisted at Olduvai into Upper Bed IV (ca. 1.1 mya), and the term is not commonly also used for sites under one million years old. With the emergence of the Acheulean between 1.7 and 1.5 mya, contemporaneous non-handaxe/cleaver/pick industries are sometimes viewed as a possibly functional variant of the Acheulean and have sometimes been assigned to a separate "African Tayacian" or "Hope Fountain" industry (Isaac, 1977).

Glynn Isaac (1976) suggested grouping these early Oldowan and Developed Oldowan sites found in Africa together under the rubric "Oldowan Industrial Complex." The term Oldowan has also been used for early sites in Eurasia, such as at Ubeidiya in Israel and at Dmanisi in the Republic of Georgia. Some researchers have preferred to use prehistorian Grahame Clark's (1961) term "Mode I industries: chopper-tools and flakes" to describe these simple technologies.

Oldowan Technology

The Oldowan Industrial Complex is characterized by simple core forms, usually made on cobbles or chunks, the resultant debitage (flakes, broken flakes, and other fragments) struck from these cores, and the battered percussors (hammerstones or spheroids) used to produce the flaking blows. Another element at many Oldowan sites is the category of retouched pieces, normally flakes or flake fragments that have been subsequently chipped along one or more edges. It has been argued (Toth, 1982, 1985; Isaac, 1997) that much of Oldowan technology can be viewed as a least-effort system for the production of sharp cutting and chopping edges by the hominin tool-makers, and that much of the observed variability between sites is a function of the quality, flaking properties, size, and shape of the raw materials that were available in a given locale.

Experiments have shown that the entire range of Oldowan forms can be produced by hard-hammer percussion, flaking against a stationary anvil, bipolar technique (placing a core on an anvil and striking it with a hammer), and, occasionally, throwing one rock against another. The typological categories commonly applied to Oldowan cores (Mary Leakey's "heavy-duty tools," outlined below) can be viewed as a continuum of lithic reduction with the intent of producing sharp-edge cutting and chopping tools (Toth, 1985; Schick & Toth 1993). In this view, many of the cores and so-called 'core tools' found in Oldowan assemblages may not have been deliberately shaped into a certain form in order to be used for some purpose; rather their shapes may have emerged as a byproduct of producing sharp cutting flakes (discussed further below in the section on "Recent Trends in Oldowan Research").

Classification of Oldowan Artifacts

Early systems for classifying "pebble tool" industries include Movius (1949), Van Riet Lowe (1952), and Ramendo (1963). Below we have outlined some examples of classification systems applied more recently to Oldowan or Mode 1 industries by some researchers in the field, to illustrate various approaches to classifying early stone artifact assemblages.

Classification System: Pierre Biberson

Pierre Biberson (1967) presented a typological system for classifying pebble tools (*galets aménagés*) from the Maghreb and the Sahara of North Africa. This system was as follows:

Unifacial forms:

I.1 Unifacial: Single scar: on end of cobble

I.2 Unifacial: Single or multiple concave scars: on any part of cobble

I.3 Unifacial: Two or more scars: on end of cobble

I.4 Unifacial: Two or more scars: on side of cobble

I.5 Unifacial: Two or more scars: on end of chopper; stepped flaking

I.6 Unifacial: Pick-like pointed form on cobble: less invasive flaking; point has a trihedral cross-section section

I.7 Unifacial: Pick-like pointed form on cobble: more invasive flaking, with intersecting scars creating a third edge; point has trihedral cross-section

I.8 Unifacial: Multiple scars: on most of side of cobble; flaked edge is curved

Bifacial forms:

II.1 Bifacial: Single scar on each face: intersecting on same end of cobble

II.2 Bifacial: At least two scars from each face: one face is struck from end of cobble; other face is struck from side of cobble, producing an edge along the side of cobble

II.3 Bifacial: Single scar on each face, oblique to each other, on end of cobble; partially intersecting scars

II.4 Bifacial: Single scar first struck from one face; multiple scars then removed from the other face (using the initial scar as striking platform) to create narrowly-developed cutting edge

II.5 Bifacial: Multiple scars first struck from one face; intersected by large single removal from other face

II.6 Bifacial: Single large scar from one face; multiple scars then removed from the other face to create well-developed cutting edge

II.7 Bifacial: Sinuous edge on short amount of the cobble circumference; a series of flakes is first removed from one face; these scars then used as striking platforms to remove a series from the other face

II.8 Bifacial: Single large scar removed from one face; multiple scars then removed from opposite face, using original scar as striking platform; original face then re-flaked

II.9 Bifacial: Pick-like pointed form made on end of cobble by intersection of two bifacial edges

II.10 Bifacial: Sinuous cutting edge produced by alternate flaking: first removing a flake from one face, then using that scar as a striking platform to remove flake from other face, etc. the cobble being turned over after each removal

II.11 Bifacial: Sinuous cutting edge along side of oblong cobble; multiple scars on each face

II.12 Bifacial: Regular flaking around much of circumference to produced fan-shaped form

II.13 Bifacial: Flaking around much of circumference, with the intersection of two edges (not forming a distinct point); with a cortical unflaked edge as well

II.14 Bifacial: Pebble or large flake flaked around entire circumference to produce a thick, biconvex discoidal core

II.15 Bifacial: Made on large cortical flake; flaked around part of circumference, with striking platform intact

II.16 Bifacial: Wedge flaked flake ("orange quarter") struck from core; dorsal face shows previous flake removals and cortex along one side of flake; cortical striking platform

Polyfacial forms:

III.1 Polyhedral: Appreciable cortex on core; unordered flaking; subspherical shape

III.2 Polyhedral: Less cortex on core; multidirectional flaking over core with one sinuous, bidirectional cutting edge

III.3 Polyhedral: multidirectional flaking over entire surface of cobble, polyhedric and subspherical ("faceted stone ball")

III.4 Polyhedral: Multidirectional flaking, pyramidal shape, flaked to a point; pick-like form on cobble

III.5 Polyhedral: One bifacially flaked sinuous edge; flakes then removed at right angle from one end of that edge (in a third direction)

III.6 Polyhedral: Unifacial or bifacial main cutting edge, with flakes removed from each end of the main edge perpendicular to the main axis of that edge

Although this system is not much used today, it is interesting that Biberson did look closely at the patterning and sequencing of flakes that were detached from cores. Aspects of this approach have been revived recently in some Oldowan analyses, with detailed drawings of artifacts indicating the direction and/or sequence of flake removals on Mode 1 cores (e.g. de la Torre *et al.*, 2003; de Lumley *et al.* 2005; de Lumley & Beyene, 2004).

Classification System: Mary Leakey

The first typological system widely used for the Oldowan, and still the one most commonly used today, was developed by Mary Leakey (1971) in her work at Olduvai Gorge. She made it clear that some of her types, e.g. choppers and scrapers, were not necessarily functional classes, but morphological ones. On the other hand, it seems clear that she felt that many of the core forms were tools in their own right and exhibited damage that she interpreted as utilization.

Leakey classified the Early Stone Age, non-Acheulean, materials in Beds I and II at Olduvai Gorge as follows:

1. Heavy-duty tools (greater than 5 cm or ~2 in maximum dimension).

   a. Choppers: cores, usually made on water-worn cobbles with a flaked edge around part of their circumference. Leakey subdivided this group into five types: side, end, two-edged, pointed, and chisel-edged.

   b. Discoids: cores, usually made on flat cobbles or thick flakes, with a flaked edge around most or all of their circumference.

   c. Polyhedrons: heavily reduced cores comprised of three or more edges.

   d. Heavy-duty scrapers: thick cores with one flat surface intersecting steep-angled flake scars.

   e. Spheroids and subspheroids: more or less spherical stones showing signs of flaking and/or battering.

   f. Proto-bifaces: artifacts intermediate in morphology between a chopper and an Acheulean biface (handaxe).

2. Light-duty tools (less than 5 cm in maximum dimension, usually retouched forms made on flakes or flake fragments).

   a. Scrapers: pieces that have been retouched along a side or end. Leakey subdivided these into six types: end, side, discoidal, perimetal, nosed, and hollow.

   b. Awls: pieces that have been retouched to form a point.

   c. Outils écaillés ("scaled tools"): pieces with flakes detached from opposite ends. Some of these are almost certainly bipolar cores for flake production.

   d. Laterally trimmed flakes: flakes with more casual and uneven retouch.

   e. Burins: rare forms with a flake detached along the thickness of the edge. In later time periods, these are known to be engraving tools.

3. Utilized artifacts

   a. Anvils: stones with pits in them that suggest their use as an anvil in stone tool manufacture.

   b. Hammerstones: stones (often cobbles) that have battered areas that suggest they have been used as a percussor or hammer in stone tool manufacture.

   c. Utilized cobblestones, nodules and blocks: pieces of stone with some damage to edges (chipping or rounding) that suggests their use as a tool.

   d. Heavy-duty flakes and light-duty flakes: flakes with some chipping or rounding of their edge

4. Debitage: flakes, broken flakes, and other fragments produced through stone knapping but which show no further modification (neither retouch nor utilization damage).

5. Manuports: Unmodified stones that appear to be found outside their natural, geological context and are assumed to have been carried to the site by hominins.

Classification System: Glynn Isaac

Another approach to classifying Oldowan sites, essentially a simplified version of Leakey's typology, was proposed by Isaac (1997) based on the study of the Koobi Fora materials. He divided the archaeological materials as follows:

1. Flaked pieces (cores and "tools") (FPs):

   a. Choppers

   b. Discoids
      i. regular
      ii. partial
      iii. elongate

   c. Polyhedrons

   d. Core scrapers
      i. Notched
      iii. Denticulate
      iii. Side (lateral)
      iv. Short (equidimensional)
      v. End (terminal)
      vi. Arcuate
      vii. Double-sided
      viii. Nosed
      ix. Pointed
      x. Perimetal

   e. Flake scrapers
      i. Notched
      ii. Denticulate
      iii. Side
      iv. Short
      v. End
      vi. Arcuate

   vii. Double-sided
  viii. Nosed
   ix. Pointed
   x. Perimetal

2. Detached pieces or debitage (DPs)
  a. Whole flakes
  b. Flake fragments (snaps, splits)
  c. Angular fragments
  d. Core fragments

3. Pounded pieces (PPs)
  a. Hammerstones
  b. Battered cobbles
  c. Anvils

4. Unmodified pieces (UNs)

Classification System: Henry de Lumley

De Lumley's approach to Oldowan industries in Africa and Eurasia (e.g. de Lumley & Beyene, 2004; de Lumley et al., 2005) makes a clear typological distinction between "pebble tools" (*galets aménagé*) and cores, the latter constituting part of his "debitage" category, along with flakes, fragments and other flaking debris.

1. Whole pebbles (unmodified)

2. Fractured pebbles

3. Blocks (unmodified)

4. Percussors

5. Pebble tools (*galets aménagé*)
  a. With a single concave flake scar
  b. Chopper (unifacial)
  c. Double chopper
  d. Rostro-carinate ("beak-like")
  e. Chopper associated with a rostro-carinate
  f. Chopping-tool (bifacial)
  g. Chopping-tool associated with a chopper
  h. Chopping-tool associated with a rostro-carinate

6. Debitage
  a. Cores (nucléus)
    i. Unifacial cores
     (a). Unidirectional
     (b). Bidirectional
     (c). Multidirectional
    ii. Bifacial cores
     (a). Unidirectional
     (b). Bidirectional
     (c). Multidirectional
    iii. Multifacial globular cores
    iv. Prismatic cores
    v. Atypical cores
    vi. Casual cores (ébauchés)
    vii. Core fragments
  b. Flakes
    i. Type 1: Flakes with total cortical surface
    ii. Type 2: Flakes with mostly cortical surface

    iii. Type 3: Flakes with residual cortex
    iv. Type 4: Flakes with no cortex
     Note - Flakes are further subdivided by eight types of platforms: cortical, smooth (lissé), dihedral or faceted, linear or punctiform, removed (öté), nul, absent, and indeterminate.

  c. Small flakes
  d. Debris

In this system of classification, retouched pieces are not considered as a separate category, but rather are noted within each lithic class (e.g. fractured pebbles, pebble tools, cores, debris, or flakes). Retouch is classified into categories such as shallow, steep, flat, invasive, burin-like, etc. Also note that cores that do not fit into his "pebble tool" categories are classified with debitage, effectively sorting cores that produced flakes into two separate classes, either as tools or as waste or byproducts.

Classification System: Nicholas Toth

An alternative system of describing the predominant technological patterns at Oldowan sites was suggested by Toth (1982, 1985) based upon his study and experimental replication of Oldowan sites at Koobi Fora, East Turkana, Kenya. This classification system allows the researcher to classify and describe Oldowan artifacts based upon major aspects of the technological operations involved in making them.

It first divides the artifact population into major categories (cores, retouched pieces, flakes, flake fragments, chunks, and percussors). It then classifies cores and retouched pieces according to dominant technological operations involved in their manufacture:

1. the blank form used (a cobble, a flake or flake fragment, or other indeterminate form)

2. the mode of flaking (unifacial, bifacial, polyfacial, or some combination thereof)

3. the extent of flaking (partial circumference, total circumference). In practice, the amount of circumference flaked is also estimated to the nearest 10%.

For instance, cores are divided first based upon the original form of the material flaked. Cores made on cobbles are then classified according to the mode of flaking, and then with regard to how extensively the form was flaked. For cores on flakes, a similar classification is made according to the mode of flaking (and also with regard to whether unifacial flaking was on the dorsal or ventral surface of the flake), and then again according to how extensively it is flaked. Retouched pieces are also classified according to mode and extensiveness of flaking.

Whole flakes are classified into one of seven flake types, according to the technological information they hold on their platforms and dorsal surfaces with regard to prior flaking of the core. This system of flake classification is similar to that of Villa (1983), a simplified system based on Tavoso (1972), in turn modified after the system of de Lumley (1969).

Flake fragments are classified according to the portion of the flake represented (split, snap, or indeterminate). Angular fragments (probable fragments of flakes whose portion of the entire flake cannot be identified with certainty) and more massive chunks represent final categories of miscellaneous debris resulting from conchoidal fracture of stone.

This technological classification system is as follows:
1. Cores and retouched pieces
    a. Cores
        i. Made on cobbles
            1. Unifacially flaked
                a. Partial circumference
                b. Total circumference
            2. Bifacially flaked
                a. Partial circumference
                b. Total circumference
            3. Unifacially and bifacially flaked
                a. Partial circumference
                b. Total circumference
            4. Polyfacially flaked
                a. Partial circumference
                b. Total circumference
        ii. Made on large flakes/flake fragments
            1. Unifacially flaked on dorsal surface
                a. Partial circumference
                b. Total circumference
            2. Unifacially flaked on ventral surface
                a. Partial circumference
                b. Total circumference
            3. Bifacially flaked
                a. Partial circumference
                b. Total circumference
            4. Unifacially and bifacially flaked
                a. Partial circumference
                b. Total circumference
        iii. Made on indeterminate blanks
            1. Bifacially flaked
                a. Partial circumference
                b. Total circumference
            2. Polyfacially flaked
    b. Retouched pieces (retouched pieces are defined as retaining flake scars that are normally less than 2 cm long, suggesting edge modification rather than flake production)
        i. Unifacially flaked on dorsal surface
            1. Partial circumference
            2. Total circumference
        ii. Unifacially flaked on ventral surface
            1. Partial circumference
            2. Total circumference
        iii. Unifacially flaked on dorsal and ventral surface
            1. Partial circumference
            2. Total circumference
        iv. Bifacially flaked
            1. Partial circumference
            2. Total circumference
2. Flakes and fragments (debitage)
    a. Whole flakes, classified into flake types
        i. Flake type I: Cortical platform, cortical dorsal surface
        ii. Flake type II: Cortical platform, partially cortical dorsal surface
        iii. Flake type III: Cortical platform, noncortical dorsal surface
        iv. Flake type IV: Noncortical platform, cortical dorsal surface
        v. Flake type V: Noncortical platform, partially cortical dorsal surface
        vi. Flake type VI: Noncortical platform, noncortical dorsal surface
        vii. Flake type VII: Indeterminate whole flake

Flakes are further subdivided on the basis of other attributes; for example, for flakes with partially cortical dorsal surfaces, it is noted whether the cortex is on the side or bottom of a flake (or some other pattern).

    b. Flake fragments
        i. Split flakes
            1. Left split
            2. Right split
        ii. Snapped flakes
            1. Proximal
            2. Mid-section
            3. Distal
        iii. Angular fragments (indeterminate part of a flake)
    c. Chunks (miscellaneous fracture from flaking, usually more massive than angular fragments)
3. Percussors
    a. Battered hammerstones
    b. Battered subspheroids/spheroids

This classification system takes a *technological* view of stone fracture at archaeological sites and makes no assumptions about end products or 'mental templates' in the mind of the prehistoric tool-maker or of the archaeologist (e.g., whether a core is a deliberate 'core tool' produced for or used in some task). This system effectively classifies assemblage components into pieces that have been flaked, the resulting flaking debris or debitage, and percussors used to flake cores. Thus, this classification system focuses on the technological operations evident in a lithic assemblage without dwelling on elaborate subclasses based on details of artifact morphology. It can thus reveal useful information regarding the actual patterns of technological operations involved in the manufacture of the artifact assemblage at hand.

Conclusion

In sum, as of yet there is no standardized system of classifying Oldowan lithic assemblages. We would argue that a system that describes the general patterns of lithic reduction (e.g. Isaac and Toth) allows for less subjective categorization of stone artifacts. However, as use of Leakey's typology is so widespread (sometimes with modification), it may be useful to classify an assemblage according to this system in addition to any other classification system employed. Leakey did not make a distinction between "heavy-duty tools" and "cores," so presumably most cores are put in the tool category in her system. Very casual and minimally flaked cores would probably be assigned to her "utilized cobblestone, nodules and blocks" category.

A major meeting of minds at the Stone Age Institute in the fall of 2005, focusing on new approaches to the Oldowan, was organized in order to address many of these issues regarding Oldowan technology and typology. One of the aims of this conference, involving many of the major researchers is this field, was improving standardization of the ways that we classify and analyze early stone artifact assemblages so that we might make more realistic comparisons and contrasts between different Oldowan occurrences.

MAJOR OLDOWAN SITES AND THEIR CONTEXT

Overview

The earliest identified stone artifacts presently date back to approximately 2.6 mya at Gona in Ethiopia. Although this date could very possibly be pushed back with future field work, this 2.6 mya date currently establishes the beginning of the known archaeological record. It is more difficult to say when the Oldowan ends, however, since many stone industries from around the world retained a similar simple, unstandardized technology, even into Holocene times. In this chapter, we are primarily considering the time period of approximately 2.6 to 1.4 mya, after which time Acheulean sites become more common on the African continent.

The time period between 3.0 and 2.0 mya was a very interesting one: it marked a phase of global climatic cooling and drying, witnessing the spread of grasslands on the African continent, major turnovers in fauna (extinctions and new speciation events), the emergence of both the robust australopithecines and the genus

Homo, and as noted, the first clear evidence of protohumans making and using stone tools and modifying animal bones. Moreover, these stone artifacts and animals bone were being deposited in densities sufficiently large to be identified during survey and to merit excavation and detailed analysis.

Most Oldowan sites are located along stream courses (fine-grained floodplains and coarser-grained channel), deltas, or lake margins. The sites in finer-grained floodplain, deltaic and lake-margin deposits normally exhibit less geological disturbances and are better candidates for examining behavioral and spatial patterning. In South Africa, the Oldowan sites are found in limestone caves infillings. The archaeological materials may have washed into these deposits from the surrounding landscape, or they may have been carried to or flaked at the site by the hominins themselves. There is evidence that some of the early hominin skeletal remains deposited in these caves had been killed and eaten by large carnivores as well.

The Oldowan sites in Africa document the earliest archaeological traces yet known, and show the development of stone technologies from their earliest occurrences in East Africa (by 2.5 to 2.6 mya) and their subsequent appearance in southern Africa by about 2 mya and in northern Africa by at least 1.8 mya. Although some studies have suggested a stage of early lithic industries technologically more 'primitive' than the Oldowan (often referred to as a "pre-Oldowan" industry, discussed further below), the earliest stone artifact assemblages from Gona, Ethiopia, establish a very well-documented lithic technology that shows that these toolmakers had mastered the basic skills to flake lava cobbles efficiently. It is true, however, that assemblages with appreciable frequencies of retouched flakes become common only after 2.0 mya.

The key technological elements of the Oldowan include battered percussors (hammerstones and spheroids/subspheroids); a range of core forms made on cobbles, chunks, and larger pieces of debitage; a range of debitage (whole flakes, split and snapped flakes, and angular fragments); and sometimes simple retouched pieces. The raw materials used generally reflect what is available and suitable for flaking in a given region: volcanic rocks such as lavas and ignimbrites are common, and in many areas basement quartzes and quartzites were also common. In North Africa, as at Ain Hanech, fine-grained limestones were also a major source of raw material. At some sites, cherts/flints were also locally available in larger quantities (e.g. some Bed II sites at Olduvai and at Ain Hanech), but generally their frequency is rather low. Different site assemblages often show major differences in their composition in terms of proportion of cores to debitage, proportion of retouched pieces, types of raw materials used, types and range of core forms produced, and the overall size distribution of cores and debitage. This could be due in part to hominin behavior, but it could also be strongly influenced by geological site formation processes, such as water action separating out different classes and sizes of artifacts.

Much of the variation among Oldowan assemblages at different sites is likely the result of use of different raw materials with different initial shapes, sizes, flaking qualities and characteristics, proximity to raw material sources, proximity to water, duration of occupation and/or reoccupation, and functional needs or constraints. Some differences among assemblages may also be attributed to technological norms that may have developed within groups of tool-makers (e.g. a preponderance of unifacial flaking of cobbles, or the Karari core scrapers made by the removal of flakes from the dorsal surface of thick cortical flakes).

When preservation is favorable, mammalian (and sometimes reptilian and avian) fauna is preserved at Oldowan sites. The mammalian remains usually have a wide representation of bovids, equids, and suids. Sometimes large mammals, such as elephant and hippopotamus are also found. These remains are often fragmentary and at some sites (e.g. FLK Zinj at Olduvai, FxJj 50 at East Turkana, and Sterkfontein Member 5), there is clear evidence of hammerstone fracture of limb bones and cut-marks from sharp-edged stone artifacts. These patterns will be discussed in more detail below.

The hominins associated near or at Oldowan sites (discussed in greater detail below) include the robust australopithecines (*A. aethiopicus*, *A. boisei*, and *A. robustus*), other australopithecines such as *A. garhi*, and larger-brained, often more gracile and smaller-toothed forms attributed to early *Homo*. At present, it cannot be established with certainty which of these species were the principal Oldowan tool-makers and tool-users. It is possible that multiple taxa could have had flaked stone technologies. It is arguable, however, that the evolving *Homo* lineage exhibits reduction of strong biological adaptations in term of the size and masticatory power of its jaws and teeth, along with a relatively rapid overall increase in brain size, which likely points to a shift toward an adaptation based more and more upon technological means and less and less upon strictly biological means. As stone technologies continue after the demise of the australopithecines, with *Homo* continuing to evolve and spread afterwards, it is certain that the *Homo* lineage had firmly incorporated stone technologies within its behavioral and adaptive repertoire.

In East Africa, the majority of Oldowan sites are found in fluviatile and sometimes lake margin environments (common places of sedimentation as well as sources of raw materials for the stone artifacts). The Eastern Rift Valley of Africa, in particular, provides an exceptionally favorable environment for the preservation of early hominin activity areas. Numerous depositional basins accumulated here during the Plio-Pleistocene, with subsequent tectonic uplift and erosion exposing these ancient deposits. Volcanic ash deposits and lavas have yielded precise radiometric dates, and

some of these ashes also have been chemically correlated from one site to another over substantial distances in the region, providing additional means to place sites within a regional chronology. The alkaline chemistry of many of these volcanic eruptions played a major role in the preservation and mineralization of the fossil fauna as well.

In South Africa, the Plio-Pleistocene Oldowan sites are all found in karstic limestone cave infillings, and as such may represent an amalgam of slope wash processes in the surrounds of the cave, carnivore transport of bone, and hominin behavior at the site. It has been suggested by Brain (1981) that these caves may have also been sleeping quarters for hominins and baboons, which might explain the extraordinarily high numbers of these taxa in the deposits.

Here we will outline the types of evidence found at the major Oldowan localities that have yielded sites which have been excavated, analyzed, and published. We will first consider the East African evidence, all sites located in the Great Rift Valley, then examine the evidence from Southern Africa found in the infillings of karstic limestone caves, and finally consider the North African evidence.

East African Localities

Gona, Ethiopia

Age: 2.6-2.5 mya; also sites at 2.2-2.1 mya

Geographic/geological setting: The Gona sites are located in overbank floodplain deposits of the ancestral Awash River.

Key sites: EG10, EG12, OGS 6, OGS 7, DAN 1, DAS 7 (2.6-2.5); DAN 2 (2.2-2.1 mya)

Raw materials: Lavas, especially trachytes; ignimbrites; some volcanic cherts (vitreous volcanic rock)

Nature of industries: Simple cores (unifacial and bifacial choppers, etc. with very few retouched flakes)

Other associated remains: Some surface cut-marked bone

Key publications: Roche & Tiercelin, 1980; Harris, 1983; Semaw, et al., 1997, 2003; Semaw, 2000; Stout et al., 2005

Hadar, Ethiopia

Age: 2.3 mya

Geographic/geological setting: Fine-grained river floodplain deposits

Key sites: AL-666; AL 894

Raw materials: Basalt and chert

Nature of industries: Simple cores and debitage; little retouch on flakes

Other associated remains: An early *Homo* maxilla (A.L. 666) was found near the *in situ* artifacts; fossil mammalian remains included *Theropithecus* (baboon) teeth, a bovid (*Raphicerus*) horn core, and murid mandible fragments

Key publications: Hovers, 2003; Hovers et al., 2002; Kimbel et al., 1996

Middle Awash, Ethiopia

Age: 2.5 mya; 1.5-1.3 mya

Geographic/geological setting: fluviatile sands and interbedded volcanic ashes

Key sites: Bouri (Hata Beds); Bodo (Bod A5 and A6)

Raw materials: Lava and chert at Bodo

Nature of industries: The Bouri peninsula (Hata Beds) has cut-marked and broken mammalian bones suggesting stone tool-using hominins at 2.5 mya (essentially contemporaneous with the Gona artifacts); at Bodo (A5 and A6) surface Oldowan artifacts have been found at 1.5-1.3 mya

Other associated remains: At Bouri (Hata Beds), the cranium and holotype of *Australopithecus garhi* was found at 2.5 mya

Key publications: Asfaw et al., 1999; De Heinzelin et al., 1999; de Heinzelin et al., 2000

Konso Gardula, Ethiopia

Age: 1.7- less than 1 mya

Geographic/geological setting: Riverine sand and silt deposits in a palaeoenvironment reconstructed as dry grassland

Key sites: KGA 3, 5 and 7-12

Raw materials: Basalt, quartz, quartzite, and silicic volcanic rock

Nature of industries: The earliest Acheulean (characterized by crude bifaces and trihedral picks made on a range of blanks (cobbles, blocks, and large flakes) and possibly contemporaneous Oldowan industries at ca. 1.7 mya

Other associated remains: A cranium and mandible (KGA 10-525) of *A. boisei* and a mandible of *Homo erectus* at ca. 1.4 mya

Key publications: Asfaw et al., 1992; Suwa et al., 1997

Melka Kunture, Ethiopia

Age: 1.7-1 mya

Geographic/geological setting: Primarily fluviatile deposits

Key sites: Garba, Gomboré, Karre

Raw materials: Lava and quartz

Nature of industries: Oldowan; later Acheulean and Developed Oldowan

Other associated remains: From the Oldowan levels, an early *Homo* child mandible at Garba IV and a partial hominin humerus from Gomboré 1B

Key publications: Chavaillon et al., 1979

Fejej, Ethiopia

Age: 1.96 mya

Geographic/geological setting: Fluviatile sediments

Key sites: FJ-1, in level C1

Raw materials: Mostly quartz and basalt

Nature of industries: Simple cores and debitage

Other associated remains: Hominin premolar and two molars, attributed to early *Homo*; distal humerus fragment, attributed to *A. boisei*

Key publications: Asfaw et al., 1991; de Lumley & Beyene, 2004

Omo Valley, Ethiopia (Shungura Formation)

Age: 2.3-2.4 mya

Geographic/geological setting: River floodplain deposits

Key sites: Omo 71, Omo 84 Omo 57, Omo 123, FtJi1, FtJi2, FtJi5

Raw materials: Primarily quartz

Nature of industries: Simple cores and debitage, bipolar technique evident

Other associated remains: *A. boisei* and early *Homo* from deposits (not at site locales)

Key publications: Chavaillon, 1970; Chavaillon & Chavaillon, 1976; Howell et al., 1987; Merrick, 1976

East Turkana (Koobi Fora), Kenya

Age: 1.9 to 1.3 mya

Geographic/geological setting: River floodplain, river channel, and deltaic deposits

Key sites: In the KBS Member, FxJj 1 (Oldowan); FxJj 3 (Oldowan); FxJj 10 (Oldowan); FxJj 11 (Oldowan); FxJj 38 (Oldowan); in the Okote Member, FxJj 17 (Oldowan); FxJj 50 (Oldowan); FxJj 16 (Karari); FxJj 18 complex (Karari). FxJj 20 complex (Karari); FwJj 1 (Karari).

Raw materials: Primarily basalts, also ignimbrite, chert, quartz

Nature of industries: Two variants of the Oldowan Industrial Complex identified, the Oldowan Industry - simple cores and debitage with a rarity of retouched forms - in both the KBS Member and the overlying Okote Member; and the Karari Industry - numerous core scrapers and more prevalent retouched pieces – in the Okote Member.

Other associated remains: Fauna present at a number of sites; overall fauna primarily larger mammals from grassland, bush, and riverine forest habitats; fauna showing cut-marks and hammerstone fracture at FxJj 50; cut-marked bones at GaJi 5, c. 1.6 my, a deltaic locality with no stone artifacts; a large number of hominin fossils stratigraphically associated with the Oldowan KBS Member and the Okote Member have been attributed to a number of taxa, including *A. boisei*, *Homo* sp., *Homo rudolfensis*, *Homo habilis*, and *Homo erectus/ergaster*.

Key publications: Isaac, 1997; Wood, 1991

West Turkana (Nachukui Formation), Kenya

Age: 2.34 mya (Lokalalei); also later sites to 1.6 mya

Geographic/geological setting: Palaeosols in a river floodplain

Key sites: Lokalalei 1 and Lokalalei 2c

Raw materials: lavas

Nature of industries: Lokalalei 1 contains simple cores and debitage, while the slightly younger Lokalalei 2c shows heavier reduction of cores and considerable refitting of stone artifacts; the authors argue that site 2c, with finer-grained lavas, shows a level of technological skill and complexity in the Oldowan that is unknown elsewhere at this time

Other associated remains: A right lower molar of a juvenile hominin attributed to early *Homo* was found at the same stratigraphic level and close to the archaeological site Lokalalei 1; cranium of *A. boisei* (KNM WT 17400) at 1.7 mya

Key publications: Kibunji et al., 1992; Roche & Kibunjia 1994; Roche et al., 1999; Brown & Gathogo, 2002; Delagnes & Roche, 2005

Chesowanja (Chemoigut and Chesowanja Formations), Kenya

Age: Approximately 1.42 mya (dated basalt separating Chemoigut and Chesowanja Formations

Geographic/geological setting: Fluviatile deposits on a saline lake margin

Key sites: GnJi 1/6E in earlier Chemoigut Formation; GnJi 10/5 in overlying Chesowanja Formation

Raw materials: Mostly lavas

Nature of industries: Oldowan forms such as scrapers, choppers, polyhedrons, flakes and fragments

Other associated remains: Partial cranium of *Australopithecus boisei* (KNM-CH 1); also additional cranial fragments of *A. boisei* (KNM-CH 304) in the Chemoigut Formation; bovids, equids, hippotamus, and crocodile; burnt clay at GnJi 1/6E

Key publications: Harris & Gowlett, 1980; Gowlett et al., 1981

Kanjera, Kenya

Age: 2.2 mya

Geographic/geological setting: Open, grassy habitat in fluvial, swamp and lake flat deposits in a lake margin environment

Key sites: (Kanjera South) Excavation 1 (in Beds KS-1 and KS-2), Excavation 2 (in Bed KS-3)

Raw materials: Mostly fine-grained lava, also other igneous rock, quartzite, quartz, and chert; some raw materials non-local

Nature of industries: Oldowan cores (choppers, polyhedrons) and debitage, some retouch

Other associated remains: Diverse vertebrate fauna with a large proportion of equids, as well as *Metridiochoerus* and *Dinotherium*; partial hippopotamus axial skeleton and artifacts in Excavation 2

Key publications: Ditchfield et al., 1999; Plummer et al., 1999

Olduvai Gorge, Tanzania

Age: 1.85-1.35 mya (Beds I and II)

Geographic/geological setting: Lake margin, channel, and floodplain deposits in grassland/woodland environments

Key sites: Bed I: DK (Oldowan); FLK NN Level 4 (Indeterminate); FLK NN Levels 1-3 (Oldowan); FLK "Zinjanthropus" Level (Oldowan); FLK Upper Levels (Indeterminate); FLK North Levels 1-6 (Oldowan); Bed II: HWK East Levels 1 and 2 (Indeterminate); FLK North, clay with root casts (Indeterminate); FLK North Deinotherium Level (Indeterminate); HWK East: Sandy Conglomerate (Developed Oldowan A); FLK North, Sandy Conglomerate (Developed Oldowan A); MNK Skull Site (Oldowan); EF-HR (Early Acheulean); MNK Main (Developed Oldowan B*); FC West (Developed Oldowan B*); SHK (Developed Oldowan B*); TK (Developed Oldowan B*); BK (Developed Oldowan B*)

(*sites with bifaces; these sites would now probably be assigned to the early Acheulean)

Raw materials: Quartz/quartzite, lava, chert

Nature of industries: Oldowan and Developed Oldowan; through time, quartz/quartzite replaces lava as the predominant raw material and frequencies of artifact classes such as spheroids/subspheroids, light-duty tools (e.g. flake scrapers)

Other associated remains: Well-preserved fauna at a number of sites, including DK and FLK *Zinjanthropus*; numerous hominin remains in Beds I and II of *Australopithecus boisei*, *Homo habilis*, and *Homo erectus*; direct association with Oldowan at FLK *Zinjanthropus* (OH 5: *Australopithecus boisei* and OH 6: *Homo habilis*) and FLK NN Level 3 (OH 7 and 8: *Homo habilis*)

Key publications: Leakey, 1971, 1975; Bunn & Kroll, 1986; Peters & Blumenschine, 1995; Blumenschine & Peters, 1998

Peninj, Tanzania

Age: 1.6-1.4 mya

Geographic/geological setting: Channels in a deltaic environment on the edge of the proto-Lake Natron

Key sites: ST Site Complex (cluster of 11 sites in same palaeosol)

Raw materials: Basalt, nephelinite, quartz

Nature of industries: Simple cores, retouched flakes, debitage,

Other associated remains: Well-preserved fauna with numerous cut-marks and percussion fractures; mandible of *A. boisei* from contemporaneous deposits; Oldowan occurrences contemporaneous with nearby early Acheulean sites

Key publications: Dominguez-Rodrigo et al., 2002; de la Torre et al., 2003; de la Torre & Mora, 2004.

Nyabusosi, Uganda (Western Rift)

Age: 1.5 mya

Geographic/geological setting: Lacustrine sands

Key sites: NY 18

Raw materials: Mostly quartz, some chert

Nature of industries: Choppers, minimally-worked cobbles, retouched pieces (notches and denticulates, etc.), and debitage (many flakes without cortex)

Other associated remains: Remains of *Elephas*, *Hippopotamus*, *Kolpochoerus* and *Phacochoerus* (pigs), *Kobus*, *Redunca*, and *Pelorovis* (buffalo) are found in the same formation

Key publications: Texier, 1993, 1995

South African Localities

Sterkfontein, South Africa

Age: Approximately 2 to 1.4 mya

Geographic/geological setting: Karst cave breccias

Key sites: Member 5

Raw materials: Quartz and quartzite

Nature of industries: In Sterkfontein East, simple Oldowan cores, flakes and fragments, a few retouched pieces; also Acheulean in Sterkfontein West deposits between 1.7 and 1.4 mya

Other associated remains: Hominin ulna, 3 teeth of *A. robustus*; early *Homo* (STW 53) with cut marks on zygomatic

Key publications: Field, 1999; Kuman, 1994, 2005; Pickering et al. 2000

Swartkrans, South Africa

Age: 1.8-1.0 mya

Geographic/geological setting: Karst cave breccias

Key sites: Member 1 (ca. 1.8-1.5 mya), Member 2 (ca. 1.5-1.0 mya), and Member 3 (ca. 1.5-1.0 mya)

Raw materials: quartz, quartzite, chert

Nature of industries: simple core forms, debitage, and some retouched pieces

Other associated remains: A wide range of fossil mammals are found in these breccias, including a wide range of artiodactyls, also hominins, baboons, carnivores (*Panthera, Euryboas, Crocuta, Canis, Hyaena,* etc.), hyrax, horse, and porcupine; hominins include *Australopithecus (Paranthropus) robustus*: SK 46 (cranium); SK 48 (cranium); SK 79 (cranium); SK 876 (mandible); SK 23 (mandible); SK 6 (mandible); SK 12 (mandible); SK 80 (pelvis); SK 3155 (pelvis); SK 97 (proximal femur); SK 82 (proximal femur); *Homo ergaster*: SK 847 (cranium); SK 15 (mandible)

Key publications: Brain, 1981; Clark 1991; Field, 1999; Kuman *et al.*, 2005.

Kromdraai, South Africa

Age: 2.0- 1.0 mya

Geographic/geological setting: Cave breccias in a grassland/woodland karstic environment

Key sites: Member A (some artifacts also in Member B).

Raw materials: Primarily quartz, also quartzite and some chert

Nature of industries: A small assemblage of 99 artifacts at Kromdraai A, mostly simple cores and flakes, with one relatively large flake (more than 10 cm long) and two subspheroids; two artifacts at Kromdraai B

Other associated remains: Remains of *A. robustus* have been found in Member B, which also contains a more closed, humid-adapted fauna than that of Kromdraai B, which is typical of a drier, more open habitat

Key publications: Kuman et al. 1997, 2005; Field, 1999

North African Localities

Ain Hanech and El-Kherba, Algeria

Age: ca. 1.8 mya

Geographic/geological setting: The archaeological occurrences are in sandy floodplain silts overlying a cobble conglomerate

Key sites: Ain Hanech and El-Kherba

Raw materials: Fine-grained limestone cobbles and flint pebbles.

Nature of industries: Limestone cores including choppers, discoids, polyhedrons, and spheroids

("faceted balls") and associated debitage; also, in flint, small cores made on pebbles, retouched flakes (scrapers, denticulates, notches), and debitage,

Other associated remains: Mammalian remains including gazelle, caprids, *Equus* (horse), and *Pelorovis* (buffalo), and *Kolpochoerus* (pig)

Key publications: Sahnouni et al. 1996, 1997, 2002; Sahnouni & de Heinzelin 1998; Sahnouni, 2005

Casablanca Sequence, Morocco

Although there have been claims that prehistoric sites in this region are over 1.5 million years old, it now appears that the earliest of these archaeological sites are no older than 1.0 million years old. As such, these occurrences are outside the scope of this chapter, but for further information the reader might consult Raynal et al., 2002.

CONTEMPORARY HOMININ TAXA

Overview

A number of early hominin taxa appear to be contemporary with the earliest stone tools of Africa during the time span of the Oldowan between 2.6 and 1.4 mya. Paleoanthropologists continue to debate just how many species are represented during this time, with "lumpers" favoring fewer species and "splitters" advocating more species.

Hominins contemporary with the earliest stone tools are generally placed in one of two genera, either *Australopithecus* or *Homo* (though some researchers place the later australopithecines in the genus *Paranthropus*). The earliest well-represented bipedal hominins tend to be the smaller-brained australopithecines, *A. afarensis* in East Africa and *A. africanus* in South Africa. Neither of these taxa is presently associated with flaked stone artifacts: *A. afarensis* precedes the first appearance of stone tools in East Africa, and *A. africanus*, though overlapping in time with stone tool sites in East Africa, is not yet found in association with archaeological materials in South Africa.

There are at least two taxa contemporary with the very earliest stone tool sites in East Africa: *A. garhi* and *A. aethiopicus*, both relatively small-brained, the latter more robust in features such as sagittal cresting and size of the cheek teeth. The presence of such relatively robust features in the cranium and teeth is observed in later australopithecines in *A. boisei* in East Africa and *A. robustus* in South Africa, until the demise of these lineages by approximately 1.2 to 1.0 mya.

Coexisting with these australopithecines, starting by at least 2.3 mya, are taxa attributed to the genus *Homo*. Finds attributed to early members of this genus (generally to Homo sp. between 2.4 and 2.0 mya), show features, particularly in their reduced dentition and somewhat larger cranial capacity (and probably a higher brain/body encephalization quotient or EQ), that distinguish them from contemporary australopithecines, linking them evolutionarily with later developments in the *Homo* lineage but not presenting features in fossil finds sufficient to produce an individual species diagnosis. Species designations among the *Homo* taxa include the earlier forms, *H. habilis* in East and South Africa and *H. rudolfensis* in East and Central Africa, and subsequently *H. ergaster/erectus*.

Phylogenetically, it is possible that the major evolutionary lineage that led to modern humans could have been *A. afarensis* to *A. garhi*, to *H. habilis*, to *H. ergaster/erectus*, and ultimately to modern humans. The lineages that led to the robust australopithecines could have been *A. afarensis* to *A. aethiopicus* to *A. boisei* in East Africa, and *A. afarensis* to *A. africanus* to *A. robustus* in South Africa.

Catalog of Hominin Fossil Taxa

The following inventory of Plio-Pleistocene hominin taxa presents the forms present during the time range of the earliest archaeological occurrences (ca. 2.6-1.4 mya). This catalog includes each taxon's known time range, key sites and fossils, major anatomical characteristics, associated archaeology, and other considerations such as possible phylogenetic status. Useful reviews of these hominin forms in their evolutionary context include Aiello & Dean, 1990; Boaz & Almquist, 1999; Bilsborough, 1992; Boyd & Silk, 1997; Campbell & Loy, 1996; Day 1986; Delson et al., 2000; Johanson & Edgar, 1996; Klein, 1999; Lewin & Foley, 2004; and Wolpoff, 1999.

Australopithecus garhi

Time range: ca. 2.5 mya

Key sites: Bouri (Middle Awash), Ethiopia

Key fossils: Bouri: BOU-VP-12/130 (partial cranium with upper dentition); BOU-VP-12/87 (crested cranial vault); BOU-VP-17/1 (mandible); BOU-VP-12/1 (partial humerus); BOU-VP-35/1 (partial humerus); BOU-VP-11/1 (proximal ulna); BOU-VP-12/1A-G (partial femur and forearm elements)

Anatomical characteristics: Small braincase; prognathic lower face; large anterior and posterior dentition; postcrania suggest a humanlike humerus/femur ratio and an apelike humerus/ulna ratio

Cranial capacity: ca. 450cc (one specimen)

Associated archaeology: Cut-marked and percussion fractured bones at Bouri; roughly contemporaneous with the earliest Stone Age sites at Gona

Other: Asfaw *et al.* (1999) have suggested that this taxon, contemporaneous with *Australopithecus (Paranthropus) aethiopicus* in East Africa and *Australopithecus africanus* in South Africa, may be ancestral to the genus *Homo*. Contemporaneous isolated non-robust dentition from the Omo may be from this taxon as well. This taxon could well be responsible for the stone tools at Gona at 2.6 mya.

Australopithecus africanus

Time range: 3.0-2.2 mya

Key sites: Taung, Sterkfontein, and Makapansgat, South Africa

Key fossils: Taung: Taung child (juvenile cranium, mandible); Sterkfontein: STS 5 (cranium); STW 505 (cranium); STS 71 (cranium, mandible); STS 36 (mandible); STS 52 (partial cranium, mandible); STS 14 (partial skeleton); STS 7 (partial scapula, humerus); Makapansgat: MLD 37/38 (cranium); STS 14 (vertebral column, rib fragments, pelvis; partial femur)

Anatomical characteristics: Small braincase; prognathic lower face; no sagittal cresting; large incisors, premolars, and molars; more ape-like limb proportions

Cranial capacity: ca. 440cc (range 430-520cc)

Associated archaeology: Unknown from South African cave deposits in which this taxon is found, but *A. africanus* is contemporaneous with the earliest stone tools in East Africa

Other: Bone tools attributed to *A. africanus* by Dart (1957) are now believed to represent animal bones modified by non-hominin agents. Some palaeoanthropologists argue that this taxon is ancestral to the genus *Homo*; others argue that it is ancestral to the South African robust australopithecine *A. robustus*.

Australopithecus (Paranthropus) aethiopicus

Time range: ca. 2.5 mya

Key sites: West Turkana, Kenya; Omo Shungura Member C, Ethiopia

Key fossils: West Turkana: KNM-WT 17000 (cranium without dentition); Omo 18 (mandible)

Anatomical characteristics: Small braincase; sagittal cresting in males; very prognathic lower face; large cheek teeth with thick enamel

Cranial capacity: 410cc (one specimen)

Associated archaeology: Unknown, but contemporaneous with the earliest stone tools at Gona in Ethiopia

Other: Some palaeoanthropologists argue that this taxon may be ancestral to both the later East and South African robust australopithecines; others argue that it is ancestral only to the East African variant *A. boisei*.

Australopithecus (Paranthropus) boisei

Time range: 2.3-1.2 mya

Key sites: Olduvai Gorge and Peninj (Natron), Tanzania; East Turkana and West Turkana, Kenya; Omo and Konso Gardula, Ethiopia; Malema, Malawi

Key fossils: Olduvai: OH 5 (cranium); East Turkana: KNM-ER 406 (cranium); KNM-ER 732 (cranium); KNM-ER 729, 1477, and 3230 (mandibles); West Turkana: KNM-WT 17400 (cranium); Chesowanja: KNM-CH 1 (partial cranium); Konso Gardula: KGA 10-525 (cranium and mandible); Peninj (mandible); Omo L.a-125 (mandible)

Anatomical characteristics: relatively small braincase; males with pronounced sagittal cresting; hyper-robust posterior dentition with reduced anterior dentition; broad and dished face with large, flaring zygomatic arches

Cranial capacity: c. 520cc; range 500-530cc

Associated archaeology: Contemporaneous with Oldowan and/or early Acheulean sites at East Turkana, West Turkana, Olduvai, Peninj, Konso Gardula, and Chesowanja. In direct association with Oldowan artifacts at FLK Zinj site, Olduvai

Other: Both *A. boisei* and early *Homo* are contemporaneous

Australopithecus (Paranthropus) robustus

Time range: 2.0-1.0 mya

Key sites: Swartkrans, Kromdraai, and Drimolen, South Africa

Key fossils: Kromdraai: TM 1517 (cranium, mandible); Swartkrans: SK 48 (cranium); SK 46 (cranium); SK 79 (cranium); SK 876 (mandible); SK 23 (mandible); SK 6 (mandible); SK 12 (mandible); SK 80 (pelvis); SK 3155 (pelvis); SK 97 (proximal femur); SK 82 (proximal femur); Drimolen: DNH 7 (cranium and mandible); DNH 8 (mandible)

Anatomical characteristics: Relatively small braincase; large molars and premolars with thick enamel; sagittal crest in males; broad and dished face

Cranial capacity: ca. 530cc; range 450-550cc

Associated archaeology: Associated with Oldowan

artifacts in the cave breccias at Swartkrans and Kromdraai

Other: *A. robustus* is contemporaneous with early *Homo* at Swartkrans and Drimolen, which makes it difficult to ascribe the Oldowan artifacts at these sites to a specific hominin

Early Homo sp.

Time range: 2.4-2.0 mya

Key sites: Baringo (Chemeron), Kenya; Hadar, Ethiopia

Key fossils: Baringo (Chemeron): KNM-BC 1 (temporal fragment); Hadar: AL-666-1 (maxilla); also possibly isolated teeth from the Omo Valley, Ethiopia

Anatomical characteristics: non-robust teeth relative to *A. boisei*

Cranial capacity: Unknown

Associated archaeology: Associated with Oldowan artifacts at Hadar; contemporaneous with Oldowan sites at Omo, West Turkana, etc.

Other: These fossils are primarily fragmentary jaws and isolated cranial fragments; there are no crania complete enough to estimate cranial capacity

Homo habilis

Time range: 2.0-1.6 mya

Key sites: Olduvai Gorge, Tanzania; East Turkana, Kenya; Sterkfontein, South Africa

Key fossils: Olduvai: OH 7 (partial cranium and mandible); OH 24 (cranium); OH 13 (partial cranium, maxilla, mandible); OH 8 (foot); OH 62 (partial skeleton); East Turkana: KNM-ER 1813 (cranium); KNM-ER 1805 (cranium); possibly Sterkfontein: STW 53 (cranium)

Anatomical characteristics: somewhat larger braincase than australopithecines; moderate brow ridge and moderately prognathic face; no sagittal cresting; reduced dentition relative to australopithecines; longer arms and shorter legs relative to modern humans

Cranial capacity: ca. 630cc; range 510-650cc

Associated archaeology: Associated with Oldowan artifacts at Olduvai and East Turkana; directly associated with Oldowan artifacts at FLK Zinj (OH 6: isolated teeth) and FLK NN Level 3 (OH 7: fragmentary cranium, mandible; and OH 8: clavicle, hand and foot bones)

Other: Many palaeoanthropologists believe that *H. habilis* is the best candidate for the ancestor of later *Homo* taxa.

Homo rudolfensis

Time range: 2.4 -1.7 mya

Key sites: East Turkana, Kenya; Uraha (Chiwondo), Malawi

Key fossils: East Turkana: KNM-ER 1470, 1590, and 3732 (crania); KNM-ER 992 and 1802 (mandibles); Olduvai: possibly OH 65 (maxilla); Uraha: UR 501 (mandible)

Anatomical characteristics: Braincase significantly larger than australopithecines; large maxilla; large premolars and molars; flat face; no brow ridge; no sagittal cresting

Cranial capacity: ca. 725; range 625-800cc

Associated archaeology: Contemporaneous with Oldowan sites in East Africa (e.g., KBS Member at East Turkana)

Other: Some palaeoanthropologists prefer to group these fossils with *Homo habilis*; others feel that the facial architecture, larger braincase, and larger dentition of this group warrant its own taxon.

Early Homo ergaster/erectus (considering fossils older than ca. 1.4 mya)

Time range: 1.8-less than 1.0 mya (in Africa)

Key sites: East Turkana and West Turkana, Kenya; Olduvai Gorge, Kenya; Swartkrans, South Africa; Dmanisi, Republic of Georgia

Key fossils: East Turkana: KNM-ER 3733 (cranium); KNM-ER 3883 (cranium); West Turkana: KNM-WT 15000 (skeleton with skull); KNM 730, 820, 992 (mandibles); Olduvai Gorge: OH 9 (cranium); Swartkrans: SK 15 (mandible), SK 847 (cranium); Dmanisi: D2280, D2282, D2700, D3444 (crania)

Anatomical characteristics: larger braincase that australopithecines and habilines; more modern human limb proportions

Cranial capacity: ca. 875cc; range 650- 1067cc

Associated archaeology: Contemporaneous with Oldowan and/or early Acheulean sites at East Turkana and West Turkana

Other: Some palaeoanthropologists view *Homo ergaster* (e.g., KNM-ER 3733, 3883, KNM-WT 15000) as a separate taxon (and a better candidate for modern human ancestry) from *Homo erectus* (e.g., OH 9). The lower end of the cranial capacity range here is based on the Dmanisi specimens. The date of the earlier Javanese *Homo erectus* materials is controversial; some favor a date of around 1.7 mya, while others argue for a date of less than 1.0 mya

CURRENT ISSUES OF CONTENTION

A number of subjects in the archaeology of human origins have generated interesting and vigorous debates over the past few decades. Here we will review key areas of contention between researchers and refer the reader to key literature on these issues.

Existence of a "Pre-Oldowan"?

Some archaeologists, notably Roche (1989), Piperno (1993), and de Lumley (de Lumley & Beyene, 2004; de Lumley *et al.*, 2005) have suggested in the past that lithic industries prior to 2.0 mya (Gona, West Turkana, Omo) exhibit less skill in knapping than those younger than 2.0 mya, such as those in the Okote Member at East Turkana and in Beds I and II at Olduvai Gorge. However, excavations in recent years at Gona in Ethiopia (Semaw, 1997, 2000; Semaw *et al.*, 1997) have shown that at the oldest archaeological sites yet known, dated to between 2.6 and 2.5 mya, flaked lava cobbles exhibit a surprising level of skill and control of flaking, suggesting that the moniker "Pre-Oldowan" might be dropped.

Since then, Roche et al. (1999) have found well-made Oldowan artifacts from West Turkana dated to 2.3 mya. They suggest that, in view of this West Turkana evidence from 2.3 mya, that sites between 2.6 and 2.0 mya are variable, with artifacts in some assemblages appearing more "sophisticated" and those in others appearing more "crude." However, it is very likely that much of this variation may be due to differences in the quality and flakability of the raw materials used in different assemblages (e.g., the trachytic lava cobbles at Gona flake much more easily flaked than many East African basalts, making it easier to remove numbers of large flakes from cores), rather than profound differences in the skill, control or cognitive abilities of the knappers.

Experimental archaeological work with modern African apes (this volume) shows that bonobos (pygmy chimpanzees) with substantial training still produce Oldowan-like artifacts, but their assemblages of artifactual products can be shown to exhibit less skill (seen in numerous attributes of the cores and flakes) than is found at early Oldowan sites (see the chapter in this volume). If such artifactual assemblages were to be found in the prehistoric record prior to 2.5 mya, and this appeared to be a consistent pattern, there may then be some justification to separate these assemblages from the classic Oldowan.

It does seem true that Oldowan sites older than 2.0 mya tend to have few or no retouched flakes in their lithic assemblages. This pattern begins to change after 2.0 mya, with retouched forms (light-duty scrapers, awls, etc.) becoming more prevalent at sites at Olduvai Gorge and also, over time, at East Turkana.

Who Were the Oldowan Tool-makers?

As noted above, there were probably at least nine hominin taxa present in Africa during the major time period of the Oldowan, between 2.6 and 1.4 mya. These included *Australopithecus garhi, A. africanus, A. aethiopicus, A. boisei, A. robustus, Homo habilis, H. rudolfensis, and H. ergaster/erectus*. At our present state of knowledge, all of these taxa, with the exception of *A. africanus*, were contemporaneous with nearby Oldowan sites.

At a number of localities, hominin fossils are found in geological strata contemporaneous with archaeological materials in that region. These localities include Hadar (early *Homo*), Middle Awash (*A. garhi*), Konso Gardula (*A. boisei, Homo ergaster/erectus*), Melka Kunture (early *Homo*), Fejej (early *Homo, A. boisei*), Omo (*A. boisei*, early *Homo*), East Turkana (*A. boisei, H. habilis, H. rudolfensis, H. ergaster/erectus*), West Turkana (early *Homo*), Chesowanja (*A. boisei*), Olduvai Gorge (*A. boisei, H. habilis, H. erectus*), Peninj (*A. boisei*), Sterkfontein (*A. robustus*, early *Homo*), Swartkrans (*A. robustus, H. ergaster*), and Kromdraai (*A. robustus*). At a few sites, notably Olduvai FLK Zinj, the fossil remains of *A. boisei* and *H. habilis* were found in direct association with a discrete Oldowan archaeological horizon. In view of the fact that diverse mammalian fauna are often in association with these Oldowan occurrences, it cannot be demonstrated clearly, based on these associations, that hominins found at or near archaeological horizons are necessarily the tool-makers. If hominin remains are indeed functionally associated with artifacts at a given locality, it must be considered that they could represent *either* the dinner or the diner. The only known Oldowan-age hominin fossil exhibiting identifiable cut-marks is the STW 53 cranium, attributed to early *Homo* (Pickering et al., 2000).

Mary Leakey (1971) argued that *H. habilis* was the principal Oldowan tool-maker, relegating *A. boisei* to a minimal role in early lithic technology (perhaps a tool-user responsible for minor modification to artifacts). Many paleoanthropologists still favor the scenario of the genus *Homo* being responsible for many or most of the Oldowan archaeological occurrences, since this genus exhibits marked brain expansion and tooth reduction over time after the onset of stone tools, while the robust australopithecines exhibit less encephalization and little if any reduction (and possible increase) in size of cheek teeth and chewing musculature over time, until their extinction by 1.0 mya.

A different scenario has been presented by Susman (e.g., Susman, 1991; Grine & Susman 1991), who proposed that the robust australopithecines may have been the first makers of stone tools and may have relied heavily upon technology in their adaptation, especially in the processing of plant resources. He argued that hand bones attributed to *A.* (*Paranthropus*) *robustus* from South Africa exhibit human-like morphology for precision grasping.

It is possible that all of the hominin taxa contemporary with the Oldowan had capacity for and some involvement in use of stone technology. It is also possible that there may have been marked variability between different taxa, and possibly substantial variability among populations of a single taxon, with regard to involvement in stone tool-making or tool-using. What is clear, however, is that by 1.0 mya, only representatives of the genus *Homo* survived (*H. erectus* or *H. ergaster*), all of the australopithecines had gone extinct, and stone tool-making continued not only in Africa but also in areas of Eurasia into which *Homo* populations had spread. This implies that the *Homo* lineage had significant involvement in stone tool manufacture and that over time this behavior maintained a relatively consistent role in its adaptation.

What Were Oldowan Tools Used For?

The function of palaeolithic stone artifacts and the overall adaptive significance of human technology are questions that have perplexed prehistorians for over a century and a half. Although the use of ethnographic analogy to associate recurrent stone artifact forms with known functions in recent times can yield possible clues to the prehistoric uses of early stone artifacts, it does not clearly identify which Oldowan artifacts were actually utilized as tools, or what they were used for.

Experimental functional studies can, at the very least, identify the functional capabilities of different artifact classes (based on such criteria as shape, edge sharpness, weight, raw material, etc.) for a range of different tasks that might have been carried out in Early Stone Age times (Jones, 1981; Toth, 1982, 1985; Tactikos, 2005). Such tasks might include stone tool manufacture, animal butchery (hide slitting, gutting, dismembering, defleshing, marrow/brain processing), nut-cracking, simple wood-working, digging, hide-working, and manufacture of simple containers. Efficiency experiments performing such tasks with a range of Oldowan artifact forms yield valuable information as to the relative efficiency of each tool for a given task.

Experiments in using the range of Oldowan artifacts for various tasks have highlighted the possible importance of an artifact type within Oldowan assemblages whose usefulness may have been underestimated in archaeology, namely, the flake. A comprehensive experimental study of Oldowan artifact function indicates that sharp flakes and flake fragments are enormously useful in cutting operations, particularly in various aspects of animal butchery (skinning, defleshing, dismembering, etc.). Thus, flakes may not simply represent debitage or "waste," but might rather represent a central component of the Oldowan toolkit (Toth, 1982, 1985, 1987b).

Such experiments can also show the relationships between *processes* (e.g., skinning, dismembering and defleshing) and resultant *products* that may have visibility in the prehistoric record (e.g., striations identified as cut-marks on bones or distinctive use-wear polishes on stone tool edges). Thus, we can identify key signatures or "smoking barrels" that can give corroborative evidence of the prehistoric functions of artifacts.

The class of battered cobbles or other pieces of stone (hammerstones, battered subspheroids or spheroids) strongly suggests that these objects were primary tools used as hammers to flake cores, and as such indicates that Oldowan hominins by 2.6 mya were using tools to make other tools, a pattern that is rare if not absent in the technological repertoires observed among modern apes in the wild. Such functional experiments can then associate a range of possible tasks which would be efficiently performed with a given artifact type.

A very valuable functional signature has been found in cut-marks on animal bones at Oldowan archaeological sites, which indicate that early hominins were skinning, disarticulating, and defleshing carcasses of small, medium, and large mammals (Blumenschine, 1986; Bunn ,1981; Dominguez-Rodrigo, 2002; Egeland et al., 2004; Monahan, 1996; Monahan & Dominguez-Rodrigo, 1999; Oliver, 1994; Pickering, 2001; Pickering and Domínguez-Rodrigo, this volume; Potts & Shipman, 1981). Experiments as well as actualistic studies of carnivore damage to bones have helped differentiate between cut-marks made with stone tools and tooth-marks made by carnivores (Bunn, 1981; Potts & Shipman, 1981). The majority of Oldowan cut-marks tend to be sets of parallel striae, suggesting that sharp-edged flakes, as opposed to unifacially or bifacially retouched edges, were primary artifacts used as butchery knives (see Toth, 1985:112).

Many of these studies have also shown that mammal long bones sometimes exhibit hammerstone percussion marks (e.g., see Blumenschine & Selvaggio, 1988), fracture patterns showing spiral fracture with discrete points of percussion (from contact with stone hammers or anvils), flake scars on bones, and occasional bone flakes. These patterns are consistent with the use of simple stone hammers, and possibly anvils, as percussors for marrow processing.

Another 'smoking barrel' so to speak, or positive evidence for stone tool function, is provided by microwear studies, or analysis of microscopic modification that can develop on stone tool edges during their use (Keeley, 1980). Microwear, or wear-patterns in the form of striations, polishes, and edge damage and modification (chipping, rounding, etc.) has been shown experimentally to develop on the edges of stone tools (primarily in fresh, fine-grained siliceous raw materials) in the process of their use for different functions. Although most raw materials used in the manufacture of early stone tools (lavas, quartz, quartzite, etc.) have not proven to be very amenable to such analysis, a study of a limited sample of more rare siliceous artifacts (mostly in chert, also ignimbrite) from sites in the Okote

Member at East Turkana have revealed a fairly diverse set of prehistoric activities dating to approximately 1.5 mya (Keeley & Toth, 1981; Keeley in Isaac, 1997:396-401).

Examination of a sample of 56 artifacts (mostly flakes and flake fragments, including a few with retouch, along with one pebble core "bifacial chopper") from 9 Okote and KBS Member sites found unequivocal and interpretable microwear traces on nine artifacts from five of the Okote Member sites. This sample of 56 artifacts represents the majority of suitable-looking specimens from Koobi Fora assemblages excavated at that time. Subsequent examination of an additional sample of 39 specimens, including a small chert core and an ignimbrite 'scraper' form did not show definite microwear traces.

These nine artifacts show a fairly diverse range of activities for the size of the sample and, interestingly, the working of both plant and animal materials: "four meat or butchering knives, two soft-plant knives, two woodscrapers (one of which had traces of use as a saw on another edge), and one wood saw" (Keeley in Isaac, 1997:399). As plant processing in Oldowan times is a nearly invisible activity due to the general lack of preservation of macroscopic plant remains, this microwear evidence provides an invaluable window into this aspect of early hominin adaptation. The plant-cutting knives show classic 'sickle gloss,' indicating the gathering or processing of soft plants, whether for food, bedding, or other purposes. (Experimentation by Toth has demonstrated that this gloss forms much more quickly on African savanna grasses than on temperate European ones).

The wood scraping and sawing would appear to indicate the shaping of wood, presumably to make other tools such as spears, digging sticks, etc. Microwear evidence for cutting meat found on four of the nine artifacts (from two sites, FxJj 50 and FxJj 20) corroborates indications from cut-marked bone, and, interestingly, two of the artifacts showing meat-cutting polish at FxJj 50 were found less than a meter away from a cut-marked bone (Keeley in Isaac, 1997:401).

Archaeologists studying Oldowan occurrences (and Palaeolithic archaeologists in general) are aware that a great deal of prehistoric tool use is, at present, invisible in the archaeological record, and what we are sampling is the 'tip of the iceberg,' but hopefully a representative sample of common tool-using activities. It is likely that future techniques will be developed to gain a much better understanding of the functions of ancient tools (for example, higher-resolution organic residue studies, even possibly DNA residues of great antiquity).

What Was the Nature of Oldowan Sites?

Explaining *how* and *why* early hominins collected, and then concentrated at discrete focal points on the landscape, lithic raw material (sometimes brought in from multiple distant sources), and also, at some sites, presumably animal bones as well, are important issues in human origins studies. Thus, the nature of variability between sites (for instance, in terms of artifact and fossil densities, the nature of artifact assemblages, spatial patterning at the site, vertical dispersion of materials, geological context, associated fauna, and so on), and possible explanations for such variability have been topics of great interest and concern in early stone age studies.

Mary Leakey (1971) divided the sites at Olduvai Gorge into: a) Living floors (limited vertical dispersion); b) Butchering or kill sites (associated with the skeleton of a large animal or group of animals) c) Sites with diffused material (significant vertical dispersion); and d) River or stream channels (artifacts incorporated in gravel deposits). Leakey interpreted her "living floors" as ancient camps of Oldowan hominins.

Glynn Isaac (1971) developed a classification of sites according to their proportions of stone artifacts relative to bone remains. The major categories in this classification were: a) "Camp" or occupation sites (high density of both stone and bone); b) Quarry or workshop sites (high density of stone, low density of bone); c) Kill or butchery sites (high density of bone, low density of stone); and d) Transitory camps (low density of both bone and stone). Isaac (1978) went on to argue that the occupation sites were early examples of "home bases" where early hominins shared food resources. He also postulated more hunting and scavenging and a sexual division of labor among these hominin social groups. Isaac (1984) later replaced the term "home base" with the term "central place foraging" areas to denote sites where hominins were concentrating stones and bones without the necessary (but possible) corollaries of food-sharing and division of labor.

Lewis Binford (1981) first suggested that early Oldowan sites simply represented places where hominins were doing marginal scavenging at places where carnivores collected and consumed animal carcasses. Later, he revised this interpretation and argued that Oldowan hominins were marginal scavengers, but may have collected bones on the landscape and processed them for marrow and relict meat (Binford, 1987). He did not think that such attributes as food sharing or a sexual division of labor were necessary for these tool-making populations.

Richard Potts (1984, 1988) suggested that Oldowan sites were "stone caches" where hominins stored materials for later use. In this model, hominins were transporting stone and depositing it in concentrations away from the original sources, thus creating "caches" of raw materials on their landscapes; if a need for stone emerged, they would have then gone to the nearest source. Potts argued that this model was energetically efficient, giving these early hominins an adaptive edge over other groups that did not cache.

Robert Blumenschine (1986, 1988) argued that many Oldowan sites represent the scavenging behavior of early hominins. In this model, hominins were accessing parts of carcasses left behind by carnivores such as large cats or cached in trees by leopards. This model posits that scavenging opportunities, including marrow extraction of felid kills abandoned in riparian woodlands, would have been markedly greater during the dry season, and thus the Oldowan archaeological record may have a built in bias for representation of dry season activities with little representation of hominin behaviors during the rainy seasons (Blumenschine, 1986).

Kathy Schick (1986, 1987a, 1987b) suggested that many Oldowan sites, particularly those with high densities of artifacts and bones, represented favored places for early hominins, likely due to proximity to resources such as food and water as well as amenities such as trees that could provide shade or escape from predators, and possibly sleeping quarters. This model of site formation proposes that hominins were repeatedly or habitually carrying stone around the landscape, with disparity between stone brought in and that taken away resulting in a range of different sites, from dense concentrations to relatively thin scatters of artifacts. At especially favorably located places, a range of food processing and feeding behaviors probably occurred, some in conjunction with stone tool-making and tool-using.

Clearly at the larger sites, the amount of lithic material brought to the site exceeded the amount taken away, resulting in the accumulation of sometimes thousands of artifacts and hundreds of kilograms of material. Some of these sites could also have served as *de facto* depots of raw material for re-use, i.e., materials discarded and left behind by hominins, not deliberately stored or "cached" (Schick, 1986:167), but which could have been tapped into by the same hominins or other individuals or groups for use at some later time, if the lithic material were not buried or obscured by vegetation or sedimentation (Schick, 1986:163-169).

It seems advisable to keep in mind that many early stone age sites could differ greatly in terms of what they represent about early hominin behaviors: They may be 'capturing' different subsets of the overall behavioral repertoire of early hominins over time and space. Sites could vary from one another, for instance, in terms of any of the following variables, each of which could potentially impact the observable archaeological patterns left behind in their wake:

- the numbers of individuals active at the locale

- composition of the group by age or sex

- the kind and variety of on-site activities pursued, e.g.,
  - food consumption
  - meat or marrow processing
  - plant processing
  - stone tool-making
  - manufacture or use of tools in other substances such as wood

- food-sharing
- sleeping
- the amount and kinds of food resources available at or near the site or brought to the location
- proximity of the site to stone resources for tool-making
- duration of occupation and frequency of reoccupation of the site locale
- proximity of the site to amenities (such as water, shade, trees or other havens from predators)
- seasonal constraints and opportunities
- available environments and microhabitats
- arbitrary differences (or 'cultures' with a small "c") among different groups
- evolutionary change in hominin behaviors over time

Thus, search for a monolithic behavioral model for the formation of early sites is likely unrealistic, but we are becoming better armed to wrestle with the more particulate questions regarding behavioral processes evident at a particular site. The past few decades have witnessed a trend toward increasingly sophisticated and self-critical modeling of the dynamics of site formation, and preliminary models and observations are being constructed based on a wealth of valuable experimental and actualistic research.

Oldowan Hunters or Scavengers?

As recently as the 1950's through the 1970's, many paleoanthropologists as well as popular authors emphasized the role of hunting in early human evolution (see, for example, Ardrey, 1976; Dart, 1953; Lee & DeVore, 1968). Binford (1981) provided a direct challenge to the assumption that hunting constituted a major component of subsistence activities of early hominins, and argued for scavenging as a major means of acquiring meat resources among these hominins. At the same time, Brain's (1981) examination of the faunal remains from the Transvaal cave accumulations in South Africa concluded that the australopithecine bone breccias were not the result of hominin predation, but rather primarily the accumulation of carnivores such as leopards and hyenas.

With the identification of animal bones with stone tool cut-marks and hammerstone percussion damage and fracture at Olduvai Gorge sites such as FLK Zinj, BK, MNK Main, and HWK East Level 1-2; at the ST Site Complex at Peninj (Natron); at FxJj 50 at Koobi Fora; and subsequently also at Swartkrans Member 3 and Sterkfontein Member 5 in South Africa, the question as to how early hominins were procuring animal resources has come under very close, active scrutiny. In the past few decades a great deal of paleoanthropologi-

cal research has centered on whether the faunal remains found at Oldowan archaeological sites show patterns indicating the relative involvement of hominins and various carnivores in accumulating, modifying, and accessing food resources from archaeological faunal remains.

At the present time, there are two major schools of thought regarding hominin-modified bones at such Oldowan sites:

1. That these faunal remains represent scavenging behavior on the part of Oldowan hominins, and that the major source of meat/marrow for these hominins was from carcasses largely consumed and left behind by predators such as large cats. This perspective has been forwarded by Blumenschine (1986, 1987, 1989); Selvaggio (1994), and Calpaldo (1995).

2. That these faunal remains represent predation or at least primary access (e.g., confrontational scavenging) to carcasses on the part of Oldowan hominins, with presumably a great deal more meat and other animal resources available to the tool-makers than in the scavenging scenario. This perspective has been forwarded especially by Bunn, Pickering, and Domínguez-Rodrigo (Bunn et. al., 1980; Bunn, 1982, 1983, 1994; Bunn & Kroll, 1986; Pickering, 1999, 2001; Domínguez-Rodrigo, 2002; Domínguez-rodrigo & Pickering, 2003). (See Pickering and Domínguez-Rodrigo, this volume, for further discussion and review of this issue).

Since Oldowan hominins were almost certainly opportunistic omnivores, and at any given time there may have been more than one species making and using stone tools, it would not be surprising if a wide range of behavioral and subsistence patterns are ultimately identified in the early archaeological record. Furthermore, these patterns could have varied seasonally, regionally, environmentally, temporally, and among different groups or populations. Hopefully fine-grained taphonomic analysis of greater numbers of Oldowan faunal assemblages in a variety of situations and environments will potentially exhibit patterning that might yield insight into such variability in hominin subsistence behaviors.

Causes for Encephalization in the Genus *Homo*?

The fossil evidence indicates that some taxa of hominins exhibited larger brains and probably higher brain to body ratios (EQ's or Encephalization Quotients) than earlier taxa (the earlier australopithecines) and contemporaneous taxa (the later robust australopithecines) by at least 2.0 mya and perhaps earlier. These larger-brained forms are conventionally put into the genus *Homo* (*H. habilis*, *H. rudolfensis*, *H. ergaster/erectus*). It is interesting and perhaps significant to note that, at the present time, we do not yet have fossil evidence of such encephalization, nor evidence of a profound reduction in the size of jaws and cheek teeth, in hominins contemporaneous with the very earliest Oldowan sites around 2.5-2.6 mya (*A. garhi, A. aethiopicus*, and *A. africanus*).

A great deal of recent debate has centered on the various causal factors that might be responsible for, or involved in, this encephalization in the genus *Homo*. Hypotheses have revolved around such factors as social intelligence, tool manufacture and use, and changes in hominin diet. Here we will review some of the major hypotheses regarding brain expansion in the human lineage.

The Social Brain Hypothesis

Primatologist Robin Dunbar (1992, 1993) has found that neocortex ratio (the ratio of neocortex size to overall brain size) in primates (and also carnivores) is correlated with group size. Group size is taken as a general index or proxy of social complexity, with primate species living in larger social groups typically having more complex social interactions than do those living in smaller groups. A larger neocortex ratio would presumably allow for a higher level of social intelligence necessary to negotiate the more complex networks of interactions and relationships in larger groups.

Dunbar suggests that the process of neocortical encephalization in the human lineage allowed for larger group sizes (for modern human foragers, the prediction would be about 150 individuals), the selective forces including clearer "theory of mind" (the cognitive ability to understand the beliefs and desires of others) as well as better communication skills that would ultimately lead to modern human language. As early hominins became more socially complex, larger neocortical areas would have evolved in tandem with larger social group sizes. Presumably, once set forth, neocortical encephalization could then also have been selected for due to other reasons as well, as it would have conveyed greater overall intelligence for use not only in social groups, but also in foraging behaviors, in planning or timing of various activities, or in tool manufacture and use. The theory of "*Machiavellian Intelligence*" (Humphrey, 1976; Byrne & Whiten, 1988) is a similar perspective that also emphasizes primate social interaction, politics, "theory of mind," deception, and intelligence.

The Symbolic Hypothesis

Neuroscientist and evolutionary anthropologist Terrence Deacon (1997) has suggested that the near-synchronous appearance of encephalization, stone tools, hunting and butchering, reduction in sexual dimorphism, and probable male provisioning, pair-bonding, and mating contracts, are interrelated features correlated with the rise of symbolic thought and communication starting in early *Homo*. In Deacon's framework, a symbol represents "… some social convention, tacit agreement, or explicit code which establishes the rela-

tionship that links one thing to another" (1997, p. 71). In his hypothesis, key results of this early enhancement in symbolic thought and communication (at first, use of simple gestures, vocalizations, activities and objects, possibly highly ritualized) would ultimately include improvements in sharing knowledge about the environment and in manipulating and negotiating with other individuals.

The Tool-Making Hypothesis

Since the time of Darwin, it has long been hypothesized that tools constitute a defining characteristic of what it is to be human. Tools have often been taken not only to represent a hallmark of the human lineage but also a major impetus for the brain encephalization in human evolution. Although in recent years we have increased our knowledge and appreciation of tool use and even occasional tool manufacture by other species, the profound technological adaptation accomplished by the human species still stands out as a remarkably significant departure from the rest of the animal world. Washburn (1960) proposed a "*biocultural feedback*" model for the coevolution of human genetic evolution and human cultural evolution (including tools). In this feedback loop, the evolution of culture and tools in our lineage would have led to selection for genetic and biological foundations for these behaviors (including intelligence), leading to more complex cultural adaptations, and so on. This idea is echoed in sociobiologists Charles Lumsden and Edward O. Wilson's (1983) "*gene-culture coevolution*" model.

A number of researchers have emphasized how the role of technology in our adaptive strategy may have contributed to the increased intelligence and encephalization in the human lineage. Kathleen Gibson (1986) has suggested an "*extracted foods*" hypothesis, arguing that primates that exploit foods which are difficult to extract and process tend to be more intelligent and encephalized. In primates neocortical size is correlated with "…the complexity and variety of the sensorimotor coordinations needed for the finding and processing of foods" (Gibson, 1986:100), and this pattern is even more exaggerated in human evolution, with tools and technology allowing for even more efficient extraction.

We have argued that, although there is nothing inherent in tool-making that would lead to encephalization, it is through tool-making and tool-use that early hominins were able to expand their diet breadth and increase the quality of their diets (Schick & Toth, 1993). By creating synthetic "organs" (a phenomenon we called "*techno-organic evolution*"), hominins were gradually able to enter the niches of other animals such as predatory and scavenging carnivores, suids, and insect-eating mammals, increasing their survivability and reproductive success. The combination of tool-making and tool-use, leading to expansion of diet breadth, increase in diet quality, increase in social complexity, and rise of more predatory behavior, and the cumulative impact of these adaptations on reproductive success, would have driven encephalization over time.

The Expensive Tissue Hypothesis

Anthropologist Leslie Aiello and Peter Wheeler (1995) have suggested that animal species tend to have as large a brain as their metabolism can support. In order to allow for evolutionary brain expansion, there must be a novel way to reallocate expenditures within their overall metabolic budget. In modern humans, the brain (a very 'expensive' tissue) comprises about 2% of body weight but consumes about 20% of the body's metabolic budget. Larger brain/body size proportions are normally associated with higher intelligence, which could increase evolutionary fitness through improving a species' adaptation by making them more efficient foragers and social animals.

In modern humans, the brain utilizes a significantly higher proportion of the metabolic budget than in other primate species, while the budget for the human gut (the gastro-intestinal tract) is significantly reduced relative to most other primates. In effect, comparing humans with other primates, the human brain has increased evolutionarily in terms of its size and its metabolic budget at the expense of the gut, which has undergone a corresponding decrease in its size and energy budget.

In early hominins, it is hypothesized that this shift towards encephalization would have been correlated with a reduction in the hominoid-like gut (a larger size necessary for digesting and detoxifying a high vegetable diet). The size of the gut is largely tied to the kinds of foods a species consumes, in terms of how digestible they are and how much quantity must be consumed to meet nutritional requirements. Species with "lower quality" diets, such as herbivores, tend to consume larger quantities of less digestible foods, requiring a larger gut. Conversely, species with "higher quality" diets, such as carnivores, tend to consume lesser quantities of more digestible foods, requiring a smaller gut (especially a smaller stomach but proportionally longer small intestine to absorb nutrients). The evident gut reduction in the human lineage could have been made possible then by increasing diet quality (especially the proportion of proteins and fats) through more omnivorous behavior (such as scavenging or hunting, or focusing on invertebrate foods such as insects). Over time, the use and reliance of tools and technology could greatly enhance acquisition and processing of a higher-quality (more easily digested) diet and thus allow further encephalization and further gut reduction.

Presence of Fire at Oldowan Sites?

At some point in human evolution, hominins likely learned to maintain natural fires (started by lightning strikes, volcanic eruptions, spontaneous combustion, etc.), and presumably much later in time learned to manufacture fire, probably through some friction technique such as a hand drill, bow drill, fire saw, etc., or by a percussion technique such as striking a flint against a pyrite. In general, the recurrent and presumably habitual use of fire at archaeological sites occurs only in the late Acheulean and Middle Palaeolithic/Middle Stone Age in the last 250,000 years. By that time, palaeolithic peoples almost certainly knew how to manufacture and maintain fires and hearths.

Intriguingly, there are a few Oldowan sites where evidence such as reddened and baked sediments or thermally altered stone or bone suggests the presence of fire. The major issue here is whether the fire was directly associated with hominin behavior, i.e., whether the hominins started or maintained the fire, or whether these are natural fires that swept through an area where hominins had left materials behind. There are no known hearth structures associated with Oldowan sites (pits, rings of stone, etc.), so establishing hominin control or manufacture of fire is even more challenging.

At Swartkrans Member 3, associated with Oldowan tools and the remains of *A. robustus*, several hundred pieces of fossil mammal bone (within a bone assemblage of nearly 60,000 pieces) have been interpreted to show signs of exposure to fire (i.e., to temperatures achieved in experimental campfires, or from approximately 650° C to a maximum of 860° C) (Brain & Sillen, 1988; Brain, 1993; Sillen & Hoering, 1993). The evidence for burning included fossil bone discoloration (buff, to dark brown or carbonized, to calcined), thin section analysis (showing cracks and other changes in structure) (Brain, 1993), and chemistry (carbon-containing char, altered fats, etc.) (Sillen & Hoering, 1993). Due to the distribution of these bones throughout much of the depth of deposit at Swartkrans (approximately six meters of deposit), it has been inferred that hominins may have tended fires repeatedly during the time of deposition.

At Koobi Fora (East Turkana), other evidence has been inferred to indicate presence of fire at an Oldowan occurrence. At the FxJj 20 Complex (at sites 20 Main and 20 East), some reddened, oxidized patches of sediment, two at 20 Main and three at 20 East, have been observed within the deposit at the approximate level of artifact horizons (Harris, 1978; Bellomo, 1993; Bellomo & Kean, 1997). These apparently burned patches are less than one meter in diameter and at least 5 cm in depth. Magnetic anomalies at FxJj 20 East roughly correlate with these reddened areas, presumably due to heating and localized alteration of the magnetic properties of the sediment (Bellomo & Kean, 1997). In addition, several chert artifacts show reddening and sometimes surface crazing and pot-lid fracturing that seem to suggest thermal alteration; one of the reddened pieces at FxJj 20 East refits to a set that does not show this color change. Burned clay has also been found at Chesowanja, Kenya, associated with Oldowan materials (Gowlett *et al.*, 1981).

These curious occurrences may indicate hominin

use of fire; however, it is not clear to us that natural processes can be completely ruled out at as a factor in the apparent burning in these instances. Glynn Isaac (pers. comm.) found that in his surface scatter study ("scatter between the patches") in the Okote Tuff complex (in which the FxJj 20 Complex is located), about one in every three surface samples yielded baked sediment fragments. This might well represent evidence of burning from bush fires in the region which also swept across the site areas at FxJj 20. For example, there is burned bone at the non-hominin site of Langebaanweg in South Africa from the Pliocene, dated to 5 mya (Hendey, 1982). Natural bush fires are relatively common occurrences in dry season conditions and can ignite bushes and trees which can burn for longer periods, and at very high temperatures, after grasses have been consumed. Until more Oldowan sites are excavated and show a clear, consistent pattern of burning, and one that is spatially discrete and stands out from 'background' burning, it is difficult to say with certainty how involved Oldowan hominins were with regard to the use of fire.

Do Chimpanzees in the Wild Produce Oldowan Sites?

Beginning in the 1960's, it became clear that modern chimpanzees in the wild made and used tools for a variety of tasks (see Goodall, 1986, for an overview of her observations of tool use among the Gombe chimpanzees). The cultural and technological patterns among different populations of chimpanzees are discussed in detail by McGrew (1992, 2004). Although the objects initially identified in chimpanzee tool-use were largely organic materials with little chance of preservation (such as twigs, grasses, leaves, etc.), later observations of chimpanzee use of stone hammers and anvils in nut-cracking activities in West Africa added materials with potential archaeological visibility to the tool-using repertoire of modern chimpanzees (Boesch & Boesch, 1983, 1984).

Beginning in the 1980's, researchers began also to look at the material culture and activities of chimpanzees from a more archaeological perspective, and to discuss the spatial distribution of materials used in different activities, density of such materials per unit area, possible ape 'mental maps' of resource locations (raw materials for hammers as well as nut resources), and optimization of transport of materials used for tools. This important research has focused on nut-cracking behavior (e.g., Boesch & Boesch, 1983, 1984, 1990, 1993; Boesch & Boesch-Achermann, 2000; Mercader *et al.*, 2002), as well as the location of nests and feeding debris (Sept, 1992b).

This research has added an exciting dimension to studies of the Oldowan. First, it has enhanced and refined our appreciation of continuities between the behaviors of the extant apes and those of early tool-

making hominins, essentially recognizing the potential of a "chimpanzee archaeology." In the primate world, tool-making, not to mention tool-using, is not an exclusively "human" or even an exclusively protohuman domain. The kinds of tool-making and tool-using behaviors we observe in modern apes gives us a valuable window into the possible range of tool use in our ancestors before percussion-fractured stone tools appear in our ancestry, as well the potential continued use of organic tools after the advent of stone tools. This research has also provided useful information for modeling the dynamics of site formation processes on the landscape, particularly concerning the interplay between tool-using behaviors and the build-up of potential archaeological residues.

If comparisons are made, however, between residues from chimpanzee activity areas and Oldowan lithic assemblages, it is imperative that the comparisons are valid and precisely evaluate comparable classes of material. For instance, it is not valid to make an "apples-to-oranges" comparison between, on the one hand, stone assemblages clearly showing conchoidal fracture through precise percussive blows, and, on the other hand, fragments or shards of crumbling or disintegrating stones or bedrock. The latter material is not characteristic of any Oldowan sites or Oldowan lithic assemblages yet known.

Mercader *et al.* (2002) carried out such a comparison in their study of stone debris excavated from an area reported to have been used over a number of years by chimpanzees for nut-processing (the "P100 site" in the Taï forest of Côte d'Ivoire). On the basis of this study, they argued:

> "Thus, chimpanzees engage in cultural activities that leave behind a stone record that mimics some Oldowan occurrences and invite us to speculate whether some of the technologically simplest Oldowan sites could be interpreted as nut-cracking sites or, more generally, if some subsets of Oldowan artifacts from the more sophisticated Oldowan assemblages could be interpreted as evidence of hard-object feeding by early hominins" (Mercader *et al.*, 2002:1455).

We would strongly disagree with the notion that this stone debris, presumably (though not observably) produced as an incidental, unintentional by-product of nut-cracking, can be meaningfully compared to the stone assemblages found at early Oldowan sites. It does *not* mimic Oldowan occurrences. From our examination of a sample of Mercader *et al.*'s stone material (shown at the Paleoanthropology Society Meeting in Denver in 2002), it appeared to us that the great majority of the ostensible "stone assemblage" (most of which is classified by the authors as "microshatter") would not merit assignment to a conchoidally-fractured, or even clearly artifactual, class at excavated Oldowan sites.

The authors claim that these stone debris "fall within the size spectrum and morphological parameters observed in a subset of the earliest known hominin technological repertoires" (Mercader *et al.*, 2002:1455). The argument that this material is like Oldowan materials because it falls within a similar size range is a *non sequitur*. In this issue, size does not matter: When stone is flaked, it fractures conchoidally producing many small, conchoidally-fractured pieces, but stone can also crumble and weather into small pieces that are not conchoidally-fractured.

This brings us to the second element in their stated criteria, i.e. that the Tai P100 stone debris falls within the "morphological parameters" of early stone technologies, which is not the case. The bulk of the Taï material does not show critical morphological parameters of flaking debitage, and thus the overall Taï 'assemblage' of stone material does not show salient characteristics of an Oldowan artifact assemblage. A basic flaw in this comparison is that, as de la Torre (2004:455) has noted, Mercader *et al.* "do not include a detailed and systematic analysis of the artifacts in question, and when this is done (see, e.g. Toth *et al.* 1993, Schick *et al.* 1999) the qualitative differences between the archaeological and ethological samples are always more important than their formal similarities."

In fact, for the most part, the P100 materials presented to us had the appearance of weathered or disintegrating rock. Whether disintegration happened "in place" due to weathering processes, or whether hammering activities were responsible or perhaps helped it along, is unclear. Largely missing are flakes with distinct bulbs of percussion, distinct platforms, and clear dorsal flake scars, as well as the cores with points of hammerstone impact, negative bulbs and scars, etc., clearly observable in Oldowan artifact assemblages. Whether this stone debris resulted directly from nut-processing activities is an interesting question that remains to be investigated and verified. However, forcing such debris into arbitrary "artifact classes" does not make them comparable to Oldowan artifacts.

At Oldowan archaeological sites, there is no question that the great majority of the stone flaking is intentional, controlled, and shows a basic sense of skill in lithic reduction. It is clearly organized in a manner to efficiently produce flakes from cores, creating sharp cutting and chopping edges (which are extremely difficult to find in nature) as well as a class of pounding/battering tools (e.g. hammerstones). Cut-marked animal bones and bone shafts showing hammerstone striae and fracture patterning, as well as the small sample of microwear polishes we have on Oldowan tools, make it clear that such sharp edges and percussors were used at times to process large animal carcasses. We would agree with de la Torre who, based on his analysis of Omo 57 and Omo 123, has asserted that the "small size of the Omo artifacts does not, as has been argued, make them similar to what chimpanzees could produce by crushing stones. On the contrary, it shows that the hominins had

the technical knowledge and the manual precision required to produce flakes from minute fragments" (de la Torre, 2004:455-456).

In short, there has been no convincing evidence yet presented that chimpanzees in the wild have produced a lithic assemblage truly comparable to those identified, excavated and analyzed at Oldowan archaeological sites. The collection of stone debris reported from Tai bears no resemblance to an Oldowan assemblage in terms of the salient technological characteristics of early stone artifacts. This view is shared by a number of colleagues in our discipline, including Mohamed Sahnouni, Sileshi Semaw, and Tim White (all pers. comm., with permission). On the other hand, this research shows that chimpanzee nut-cracking behavior has the potential for "archaeological visibility" in the prehistoric record.

To identify potential "chimpanzee archaeology," it will be necessary and useful to have a critical, detailed description of materials altered by chimpanzee activities, and to verify the link between the materials and the activities. "Shatter" material such as that identified at the Taï P100 site should be analyzed and described accurately as to its salient characteristics, clearly noting differences from the conchoidally-fracture debitage produced in stone artifact manufacture. The raw materials of such shatter should be assessed as well, to see if these are consistent with the raw materials of the hammer and anvil components of the stone debris. If some materials can *truly* and *convincingly* be classified as cores or as conchoidally-fractured flakes and flake fragments, this (likely very small) sample should be identified and clearly presented in photographs or drawings and subjected to archaeological attribute analysis, not simply placed in "artifact-like" categories. Some materials might be able to be classified as cobble fragments, but classification as hammer or anvil fragments would require good evidence in terms of distinct battering marks.

In such an analysis, it would also be important to recognize and acknowledge any possible 'ringers.' For instance, it might be expected that an occasional flake might be found on the landscape that may represent low-density, archaeological background material from human activities on the landscape, and which may well stand out from the other debris, perhaps in terms of its weathering or an unusual or higher-quality raw material.

For primatologists in the field, some of whom have asked us what sorts of materials and conditions would be helpful to explore the issue of possible "chimpanzee archaeology," and how best to identify, distinguish and describe such residues, we suggest that it would be ideal to undertake the following procedures:

- To retrieve stone material from sites where chimpanzee nut-processing has been observed in real time, with observations of the types of materials used for hammers or anvils;

- To conduct controlled experiments in nut-processing with similar rocks from that region;

- To describe debris resulting in each situation with a neutral, critical eye in order to develop a better sense of real characteristics of the residues that result from the nut-cracking process;

- To be diligent in refraining from applying archaeological classification or artifact terminology (e.g. "platforms," "flakes," "flake fragments," etc.) unless absolutely justified by clear evidence of characteristics of conchoidal fracture;

- To be aware of the possible presence of some archaeological background "noise," or chance presence of some stone artifactual material from past human occupation in the area (very possibly in materials other than the local bedrock or the nut-cracking hammers and anvils);

- To compare and contrast nut-processing debris from the natural weathering and disintegration of the rocks available in the region, as rocks can disintegrate and crumble from weathering processes, affected also by internal bedding characteristics and flaws. It would be very useful to excavate samples of disintegrated stones or bedrock in the region away from nut-processing areas to see if many of the features found at the P100 Taï site or established chimpanzee nut-cracking sites are also be found in a non-ape context. In fact, we have recently analyzed a sample of naturally disintegrating Franciscan rock from the San Francisco region that exhibits size characteristics very similar to that of the excavated Taï P100 sample.

We and many of our paleoanthropological colleagues would welcome and value such critical investigations and analyses as important contributions to understanding stone residues that might be produced by chimpanzees. We look forward to such approaches in the future, and to the development of criteria to contrast and compare chimpanzee activity residues and the Plio-Pleistocene archaeological record.

BEYOND TYPOLOGY: RECENT TRENDS IN OLDOWAN RESEARCH

During the past few decades, a number of new approaches have been developed, many of them actualistic, that have been usefully applied to Oldowan studies. Here we will review some of the major approaches that have expanded our knowledge of the patterning, complexity, and context of Oldowan hominin behaviors.

Experimental Artifact Replication and Use

Experiments in making and using prehistoric stone artifacts, as well as have become an increasingly common approach to early stone age artifact assemblages. Such experimental approaches can address a number of important archaeological and paleoanthropological questions, including:

1. What *techniques of manufacture* were employed (e.g., direct freehand percussion, anvil technique, bipolar technique, throwing against an anvil, etc.)?

2. What *strategies or methods* were employed by Oldowan hominins? Can we diagram a clear reduction pattern (or *chaine operatoire*) from the unmodified raw material to the resultant archaeological flaked and battered artifacts?

3. What is the *relationship between artifact type and raw material type*? Do certain artifact forms tend to be made in certain raw materials? If so, might this result as a byproduct, with the nature of the raw material influencing patterns of fracture and modification, or is there good reason to invoke intentional selection of certain raw materials for the manufacture of certain artifact forms?

4. What are the *functional attributes for a certain artifact class in a given raw material*? For example, what Oldowan artifact classes are best for bone-breaking, animal disarticulation, or wood-working? How long can a given tool be used for a given function before it needs to be discarded or resharpened?

Casual experiments in making and/or using Oldowan types of tools were carried out in the 1960's by such prehistorians as J. Desmond Clark and Louis Leakey. A number of more detailed replication and use studies have since yielded insights into the manufacture and potential use of Oldowan artifacts (e.g., Jones, 1980, 1981, 1994; Toth, 1982, 1985, 1987b, 1991, 1997; Schick & Toth, 1993; Sahnouni *et al.*, 1997; Ludwig, 1999; Tactikos, 2005; Braun *et al.*, 2005a).

Some of the major observations that have emanated from these experimental studies have included that:

1. Direct, hard-hammer percussion was a major technique in the Oldowan, with bipolar technique also being used at some sites

2. Early tool-making hominins could be very dexterous and coordinated in reducing stone, sometimes reducing cores to a small size and directing blows of percussion in a skilled, controlled way

3. Many of the Oldowan "core tool" forms may simply be least-effort residual cores resulting from flake production, and that the final form of the core may be the product of the raw material type, size, and shape of the blank (cobble or chunk), and the extent of flaking (Toth, 1982, 1985, 1987b). Many of these Oldowan core forms grade into each other

(e.g, with continued flaking, choppers can transform into discoids or even polyhedrons).

4. There may be some indications of simple lithic "traditions" in the Oldowan that have a cultural (i.e. "learned") component. The predominance of unifacial flaking of cobbles at the Gona sites of EG 10 and EG 12, and at the Koobi Fora site of FxJj50 suggest such a component, as does the predominance of unifacially reduced thick flakes ("core scrapers" or "Karari scrapers") at a number of Koobi Fora sites along the Karari escarpment in the Okote Member (Toth, 1997; Ludwig, 1999).

5. There is a discrepancy between the cores/retouched pieces at many Oldowan sites and the experimentally-predicted debitage patterns. Often later stages of core reduction are preferentially represented at Oldowan sites, suggesting that tool-making hominins were testing cobbles and partially reducing cores "off-site", and transporting partially-flaked cores to sites for further reduction (Toth, 1982, 1985, 1997). This observation is also corroborated by refitting studies (see below).

One example of how experimentation can shed light on a palaeolithic problem can be seen in the class of battered artifacts called spheroids and subspheroids. For decades there has been considerable speculation as to what these enigmatic artifact types represent, with some ideas focused on their having been fashioned and shaped for some specific purpose or function. Various suggestions have included thrown missiles used in hunting or defense, hafted bolas stones, club heads, or some sort of plant processing tool (Willoughby, 1985). Experiments conducted in quartz/quartzite (Schick & Toth, 1994; Jones, 1994) however, demonstrate that these battered, rounded and spherical forms can be unintentionally produced after a few hours by using these stones as hammerstone percussors when flaking Oldowan cores.

These experimental observations were then tested against the archaeological sites in Beds I and II at Olduvai Gorge. Early in this Bed I to Bed II sequence, sites show relatively high percentages of lava in the "heavy-duty tool" categories and relatively little quartz, and these same sites contain low proportions of spheroids versus the numbers of hammerstones. Progressing upward through this Olduvai sequence, Oldowan and Developed Oldowan sites exhibit increasing greater percentages of quartz/quartzite (versus lava) in the "heavy-duty tool" categories, and these sites also exhibit increasing numbers of spheroids/subspheroids versus hammerstones in their assemblages. That is, as a greater emphasis develops on quartz rather than lava in producing cores or core tools, quartz spheroids and subspheroids become increasingly more common, and lava hammerstones less common. This pattern is predictable and readily understood in light of the experimental study of spheroid production: as quartz utilization increases over

time for artifact manufacture at Olduvai, quartz is used correspondingly more often as hammerstones, resulting in battered quartz forms such as spheroids and subspheroids representing well-used quartz hammerstones (Schick & Toth, 1994).

Another example of how experimental research can lend insight into unusual or puzzling artifact forms can be seen in an investigation into another type of spheroid, the "faceted spheroid." These oddly-shaped artifacts, shaped into polyhedral, nearly spherical forms but with angular facets from flake scars around most of all of their surface, have presented provocative questions as to whether they are themselves tools or not, and, if so, why would they have been deliberately shaped in this way? Experiments have now shown that faceted spheroids may also represent an artifact type whose morphology results as a byproduct rather than through deliberate shaping per se.

Experiments in flaking limestone cores with a hammerstone have shown that, in their later stages of flaking, these cores can develop a faceted and nearly spherical form, like those artifacts often referred to as a faceted balls ("*boules à facettes*"), polyhedric balls ("*boules polyédriques*"), or faceted spheroids within Early Palaeolithic assemblages. Such faceted spheroids can develop when flaking cores in certain materials such as limestone that allow flaking to proceed until very obtuse angles are achieved on the core as it approaches exhaustion and further flaking becomes very difficult (Sahnouni *et al.*, 1997). Thus, faceted spheroids may represent exhausted cores resulting as a byproduct from extensive flake production from certain raw materials such as limestone, rather than tools purposefully or deliberately "shaped" into this form.

Experiments in Oldowan stone artifact manufacture have also been conducted to examine physiological and biomechanical patterns that pertain to human evolutionary questions. These include studies of brain activity using positron emission tomography or PET (Stout et al., 2000 and this volume; Stout, this volume), kinesiological and biomechanical patterns (Dapena *et al.*, this volume), and hand and arm muscle activity (Marzke *et al.*, 1998),

Experiments in Site Formation Studies

Understanding both the behavioral and geoarchaeological processes of archaeological site formation can allow prehistorians to tease apart which patterns are the probable result of hominin behavior, other biological agents (carnivore or rodent modification of bones, bioturbation, etc.), geological processes (water action through stream flooding or wave action, sediment compaction, etc.). For example, before detailed spatial analysis is done to look for discrete behavioral patterns, it would be useful to know if geological processes have seriously reworked stone artifacts and animal bones to make such an analysis meaningless from a behavioral perspective. Isaac conducted pioneering exploratory experiments looking at geological site formation processes that can be involved in Oldowan sites (Isaac & Keller, 1967). A detailed study was conducted by Schick (1984, 1986, 1987a, 1987b, 1991, 1997) who set out a large number of facsimiles of Oldowan artifact and fossil concentrations, monitored their transformation by natural processes (particularly flood waters), and subsequently excavated and analyzed the remaining sites, and in addition conducted more controlled flume or laboratory channel studies.

Actualistic studies into site formation processes can yield important clues to evaluate the *degree* of disturbance of a site, rather than addressing disturbance in an unrealistic, binary, "either-or" scenario (Schick 1986, 1987a). Site formation experiments have made it very clear that we should discard the bimodal categories of "primary" context versus "secondary or disturbed" context. Rather, site disturbance occurs along a continuum and a detailed analysis of site patterns can yield valuable evidence as to how disturbed a site might be and clues as to the nature of that disturbance. This analytical procedure naturally makes an assessment of the sedimentary context of the archaeological deposit and the probable nature and energy level of depositional forces (channel flow, overbank floods, slope wash, etc.). Importantly, it also then analyzes characteristics of the artifact (and bone) assemblage contained within the sediment for clues as to the level and nature of disturbance during burial, as the sedimentary substrate can both overestimate and underestimate the energy of the depositional episode.

Criteria used as evidence of behavioral and geological site formation processes include:

1. Sediment particle size and sorting

2. Assemblage composition, including:

    i. Debitage size distribution

    ii. Relative proportion of debitage versus core forms

    iii. Proportion of micro-debitage (sampled at least, with very fine screen size and wet sieving)

    iv. 'Technological coherency,' or whether the debitage composition matches or is predictable from the cores present in terms of expected flake types and numbers, raw material, etc.

3. Refitting of stone and bone

4. Spatial patterns of assemblage components (cores, debitage size classes, conjoining sets) that might indicate on-site hominin behaviors (tool-making, tool-using) or rather disturbance by floods or other processes before or during the process of burial

5. Fabric of the artifact and bone deposit, such as imbrication, orientation, or dip, which might indicate disturbance by water

6. Artifact and bone condition, such as abrasion, physical or chemical weathering

Application of such criteria to archaeological site assemblages can help evaluate the probable degree of disturbance of sites during the process of burial within a unit of sediment. Ideally, such fine-grained assessment of the nature of site assemblages and their spatial distribution can help identify those sites with better 'behavioral integrity,' i.e., those that might bear more direct indications of on-site hominin behaviors (Schick, 1992, 1997).

Consideration of Raw Material Selection and Transport

Other aspects of lithic technology that have become areas of interest in their own right concern hominin selection of raw material for tool production and the transport of this raw material and manufactured stone artifacts to and from site localities where artifact manufacture and/or tool use took place. Such studies shed light on hominin tool-related behaviors in a larger framework, both spatially and chronologically, than merely what has occurred at a particular archaeological site. In effect, research into raw material selection and the transport of raw materials and artifacts bring into focus the larger environment in which the hominins were moving, living and adapting and allows us to appreciate aspects of hominin behaviors and the choices they made not just from "on-site" but also "off-site" archaeology.

Studies of the selection of raw materials for stone artifacts have obvious importance in view of:

- the impact that different raw material types can have on stone tool manufacture and its products

- if strong selectivity in use of raw materials can be demonstrated, possible implications with regard to hominin cognitive abilities and their familiarity and experience in tool-making activities

- potential insights into larger-scale hominin movements across the landscape from raw material sources to sites where artifacts were manufactured or discarded

- possible impact that a site's "distance-from-raw-materials" might have on lithic reduction patterns and the artifacts produced (for instance, whether local sources might yield larger, less heavily reduced cores than more heavily 'curated' materials from more distant sources, etc.)

- possible evidence for "opportunism" versus greater planning depth among hominins in their use of very local or more distant resources

Geologist Richard Hay carried out an assessment of

the sources of raw materials at Olduvai Gorge in Beds I and II (Hay, 1971, pp. 17-18; Hay 1976:182-186) showing that tool-making hominins regularly transported rock to sites from sources a few kilometers away. He found that the "majority of artifacts at all excavated sites in Beds I and II are made of materials obtainable within a distance of 4 km, and at most sites are less than 2 km from possible sources" (Hay, 1976:183). Hay estimated that the transport distances of quartz/quartzite (from the basement outcrops at the Naibor Soit inselberg on the north side of the lake basin) and of lavas (from highlands to the south and east of the lake basin) would generally have been in the order of several kilometers. Notably, however, larger site assemblages normally also contain some materials obtained from more distant sources, at least 8 to 10 km away, and the proportion of such distant raw materials increases over time at sites in the Olduvai sequence to at least early Bed III times. Very few artifacts are made in 'exotic' materials from sources completely outside the basin (Hay, 1976:183).

Hay found that most of the lavas used in Bed I and lower Bed II probably came from the volcano Sadiman (in the volcanic highlands to the southeast of the Olduvai basin) in the form of rounded cobbles found in streams on the north side of the volcano. He argued that abundant use of the Sadiman-type lava versus other available lavas at Oldowan and Developed Oldowan A sites may have been due to preference for its "dense, homogenous nature" (Hay, 1976:183), but also noted that use of Sadiman lava dropped by Developed Oldowan B times, a trend which continue upward in the sequence. Use of a phonolite from Engelosin, a volcano to the north of the basin, is found first in Bed II sites situated from 9 to 11 km away from this source (Hay, 1971), and use of a gneiss of Kelogi type (outcropping in inselbergs near the west edge of the Side Gorge) is found in Bed I and Bed II sites at least 8 to 10 km from the nearest outcrops (Hay, 1976:184). Chert was available at the lake margin in Bed II times, but even this local material appears often to have been transported to other locales where it was worked and/or utilized (Stiles *et al.*, 1974), and artifacts made in local basin cherts "are abundant only within 1 km of probable source areas" (Hay, 1976:185).

The chert-bearing exposures that were available to hominins at site MNK "Chert Factory Site" differ in oxygen isotope composition and in size to the chert artifacts found at that site, suggesting that hominins were transporting larger lithic materials there from some other chert source(s) (the cherts flaked at the site having formed in lower-salinity water than the local chert, possibly further south in the lake basin). Furthermore, at site HWK some 1.3 km northeast of MNK, it appears that many of the chert flakes found at HWK were brought in from such a quarry site already flaked, as they seem to represent a selected size without much small debitage (although this could also represent

fluvial winnowing of the smaller size fraction from the site.)

Beginning in Bed II of Olduvai (the "Developed Oldowan"), tool-making hominins began to concentrate on working quartz and quartzite rather than lava at many sites (Leakey, 1971). Whether this represents intentional selection of these materials because of their hardness and sharpness, or whether this reflects difference ranging patterns or some other factor, remains unclear. In any case, this de facto 'preference' for quartz in fact increases over time throughout Bed II times, as quartz often represents the vast majority of the "heavy-duty tools," "light-duty tools," and debitage (Schick & Toth, 1994).

At Koobi Fora, Oldowan hominins usually selected raw materials in proportions that were generally available in the nearby gravel deposits (Toth, 1997), although they clearly avoided clasts that were highly vesicular or badly weathered. The occasional chert artifacts with a very diagnostic and distinctive color also suggest that some high-quality flakes may have been transported as individual artifacts some distance. The low proportion of early stages of reduction at many sites can partially be explained by testing raw materials out at gravel sources before transporting them for further reduction.

Analysis of the artifact assemblages at Kanjera in Kenya indicate that most of the artifacts were manufactured from fine-grained igneous rocks that were locally available. Nevertheless, approximately 15% of the artifacts were manufactured from raw materials that were not immediately local to the site (quartzite, chert, vein quartz and quartz porphyry) that were apparently brought in from more remote sources (Plummer et al., 1999). Recent research by Braun et al. (2005b) has been investigating the relationship between the mechanical properties of raw material and how it relates to artifact form and function.

More recently, research at Gona has indicated that at some very early Oldowan sites, dating to between 2.6 and 2.2 mya have rock types in higher than expected frequencies compared to frequencies in contemporaneous cobble gravels there (Stout et al., 2005). High-quality, fine-grained trachytes cobbles appear to have been preferentially selected as raw materials at sites EG13, DAS7, DAN1, OGS6, and DAN2d. In addition, most Gona sites (OGS7, DAS7, DAN1, OGS6, and DAN2d) had higher than expected frequencies of aphanitic volcanic rock (so fine-grained that individual crystals cannot be seen with the naked eye), and vitreous volcanic rock ("volcanic chert") in higher than expected frequencies. This suggests that the earliest known hominin toolmakers in the world already showed some discrimination in selecting higher-quality raw materials. Interestingly, similar selectivity is also known among certain birds in their choice of gizzard stones, presumably using visual clues such as polish, luster, and color to identify harder, finer-grained stones of chert.

Refitting Studies and Spatial Analysis

The ability to refit flaked lithic materials (and sometimes fractured bone) back together can yield very valuable information about early hominin behavior, technology, and site context. Refitting of stone artifacts can be used to:

1. Give a "blow-by-blow" account of core reduction/flake production at an archaeological site; this may give important information regarding the decision-making and possible technological patterns or strategies of Oldowan tool-makers.

2. Identify what stages of flaking are represented:

   - Do the refits form a complete cobble or chunk (complete reduction)?

   - Were cores brought in partially flaked?

   - Were cores (or select debitage) apparently carried away from the excavated areas?

3. Show whether the refitting sets have any special patterning, e.g., was the flaking done in a discrete place, or was the core moved around the site as reduction progressed?

4. Show what type of core/retouched piece may have been produced at a site and then taken away from the flaking area (or away from the site altogether) ("phantom" artifact), by making a mold of the empty space produced by refitted flakes/fragments.

5. Help assess the geoarchaeological context of the site, for instance, whether the spatial patterns are due to hominin behavior or have been altered appreciably by depositional forces such as stream or wave action, and also whether significant vertical dispersion of artifacts (as well as bones) has occurred after deposition (e.g. through bioturbation by roots, rodents, etc.).

6. Assess whether there is more than one temporal bout of hominin activity at a given site (e.g. are refitted sets found at different horizons offset vertically or microstratigraphically, suggesting separate flaking episodes spaced through time?).

To date, refitting has been successfully and systematically employed at a number of Oldowan localities in the greater Turkana basin. These include sites at Koobi Fora at East Turkana, where refitting at a minimum of eight different localities has revealed technological patterns of refitting stones, some of which show minimal disturbance by depositional forces as well as spatial configurations of both stone artifacts and broken bones that have been refitted (Bunn *et al.*, 1980; Toth, 1982, 1985; Kroll & Isaac, 1984; Schick, 1984, 1986; Kroll, 1997). Some Koobi Fora sites, such as FxJj 3, FwJj 1, FxJj 20E, FxJj 50, and FxJj 64, exhibit refitting patterns

that indicate discrete knapping areas, suggesting minimal geological reworking of lithic materials after hominin discard. At two of these sites, FxJj 50 and FxJj 64, there are also spatial concentrations of refitted animal bones exhibiting cut-marks and hammerstone fracture, suggesting that these were areas of meat and marrow consumption (Kroll, 1997).

Refitting has also been accomplished with great success in a lithic assemblage from Lokalalei 2C (LA 2C) at West Turkana, showing technological patterns at the site and spatial configurations of on-site flaking episodes dating to approximately 2.34 mya (Roche *et al.*, 1999). At the Lokalalei site, approximately 10% of an assemblage of 2583 surface and excavated artifacts has been refitted (and 20% of the excavated assemblage of 2067 artifacts) to at least one other artifact. The refitting sets show knapping of over 60 cobbles at the site, most of these showing a few flakes struck from a core, sometimes with the core included. A few refitting sets, however, show a fairly large series of between 10 and 20 flakes removed from the core. The authors suggest that the technological patterns exhibited by the refitting at Lokalalei 2C demonstrate greater cognitive abilities and motor skills than they had previously attributed to Pliocene tool-makers, and conclude that, therefore, early Oldowan sites exhibit diversity in their technological patterns that likely represent differences among hominin groups in cognitive and motor skills (Roche *et al.*, 1999).

However, the core forms at Lokalalei 2C do not necessarily exhibit more 'skilled' flake removal than at many other Pliocene localities, even at occurrences at Gona dating to 2.6 mya. The Gona sites, for instance, show a great deal of control, precision and coordination in consistent unifacial flaking and bifacial flaking of cores (Semaw et al., 1997, 2003, and Semaw, this volume). Clearly, however, the cores at Lokalalei 2C show on average more extensive core reduction than most Oldowan sites predating 2 mya. It may be that the Lokalalei 2C site differs more in degree than in kind: most of the cores are fairly extensively flaked, and, moreover, the refitting evidence provides more detailed information about flake removals and core manipulation than is normally available at early Oldowan sites. Of course, extent of core reduction might be expected to vary for any of a number of reasons, including, for instance, access to raw materials, competition within the hominin group for easily flaked cores (e.g. due to cobble shape), size and quality of the cobble blank, reoccupation and reuse of the site and its materials, demand or need for tools, etc., and it may not necessarily reflect tool-making skill.

The notion that hominin flaking at the Lokalalei 2C site is more 'elaborate' or 'sophisticated' than at other Pliocene localities, and that this reflects some profound difference in hominin abilities between sites, has not yet been demonstrated to the satisfaction of many Palaeolithic archaeologists. Much of Oldowan lithic technological variability may be due to differences in flaking qualities of the raw materials used and extent of core reduction at different sites for any of the reasons mentioned above, rather than firm differences in hominin technical abilities in their stone flaking. Such differences in abilities may certainly exist (especially considering the fact that multiple hominin taxa appear to have coexisted), but more work needs to be done to demonstrate real differences in technological 'skill' per se.

Taphonomic Studies

As previously noted in the section above on "Oldowan Hunters or Scavengers?," palaeoanthropologists have become increasingly critical in their approach to faunal materials found at archaeological sites. Over the past few decades, great strides have been made in developing criteria to identify diverse site formation processes involved in the concentration, modification, and preservation of animal remains. A range of diverse actualistic and experimental studies have greatly enlarged our understanding of the potential role of diverse agencies (hominins, different carnivores, rodents, etc.) in building faunal concentrations and provided valuable criteria to help infer the relative impacts of these different agencies. Application of such studies to Oldowan faunal assemblages is exemplified in the chapter by Pickering and Domínguez-Rodrigo in this volume, Pickering (2002), Brain (1981), and Bunn & Kroll (1986), .

The diverse range of criteria of interest to taphonomists now include, in addition to more traditional zooarchaeological analyses (such as identification of taxon, element, age, sex, left/right siding), a range of other features of bone assemblages that can potentially give information regarding the agent(s) of its accumulation and modification. These criteria include aspects of bone fragmentation (completeness/degree of fragmentation, element portion, shaft circumference, fracture patterning, etc.), more discrete evidence of hominin-induced modification (cut-marks, hammerstone striae, bone flakes and flake scars), evidence of other biological agents of modification (carnivore gnawing, rodent gnawing, root marks, trampling), other types of chemical or mechanical modification (acid etching, sediment abrasion, weathering), and signs of modification by heating or burning.

In addition, certain types of modification (e.g. polishing, patterned flaking or shaping) may suggest that bone was modified during manufacture and/or use as a tool. The increased use of optical and electron microscopy in examining striae, pits, depressed fractures has become a standard approach among zooarchaeologists and taphonomists. Image analysis systems are beginning to be being employed to help image, quantify, and analyze patterns of modification on bone surfaces.

Dietary Studies

Besides the information from faunal remains, several other lines of evidence can help yield clues pertaining to the diets of early hominins, what food items may have been part of this diet, and how early technology might have enhanced the acquisition and processing of such food items. Using modern primates and hunter-gatherers as models, most paleoanthropologists believe the bulk of the diet of early hominin populations consisted of plant resources such as berries, fruits, nuts, leaves, pith, flowers, shoots, seeds, and gum, as well as underground resources such as roots, tubers, corms and rhizomes. This plant component would have likely been supplemented by animal resources such as insects, eggs, small reptiles (e.g., lizards, tortoises, and snakes), amphibians, mollusks, and fish, as well as larger animals.

Systematic survey of modern environments thought to be similar to those of Oldowan sites (e.g. grassland, woodland mosaics with riverine forest and/or lake margin habitats) can enable researchers, based on hunter-gatherer and primate analogs, to assess the distribution and density of different edible resources and to model different hominin foraging patterns. Such studies have been done by Peters & O'Brien (1981), Sept (1984, 1992a), Vincent (1984), and Copeland (2004). Recently, the importance of underground storage organs (USO's) such as roots, tubers, rhizomes, and corms has been stressed by Laden & Wrangham (2005), showing that many of the Plio-Pleistocene sites yielding early hominin remains also contain cane rat fossils, animals that specialize in the feeding of such underground food resources.

Tooth-wear on Plio-Pleistocene hominins from can also show patterns of wear that may be indicative of dietary patterns (Grine, 1986; Grine & Kay, 1988). Recently, the teeth of *A. africanus* and *A. (P.) robustus* were re-analyzed using dental microwear texture analysis (scale-sensitive fractal analysis) (Scott et al., 2005). The results suggested that the microwear textures of *robustus* were more complex, and more variable in complexity, than *africanus*, suggesting that the *robustus* diet included hard, brittle foods, analogous to the diet of capuchin monkeys, which includes hard, brittle seeds. The dental microwear of *africanus*, on the other hand, suggested tougher foods, analogous to the diet of howler monkeys that includes tough leaves. The authors conclude that "early hominin diet differences might relate more to microhabitat, seasonality or fall-back food choice than to oversimplified, dichotomous food preferences" (Scott et al., 2005, p. 694).

A number of studies have investigated the stable carbon isotope ratios ($^{13}C/^{12}C$) in tooth enamel of fossil and modern animals (including hominins, baboons, and other mammals) in order to discern relative proportions of C_3 to C_4 resources in their respective diets (Lee-Thorpe & van der Merwe, 1993; Lee-Thorpe et al.,

1994; Sponheimer et al., 2005a). Initial studies found a dietary distinction between, on the one hand, robust australopithecines along with three fossil baboon species (two species of *Papio* and one species of *Parapapio*), and, on the other hand, the *Theropithecus dartii* baboon (Lee-Thorpe & van der Merwe, 1993; Lee-Thorpe et al., 1994). This isotopic evidence appeared to indicate that the robust australopithecines and the three baboons with similar signatures were consuming relatively high amounts of C_3 plants such as tubers, roots, corms, fruits and nuts, while the *Theropithecus* baboon's carbon isotope ratios indicated a diet involving large amounts of C_4 foods, presumably grasses. Another interesting difference in C_3 versus C_4 isotopes was observed between two different felids, with lions apparently preying largely on C_4-rich grazers, but with leopards preying more on animals with high C_3 concentrations, perhaps including the C_3-rich robust australopithecines and baboons (Lee-Thorpe & van der Merwe, 1993). This earlier research indicated that robust australopithecines, in addition to their consumption of C_3 foods, also took in an appreciable quantity of C_4 foods (25 to 30% of their diet), either in the form of grasses or, perhaps, grass-eating animals (Lee-Thorpe et al., 1994).

Recently Sponheimer et al. (2005a) reported that carbon isotope evidence in tooth enamel of South African australopithecines indicates that gracile forms (*Australopithecus africanus* from Sterkfontein) and robust forms (*Australopithecus* or *Paranthropus robustus* from Swartkrans and Kromdraai) both had a strong reliance on non-C_3 foods, with C_4 resources evidently comprising 35 to 40% of their diet. This strongly contrasts with modern chimpanzee data, which indicates that these modern apes, living in tropical forests, have an essentially pure C_3 diet. Sponheimer et al. (2005a) suggest that other C_4 foodstuffs, besides such foods as sedges and termites, were probably a part of the gracile and robust australopithecine diet. Important food items the authors suggest could have contributed to the C4-component of australopithecine diet include grass seeds and roots, succulents (e.g. euphorbias and aloes), and various animal foods (e.g., grasshoppers, bird eggs, rodents, lizards, and young mammals such as antelope).

Oxygen isotope data suggest that the later, robust australopithecines lived in a more arid environment than the earlier, gracile form (Sponheimer et al., 2005a). This pattern may reflect changing environments over time, from wetter woodlands to more arid grasslands between 2.5 and 1.8 mya. Alternatively, ecological differences between the taxa could also have contributed to the oxygen isotope differences. Sponheimer et al. (2005a) make the interesting observation, however, that the proportion of C4 dietary resources remains much the same between the two taxa despite the oxygen isotope evidence for marked environmental change over time.

Investigation of another isotopic ratio, that of strontium to calcium (Sr/Ca), has provided additional information regarding early hominin dietary components (e.g., Sillen, 1992; Sillen et al., 1995; Sponheimer et al., 2005b). As strontium levels have been observed to be generally higher in herbivores than faunivores, Sr/Ca levels have sometimes been used as an indication of the relative degree of herbivory vs. carnivory among different animals. A study of the Sr/Ca ratio in tooth enamel in fossil fauna from the Sterkfontein Valley, fossil South African hominins (*Australopithecus africanus* and *Australopithecus/Paranthropus robustus*), and modern African mammals in Kruger National Park shows relatively high Sr/Ca ratios in both hominin taxa, more within the range of most grazing animals). This study also showed a higher ratio as well as a greater range of Sr/Ca levels among samples in the gracile form (*A. africanus*) than in the robust form (*A./P. robustus*) (Sponheimer et al., 2005b).

This relatively high strontium level for both hominin taxa here (higher than carnivores and leaf-eating browsers in the modern Sterkfontein Valley, the latter showing the lowest Sr/Ca ratios) appears to contradict a lower value for *Paranthropus* found in a earlier study, which was taken to indicate a more omnivorous diet (including significant intake of some sort of animal foods) for this robust form (Sillen, 1992). Sponheimer et al. (2005b) reject insectivory as a likely cause of the relatively high Sr/Ca ratios in these australopithecine taxa, for while insectivores also have high Sr/Ca, another ratio, Ba/Ca, is high in insectivores but very low in the australopithecines here. The combination of high Sr/Ca and low Ba/Ca are noted to have been found in mole rats and warthogs, so that consumption of roots and rhizomes, an important component of the diet of these animals, is suggested as a possible cause of the isotope patterns observed (Sponheimer et al., 2005b).

Interestingly, early *Homo* at Swartkrans (e.g. SK 847) has elevated Sr/Ca ratios compared to *A. robustus* (Sillen et al., 1995). It was suggested in this study that early *Homo* may have been exploiting geophytes, such as bulbs of edible lilies of the genus *Hypoxis*, which are available locally in the Transvaal today; alternatively, they may have been consuming animals with an elevated Sr/Ca ratio such as hyraxes.

Isotopic research has shown considerable promise as a tool for deciphering aspects of the diet of prehistoric animals, including various forms of hominin taxa as well as other mammals. At the present stage of development of this field, however, it seems advisable to keep in mind that:

- a relatively small sample of fossil specimens has thus far been subjected to analysis

- there is likely more to learn regarding diagenetic processes impacting isotopic signatures (isotope content in living specimens is often different from that in similar fossil taxa)

- there may also be more to learn about the influence of diet earlier versus later in life, seasonal changes in diet, etc., on the isotopic signatures contained in different structural elements

- our reference sample needs to be enlarged for modern analogue species with known diet

- we should expand our knowledge of the range of variation in the diet of modern species in different environments, with different sets of available foods, etc., and the effect of such differences on isotopic signatures

- similar isotopic signatures might be obtained from diets with quite different food profiles, so that we may need additional analytical procedures to tease these apart (as in the above case involving the Sr/Ca ratio, in which the Ba/Ca ratio might help distinguish between different diet compositions).

It appears that, at the present stage of development of this line of research, each study presents some new intriguing pattern, though an overall synthesis of results is not yet attainable. Researchers are becoming increasingly aware, however, of the complexities involved in paleoisotopic studies, and undoubtedly further research along these lines will help further elucidate aspects of the diets of fossil hominins and other animals.

Landscape Archaeology

Most conventional Oldowan archaeological fieldwork at open-air sites has consisted of locating high-density surface occurrences on an eroding stratigraphic outcrop and subsequently conducting excavations at these localities to recover dense and informative *in situ* materials. Another approach, called landscape archaeology (earlier called study of "scatters between the patches" and subsequently referred to as "scatters and patches analysis" (Isaac & Keller, 1967; Isaac, 1981, 1997:9; Isaac et al., 1981; Stern, 1991), attempts to understand distributions of archaeological materials (artifacts and fossil bones) on the paleolandscape by examining their presence, nature, and densities along erosional outcrops at a given stratigraphic horizon. These surface materials are thus regarded as a sign or proxy of the *in situ* materials buried within the sedimentary units in these areas and which might represent or give some indication of landscape use within a single "unit" of time (although this "unit" is . Such materials are then examined for lithic technological patterns (e.g. raw materials, artifact types, ratio of cores to debitage, flake types as an indication of earlier or later stages of flaking, etc.) as well as the nature of the fossil remains.

Some of the interesting patterns that have emerged from this work are:

1. There often tends to be a co-occurrence on the landscape of the peaks of higher densities of both fossil bone and stone artifacts. This suggests that either early tool-using hominins were focusing on areas of their landscapes where animals were congregating and/or dying, or the hominins were major agents of collection and concentration of animal remains in areas where they also discarded large numbers of artifacts. In addition, bone preservation tends to be favored in fluvial floodplain environments with rapid burial and deposition. (Isaac, pers. comm., Stern, 1991:343).

2. That the *majority* of surface artifacts in these surveys are found in very low-density occurrences, rather than the dense concentrations or "sites" that archaeologists tend to focus on and excavate (Isaac, pers. comm., Stern, 1991). This suggests that much of the manufacture and/or use and ultimate discard of lithic materials occurred away from the anomalous dense concentrations. The implications of this pattern are that archaeologists may be missing important aspects of the overall behavioral repertoire of early hominins if they concentrate exclusively on the larger-scale concentrations of artifacts and bones for excavation, and that they should also pay special attention to lower-density "mini-sites" or even "single bout of activity" areas (e.g. Marshall, 1997: 220-223; also see Isaac *et al.*, 1981, "Small is informative..." and Isaac 1981, "Stone Age Visiting Cards").

More recently, landscape archaeology has been applied systematically to the lowermost Bed II at Olduvai Gorge where a single stratigraphic horizon may be traced laterally for some distance and thus provide a window into patterns of hominin activities over the larger landscape (Blumenschine & Masao, 1991; Peters & Blumenschine, 1995, 1996; Blumenschine & Peters, 1998). Applications of this approach have been made as well to the East Turkana (Koobi Fora) archaeology (Stern, 1991; Rogers, 1996, 1997).

Placement of sites on the landscape relative to paleoecological variables such as climate (wetter or dryer climate, seasonality, etc.) and paleoenvironments (lake margin, alluvial plain, alluvial fans, riparian corridors, etc.) has been a major concern in a number of these studies (Peters & Blumenschine, 1995, 1996; Rogers et al, 1994; Blumenschine & Peters, 1998). Such research has also incorporated actualistic studies of modern analog landscapes in order to compare and contrast potential distribution of resources (e.g., drinking water, raw materials for tools, plant foods such as fruits or underground storage organs, animal foods such as scavengeable carcasses, trees and shrubs for shade or refuge) in the paleolandscape relative to the locations of lower-density scatters and the excavated, higher-density localities (Peters & Blumenschine, 1995).

Such studies have not yet generated overarching conclusions regarding Oldowan hominin activity variation across Plio-Pleistocene paleolandscapes, and of course such activity variation may well be inextricably linked to the constraints and possibilities within a particular sedimentary basin. Stern (1991) has also cautioned that sedimentary layers used for landscape archaeological studies, such as the lower Okote Member at Koobi Fora, can represent palimpsests developed over tens of thousands of years, and thus can be viewed as 'contemporaneous' in only a very special, very broad sense.

Nevertheless, landscape archaeological research can help build predictive models as to the kinds of stone and bone assemblages might be found within different components of the environment, and these predictions can be tested in future research. Further application of the landscape archaeological approach to Oldowan occurrences in a variety of sedimentary basins and critical synthesis of the results of different such studies are likely to provide interesting and useful information regarding the larger picture of Oldowan hominin activities across their paleoenvironmental landscape.

CONCLUSION

Some seventy years after Louis Leakey first proposed the concept of an Oldowan, dozens of sites in East, Central, South, and North Africa have produced evidence of early hominin tool-makers between 2.6 and 1.4 mya. The earliest Oldowan sites are found shortly before the first evidence of the emergence of the genus *Homo*, a lineage characterized through time by reduced jaws and teeth and increased encephalization relative to the robust australopithecines. Although we cannot presently establish which Plio-Pleistocene hominins were the predominant tool-makers, it appears that by 1.0 mya the robust australopithecines had gone extinct, making *Homo* the only hominin genus to survive.

Oldowan lithic technology was simple, characterized by battered percussors, cores made on cobbles and chunks, flakes and fragments, and sometimes retouched flakes. Although this technology was simple, many sites show considerable skill in removing large, sharp flakes from cores. When preservation is ideal, Oldowan assemblages are associated with cut-marked and broken animal bones that indicate these early tool-makers were processing animal carcasses as part of their overall adaptive strategy. Sites become more numerous after 2.0 mya, and by c. 1.7-1.4 mya new technological elements begin to appear, such as the consistent production of large (> 15 cm) flakes and the manufacture of picks, cleavers, and handaxes that heralds the beginnings of the Acheulean Industrial complex.

Some of the major conclusions of Oldowan research over the past half century have included:

- The Oldowan Industrial Complex is characterized by simple stone technologies that include percussors (hammerstones and spheroids), cores (e.g., choppers, discoids, heavy-duty scrapers, polyhedrons), unmodified flakes, sometimes retouched flakes, and other debitage (snapped flakes, split flakes, angular fragments, chunks).

- The earliest Oldowan sites are known from Gona, Ethiopia at c. 2.6 mya. The Oldowan is contemporaneous with early Acheulean indus-

tries (starting at c. 1.7-1.4 mya), and Oldowan-like occurrences continue to be found in later phases of prehistory. The term is especially used in Africa, but has also been applied to sites in Eurasia. Defining the end of the Oldowan is somewhat arbitrary, but usually the term is not used for industries less than c. 1 mya.

- Oldowan sites are found at open air localities along streams and lake margins in East Africa (Gona, Hadar, Middle Awash, Konso Gardula, Melka Kunture, Fejej, Omo, East Turkana, West Turkana, Chesowanja, Kanjera, Olduvai Gorge, Peninj, and Nyabusosi), cave deposits in the Transvaal region of South Africa (Sterkfontein, Swartkrans, Kromdraai), and open air sites along streams in North Africa (Ain Hanech and El-Kherba).

- Hominin taxa contemporaneous with Oldowan sites between 2.6 and 1.4 million years ago include *Australopithecus garhi, Australopithecus africanus, Australopithecus (Paranthropus) aethiopicus, Australopithecus (Paranthropus) boisei, Australopithecus (Paranthropus) robustus*, early *Homo* sp., *Homo habilis, Homo rudolfensis*, and *Homo ergaster/erectus*.

- Although it is not clear which hominin taxa were responsible for the manufacture and use of Oldowan stone artifacts, many anthropologists believe the genus *Homo*, with a larger brain and reduced jaws and dentition over time, was a more likely dedicated tool-maker and tool-user. After about 1 mya, the robust australopithecines (and earlier forms of *Homo*) were extinct, with only *Homo ergaster/erectus* and their descendants that carry on the stone tool-making tradition.

- Although some prehistorians suggest that some early sites should be assigned to a "Pre-Oldowan," others feel that the ranges of technologies exhibited at early sites all fall within the Oldowan Industrial Complex, with some sites exhibiting more heavily reduced cores and more retouched flakes than other sites.

- Oldowan tools were clearly used in animal butchery (meat-cutting/bone fracture) based on bone modification and lithic microwear. Microwear has also suggested that hominins used stone tools for a range of other functions, including woodworking and cutting of soft plant material. Much of the evidence from cut marks and microwear suggest that sharp, unmodified flakes were a major part of the technological repertoire of Oldowan hominins, along with stone hammers used in knapping and to break bones.

- Theories of how Oldowan sites formed include camps or home bases, scavenging stations, stone caches, or more generic favored places. It is likely that a range of explanations is required to explain Oldowan occurrences through time, space, and environmental settings, and that different sites may have served different functions.

- Current debate about hominin foraging strategies has divided archaeologists into two main camps: those that favor a scavenging model, and those that favor a hunting/primary access model. It is possible that aspects of both models characterize early hominin procurement of animal resources, again, at different times, places and environments. Further research should clarify this picture.

- Theories to explain encephalization in the genus *Homo* include social complexity, the rise of symbolic behavior, tool-making, and higher-quality diet. It is likely that this phenomenon of accelerated brain expansion in the human lineage was due to the ability of hominins to access higher-quality food resources through the use of technology, which allowed for a decreased gut size and increased brain size.

- Evidence for fire is found at several Oldowan sites, notably Swartkrans Member 3 in South Africa and the FxJj 20 Complex at Koobi Fora, Kenya. Although hominins may have maintained fire at these sites, the possibility of natural fires modifying bones and lithic artifacts cannot be ruled out.

- Although modern chimpanzees nut-cracking behavior may produce battered and pitted stones and occasional stone fracture or disintegration, these phenomena are not really comparable to Oldowan sites. Oldowan lithic technology shows clear, deliberate, and patterned flaking of stone.

- Recent trends in Oldowan research have included experimental artifact replication and use, site formation studies, studies of raw material selection and transport, refitting and spatial analysis, taphonomic studies, dietary studies including chemical analysis of isotopic signatures in fossil bone, and landscape archaeology.

It is likely that many new Oldowan occurrences will be discovered in this century and that a range of new theoretical and methodological approaches will be applied to the earliest Palaeolithic record. These new lines of evidence should give us a clearer understanding of the complexity of the Oldowan archaeological record and a greater appreciation of the range of adaptive behaviors in the emergent tool-making and tool-using hominins that ultimately led to the modern human condition.

REFERENCES CITED

Aiello, L. & Dean, C. (1990). An Introduction to Human Evolutionary Anatomy. London: Academic Press.

Aiello, L.C. & Wheeler, P. (1995). The expensive-tissue hypothesis. Current Anthropology 36(2):199-221.

Ardrey, R. (1976). The Hunting Hypothesis . New York: Athenaeum.

Asfaw, B., Beyene, Y., Semaw, S., Suwa, G., White, T. & WoldeGabriel, G. (1991). Fejej: a new palaeoanthropological research area in Ethiopia. Journal of Human Evolution 20:137-143.

Asfaw, B., Beyene, Y., Suwa, G., Walter, R.C., White, T.D., WoldeGabriel, G. & Yemane, T. (1992). The earliest Acheulean from Konso-Gardula. Nature 360:732-5.

Asfaw, B., White, T., Lovejoy, O., Latimer, B. & Simpson, S. (1999). Australopithecus garhi: A New Species of Early Hominid from Ethiopia. Science 284(5414):629-34.

Bellomo, R.V. (1993). A methodological approach for identifying archaeological evidence of fire resulting from human activities. Journal of Archaeological Science 20:525-53.

Bellomo, R.V. & Kean, W.F. (1997). Appendix 4: Evidence of hominid-controlled fire at the FxJj 20 site complex, Karari Escarpment. In: Koobi Fora Research Project, Vol. 5: Plio-Pleistocene Archaeology (G.L. Isaac, Ed.), pp. 224-236. Oxford: Clarendon Press.

Biberson, P. (1967). Galets Aménagés du Maghreb et du Sahara: Types I.1-I.8, II.1-II.16, III.1-III.6. In: Congres Panafrican de Préhistoire et d'Études Quaternaires. Paris: Arts et Métiers Graphiques.

Bilsborough, A. (1992). Human Evolution. Glasgow, UK: Blackie Academic & Professional.

Binford, L. R. (1981). Bones: Ancient Men and Modern Myths. New York: Academic Press.

Binford, L.R. (1987). Searching for camps and missing the evidence? Another look at the Lower Paleolithic. In: The Pleistocene Old World (O. Soffer, Ed.), pp. 17-31. Plenum Publishing Corporation.

Blumenschine, R.J. (1986). Early Hominid Scavenging Opportunities. Oxford: British Archaeological Reports.

Blumenschine, R.J. (1987). Characteristics of an early hominid scavenging niche. Current Anthropology 28(4):383-407.

Blumenschine, R.J. (1988). An experimental model of the timing of hominid and carnivore influence on archaeological bone assemblages. Journal of Archaeological Science 15:483-502.

Blumenschine, R.J. (1989). Man the scavenger. Archaeology 42:26-32.

Blumenschine, R.J. & Masao, F.T. (1991). Living sites at Olduvai Gorge, Tanzania? Preliminary landscape archaeology results in the Basal Bed II Lake Margin Zone. Journal of Human Evolution 21:451-62.

Blumenschine, R.J. & Peters, C.R. (1998). Archaeological predictions for hominid land use in the paleo-Olduvai Basin, Tanzania, during lowermost Bed II times. Journal of Human Evolution 34:565-607.

Blumenschine, R.J. & Selvaggio, M.M. (1988). Percussion marks on bone surfaces as a new diagnostic of hominid behavior. Nature 333:763-765.

Boaz, N.T. & Almquist, A.J. (1999). Essentials of Biological Anthropology. New Jersey: Prentice-Hall, Inc.

Boesch, C. & Boesch, H. (1983). Optimisation of nut-cracking with natural hammers by wild chimpanzees. Behavior 83:265-286.

Boesch, C. & Boesch, H. (1984). Mental map in Wild Chimpanzees: an analysis of hammer transports for nut cracking. Primates 25(2):160-70.

Boesch, C. & Boesch, H. (1990). Tool use and tool making in wild chimpanzees. Folia Primatologica 54:86-99.

Boesch, C. & Boesch, H. (1993). Different hand postures for pounding nuts with natural hammers by wild chimpanzees. In: Hands of the Primates , (H. Prueschoft & D.J. Chivers, Eds.), pp. 31-43. Vienna: Springer-Verlag.

Boesch, C. & Boesch-Achermann, H. (2000). The Chimpanzees of Tai Forest: Behavioural Ecology and Evolution. New York: Oxford University Press.

Boyd, R. & Silk, J. B. (1997). How Humans Evolved. New York: W. W. Norton & Company.

Brain, C.K. (1981). The Hunters or the Hunted? An Introduction to African Cave Taphonomy. Chicago: The University of Chicago Press.

Brain, C.K. (1993). Swartkrans: A Cave's Chronicle of Early Man. Pretoria: Transvaal Museum.

Brain, C.K. & Sillen, A. (1988). Evidence from the Swartkrans Cave for the earliest use of fire. Nature 336(6198):464-6.

Braun, D.R., Tactikos, J.C., Ferraro, J.V. & Harris, J.W.K. (2005a). Flake recovery rates and inferences of Oldowan hominin behavior: a response to Kimura 1999, 2002. Journal of Human Evolution 48(5):525-31.

Braun, D., Plummer, T., Ferraro, J., Bishop, L., Ditchfield, P., Potts, R., & Harris, J. (2005b). Oldowan technology at Kanjera South, Kenya: the context of technological diversity. Paper presented at Paleoanthropology Society Meetings, Milwaukee.

Brown, F.H. & Gathogo, P.N. (2002). Stratigraphic relation between Lokalalei 1A and Lokalalei 2C, Pliocene archaeological sites in West Turkana, Kenya. Journal of Archaeological Science 29:699-702.

Bunn, H.T. (1981). Archaeological evidence for meat-eating by Plio-Pleistocene hominids from Koobi Fora and Olduvai Gorge. Nature 291(5816):574-7.

Bunn, H.T. (1982). Meat-Eating and Human Evolution: Studies on the Diet and Subsistence Patterns of Plio-Pleistocene Hominids in East Africa. Ph.D. Dissertation. University of California, Berkeley.

Bunn, H.T. (1983). Evidence on the diet and subsistence patterns of Plio-Pleistocene hominids at Koobi Fora, Kenya, and Olduvai Gorge, Tanzania. In: Animals and Archaeology, Vol. 1: Hunters and Their Prey, (J. Clutton-Brock & C. Grigson, Eds.), pp. 21-30. Oxford, England: British Archaeological Reports.

Bunn, H.T. (1994). Early Pleistocene hominid foraging strategies along the ancestral Omo River at Koobi Fora, Kenya. Journal of Human Evolution 27:247-266.

Bunn, H.T. & Kroll, E.M. (1986). Systematic butchery by Plio/Pleistocene hominids at Olduvai Gorge, Tanzania. Current Anthropology 27(5):431-52.

Bunn, H., Harris, J.W.K., Isaac, G.L., Kaufulu, Z., Kroll, E., Schick, K., Toth, N. & Behrensmeyer, A.K. (1980). FxJj50: an Early Pleistocene site in Northern Kenya. World Archaeology 12(2):109-44.

Byrne, R. & Whiten, A. (Eds.) (1988). Machiavellian Intelligence: Social Expertise and the Evolution of Intellect in Monkey, Apes, and Humans. New York: Clarendon Press.

Campbell, Bernard G. & Loy, James D. (1996). Humankind Emerging. Rhode Island: Harper Collins College Publishers.

Capaldo, S.D. (1995). Inferring Hominid and Carnivore Behavior From Dual-Patterned Archaeofaunal Assemblages . Ph.D. Dissertation. Rutgers, The State University of New Jersey.

Chavaillon, J. (1970). Découverte d'un niveau oldowayen dans la basse vallée de l'Omo (Ethiopie). Bulletin de la Société Préhistorique Française 67:7-11.

Chavaillon, J. (1976). Evidence for the technical practices of early Pleistocene hominids, Shungura Formation, Lower Omo Valley, Ethiopia. In: Earliest Man and Environments in the Lake Rudolf Basin, (Y. Coppens, F.C. Howell, G.L. Isaac & R.E.F. Leakey, Eds.), pp. 565-573. Chicago: University of Chicago Press.

Chavaillon, J. & Chavaillon, N. (1976). Le paléolithique ancien en ethiopie caractères techniques de l'Oldowayen de Gomboré I a Melka-Konturé. In: Les Plus Ancienne Industries en Afrique, UISPP, IX Congrès: pp. 43-69.

Chavaillon, J., Chavaillon, N., Hours, F. & Piperno, M. (1979). From the Oldowan to the Middle Stone Age at Melka-Kunturé (Ethiopia). Understanding cultural changes . Quaternaria 21:87-114.

Clark, G. (1961). World Prehistory. Cambridge: Cambridge University Press.

Clark, J.D. (1991). Stone artifact assemblages from Swartkrans, Transvaal, South Africa. In: Cultural beginnings: approaches to understanding early hominid lifeways in the African savanna (J.D. Clark, Ed.), pp. 137-158. Bonn: Dr. Rudolf Habelt GMBH.

Copeland, S. (2004). Paleoanthropological Implications of Vegetation and Wild Plant Resources in Modern Savanna Landscapes, with Applications to Plio-Pleistocene Olduvai Gorge, Tanzania. Ph.D. Dissertation. New Brunswick: Rutgers University.

Dart, R.A. (1953). The predatory transition from ape to man. International Anthropological and Linguistic Review 1(4):201-18.

Dart, R.A. (1957). The Osteodontokeratic Culture of Australopithecus prometheus. Pretoria: Transvaal Museum.

Day, M.H. (1986). Guide to Fossil Man. Chicago: The University of Chicago Press.

de Heinzelin, J., Clark, J.D., White, T.D., Hart, W.K., Renne, P.R., WoldeGabriel, G., Beyene, Y. & Vrba, E.S. (1999). Environment and behavior of 2.5-million-year-old Bouri hominids. Science 284:625-9.

de Heinzelin, J., Clark, J. D., Schick, K. D. & Gilbert, W. H. (Eds.) (2000). The Acheulean and the Plio-Pleistocene Deposits of the Middle Awash Valley Ethiopia. Belgium: Department of Geology and Mineralogy/ Royal Museum of Central Africa.

Deacon, T.W. (1997). The Symbolic Species: The Co-Evolution of Language and the Brain. New York: W.W. Norton and Company.

Delagnes, A. & Roche, H. (2005). Late Pliocene hominid knapping skills: the case of Lokalalei 2C, West Turkana, Kenya. Journal of Human Evolution 48(5):435-72.

Delson, E., Tattersall, I., Van Couvering, J.A., & Brooks, A.S. (Eds.) (2000). Encyclopedia of Human Evolution and Prehistory. New York: Garland.

Ditchfield, P., Hicks, J., Plummer, T., Bishop, L.C. & Potts, R. (1999). Current research on the Late Pliocene and Pleistocene deposits north of Homa Mountain, southwestern Kenya. Journal of Human Evolution 36:123-50.

Domínguez-Rodrigo, M. (2002). Hunting and scavenging by early humans: the state of the debate. Journal of World Prehistory 16(1):1-54.

Domínguez-Rodrigo, M. & Pickering, T.R. (2003). Early hominid hunting and scavenging: a zooarchaeological review. Evolutionary Anthropology 12:275-82.

Domínguez-Rodrigo, M., de la Torre, I., de Luque, L., Alcalá, L., Mora, R., Serrallonga, J. & Medina, V. (2002). The ST site complex at Peninj, West Lake Natron, Tanzania: implications for early hominid behavioral models. Journal of Archaeological Science 29:639-65.

Domínguez-Rodrigo, M., Pickering, T., Semaw, S. & Rogers, M. (2005). Cutmarked bones from Pliocene archaeological sites at Gona, Afar, Ethiopia: implications for the function of the world's oldest stone tools. Journal of Human Evolution 49:109-21.

Dunbar, R.I.M. (1992). Neocortex size as a constraint on group size in primates. Journal of Human Evolution 20:469-493.

Dunbar, R.I.M (1993). Coevolution of neocortical size, group size and language in humans. Behavioural and Brain Sciences 16:681-735.

Egeland, C.P., Pickering, T.R., Domínguez-Rodrigo, M. & Brain, C.K. (2004). Disentangling Early Stone Age palimpsests: determining the functional independence of hominid- and carnivore-derived portions of archaeofaunas. Journal of Human Evolution 47:343-57.

Field, A.S. (1999). An Analytical and Comparative Study of the Earlier Stone Age Archaeology of the Sterkfontein Valley. Ph.D. Dissertation. University of the Witwatersrand, Johannesburg.

Gibson, K.R. (1986). Cognition, brain size and the extraction of embedded food resources. In: Primate Ontogeny, Cognition and Social Behaviour, (J.G. Else & P.C. Lee, Eds.), pp. 93-103. Cambridge: Cambridge University Press.

Goodall, J. (1986). The Chimpanzees of Gombe: Patterns of Behavior. Cambridge MA and London, England: The Belknap Press of Harvard University.

Gowlett, J.A.J. (1990). 2. Archaeological Studies of Human Origins & Early Prehistory in Africa. In: A History of African Archaeology, (P. Robertshaw, Ed.), pp. 13-38. New Hampshire: Heinemann Educational Books.

Gowlett, J.A.J., Harris, J.W.K., Walton, D. & Wood, B.A. (1981). Early archaeological sites, hominid remains and traces of fire from Chesowanja, Kenya. Nature 294(5837):125-9.

Grine, F.E. (1986). Dental evidence for dietary differences in Australopithecus and Paranthropus: a quantitative analysis of permanent molar microwear. Journal of Human Evolution 15:783-822.

Grine F.E. & Kay, R.F. (1988). Early hominid diets from quantitative image analysis of dental microwear. Nature 333:765-768.

Grine, F.E. & Susman, R.L. (1991). Radius of Paranthropus robustus from Member 1, Swartkrans Formation, South Africa. American Journal of Physical Anthropology 84:229-48.

Harris, J.W.K. (1978). The Karari Industry: Its Place in East African Prehistory. Ph.D. Dissertation. University of California.

Harris, J.W.K. (1983). Cultural beginnings: Plio-Pleistocene archaeological occurrences from the Afar, Ethiopia. The African Archaeological Review 1:3-31.

Harris, J.W.K. & Gowlett, J.A.J. (1980). Evidence of early stone industries at Chesowanja, Kenya. In: Pre-Acheulean and Acheulean cultures in Africa, (R.E. Leakey and B.A. Ogot, Eds.). Proceedings of the 8th Panafrican Congress of Prehistory and Quaternary Studies, pp. 208-212.

Hay, R.L. (1971). Geologic background of Beds I and II: stratigraphic summary. In: Olduvai Gorge, Volume 3: Excavations in Beds I and II, 1960-1963 (M.D. Leakey), pp. 9-18. Cambridge: Cambridge University Press.

Hay, R.L. (1976). Geology of the Olduvai Gorge. Berkeley: University of California Press.

Hendey, Q.B. (1982). Langebaanweg: A Record of Past Life. Cape Town: South African Museum.

Hovers, E. (2003). Treading carefully: site formation processes and Pliocene lithic technology. In: Oldowan: Rather More than Smashing Stones: First Hominid Technology Workshop, Treballs d'Arqueologia, 9 (J. Martinez Moreno, R. Mora Torcal, I. de la Torre Sainz, Eds.), pp.145-158. Bellaterra, Spain: Universitat Autònoma de Barcelona.

Hovers, E., Schollmeyer, K., Goldman, T., Eck, G.G., Reed, K.E., Johanson, D.C., & Kimbel, W.H. (2002). Later Pliocene archaeological sites in Hadar, Ethiopia. Paleoanthropology Society Abstracts, Journal of Human Evolution, A17.

Howell, F.C., Haesaerts, P. & de Heinzelin, J. (1987). Depositional environments, archaeological occurrences and hominids from Members E and F of the Shungura Formation (Omo basin, Ethiopia). Journal of Human Evolution 16:665-700.

Humphrey, N.K. (1976). The social function of intellect. In: Growing Points in Ethology, (P.P.G. Bateson and R.A. Hinde, Eds.), pp. 303-317. Cambridge: Cambridge University Press.

Isaac, G.L. (1971). The diet of early man: aspects of archaeological evidence from Lower and Middle Pleistocene sites in Africa. World Archaeology 2(3):278-98.

Isaac, G.L. (1976). Plio-Pleistocene artifact assemblages from East Rudolf, Kenya. In: Earliest Man and Environments in the Lake Rudolf Basin: Stratigraphy, Paleoecology, and Evolution (Y. Coppens, F.C. Howell, G.L. Isaac & R. Leakey, Eds.), pp. 552-564. Chicago: University of Chicago Press.

Isaac, G.L. (1977). Olorgesailie: Archaeological Studies of a Middle Pleistocene Lake Basin in Kenya. Chicago: University of Chicago Press.

Isaac, G.L. (1978). The food-sharing behavior of protohuman hominids. Scientific American 238(4):90-109.

Isaac, G. L. (1981). Stone age visiting cards: approaches to the study of early land-use patterns. In: Patterns of the Past. (I. Hodder, G. Isaac, and N. Hammond, Eds.), pp. 131-155. Cambridge: Cambridge University Press.

Isaac, G.L. (1984). The archaeology of human origins: studies of the Lower Pleistocene in East Africa. In: Advances in World Archaeology, (F. Wendorf & A. Close, Eds.), pp. 1-87. New York: Academic Press.

Isaac, G.L. (Ed.) (1997). Koobi Fora Research Project, Vol. 5: Plio-Pleistocene Archaeology. Oxford: Clarendon Press.

Isaac, G.L. & Keller, C.M. (1967). Towards the interpretation of occupation debris: some experiments and observations. Kroeber Anthropological Papers 37:31-57.

Isaac, G.L., Harris, J.W.K. & Marshall, F. (1981). Small is informative: the application of the study of mini-sites and least-effort criteria in the interpretation of the Early Pleistocene archaeological record at Koobi Fora, Kenya. In: Las Industrias mas Antiguas, (J.D. Clark and G. L. Isaac, Eds.), pp. 101-119. Mexico City: X Congresso Union International de Ciencias Prehistoricas y Protohistoricas.

Johanson, D. & Edgar, B. (1996). From Lucy to Language. New York: Simon & Schuster Editions.

Jones, P.R. (1980). Experimental butchery with modern stone tools and its relevance for palaeolithic archaeology. World Archaeology 12(2):153-165.

Jones, P.R. (1981). Experimental implement manufacture and use: a case study from Olduvai Gorge, Tanzania. In: The Emergence of Man. (J.Z. Young, E.M. Jope, & K.P. Oakley, Eds.), pp. 189-195. Philosophical Transactions of the Royal Society of London, Series B. vol. 292, no. 1057.

Jones, P.R. (1994). Results of experimental work in relation to the stone industries of Olduvai Gorge. In: Olduvai Gorge Volume 5: Excavation in Beds III, IV, and the Masek Beds, 1968-1971, (By Mary Leakey, with Derek Roe), pp. 254-298. Cambridge: Cambridge University Press.

Keeley, L.H. (1980). Experimental Determination of Stone Tool Uses: A microwear analysis. Chicago: University of Chicago Press.

Keeley, L.H. & Toth, N. (1981). Microwear polishes on early stone tools from Koobi Fora, Kenya. Nature 293:464-6.

Kibunjia, M., Roche, H., Brown, F.H. & Leakey, R.E. (1992). Pliocene and Pleistocene archaeological sites west of Lake Turkana, Kenya. Journal of Human Evolution 23:431-8.

Kibunjia, M. (1994). Pliocene archaeological occurrences in the Lake Turkana Basin. Journal of Human Evolution 27:159-71.

Kimbel, W.H., Walter, R.C., Johanson, D.C., Reed, K.E., Aronson, J.L., Assefa, Z., Marean, C.W., Eck, G.C., Bobe, R., Hovers, E., Rak, Y., Vondra, C., Yemane, T., York, D., Chen, Y., Evensen, N.M. & Smith, P.E. (1996). Late Pliocene Homo and Oldowan tools from the Hadar Formation (Kada Hadar Member), Ethiopia. Journal of Human Evolution 31:549-61.

Klein, R.G. (1999). The Human Career. Chicago: University of Chicago Press.

Kroll, E.M. (1997). Lithic and faunal distributions at eight archaeological excavations. In: Koobi Fora Research Project, Vols. 5: Plio-Pleistocene Archaeology, pp. 459-543. (G.L. Isaac, Ed.). Oxford: Clarendon Press.

Kroll, E.M. & Isaac, G.L. (1984). Configurations of artifacts and bones at early Pleistocene sites in East Africa. In: Intrasite Spatial Analysis in Archaeology, (H.J. Hietala, Ed.), pp. 4-31. Cambridge: Cambridge University Press.

Kuman, K. (1994). The Archaeology of Sterkfontein - past and present. Journal of Human Evolution 27:471-95.

Kuman, K. (2005). La Préhistoire ancienne de l'Afrique méridonale: contribution des sites à hominidés d'Afrique du Sud. In: Le Paléolithique en Afrique: L'histoire la plus longue, (M. Sahnouni, Ed.), pp. 53-82. Paris: Éditions Artcom', Guides de la Préhistoire Mondiale.

Kuman, K., Field, A.S. & Thackeray, J.F. (1997). Discovery of new artifacts at Kromdraai. South African Journal of Science 93:187-93.

Laden, G. & Wrangham, R. (2005). The rise of the hominids as an adaptive shift in fallback foods: plant underground storage organs (USOs) and australopith origins. Journal of Human Evolution 49:482-498.

Leakey, L.S.B. (1936). Stone Age Africa. London: Oxford University Press.

Leakey, M.D. (1971). Olduvai Gorge, Volume 3: Excavations in Beds I and II, 1960-1963. Cambridge: Cambridge University Press.

Leakey, M.D. (1975). Cultural patterns in the Olduvai sequence. In: After the Australopithecines: Stratigraphy, Ecology, and Culture Change in the Middle Pleistocene, (K.W. Butzer and G.L. Isaac, Eds.), pp. 477-493. The Hague: Mouton Publishers.

Leakey, M.D. (1994). Olduvai Gorge Volume 5: Excavation in Beds III, IV, and the Masek Beds, 1968-1971. (With Derek Roe). Cambridge: Cambridge University Press.

Lee, R.B. & DeVore, I. (1968). Man the Hunter. Chicago: Aldine Publishing Company.

Lee-Thorp, J.A. & van der Merwe, N.J. (1993). Stable carbon isotope studies of Swartkrans fossils. In: Swartkrans: A Cave's Chronicle of Early Man, (C.K. Brain, Ed.), pp. 251-256. Pretoria: Transvaal Museum.

Lee-Thorp, J.A., van der Merwe, N.J. & Brain, C.K. (1994). Diet of Australopithecus robustus at Swartkrans from stable carbon isotopic analysis. Journal of Human Evolution 27:361-372.

Lewin, R. & Foley, R.A. (2004). Principles of Human Evolution. Oxford UK: Blackwell Publishing.

Ludwig, B.V. (1999). A Technological Reassessment of East African Plio-Pleistocene Lithic Artifact Assemblages. Ph.D. Dissertation. Rutgers, The State University of New Jersey.

de Lumley, H. (1969). Le paléolithique inférieur et moyen du Midi Méditerranéen dans son cadre géologique. Paris: Éditions CNRS.

de Lumley, H. & Beyene, Y. (Eds.) (2004). Les Sites Préhistoriques de la Région de Fejej, Sud-Omo, Éthiopie, dans leurs contexte stratigraphique et paléontologique. Paris: Éditions Recherche sur les Civilisations.

de Lumley, H., Nioradzé, M., Barsky, D., Cauche, D. Celiberti, V., Nioradzé, G., Notter, O., Zhvania, D. & Lordkipanidze, D. (2005). Les industries lithiques préoldowayennes du début du Pléistocène inférieur du site de Dmanissi en Géorgie. L'anthropologie 109:1-182.

Lumsden, C.J. & Wilson E. (1983). Promethean Fire: Reflections on the Origin of Mind. Cambridge, MA: Harvard University Press.

Marshall, F. (1997). FxJj 64. In: Koobi Fora Research Project, Vol. 5: Plio-Pleistocene Archaeology. (G.L. Isaac, Ed.), pp. 220-223. Oxford: Clarendon Press.

Marzke, M.W., Toth, N., Schick, K., Reece, S., Steinberg, B., Hunt, K., Linscheid, R.L. & An, K.-N. (1998). EMG study of hand muscle recruitment during hard hammer percussion manufacture of Oldowan tools. American Journal of Physical Anthropology 105:315-32.

McGrew, W.C. (1992). Chimpanzee Material Culture: Implications for Human Evolution. New York: Cambridge University Press.

McGrew, W. (2004). The Cultured Chimpanzee. Cambridge, UK: Cambridge University Press.

Mercader, J., Panger, M. & Boesch, C. (2002). Excavation of a chimpanzee stone tool site in the African Rainforest. Science 296:1452-1455.

Merrick, H.V. (1976). Recent archaeological research in the Plio-Pleistocene deposits of the Lower Omo, southwestern Ethiopia. In: Human Origins: Louis Leakey and the East African evidence, (G.L. Isaac & T. McCown, Eds.), pp. 461-481. Menlo Park, CA: W. A. Benjamin, Inc.

Monahan, C.M. (1996). New zooarchaeological data from Bed II, Olduvai Gorge, Tanzania: implications for hominid behavior in the Early Pleistocene. Journal of Human Evolution 31:93-128.

Monahan, C.M. & Dominguez-Rodrigo, M. (1999). Comparing apples and oranges in the Plio-Pleistocene: methodological comments on 'Meat-eating by early hominids at the FLK 22 Zinjanthropus site, Olduvai Gorge (Tanzania): an experimental approach using cut-mark data.' Journal of Human Evolution 37:789-792.

Movius, H.L. (1949). Pleistocene research: Old-World palaeolithic archaeology. Bulletin of the Geological Society of America 60:1443-56.

Oliver, J.S. (1994). Estimates of hominid and carnivore involvement in the FLK Zinjanthropus fossil assemblages: some socioecological implications. Journal of Human Evolution 27:267-294.

Peters, C.R. & Blumenschine, R.J. (1995). Landscape perspectives on possible land use patterns for Early Pleistocene hominids in the Olduvai Basin, Tanzania. Journal of Human Evolution 29:321-62.

Peters, C.R. & Blumenschine, R.J. (1996). Landscape perspectives on possible land use patterns for Early Pleistocene hominids in the Olduvai Basin, Tanzania: Part II, Expanding the landscape models. Kaupia 6:175-221.

Peters, C.R. & O'Brien, E.M. (1981). The early hominid plant-food niche: insights from an analysis of plant exploitation by Homo, Pan, and Papio in eastern and southern Africa. Current Anthropology 22(2):127-40.

Pickering, T.R. (1999). Taphonomic Interpretations of the Sterkfontein Early Hominid Site (Gauteng, South Africa) Reconsidered in Light of Recent Evidence. Ph.D. Dissertation, Department of Anthropology, University of Wisconsin, Madison.

Pickering, T.R. (2001). Taphonomy of the Swartkrans hominid postcrania and its bearing on issues of meat-eating and fire management. In: Meat-Eating & Evolution, (D. Stanford & H. Bunn, Eds.), pp. 33-51. Oxford: Oxford University Press.

Pickering, T.R. (2002). Reconsideration of criteria for differentiating faunal assemblages accumulated by hyenas and hominids. International Journal of Osteoarchaeology 12:127-141.

Pickering, T.R., White, T.D. & Toth, N. (2000). Cutmarks on a Plio-Pleistocene hominid from Sterkfontein, South Africa. American Journal of Physical Anthropology 111:579-84.

Piperno, M. (1993). The origins of tool use and the evolution of social space in palaeolithic times: some reflections. In: The Use of Tools by Human and Non-human Primates, (A. Berthelet & J. Chavaillon, Eds.), pp. 254-266. Oxford: Clarendon Press.

Plummer, T., Bishop, L.C., Ditchfield, P. & Hicks, J. (1999). Research on late Pliocene Oldowan sites at Kanjera South, Kenya. Journal of Human Evolution 36:151-70.

Potts, R. (1984). Home bases and early hominids. American Scientist 72:338-47.

Potts, R. (1988). Early Hominid Activities at Olduvai. New York: Aldine De Gruyter.

Potts, R. & Shipman, P. (1981). Cutmarks made by stone tools on bones from Olduvai Gorge, Tanzania. Nature 291:577-580.

Ramendo, L. (1963). Les Galets Aménagés de Reggan (Sahara). Libyca t. XI:43-74.

Raynal, J.-P., Sbihi Alaoui, F.-Z., Magoga, L., Mohib, A. & Zouak, M. (2002). Casablanca and the earliest occupation of North Atlantic Morocco. Quaternaire 13(1):63-77.

Roche, H. (1989). Technological evolution in early hominids. OSSA 4:97–98.

Roche, H. & Kibunjia, M. (1994). Les sites archéologiques Plio-Pléistocènes de la formation de Nachukui, West Turkana, Kenya. Comptes Rendus de l'Académie des Sciences, Paris 318(2):1145-51.

Roche, H. & Tiercelin, J.-J. (1980). Industries lithiques de la formation plio-pléistocène d'Hadar Ethiopie (campagne 1976). In: Pre-Acheulean and Acheulean cultures in Africa, (R.E. Leakey & B.A. Ogot, Eds), pp. 194-199. Nairobi: Proceedings of the 8th Panafrican Congress of Prehistory and Quaternary Studies.

Roche, H., Delagnes, A., Brugal, J., Feibel, C., Kibunjia, M., Mourre, V. & Texier, P. (1999). Early hominid stone tool production and technological skill 2.34 Myr ago in West Turkana, Kenya. Nature 399:57-60.

Rogers, M.J. (1996). Landscape archaeology at East Turkana, Kenya. In: Four million years of hominid evolution in Africa: papers in honour of Dr. Mary Douglas Leakey's outstanding contribution in palaeoanthropology (C.C. Magori, C.B. Saanane & F. Schrenk, Eds.), pp. 21-26. Darmstadt: Darmstädter Beiträge zur Baturgeschichte, Heft 6.

Rogers, M.J. (1997). A Landscape Archaeological Study from East Turkana, Kenya. Ph.D. Dissertation. New Brunswick: Rutgers University.

Rogers, M.J., Feibel, C.S., & Harris, J.W.K. (1994). Changing patterns of land use by Plio-Pleistocene hominids in the Lake Turkana Basin. Journal of Human Evolution 27:139-158.

Sahnouni, M. (Ed.) (2005). Le Paléolithique en Afrique: L'histoire la plus longue. Paris: Éditions Artcom', Guides de la Préhistoire Mondiale.

Sahnouni, M. & de Heinzelin, J. (1998). The site of Ain Hanech revisited: new investigations at this Lower Pleistocene site in Northeastern Algeria. Journal of Archaeological Science 25(11):1083-101.

Sahnouni, M., de Heinzelin, J. & Saoudi, Y. (1996). Récentes recherches dans le gisement oldowayen d'Ain Hanech, Algérie. Comptes rendus de l'Académie des sciences, Série 2 323:639-644.

Sahnouni, M., Hadjouis, D., van der Made, J., Derradji, A., Canals, A., Medig, M. & Belahrech, H. (2002). Further research at the Oldowan site of Ain Hanech, North-eastern Algeria. Journal of Human Evolution 43:925-37.

Sahnouni, M., Schick, K. & Toth, N. (1997). An experimental investigation into the mature of faceted limestone "spheroids" in the Early Paleolithic. Journal of Archaeological Science 24:701-13.

Schick, K.D. (1984). Processes of Palaeolithic Site Formation: An Experimental Study. Ph.D. Dissertation. University of California.

Schick, K.D. (1986). Stone Age Sites in the Making: Experiments in the Formation and Transformation of Archaeological Occurrences. Oxford: British Archaeological Reports.

Schick, K.D. (1987a). Experimentally-Derived Criteria for Assessing Hydrologic Disturbance of Archaeological Sites. In: Natural Formation Processes and the Archaeological Record, (D.T. Nash & M.D. Petraglia, Eds.), pp. 86-107. Oxford: British Archaeological Reports.

Schick, K.D. (1987b). Modeling the formation of Early Stone Age artifact concentrations. Journal of Human Evolution 16:789-807.

Schick, K.D. (1991). On making behavioral inferences from early archaeological sites. In: Cultural Beginnings: Approaches to understanding early hominid life-ways in the African savanna (J.D. Clark, Ed.), pp. 79-107. Bonn: Dr. Rudolf Habelt GMBH.

Schick, K.D. (1992). Geoarchaeological analysis of an Acheulean site at Kalambo Falls, Zambia. Geoarchaeology 7(1):1-26.

Schick, K.D. (1997). Experimental studies of site-formation processes. In: Koobi Fora Research Project, Volume 5: Plio-Pleistocene Archaeology, (G.L. Isaac, Ed.), pp. 244-256. Oxford, UK: Clarendon Press.

Schick, K.D. & Toth, N. (1993). Making Silent Stones Speak: Human Evolution and the Dawn of Technology. New York: Simon & Schuster.

Schick, K. & Toth, N. (1994). Early Stone Age technology in Africa: a review and case study into the nature and function of spheroids and subspheroids. In: Integrative Paths to the Past: Palaeoanthropological Advances in Honor of F. Clark Howell, (R. Coruccini & R. Ciochon, Eds.), pp.1-29. Englewood Cliffs, New Jersey: Prentice-Hall.

Schick, K.D., Toth, N., Garufi, G.S., Savage-Rumbaugh, E.S., Rumbaugh, D. & Sevcik, R. (1999). Continuing investigations into the stone tool-making and tool-using capabilities of a bonobo (Pan paniscus). Journal of Archaeological Science 26:821-32.

Scott, R.G., Ungar, P.S., Bergstrom, T.L., Grown, C.A. Grine, F.E., Teaford, M.F., & Walker, A. (2005). Dental microwear texture analysis shows within-species diet variability in fossil hominins. Nature 436:693-695.

Selvaggio, M.M. (1994). Evidence From Carnivore Tooth Marks and Stone-Tool-Butchery Marks for Scavenging by Hominids at FLK Zinjanthropus Olduvai Gorge, Tanzania. Ph.D. Dissertation. Rutgers, The State University of New Jersey.

Semaw, S. (1997). Late Pliocene Archaeology of the Gona River Deposits, Afar, Ethiopia. Ph.D. Dissertation, Anthropology Department. New Brunswick: Rutgers University.

Semaw, S. (2000). The world's oldest stone artefacts from Gona, Ethiopia: their implications for understanding stone technology and patterns of human evolution between 2.6-1.5 million years ago. Journal of Archaeological Science 27:1197-214.

Semaw, S., Renne, P., Harris, J.W.K., Feibel, C.S., Bernor, R.L., Fesseha, N. & Mowbray, K. (1997). 2.5-million-year-old stone tools from Gona, Ethiopia. Nature 385(January 23 1997):333-6.

Semaw, S., Rogers, M.J., Quade, J., Renne, P.R., Butler, R.F., Dominguez-Rodrigo, M., Stout, D., Hart, W.S., Pickering, T. & Simpson, S.W. (2003). 2.6-million-year-old stone tools and associated bones from OGS-6 and OGS-7, Gona, Afar, Ethiopia. Journal of Human Evolution 45:169-77.

Sept, J.M. (1984). Plants and Early Hominds in East Africa: A Study of Vegetation in Situations Comparable to Early Archaeological Site Locations . Ph.D. Dissertation. University of California.

Sept, J.M. (1992a). Archaeological evidence and ecological perspectives for reconstructing early hominid subsistence behavior. In: Archaeological Method and Theory. Vol. 4 (M.B. Schiffer, Ed.), pp. 1-56. New York: Academic Press.

Sept, J.M. (1992b). Was there no place like home? A new perspective on early hominid archaeological sites from the mapping of chimpanzee nests. Current Anthropology 33(2):187-207.

Sillen, A. (1992). Strontium-calcium ratios (Sr/Ca) of Australopithecus robustus and associated fauna from Swartkrans. Journal of Human Evolution 23:495-516.

Sillen, A. & Hoering, T. (1993). Chemical characterisation of burnt bones from Swartkrans. In: Swartkrans: a cave's chronicle of early man, vol. 8 (C. K. Brain, Ed.), pp. 243–9. Pretoria: Transvaal Museum Monographs.

Sillen, A., Hall, G., & Armstrong, R. (1995). Strontium calcium ratios (Sr/Ca) and strontium isotopic ratios (87Sr/86Sr) of Australopithecus robustus and Homo sp. from Swarkrans. Journal of Human Evolution 28:277-285.

Sponheimer, M., Lee-Thorp, J., de Ruiter, D., Codron, D., Codron, J., Baugh, A.T. & Thackeray, F. (2005a). Hominins, sedges, and termites: new carbon isotope data from the Sterkfontein valley and Kruger National Park. Journal of Human Evolution 48:301-312.

Sponheimer, M., de Ruiter, D., Lee-Thorp, J. & Späth, A. (2005b). Sr/Ca and early hominin diets revisited: new data from modern and fossil tooth enamel. Journal of Human Evolution 48:147-156

Stern, N. (1991). The Scatters-Between-the-Patches: A Study of Early Hominid Land Use Patterns in the Turkana Basin, Kenya. Ph.D. Dissertation. Harvard University.

Stiles, D.N., Hay, R.L., & O'Neil, J. (1974). The MNK chert factory site, Olduvai Gorge, Tanzania. World Archaeology 5: 285-308.

Stout, D., Toth, N., Schick, K., Stout, J. & Hutchins, G. (2000). Stone tool-making and brain activation: positron emission tomography (PET) studies. Journal of Archaeological Science 27:1215-23.

Stout, D., Quade, J., Semaw, S.R.M.J. & Levin, N.E. (2005). Raw material selectivity of the earliest stone toolmakers at Gona, Afar, Ethiopia. Journal of Human Evolution 48(4):365-80.

Susman, R.L. (1991). Who made the Oldowan tools? Fossil evidence for the tool behavior in Plio-Pleistocene hominids. Journal of Anthropological Research 47(2):129-151.

Suwa, G., Asfaw, B., Beyene, Y., White, T.D., Katoh, S., Nagaoka, S., Nakaya, H., Uzawa, K., Renne, P. & WoldeGabriel, G. (1997). The first skull of Australopithecus boisei. Nature 389:489-492.

Tactikos, J.C. (2005). Landscape and Experimental Perspectives on Variability in Oldowan Technology at Olduvai Gorge, Tanzania. Ph.D. Dissertation. New Brunswick: Rutgers University.

Tavoso, A. (1972). Les industries de la moyenne terrasse du Tarn à Tecou (Tarn). Bulletin du Musée d'Anthropologie Préhistorique de Monaco 18:113-144.

Texier, P.-J. (1993). NY 18, an Oldowan site at Nyabusosi, Lake Albert Basin, Toro, Uganda. In: Four million years of hominid evolution in Africa: An international congress in honour of Dr. Mary Douglas Leakey's outstanding contribution to palaeoanthropology, Arusha (Tanzania), Volume de Résumés. (C.C. Magori, C.B. Saanane & F. Schrenk, Eds.), pp. 71-72. Darmstadt: Darmstädter Beiträge zur Baturgeschichte, Heft 6.

Texier, P.-J. (1995). The Oldowan assemblage from NY 18 Site at Nyabusosi (Toro-Uganda). Comptes Rendus de l'Académie des Sciences, Paris 320:647-53.

de la Torre, I. (2004). Omo revisited: evaluating the technological skills of Pliocene hominids. Current Anthropology 45(4):439-65.

de la Torre, I & Mora, R. (2004). El Olduvayense de la Sección Tipo de Peninj (Lago Natron, Tanzania). Barcelona: Universitat Autónoma de Barcelona.

de la Torre, I., Mora, R., Domínguez-Rodrigo, M., Luque, L. & Alcalá, L. (2003). The Oldowan industry of Peninj and its bearing on the reconstruction of the technological skills of Lower Pleistocene hominids. Journal of Human Evolution, 44:203-224.

Toth, N. (1982). The Stone Technologies of Early Hominids at Koobi Fora: An Experimental Approach. Ph.D. Dissertation. University of California, Berkeley.

Toth, N. (1985). The Oldowan reassessed: a close look at early stone artifacts. Journal of Archaeological Science 12:101-20.

Toth, N. (1987a). Behavioral inferences from Early Stone Age artifact assemblages: an experimental model. Journal of Human Evolution 16:763-87.

Toth, N. (1987b). The first technology. Scientific American 255(4):112-21.

Toth, N. (1991). The importance of experimental replicative and functional studies in palaeolithic archaeology. In: Cultural Beginnings: Approaches to Understanding Early Hominid Life-Ways in the African Savanna (J.D. Clark, Ed.), pp. 109-124. Bonn: Dr. Rudolf Habelt GMBH.

Toth, N. (1997). The Artefact Assemblages in the Light of Experimental Studies. In: Koobi Fora Research Project Vol. 5, Plio-Pleistocene Archeology, (G.L. Isaac, Ed.), pp. 363-401. New York: Oxford University Press.

Toth, N., Schick, K.D., Savage-Rumbaugh, E.S., Sevcik, R.A. & Rumbaugh, D.M. (1993). Pan the tool-maker: investigations into the stone tool-making and tool-using capabilities of a bonobo (Pan paniscus). Journal of Archaeological Science 20:81-91.

Van Riet Lowe, C. (1952). The Pleistocene Geology and Prehistory of Uganda, Vol. II: Prehistory. Colchester: Authority of the Uganda Government.

Villa, P. (1983). Terra Amata and the Middle Pleistocene Archaeological Record of Southern France. Berkeley: University of California Press.

Vincent, A.S. (1984). Plant foods in savanna environments: a preliminary report of tubers eaten by the Hadza of Northern Tanzania. World Archaeology 17(2):131-47.

Washburn, S.L. (1960). Tools and human evolution. Scientific American 203(3):3-15.

Willoughby, P.R. (1985). Spheroids and battered stones in the African Early Stone Age. World Archaeology 17(1):44-60.

Wolpoff, M. H. (1999). Paleoanthropology, 2nd ed. New York: McGraw-Hill.

Wood, B. (1991). Koobi Fora Research Project (Volume 4): Hominid Cranial Remains. Oxford: Clarendon Press.

CHAPTER 2

THE OLDEST STONE ARTIFACTS FROM GONA (2.6-2.5 MA), AFAR, ETHIOPIA: IMPLICATIONS FOR UNDERSTANDING THE EARLIEST STAGES OF STONE KNAPPING

BY SILESHI SEMAW

ABSTRACT

Gona is key for understanding the earliest stages of ancestral human stone technology. Systematic investigations at Gona (1992-94) led to the discovery of EG10 and EG12, which yielded more than 3,000 surface and excavated artifacts. The artifacts are dated to 2.6-2.5 million years (Ma) by ^{40}Ar/^{39}Ar and paleomagnetic stratigraphy, and are the oldest yet documented from anywhere in the world. Thus, they offer the best opportunity for investigating the earliest stages of ancestral hominin stone technology, raw material preference and selection strategy. The evidence strongly indicates that the first toolmakers, though at the early stages of crossing over the threshold, had sophisticated control of conchoidal fracture, selected for raw materials with good flaking quality, and had remarkable skills in producing sharp-edged flakes. A handful of slightly younger artifact sites are known in East Africa including Omo and Hadar from Ethiopia, and Lokalalei from Kenya. Earlier descriptions alleged the Lokalalei 1 hominins to have been technologically less advanced, but to the contrary, excavated materials recovered from the contiguous Lokalalei 2C have shown that the toolmakers commanded excellent flaking control, further corroborating earlier observations made by Semaw et al. (1997). Bouri, from the Middle Awash has yielded the oldest cutmarked fossilized animal bones dated to 2.5 Ma, complementing the archaeology of Gona by showing unequivocally that the earliest artifacts were used for processing animal carcasses. Further, Bouri has produced *Australopithecus garhi*, probably the best candidate for making and using the earliest artifacts. The Oldowan was named for the 1.8 Ma non-standardized simple core/flake artifacts discovered from the Lower Beds of Olduvai Gorge, Tanzania. The Gona artifacts are the earliest examples of this "least effort" core/flake tradition, and it is argued here that the stone assemblages dated between 2.6-1.5 Ma group into the Oldowan Industry. The first intentionally produced sharp-edged stones made an abrupt entrance into the archaeological record by 2.6 Ma, and the same patterns of stone manufacture persisted for over a million years with little change suggesting a "technological stasis" in the Oldowan.

KEY WORDS:

Gona, Earliest stone tools, Late Pliocene stone assemblages, Oldowan Industry

INTRODUCTION

Continued systematic archaeological investigations of the major East African Late Pliocene sites are shedding further light on the initial appearance and the earlier stages of stone technology, and the adaptive role tools played in the lives of ancestral hominins. The major sites including Gona, Bouri and Hadar are located within the main Afar Rift of Ethiopia (Kimbel et al., 1996; Semaw, 2000; Semaw et al., 1997; Asfaw et al., 1999; de Heinzelin et al., 1999), Omo, in the southern part of the Ethiopian rift (Chavaillon, 1976; Merrick, 1976; Howell et al., 1987) and Lokalalei, at West Turkana, in northern Kenya (Kibunjia, 1994; Kibunjia et al., 1992; Roche et al., 1999). The stone assemblages and contextual data from these sites are providing major insights on the beginnings of stone technology and ancestral hominin tool use behavior. The earliest and the most informative of these are the stone artifacts exca-vated from the two East Gona sites of EG10 and EG12. Bouri is dated to 2.5 Ma, and Omo, Lokalalei and Hadar are from slightly younger deposits dated between 2.4-2.3 Ma.

Archaeological reconnaissance survey of the Gona deposits began in the 1970's, and initial fieldwork showed the presence of a low density of artifacts east and west of the Kada Gona River (Roche *et al.*, 1977, 1980; Harris, 1983; Harris & Semaw, 1989). Extensive and systematic investigations of the two sites between 1992-94 produced more than 3,000 surface and exca-vated artifacts. Based on a combination of radioisotopic (^{40}Ar/^{39}Ar) and paleomagnetic dating techniques, EG10 and EG12 were firmly dated between 2.6-2.5 Ma and the stone assemblages are the oldest yet known from anywhere in the world (Semaw, 2000; Semaw *et al.*, 1997). The EG10 and EG12 stone artifacts were deposited in fine-grained sediments and excavated with-in a primary geological context, therefore, offering the best opportunity for investigating the stone manufacture techniques and skills of the first toolmakers, and for understanding the overall behavioral repertoire of Late Pliocene hominins. Analysis of the EG10 and EG12 artifacts show that the first toolmakers had sophisticat-ed understanding of the mechanics of conchoidal frac-ture on stones, that they selected for appropriate size and fine-grained raw materials with good flaking quali-ty, and commanded superior control in stone working techniques than previously recognized. Based on the remarkable knapping skills shown at EG10 and EG12, Semaw *et al.* (1997) suggested that Late Pliocene hominins had a clear mastery and sophisticated under-standing of stone flaking techniques comparable to Early Pleistocene Oldowan toolmakers. The recently excavated LA2C assemblages of West Turkana (Roche *et al.*, 1999) further corroborate this suggestion. The Gona assemblages are the earliest examples for the "least effort" core/flake Oldowan technology (Toth, 1982, 1985, 1987), which lasted between 2.6-c.1.5 Ma. This mode of stone manufacture persisted for over 1 million years with little change suggesting "technologi-cal stasis" in the earliest stone industry (Semaw *et al.*, 1997). The makers have yet to be identified at Gona, but the recent discovery made from the nearby contempo-rary site of Bouri, in the Middle Awash, indicates that *Australopithecus garhi* (2.5 Ma) may be the best candi-date for inventing and utilizing the earliest sharp-edged stone implements (Asfaw *et al.*, 1999). Additionally, the fossilized animal bones associated with *Australopithecus garhi* bear evidence of cutmarks showing that the first stone tools were used for activities related to animal butchery (de Heinzelin *et al.*, 1999). Standardized artifacts made with predetermined shape and symmetry and characterized by large handaxes and cleavers appeared for the first time by c. 1.5 Ma with the advent of the Acheulean tradition in Africa (Isaac & Curtis, 1974; Gowlett, 1988; Asfaw *et al.*, 1992;

Dominguez-Rodrigo *et al.*, 2001).

Detailed descriptions of the Gona stone assem-blages and their geological context are provided in this chapter. In addition, the chapter offers the background archaeological information for comparing with the results currently available from the analysis of the first experimental replicative stone knapping conducted on water-worn ancient river cobbles sampled from Gona. The river cobbles were sampled from the conglomerate probably used as the source of the same raw materials used by the first toolmakers. The cobbles were brought to the U.S. under a permission granted by the Authority for Research and Conservation of Cultural Heritage (ARCCH) of the Ministry of Youth Sports and Culture of Ethiopia. The knapping experiments were carried out by Nicholas Toth and Kathy Schick (CRAFT Research Center, Indiana University), and by non-human pri-mates (chimpanzees) from the Language Research Center in Atlanta, Georgia (Toth *et al.*, this volume).

BACKGROUND: THE SEARCH FOR THE OLDEST STONE TOOLS IN AFRICA

Early Half of the 20th Century

It is useful to present here a brief overview of ear-lier investigations undertaken in search of the oldest stone tools in East Africa to provide a historical link and a background for the major research activities and archaeological discoveries recently made at several Late Pliocene/Early Pleistocene sites. Comprehensive sum-mary and details of important events on the history of archaeological research in Africa during the late 19th and the early parts of the 20th Century are provided in Gowlett (1990; see also Clark, 1976; Tobias, 1976). The initial search for early stone tools in Africa started dur-ing the late 19th Century, and the first archaeological explorations were those undertaken in the northern and southern parts of the continent within the countries which were then French and British colonies (Gowlett, 1990). The archaeological riches of the Eastern part of Africa were recognized beginning in the 1890's follow-ing artifact discoveries made by geologists who began collecting stone tools for a hobby (Gowlett, 1990). The explorations and archaeological collections made, for e.g., by E.J. Wayland in Uganda in the 1920's were among the earliest examples for the beginnings of sys-tematic investigations. In the valleys of Kagera, Muzizi and Kafu, Wayland discovered crudely made "pebble tools" which at the time were believed to be the earliest artifacts ever to be documented from the African conti-nent. These so called "pebble tools" were labeled as the "Kafuan Culture," named after one of the valleys in Uganda where these alleged artifacts were found (Wayland, 1934 in Gowlett, 1990). Subsequently, Wayland proposed an archaeological sequence in East Africa with the "Kafuan" as the original "pebble cul-ture" and the oldest in the region (Gowlett, 1990).

Louis Leakey began extensive palaeoanthropological investigations in East Africa during the 1930's, and he was responsible for outlining the initial culture-sequence of the stone artifacts found at Olduvai Gorge, in Tanzania. At the time, the Kafuan was still recognized as the earliest "pebble culture" and Wayland's ideas greatly influenced the interpretations of older archaeological assemblages from East Africa. This was clearly exemplified with the evolutionary stages proposed by Louis Leakey for the culture-sequence /culture history of the Olduvai Gorge artifacts:

> "Following upon the Kafuan culture came a culture step which Mr. Wayland calls pre-Chellean and to which I have given the name of the Oldowan culture. I should have preferred to call it 'Developed Kafuan' but Mr. Wayland holds that it is quite distinct from even the most developed Kafuan" (L.S.B. Leakey, 1936, quoted in Gowlett, 1990, p.22).

Subsequently, the artifactual authenticity of these so called "pebble tools" was challenged by a number of researchers (for e.g., Van Riet Lowe, 1957), and the "Kafuan" was finally rejected in 1959 after it was proven that the majority were broken pebbles which occurred due to geological processes instead of being the results of deliberate fashioning by ancestral humans (Bishop, 1959). The Oldowan, as originally proposed for the assemblages of the Lower Beds of Olduvai Gorge, is still valid for classifying the earliest stone tools dated roughly between c.2.6-1.5 Ma (L.S.B. Leakey, 1936; M. D. Leakey, 1971; Semaw, 2000; Semaw *et al.*, 1997). The Oldowan artifact tradition was widespread between 2.0-1.5 Ma and well-documented from several sites distributed across Africa. In North Africa, Oldowan artifacts traditionally referred to as "pebble tools" (galets aménagés) are known from Ain Hanech in Algeria (Balout, 1955; Sahnouni, 1998; Sahnouni and de Heinzelin, 1998) and the Sidi Abderrahman from the Casablanca sequence in

Figure 1

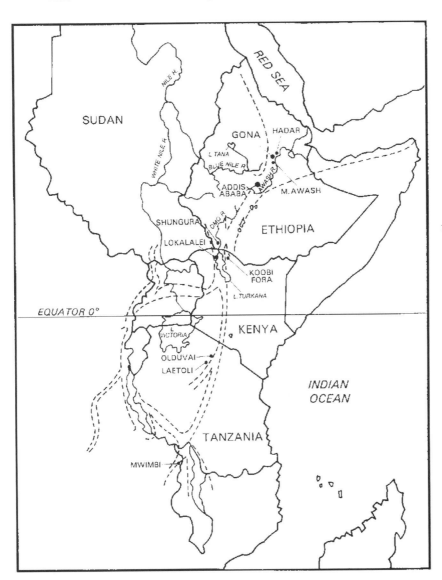

1. A map showing the Late Pliocene archaeological sites in East Africa.

Morocco (Biberson, 1961; Clark, 1992). The major East African sites with the Oldowan include Melka Kontouré, Middle Awash, Gadeb and Fejej from Ethiopia (Chavaillon *et al.*, 1979; Clark & Kurashina, 1979; Clark *et al.*, 1984, 1994; Asfaw *et al.*, 1991), Koobi Fora and Chesowanja from Kenya (Isaac, 1976a & 1976b; Isaac & Isaac, 1997; Harris, 1978; Gowlett *et al.*, 1981) and Nyabusosi from Uganda (Texier, 1995). Early Pleistocene Oldowan sites from Southern Africa include Swartkrans Members 1 and 2 (Brain *et al.*, 1988), Sterkfontein Member 5 (Kuman, 1994a, 1994b, 1996, 1998) and Kromdraai (Kuman *et al.*, 1997). The stone artifacts from these sites are characterized mainly by the "least effort" core/flake industry of the Oldowan tradition.

Major Late Pliocene (c. 2.6-2.0 Ma) Artifact Discoveries Made in the Last 50 Years

The systematic explorations and discoveries made by Louis and Mary Leakey at Olduvai Gorge were instrumental for placing East Africa on the global map of prehistoric studies. East African archaeology took a major turn in the early 1960's following the discovery of Zinjanthropus (*Australopithecus boisei*), which brought great focus and attention to the field research being undertaken in that part of the continent. Newly developed dating techniques (for e.g., K/Ar, zicron fission track, etc.) clearly showed that the archaeological record of East Africa was much earlier in time than originally perceived and greatly assisted in asserting Africa's leading position in the study of human beginnings.

A large number of substantive and informative discoveries have been made through extensive field and laboratory studies conducted mainly within the last five decades. These include findings regarding the physical and behavioral evolution of ancestral hominins, systematic archaeological investigations of the initial stages on the techniques of stone knapping and their function, and intensive studies on the stratigraphy, dating and palaeoenvironments of the major East African Plio-Pleistocene archaeological sites (e.g., M.D. Leakey, 1971; Isaac, 1976a & b; Isaac & Isaac, 1997; Chavaillon, 1976; Merrick, 1976; Roche & Tiercelin, 1977, 1980; Bunn *et al.*, 1980; Walter, 1980; Toth, 1982; 1985, 1987; Harris, 1978, 1983; de Heinzelin, 1983; de Heinzelin *et al.*, 1999; Clark *et al.*, 1984, 1994; Walker *et al.*, 1986; Howell *et al.*, 1987; Kaufulu & Stern, 1987; Feibel *et al.*, 1989; Schick & Toth, 1993; Brown, 1995; Asfaw *et al.*, 1991, 1992, 1999; Kibunjia *et al.*, 1992; Kimbel *et al.*, 1996; Semaw *et al.*, 1997; Roche *et al.*, 1999).

Despite considerable progress in field research, only a handful of Late Pliocene archaeological sites are known with secure radiometric dates and a high density of excavated artifacts (Chavaillon, 1976; Merrick, 1976;

Merrick & Merrick, 1976; Howell *et al.*, 1987; Kibunjia, 1994; Kibunjia *et al.*, 1992; Roche, 1989; Roche *et al.*, 1999; Kimbel *et al.*, 1996; Semaw, 1997, 2000; Semaw *et al.*, 1997; Asfaw *et al.*, 1999; de Heinzelin *et al.*, 1999). The sites referred to in this text are shown in Figure 1. Except for Lokalalei, which is located in Kenya, all of the other Late Pliocene archaeological sites are documented in Ethiopia, and most of these are situated on the floor of the main Afar Rift. Gona is close to 2.6 Ma, Bouri c. 2.5 Ma, and the rest are from slightly younger deposits dated between 2.4-2.3 Ma. There are claims for the presence of Late Pliocene sites in the Western African Rift (Harris *et al.*, 1987, 1990), and the Malawi region (Kaufulu & Stern, 1987). However, both sites lack radiometric dates, and the ages of the Oldowan occurrences of the two regions are still problematic.

Omo, Shungura Formation, Southern Ethiopia

The first multidisciplinary scientific research team in palaeoanthropology was organized by F.C. Howell and colleagues in the early 1960's to investigate the fossiliferous deposits of the Omo, in southern Ethiopia (Howell, 1978). The new interdisciplinary approach contributed enormously to the research and set the standard currently adopted in the field, with multiple aspects of the problems to be addressed dealt with jointly by archaeologists, paleontologists, geologists, and others specializing in the various allied sub-disciplines. The stone artifacts from Upper Bed I and Lower Bed II from Olduvai were dated to 1.8 Ma by K/Ar (M.D. Leakey, 1971; Hay, 1971), and these were the oldest known until fresh archaeological discoveries were made in the Omo, within Member F of the Shungura deposits dated by K/Ar to 2.4-2.3 Ma (Chavaillon, 1976; Merrick, 1976; Merrick & Merrick, 1976; Howell *et al.*, 1987). Two groups (American and French) were involved in the archaeological excavations of the artifacts recovered from the Shungura Formation. The Omo artifacts were predominantly made of small-size quartz pebbles and mainly worked with the "bipolar" flaking technique. Thus, the artifacts were essentially smaller in size compared to the stone assemblages excavated from Olduvai, which were made of moderate size quartz and lava produced mainly by the hand-held percussion technique. Because of the simplicity of the artifacts, Chavaillon (1976) proposed the "Shungura facies" to differentiate the Omo from slightly younger assemblages of the Oldowan. Composition of the surface and excavated artifacts from Omo and the other major Late Pliocene sites in East Africa are shown in Table 1. There were no hominins directly associated with the excavated Omo artifacts, but *Australopithecus aethiopicus* and probably early *Homo* were contemporaneous with the Shungura 2.4-2.3 Ma artifacts (Howell *et al.*, 1987; Walker *et al.*, 1986; Suwa *et al.*, 1996). To date,

Table 1

Artifact Category	EAST GONA				WEST TURKANA				OMO									
	EG10		EG12		Lokalalei 1		Lokalalei 2C		FtJi1		FtJi2		FtJi5		Omo57		OMO123	
	Exc.	Surf.	Exc.	Surf.	Exc.	Surf.	Exc.	Surf.	Exc.	Surf.	Exc.	Surf.	Exc.	Surf.	Exc.	Surf.	Exc.	Surf.
Cores	2.19	1.09	2.02	1.30	11.99	6.12	2.61	2.91	0.26					3.90		3.68	1.56	2.37
Whole Flakes	23.37	17.80	32.58	30.52	17.51	12.24	16.88	27.71	4.50	7.00	1.35	4.60	4.20	7.80	23.34	25.15	38.86	34.12
Flake Fragments	14.16	6.13	11.46	12.66	55.63	81.64	57.14	55.62	2.10	11.10	1.79	10.80		1.30				
Angular Fragments	59.12	73.50	50.79	54.22	11.99		16.06	12.02	93.30	81.85	96.90	84.60	95.80	87.00	70.00	65.03	56.45	56.31
Retouched Pieces	0.58						0.44	0.78										
Piece*															6.66	9.81	0.26	0.30
Core Fragments	0.58	1.48	2.25	1.30			0.44	0.19							6.66	6.14	2.35	6.90
Hammerstones							0.82											
Modified Pebbles					2.88		3.34	0.78										
Unmodified Pieces							2.27											
Total number of artifacts	685	1551	445	309	417	49	2067	516	375	270	223	130	24	77	30	193	767	1014

Key: Exc., = Excavated, and Surf. = Surface artifacts

Source for the Omo artifacts, Howell et al. 1987, p. 679. Omo 84 is not included because the available data seems incomplete.

* The meaning of this category is not clear, and exists only in the inventory of Chavaillon 1976 (Howell et al. 1987). The data for Omo 57 is much higher than 100%.

Source for Lokalalei 2C, Roche et al. 1999, p. 59. Broken flakes and small flakes are included together, all the core categories are included together, & broken cores are listed as core fragments. Worked pebbles and broken pebbles are included together, Unmodified pebbles are listed under Unmodified Pieces.

Source for Lokalalei 1, Kibunjia 1994, p. 164. The Lokalalei Artifacts from both the 1987 and 1991 excavations are included together. Data not yet available for the 2.3 Ma Hadar artifacts.

1. *Composition and percentages of artifact assemblages, the 2.6-2.5 Ma Gona and other Late Pliocene archaeological sites dated to 2.4-2.3 Ma (modified after Semaw, 2000).*

Australopithecus aethiopicus is the least favored by most anthropologists as the possible candidate for Late Pliocene artifact manufacture and use (but see Sussman, 1991).

Koobi Fora, East Turkana, Kenya

In the early 1970's claims were made for the presence of older artifacts dated to c. 2.6 Ma from the Koobi Fora Formation, East Turkana in northern Kenya (R.E. Leakey, 1970; Isaac, 1997). Initial radioisotopic dates yielded 2.6 Ma for the KBS tuff, and the artifacts were thought to be the oldest at the time (R.E. Leakey, 1970; Fitch & Miller, 1976; Isaac, 1997). However, the age of the KBS tuff was later revised to 1.89 Ma based on repeated K/Ar dates and comparative biochronological data gathered from Koobi Fora and the Shungura Formation in the Omo (McDougall et al, 1980; Fitch & Miller, 1976; Cooke, 1976; Isaac, 1997). The lava-dominated stone assemblages lacked spheroids and retouched pieces, artifact types known to occur within the Oldowan (mainly at Olduvai Gorge), and initially their absence at Koobi Fora was taken to imply greater antiquity, and the assemblages were assigned to the KBS Industry (Isaac, 1976a & b). Despite the similarities in the techniques of Late Pliocene-Early Pleistocene stone manufacture and use, assemblage variations could occur due to differences in the quality of the raw materials utilized (Isaac 1976a; Toth, 1982, 1985, 1987). Early on, Isaac recognized the role raw material variations could have played in influencing artifact morphol-

ogy and he accepted that the stone assemblages originally assigned to the KBS Industry could be subsumed under the Oldowan (Isaac, 1976a).

The Lokalalei Sites, West Turkana, Kenya

Beginning in the mid-1980's, archaeological survey and excavations undertaken at West Turkana, in northern Kenya revealed the presence of Late Pliocene artifacts at the site named Lokalalei 1 (GaJh5) (Roche, 1989, 1996; Kibunjia, 1994; Kibunjia *et al.*, 1992). During the mid 1990's, continued archaeological investigations of the contiguous deposits produced an additional contemporary site named Lokalalei 2C (LA2C) (Roche *et al.*, 1999). The Lokalalei sites were discovered in ancient sediments that are contemporaneous with the adjacent artifact-bearing deposits of the Omo. Lokalalei 1 was excavated in 1987 and 1991, and LA2C in 1997. Both sites were placed stratigraphically above the Kalochoro tuff, a geochemical correlate of Tuff F of the Shungura Formation and dated between 2.4-2.3 Ma (Feibel *et al.*, 1989; Brown, 1995).

Lokalalei 1 produced close to 500 surface and excavated artifacts, and the LA2C site yielded more than 2,500 artifacts (Table 1). The Lokalalei artifacts were found within fine-grained sediments deposited in floodplain settings and they were in primary context. The assemblages of both Lokalalei 1 and LA2C consisted of cores, *débitage*, and pounded pieces, stone artifacts typical of the Oldowan Industry. There are a few instances of retouched pieces from LA2C, but none were recognized from Lokalalei 1. Possible cut-marked bones were found at Lokalalei 1, but none were reported from LA2C. The artifacts earlier excavated from Lokalalei 1 were described to be poorly-made, the makers alleged to be less-coordinated, less-skilled, and lacking in manual dexterity compared to the "better-skilled" hominins responsible for making the "elaborate" artifacts known from slightly younger Early Pleistocene Oldowan sites dated between c. 2.0-1.5 Ma (Roche, 1989, 1996; Kibunjia, 1994). More than 50 cores were recovered (both surface and excavated) at Lokalalei 1 which average close to 100 mm in size, with the majority having between 1-12 flake scar counts. "About 80% of the flaking scars on these cores was [sic] characterized by step fractures and only a few instances of complete flake removals were observed" (Kibunjia, 1994, p.165). According to Kibunjia, the majority of the cores were discarded because repeated flaking attempts produced nothing but step/hinge flakes. He concluded that "factors other than raw material account for the poor technology" (Kibunjia, 1994, p. 165). Furthermore, he argued that the Lokalalei 1 artifacts were substantially different and less exhaustively worked compared to the Oldowan cores that post-date 2.0 Ma, and the assemblages were assigned to the "Nachikui facies" (Kibunjia, 1994). The "Shungura facies," which was earlier proposed for the Omo artifacts by Chavaillon (1976) was accepted as a Late Pliocene "variant." In addition, the Gona assemblages were included with the so called "technologically less- advanced" Omo and Lokalalei, and the three assemblages were assigned to "the Omo Industrial Complex" and/or to the generic "Pre-Oldowan" to differentiate them from the Oldowan (*sensu stricto*) known within the deposits dated between 2.0-1.5 Ma (Kibunjia, 1994).

The artifacts recently excavated from the adjacent and contemporary site of LA2C were described to be more "sophisticated" and the "*débitage* scheme" different from any of the Oldowan assemblages known during the Pliocene (Roche *et al.*, 1999; Roche, 2001; Roche & Delagnes, 2001). Remarkably, c. 20% of the LA2C excavated artifacts included refitting pieces. Ten different raw material types in the regions were identified as potential raw material sources, but basalt (varying from coarse to fine- grained) and phonolite were the main types used for making the LA2C artifacts. Interestingly, the coarse-grained materials were not as intensively worked as the finer ones (Roche *et al.*, 1999). The authors claim that the large number of refitting pieces from the LA2C excavations makeup for the strongest case presented for showing Late Pliocene hominin "sophistication" in artifact making techniques (Roche *et al.*, 1999) (to be discussed further below).

The Senga 5a (Western Rift), DR of Congo, and the Mwimbi Site, Malawi

There were claims for the presence of Late Pliocene artifacts in the Western Rift of Africa at two sites named Kanyatsi and Senga 5a, in the eastern part of what is now the Democratic Republic of the Congo (Harris *et al.*, 1987, 1990). The Kanyatsi site was found by Jean de Heinzelin in the early 1960's and yielded several flakes made of quartz (Harris *et al.*, 1987). During the mid-late 1980's, the area was resurveyed and further artifacts were found at Senga 5a. The site was excavated and produced a high density of quartz artifacts including cores and a large number of flaking debris. Based on associated fossilized fauna, the Senga 5a artifacts were estimated to be between 2.2-2.3 Ma. However, subsequent investigations showed that the materials were in a derived geological context and of unsubstantiated age (Harris *et al.*, 1990).

Farther to the south, field investigations in the Malawi region yielded an archaeological site named Mwimbi stratified within the Chiwondo Beds, and the site was estimated to be of Late Pliocene age (Kaufulu & Stern, 1987). Excavations at Mwimbi produced a high density of artifacts made of quartz (mainly cores and flakes) and the site was estimated between 2.2-1.6 Ma based on stratigraphically associated fauna. Additional geological/geochronological studies are critical for substantiating the age of the Mwimbi artifacts before expanding the geographical range of Late Pliocene toolmakers to include the areas south of the Omo/Turkana basin.

Hadar, Afar, Ethiopia

The Hadar study area is located east of Gona. Beginning in the 1970's, for decades the field research at Hadar was primarily focused on searching for fossil hominins, and continuous palaeontological survey in the Hadar basin resulted in the discovery of hundreds of remarkable hominin specimens attributed to *Australopithecus afarensis* within ancient deposits dated between 3.3-2.9 Ma (Johanson *et al.*, 1982; Kimbel *et al.*, 1994 and references therein). A brief archaeological exploration carried out in the early 1970's produced Acheulean assemblages from the Denen Dora area (Corvinus, 1976) and Oldowan artifacts further west of the Hadar study area at Kada Gona (Corvinus & Roche, 1976; Roche & Tiercelin, 1977, 1980).

Archaeological investigations were continued at Hadar in the mid-1990's, and a reconnaissance survey undertaken in the younger deposits of the Upper Kada Hadar Member produced Oldowan artifacts and associated fossilized animal bones. The artifacts and fauna were dated to 2.3 Ma based on the ^{40}Ar/^{39}Ar age of the BKT-3 tuff (Kimbel *et al.*, 1996). The deposits also yielded a hominin maxilla attributed to early *Homo* in stratigraphic association with the artifacts.

Bouri, the Middle Awash, Afar, Ethiopia

Palaeolithic researchers for a long time (based solely on the evidence of stone artifacts) inferred that the first intentionally created sharp-edged stones were used for processing carcasses for meat (e.g., Vrba, 1990; Harris, 1983; Pickford, 1990). However, such early hominin practice remained archaeologically unproven for quite sometime because of the lack of empirical evidence. The recent field investigations of the Late Pliocene Hata Beds of the Bouri peninsula, in the Middle Awash yielded very well-fossilized excavated animal bones bearing evidence of definite stone tool cutmarks dated to 2.5 Ma (de Heinzelin, 1999). Bouri is located c. 90 Km south of Gona, and the cutmark data complements the archaeology by providing direct evidence for the function of the oldest artifacts. The excavations at Bouri failed to produce associated *in situ* artifacts, but the cutmarks for the first time unequivocally showed that the earliest stone tools were made and used for activities related to animal butchery. Furthermore, Bouri yielded stratigraphically associated fossilized remains of a hominin named *Australopithecus garhi*, probably the best candidate for inventing and using the first stone tools (Asfaw *et al.*, 1999; de Heinzelin *et al.*, 1999).

Gona is unique for providing the earliest and most informative assemblages for studying the stone manufacture techniques of the first toolmakers. Attempts are made in this chapter to analyze the earliest assemblages from Gona in light of current understanding of Late Pliocene-Early Pleistocene human behavior, and to address issues related to technical and coordination skills, raw material preference, and acquisition strategies of the first toolmakers. The Gona evidence is compared and contrasted with the archaeological information available from slightly younger 2.4-1.5 Ma sites, and inferences on Late Pliocene hominin behavior are drawn mainly from reports published on the comparable artifacts excavated from Lokalalei 1 & LA2C and the other contemporary sites (Chavaillon, 1976; Merrick, 1976; Merrick & Merrick, 1976; Isaac, 1976a & b; Kibunjia, 1994; Kibunjia *et al.*, 1992; Roche, 1989, 1996; Roche *et al.*, 1999; Piperno, 1989; Asfaw *et al.*, 1999; de Heinzelin *et al.*, 1999).

Earlier Research at Gona

Maurice Taieb (a French geologist) was the first to recognize the palaeoanthropological importance of the deposits exposed within the Afar Rift. His geological reconnaissance survey of the late 1960's along the Awash indicated the presence of laterally extensive artifact and fossil-rich Plio-Pleistocene deposits outcropping in the areas adjacent to the main course of the Awash and its tributaries. Subsequently, extensive field research programs were initiated in the region leading to the discovery of what are now recognized as the Hadar, the Middle Awash and the Gona study areas. Decades of

fieldwork undertaken at these major sites have produced remarkable fossil hominins and archaeological materials which have provided great insights on the physical and behavioral evolution of ancestral humans and their surroundings (for e.g., Kalb *et al.*, 1982a & b, 1993; Johanson *et al.*, 1982; Clark *et al.*, 1984, 1994; White *et al.*, 1994; WoldeGabriel *et al.*, 1994; Kimbel *et al.*, 1994, 1996; Semaw, 1997, 2000; Semaw *et al.*, 1997; Asfaw *et al.*, 1999; de Heinzelin *et al.*, 1999; Haile-Selassie, 2001; WoldeGabriel *et al.*, 2001, and references therein).

During the early 1970's, G. Corvinus began archaeological survey in the Hadar deposits and documented Middle Pleistocene Acheulean artifacts (Corvinus, 1976). H. Roche later joined Corvinus and they extended explorations into the adjacent deposits exposed west of the Hadar study area. Their brief archaeological survey revealed the presence of Oldowan artifacts within the deposits exposed by the Kada Gona river, and surface artifacts were documented at localities named Afaredo-1, and Kada Gona 1, 2, 3 & 4 (Corvinus & Roche, 1976; Roche & Tiercelin, 1977, 1980). The artifacts were characterized by a low density of surface scatters located stratigraphically between two Cobble Conglomerates referred to as the Intermediate Conglomerate (*Conglomérat Intermédiaire*) and the Upper Conglomerate (*Conglomérat Supérieur*). Initially, four volcanic tuffs labeled as *Cinérites* (ashes) I-IV were recognized. Three of the *Cinérites* were later renamed as Artifact Site Tuffs (AST-1, -2 & -3) by Walter (1980). The three AST tuffs were useful for tephra chronology, but they were contaminated for radioisotopic dating techniques and have not provided absolute chronology for the artifacts (Aronson *et al.*, 1977, 1981; Walter, 1980, 1994; Walter & Aronson, 1982). Therefore, the 2.5 Ma age earlier reported for the Kada Gona artifacts was derived from estimates based on the higher stratigraphic position of the artifact-bearing sediments in relation to the BKT-2 tuff (dated to 2.9 Ma by K/Ar) known from the contiguous older deposits of the Hadar Formation (Roche & Tiercelin, 1977, 1980). The BKT-2 tuff also provided the minimum age for the *Australopithecus afarensis* specimens documented at Hadar, and the maximum age estimate for the Kada Gona artifacts (Corvinus & Roche, 1976; Roche, 1976; Roche & Tiercelin, 1977, 1980). Further, an age of 3.14 Ma was reported for the BKT-2 tuff (by Hall *et al.*, 1985) and the age of the Gona artifacts remained uncertain. The first *in situ* artifacts were systematically excavated in 1976 from a West Gona locality later renamed WG1 (Harris, 1983; Harris & Semaw, 1989). Again, the West Gona surface and excavated artifacts were low density concentrations, and the 2.5 Ma age for WG1 was an extrapolation based on the higher stratigraphic position of the artifact-bearing strata in relation to the BKT-2 tuff. Because of a moratorium passed by the government, there were no field activities in Ethiopia during the 1980's. After a long hiatus, the 1987

Gona field permit was the first to be issued by the then Ministry of Culture of Ethiopia. The deposits exposed on both sides of the Kada Gona were briefly surveyed and yielded two new archaeological localities named WG2 and WG3. The two sites were documented in close proximity to WG1, the site earlier excavated at West Gona (Harris, 1983; Harris & Semaw, 1989). The brief field survey indicated the great potential of the Gona region for future extensive palaeoanthropological investigations.

The importance of Gona was overlooked for a long time mainly because the artifact-bearing deposits investigated in the 1970's lacked radiometric dates. It was not clear from the published reports whether or not the artifacts reported by Roche *et al.* (1977, 1980) were retrieved from a sealed stratigraphic context. In addition, the artifact assemblages documented by Roche *et al.* (1977, 1980) and those excavated by Harris (1983) were relatively low density and perhaps of limited utility in characterizing the earliest assemblages. In addition, details have yet to be available for the surface artifacts from Afaredo 1, Kada Gona 1, 2, 3 and 4 of Roche *et al.* (1977, 1980).

The Gona Paleoanthropological Research Study Area: New Investigations

The Gona Palaeoanthropological Research Project (GPRP) study area is situated in the west-central part of the Afar Administrative Region of Ethiopia. The GPRP covers an area of 500 Km^2 badlands with fluvio-lacustrine fossiliferous and artifact-rich deposits. Cobble conglomerates and interbedded tuffaceous markers are prominent throughout much of the sequence. The GPRP study area stretches to the Mile-Bati road in the north and to the headwaters of the Asbole River in the south. The Hadar study area bounds Gona to the east, and exposures of the Western Escarpment of Ethiopia are the western limits of the Gona study area. The Western Margin deposits are rich with Late Miocene-Early Pliocene faunas. The largest portion of the study area to the east and to the south contain ancient sediments with a wealth of Plio-Pleistocene fossilized faunas, and stone artifacts spanning the time period between 2.6 Ma-c.500 Ka. The major rivers east to west, include Kada Gona, Ounda Gona, Dana Aoule, Busidima, Gawis Yalu, and Sifi. These rivers and associated tributaries drain the surrounding region and seasonally flow into the Awash, also cutting through the ancient sediments and exposing artifacts and fossilized fauna. The deposits outcropping in the GPRP study area are now providing windows of opportunities for systematic archaeological, palaeontological, and geological field studies. Recent field investigations (1999-2001) have produced a large number of new archaeological and palaeontological sites, including hominins sampling various critical time intervals during the Early Pliocene and the Early Pleistocene (Semaw *et al.*, 2002).

The East Gona Archaeological Localities of EG10 and EG12: An Overview

The first systematic and extensive archaeological and geological field studies at Gona were undertaken between 1992-94. Fossilized fauna and artifact rich deposits at Gona are being exposed by erosion and often washed away quickly due to the high relief and badlands topography of the region, and the 1992 survey was timely due to the very high density of freshly eroded artifacts found littering the ancient landscape. More than 12 new localities were documented on both sides of the Kada Gona river, and the two East Gona sites of EG10 and EG12 were excavated yielding more than 3,000 surface and *in situ* artifacts within fine-grained deposits securely dated between 2.6-2.5 Ma (Semaw, 1997, 2000; Semaw *et al.*, 1997). Figure 2 shows the photo of the excavations carried out at EG10. The location and stratigraphy of the major East Gona sites are shown in Figure 3a & 3b.

The EG10 and EG12 sites are situated on the eastern side of the Kada Gona c. 7 Km upstream from its confluence with the Awash River. Surface artifacts with abundant cores, flakes and fragments were also systematically sampled from several localities distributed within laterally extensive deposits exposed on both sides of the river. EG10 and EG12 were chosen for systematic excavation and detailed investigations because the two localities had the highest density of freshly eroded artifacts with cores, flakes, and flaking debris of all sizes found exposed on the surface. The three laterally extensive marker tuffs (AST-1, -2 & -3), the two additional new tuffs (AST-2.5 & AST-2.75), and the three prominent cobble conglomerates were found very well-exposed in close proximity to EG10 and EG12, aiding stratigraphic placement and age determinations of the excavated artifacts. Furthermore, the two localities were amenable for systematic excavations because of less overburden to deal with to expose the artifact-bearing horizon as well as ease of access to the sites by vehicle from the Kada Gona River.

Stratigraphy and Dating

The EG10 and EG12 artifact-bearing horizons are stratified in fine-grained sediments of floodplain context situated within and immediately overlying the AST-2 marker tuff. The Intermediate Cobble Conglomerate (ICC) is situated less than one meter below the EG10 and EG12 excavations and is prominent within the stratigraphic section exposed near both sites, and as a marker horizon, it also extends laterally within the Kada Gona and associated drainages. Geological studies indicate that the ICC was the closest source of the stone raw materials used for making the EG10 and EG12 artifacts. The stratigraphic details and the dating of the EG10 and EG12 sites are shown in Figure 3b. The Kada Gona River has exposed a c. 50 meter-thick, upward-fining stratigraphic section near EG10 and EG12. From bot-

Figure 2

2. *A photo showing view of the EG10 excavations.*

Figure 3

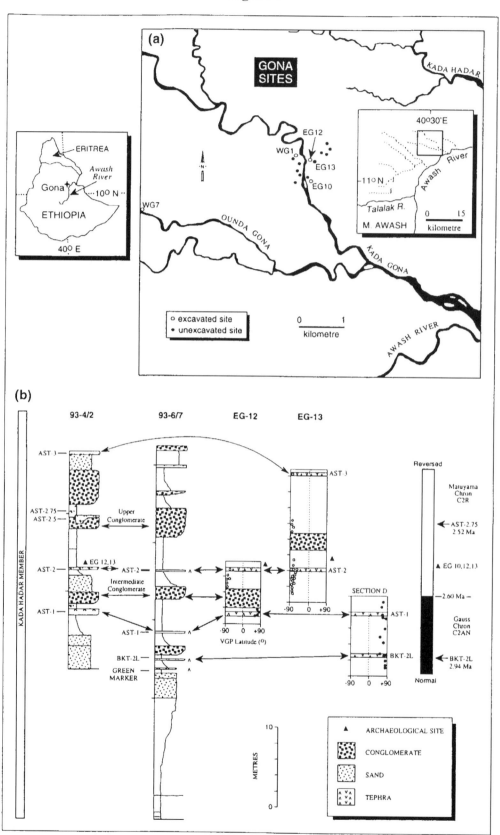

3a. *A map showing the location of EG10, EG12 and other contemporary archaeological sites of East and West Gona.*

3b. *The stratigraphy and dating of the EG10 and EG12 sites. The composite sections (93-4/2, 93-6/7) are correlated with the results of the magnetostratigraphy of EG12 and EG13. Normal polarity is shown by filled circles, and reversed by open circles. The 40Ar/39Ar dates are shown with units of the magnetic polarity timescale (MPTS) on the right (Figure modified after Semaw et al., 1997).*

Figure 4

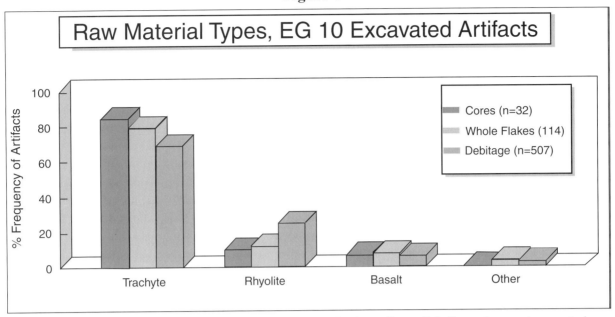

4. *The raw material types used at EG10. Note: data based on* in situ *artifacts. Only the surface and excavated cores/choppers are combined.*

tom to top, major stratigraphic layers include the Green marker tuff (a geochemical correlate of the Kada Hadar BKT-2 tuff sampled beneath EG10, and dated to 2.9 Ma), the Basal Cobble Conglomerate, the AST-1, the ICC, the AST-2, the AST-2.5, the AST-2.75, the Upper Cobble Conglomerate, the AST-3 tuff and slightly younger fossil and artifact-bearing deposits of the capping strata. Two of the new tuffs, the AST-2.5 and AST-2.75 were sampled in 1993, and the latter proved amenable for $^{40}Ar/^{39}Ar$ dating, and provided the critical data needed for resolving the age of the Gona artifacts. The AST-2.75 tuff was sampled just above locality EG13 and dated to 2.52+ 0.075 by the $^{40}Ar/^{39}Ar$ dating technique (Semaw *et al.*, 1997). The tuff was located c. 200 m north of EG10 and stratigraphically less than 5 meters above EG10, EG12 and EG13. Paleomagnetic calibrations of the sediments sampled along the stratigraphic sections exposed near EG10 and EG12 placed the Gauss-Matuyama transition, dated to 2.6 Ma (McDougall *et al.*, 1992), within the ICC. Hence, the 2.52 Ma $^{40}Ar/^{39}Ar$ date provided a minimum age for the excavated sites, and the paleomagnetic transition identified below the sites provided a maximum age for the artifacts. In addition, the paleomagnetic calibrations corroborated the 2.52 Ma minimum age derived from the $^{40}Ar/^{39}Ar$ analyses of the overlying tuff. Therefore, the EG10 and EG12 artifacts are firmly dated between 2.6-2.5 Ma, and they are the earliest documented archaeological occurrences yet identified in Africa (Semaw, 2000; Semaw *et al.*, 1997).

Raw Materials

The EG10 and EG12 artifacts were predominantly made of trachyte and rhyolite cobbles (Figure 4). The stone raw materials were acquired from nearby gravels, and the ICC was the most likely source of the cobbles used for making the artifacts. Geological studies have indicated the presence of ancient channels that carried the cobbles in close proximity to the sites. The ICC contains water-worn rounded trachyte and rhyolite cobbles suitable for making the EG10 and EG12 stone tools, suggesting that the toolmakers had to travel only short distances (c.100-200 m) to select and acquire these stone raw materials.

Moreover, preliminary geological studies have shown that the raw materials used at EG10 and EG12 were similar to the cobbles that are now eroding from the ICC. The raw materials selected and used by the hominins appear to have been fist-sized water worn rounded cobbles. The ICC is currently eroding into the Kada Gona River and being exposed near EG10 and EG12, providing a wonderful opportunity for sampling and analyzing the cobbles from present day settings. During the 1993 fieldwork, a total of 103 cobbles eroded from the ICC were picked randomly (near EG10) and sorted by raw material types. The cobble samples consisted of 48% trachyte, 24% rhyolite and the remaining were identified as chalcedony, breccia, basalt and other types (Semaw 1997). The trachytes used at Gona were of fine-grained quality varying from light grey to brown, often with inclusion of phenocrysts, and their cortical surface mostly dark brown in color. Measurements taken on the maximum dimensions of the cobbles ranged between 60-170 mm (mean 105mm, s.d. 29), size classes optimal for making the Oldowan artifacts of EG10 and EG12 (Semaw, 1997).

As is shown in Figure 4, more than 75% of the EG10 and EG12 excavated artifacts were made of trachyte, c. 20% of rhyolite, and the remaining of basalt, chalcedony and other raw materials. Comparison of the

Table 2

	EG 10		EG12	
	Surface	Excavated	Surface	Excavated
All Lithics (n)	1549	686	309	445
All Artifacts (n)	1549	685	308	445
Manuports (Unmodified Stones)	0	0	0	0
Split Cobbles	0	1	1	0
Cores/Choppers or Tools (Flaked Pieces)	1.1	2.19	1.3	2.03
Débitage (Detached Pieces)	98.9	97.81	98.7	97.97
Utilized Material (Battered & Pounded Pieces)	0.00	0.00	0.00	0.00
% Artifacts	100.00	100.00	100.00	100.00
Cores/Choppers (Flaked Pieces) (n)	17	15	4	8
Cores/Choppers	88.24	73.33	100.00	88.89
Discoids	5.88	20.00	0.00	0.00
Core Scrapers	5.88	6.67	0.00	11.11
% Total	100.00	100.00	100.00	100.00
Débitage (Detached Pieces) (n)	1532	670	304	436
Whole Flakes	18.01	24.48	30.92	33.94
Angular Flakes	74.28	60.45	54.93	51.83
Split Flakes	5.42	8.36	12.17	9.86
Snapped Flakes	0.72	3.43	0.66	2.06
Split & Snapped Flakes	0.07	2.69	0.00	0.00
Core/Cobble Fragments	1.50	0.59	1.32	2.31
% Total	100.00	100.00	100.00	100.00
Utilized Materials				
(Battered & Pounded Pieces) (n)	0	0	0	0

2. *Percentage composition of all the surface and excavated artifacts from EG10 and EG12. Data modified after Semaw (1997). The description of the artifacts follows terminologies introduced by M.D. Leakey (1971). Isaac et al.'s (1981) artifact categories are shown in parentheses.*

raw materials used at EG10 and EG12 with the types identified from the samples picked within the ICC show that trachyte was the most abundant as well as the most preferred. Recently discovered contemporary archaeological sites located c.10 Km away from the EG sites are confirming that the hominins practiced even more systematic raw material selection strategies and preference for finer-grained raw materials (like chert), further reinforcing the fact that the first toolmakers were more selective in their decision-making and stone-crafting behavior than previously recognized (personal observation.).

East Gona 10 (EG10)

EG10 is the most informative of all the East Gona sites with the highest density of surface and excavated stone artifacts. The composition of the surface and excavated artifacts from EG10 and EG12 are shown in Table 2. Fresh artifacts comprising of cores, and *débitage*

were found eroding down a slope situated above the ICC. All sizes and classes of artifacts were found freshly eroding out of fine-grained sediments indicating the great potential of EG10 for yielding a high density of *in situ* artifacts. A total of 1, 549 surface exposed artifacts (including plotted and surface-scraped) were collected from the 38m² grid established down the slope to the level of the ICC. There were several fossilized bones found on the surface, but none from the excavations, and the evidence was not adequate to make any behavioral associations between the fauna and the artifacts. A total of 13m² area was excavated following the edge of the outcrops, and 686 artifacts were recovered *in situ*. The excavated materials consisted of cores, broken flakes (with majority of split and snapped), broken cores and a high density of angular fragments. As shown in Figure 5, two artifact-bearing levels separated by c. 40 cm of nearly sterile deposits were documented at the site. The artifacts from the two levels were tightly clustered and restricted, each within a 10 cm layer of

Figure 5

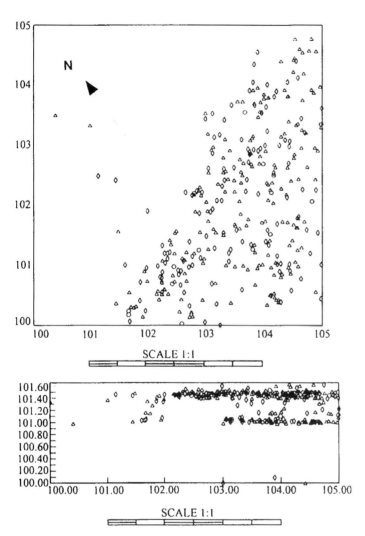

5. *Horizontal and vertical distributions of the EG10 excavated artifacts. (○ = Cores, ◇ = Whole Flakes and △ = Angular Fragments.)*

deposits, suggesting absence of vertical dispersion or minimal disturbance as a result of geological processes. The presence of two discrete levels suggests possible repeated occupation of the site, and that the area may have been favored by the toolmakers due to its close proximity to raw materials and ancient streams with fresh water.

The artifacts were excavated in fine-grained, consolidated brown clays that appear to have distinctive characteristics of swelling and cracking that occur during different moisture regimes. The paleosols also indicate marked seasonality during the time of the deposition of the artifacts. Glass shards of the contaminated AST-2 tephra were chemically identified within the artifact-bearing sediments. A majority (>99%) of the surface and excavated artifacts were exceptionally fresh, and only a fraction of the cores and flakes show edge-damage that may be attributable to utilization. There were no preferred orientations recorded for the excavated artifacts implying excellent site integrity and absence

of disturbance by water. The overall evidence is in favor of a quick burial after discard. However, the absence of utilization damage is intriguing when considering the time and energy expended for seeking out raw materials and making the artifacts.

The excavation was extended into a geological trench to the north and dug c. 3 meters down the sequence to the level of the ICC, and no artifacts were found from the geological test-trench. The latest survey (1999-2001) of EG10 has shown the presence of freshly exposed artifacts eroding out of the sediments left unexcavated, and there is high potential for the presence of further artifacts still buried within intact sediments beneath the overburden (personal observation).

East Gona EG12

Locality EG12 was discovered within the Aybayto Dora stream, c. 300 meters north of EG10. The stream flows west into the Kada Gona, also exposing Late Pliocene sediments. The stratigraphy of EG12 is similar to EG10 because the same deposits extend laterally, and very well exposed for c. 2-3 km along the Kada Gona and feeding streams. Therefore, the Green Marker, the AST-1, -2 & -3 marker tuffs and interbedded cobble conglomerates are prominently featured in the stratigraphic section exposed near EG12. The AST-2.75 tuff is located c. 100 meters east of EG12, and stratigraphically situated c. 0.5 meters above the excavation. The same artifact-bearing horizon extends east and surface artifacts were sampled at EG13, located less than 5 meters directly below the AST-2.75 tuff. Based on their stratigraphic position in relation to the $^{40}Ar/^{39}Ar$ dated AST-2.75 tuff, and the magnetostratigraphy of the nearby sediments (identified as the Gauss-Matuyama transition), the EG12 and EG13 localities are firmly dated between 2.5-2.6 Ma (Figure 3a).

At EG12, a high density of very fresh stone artifacts was found exposed on a steep slope and eroding down into the Aybayto Dora stream. The artifact-bearing horizon was located on a small flat ridge mid-way up the steep section and the area difficult to reach for excavation. An area of 26 m² was gridded over the tilted slope down the sequence to the level of the ICC and a total of 309 artifacts (including surface plotted and surface scraped) were systematically collected. An area of 9m² was excavated atop the flat ridge and yielded a total of 444 artifacts *in situ* within fine-grained sediments. The vertical and horizontal distribution of the EG12 excavated artifacts is shown in Figure 6. The artifacts were clustered within 40 cm thick well-consolidated brown clays, and their sedimentary context was very similar to EG10. The AST-2 tephra was located c. 0.5 meters below the EG12 excavated horizon. Composition of the artifacts, both surface and excavated, was similar to

Figure 6

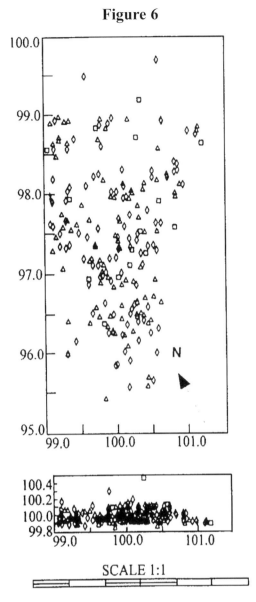

SCALE 1:1

6. *Horizontal and vertical distributions of the EG12 excavated artifacts. (○ = Cores, ◇ = Whole Flakes and △ = Angular Fragments.)*

EG10, and all of the artifacts were in mint condition. There were no fossilized bones associated with the surface or the excavated artifacts. EG12 was resurveyed during the recent rounds of fieldwork and high density concentrations of freshly exposed stone artifacts were found eroding at the site of the 1992 excavation. As was the case with EG10, it is likely that two artifact-bearing levels may be present at EG12. The surface exposed artifacts were collected, but no further excavations were undertaken due to time constraints.

Characteristics of the Gona Stone Assemblages

The EG10 and EG12 artifacts are broadly similar to the Oldowan assemblages known from archaeological sites dated between 2.4-1.5 Ma. The surface and excavated artifacts consist of a large number of unifacially-flaked cores, and *débitage* including whole flakes, and a high density of flaking debris (split and snapped flakes, and angular fragments). The composition of the EG10 and EG12 artifacts and the types of raw materials used are summarized in Table 2. Water worn rounded cobbles, mainly trachyte, rhyolite, basalt, and in some instances rare raw materials such as chalcedony and chert were used. Retouched pieces are rare, but present. There were no spherical cobbles with evidence of pitting or battering marks identified as typical hammerstones. In addition, spheroids and subspheroids, known at other Early Pleistocene sites (for e.g., at Olduvai) are totally lacking at Gona.

The Gona assemblages were classified by using the typology devised by M.D. Leakey (1971) for describing the artifacts excavated from the Lower Beds at Olduvai Gorge. None of the Gona artifacts were identified into the elaborate Oldowan tool types of M. Leakey such as proto-handaxes, awls, burins, etc. Despite the feasibility of doing so at Gona, these tool types were not included in the inventory of EG10 and EG12 because it would be unlikely for any of the Plio-Pleistocene toolmakers (2.5-1.5 Ma) to have had such functionally elaborate artifacts planned in their "toolkits."

By using a simple technological approach, Isaac *et al.* (1981) designed a system of classifications that minimizes the functional implications associated with M. Leakey's typology. Isaac *et al.* (1981) classified Plio-Pleistocene Oldowan artifacts into major categories including *Flaked Pieces* (the various choppers/ core forms and retouched pieces), *Detached Pieces* (the *débitage*), *Pounded Pieces* (the hammerstones and battered cobbles), and *Unmodified Pieces* (the manuports). Their scheme is useful, but very broad and general for detailed comparative study of Oldowan stone assemblages and for investigating possible changes in the tool manufacture behavior of Late Pliocene-Early Pleistocene hominins. Artifacts from Gona and the major Late Pliocene-Early Pleistocene sites in East Africa were analyzed and described by using the typology outlined by M. Leakey (1971), and her classifications are still vital for comparative studies of Oldowan assemblages.

The EG10 and EG12 artifacts were reanalyzed in 1999 and the descriptions of the artifacts presented here are based on the recently collected data. All of the surface and excavated cores from both EG10 and EG12 were described. All of the excavated whole flakes from EG10 and a sample of the whole flakes from EG12 were reanalyzed for this study. The brief summary of the measurements for the remaining *débitage* category (broken flakes, and angular and core fragments) is reliant on the excavated artifact data collected from EG10.

The Cores/Choppers

A majority of the cores/choppers at Gona were recovered from the surface and the excavations at EG10, and they provide the best representative samples for characterizing the stone knapping behavior of the first toolmakers, particularly those who inhabited the area around Kada Gona. The excavated cores/choppers include 15 specimens from EG10, and 7 from EG12, accounting for c. 1% of the total excavated assemblages (Table 3). Most of the EG10 cores (60% surface and 69% excavated) and close to half (44%) of the EG12 excavated cores were flaked only on one face, and the remaining bifacially worked. Only a small percentage were exhaustively-reduced, but most cobbles were flaked around much of the circumference, and the overall evidence clearly shows that the Late Pliocene/Early Pleistocene toolmakers understood conchoidal fractures on stones, sought acute angles when striking the cobbles and had excellent coordination and motor skills to successfully remove large flakes off the cores. Unifacial side choppers made of trachyte cobbles dominate the assemblages and only a few of the specimens were identified as discoids (Table 4). However, there were also

Table 3

	EG 10		EG12	
	Surface	Excavated	Surface	Excavated
No. of Cores/Choppers	17	15	4	7
Raw Materials				
Trachyte	94.12	73.33	100.00	42.86
Rhyolite	5.88	13.33	0.00	28.57
Basalt	0.00	13.33	0.00	0.00
Other	0.00	0.00	0.00	28.57
% Total	100.00	100.00	100.00	100.00
Length				
Mean	76.88	83.33	72.25	74.45
Std	6.82	10.34	2.39	8.72
Range	(64-87)	(69-105)	(69-75)	(58-93)
Breadth				
Mean	62.20	60.90	62.50	59.73
Std	6.01	9.18	3.91	8.06
Range	(51-71)	(44-80)	(57-68)	(49-77)
Thickness				
Mean	46.59	45.27	49.00	43.73
Std	9.25	12.36	2.92	7.74
Range	(31-63)	(30-69)	(45-53)	(25-53)
Total Scars				
Mean	8.76	10.27	11.00	8.91
Std	2.67	3.74	2.12	3.45
Range	(4-13)	(6-21)	(8-13)	(3-15)
Largest Scars				
Mean	45.47	48.07	52.25	45.45
Std	8.63	11.79	8.39	8.40
Range	(30-63)	(30-65)	(40-60)	(33-60)
B/L				
Mean	0.81	0.73	0.87	0.77
Std	0.05	0.08	0.00	0.09
T/B				
Mean	0.73	0.74	0.79	0.71
Std	0.08	0.16	0.00	0.14

3. *Summary of the basic attributes for the EG10 and EG12 surface and excavated cores/choppers.*

Table 4

	EG 10				EG12			
	Surface		Excavated		Surface		Excavated	
	<u>Uni</u>	<u>Bi/Multi</u>	<u>Uni</u>	<u>Bi/Multi</u>	<u>Uni</u>	<u>Bi/Multi</u>	<u>Uni</u>	<u>Bi/Multi</u>
Total no. of Artifacts	9	6	11	4	3	0	4	5
Side Choppers	3	2	8	0	1	0	4	1
End Choppers	3	0	0	0	1	0	0	0
Side & End Choppers	2	2	2	1	0	0	0	3
Discoids	1	0	1	3	1	0	0	0
Core Scrapers	0	2	0	1	0	0	0	1
% Total	60.00	40.00	68.75	31.25	100.00	0.00	44.44	55.55

4. *Composition of the unifacially and bifacially/multifacially worked EG10 and EG12 surface and excavated cores/choppers. Uni=unifacial, Bi=bifacial, and Multi=multifacial. Note: Multifacial flaking here refers to working of the core around much of the circumference.*

several multifacially worked and exhaustively reduced cores with evidence of many generations of flake removals from EG10, EG12, EG13, and other contemporary sites. The EG10 excavated cores have an average length of 83.33 mm (s.d. 10.34) and those of EG12 average 74.45 mm in length (s.d. 8.72). Detailed measurements of the basic attributes for the surface and excavated EG10 and EG12 cores are provided in Table 3. The average size of the Gona cores accords with the measurements of the cobbles sampled (in 1993) from the ICC (range between 60-170 mm, average 105mm, s.d. 29) (Semaw, 1997). The EG10 excavated cores have an average of 10 scar counts (range between 6-21 scar counts, with s.d. 3.74). The average size of the Gona cores is relatively smaller compared to the c. 105 mm average reported for Lokalalei 1 (Kibunjia, 1994). In addition, the pattern at Lokalalei 1with the number of flake scars (between 1-12) seem to correspond with the less exhaustively flaked nature of the core forms. It may be that the cores were not as heavily reduced because of the lower quality of the raw materials used at Lokalalei 1, and repeated attempts may have failed to produce workable flakes, as suggested by Kibunjia (1994).

Only two of the excavated cores/choppers from EG10 show typical hammerstone battering marks characteristic of flaking attempts, and none were recorded for the cores from EG12. As surface-exposed trachyte artifacts exfoliate because of weathering and can superficially mimic battering/pounding marks, only excavated artifacts were included in the analysis of pitting and bruising marks, and special caution was taken to distinguish actual pitting from such exfoliation. Analysis of the Gona specimens show that at least one step/hinge has been recorded on nearly 80% of the cores. The percentage frequency of the steps/hinges recorded for the Gona cores is shown in Figure 7. Some researchers attribute the high incidence of steps/hinges on cores as an indication of low level of technical skills (Kibunjia, 1994; Ludwig, 1999). Based on observation of the Gona

assemblages and experimental knapping studies, Ludwig (1999) argues for a relatively low level of skills for the Gona and Lokalalei 1 toolmakers compared, for example., to the hominins responsible for the Oldowan assemblages known from the Lower Beds at Olduvai. His experimental studies were based on argillite blanks collected from river gravels in New Jersey, and it is debatable whether or not the materials he used approximate the flaking properties of the trachyte utilized at Gona, and if direct comparisons can be made between the two different raw materials. Nonetheless, the question of how many of the steps/hinges were influenced by raw material characteristics (e.g., quality of flaking, internal flaws, or shape) vs. the level of skills of the knapper (novice/ experienced) require careful investigations, and need to be further determined by extensive knapping experiments using comparable raw materials. Furthermore, recent experimental studies show that there appear no clear relationships between the preponderance of steps/hinges in an assemblage and the level of the skill of a knapper (see Toth *et al.*, this volume).

Despite the mastery and control of flaking shown by the hominins, a majority of the EG10 cores were flaked only on one face. The preponderance of unifacial working at EG10 might be explained by the abundance of large-size cobbles readily available from the nearby ancient streams for producing flakes for immediate use as "expedient tools," or the sites may represent activity loci where the hominins acquired raw materials, did casual flaking and transported selected specimens for use elsewhere. However, all the cores made of exotic raw materials such as chalcedony and chert were heavily reduced and several examples of diminutive cores are known from EG13 and across Kada Gona at WG2 and elsewhere in the study area (Figures 8 and 9). Nonetheless, the overall evidence clearly indicates that the earliest toolmakers understood the properties of conchoidal fractures, and they have already mastered stone-on-stone flaking techniques as early as 2.6 Ma.

The Débitage

Following the criteria outlined by M.D. Leakey (1971), the whole flakes and resultant flaking debris produced during the process of reduction of the cores were classified as *débitage*. The *débitage* includes whole flakes, broken flakes, angular and core fragment often accounting for the highest percentage of the artifacts in the Oldowan. Over 97% of the total assemblages from EG10 and EG12 fall into this category. The whole flakes provide important information on the technical skills of early stone knappers and they are discussed in more detail. The flaking debris including split and snapped flakes, angular and core fragments are described in greater details in Semaw (1997), and only a brief summary of the measurements are provided here.

Artifacts with obvious platforms, diagnostic bulbs of percussion, and clear release surfaces were classified as whole flakes. A total of 110 specimens from EG10 and 58 from EG12 are included in this study. The whole flakes account for 25% of the EG10, and close to 34% of the EG12 excavated artifacts (Table 5). Like the cores, a majority (80%) of the EG10 whole flakes were made of trachyte, 11% of rhyolite and the remaining on basalt and other raw materials. EG12 also shows a similar trend with 69% of the whole flakes made of trachyte, 17% of rhyolite, and the remaining of basalt and other types of raw materials. Remarkably, most of the Gona whole flakes exhibit very prominent bulbs of percussion and show that the toolmakers practiced bold flaking and had excellent coordination and control over the core reduction processes (Figures 9 & 10). In maximum dimensions, the whole flakes from EG10 average to 42.18 mm (range between 85-20 mm, s.d. 15.56), and the EG12 average to 40.94 mm (range between 20-71 mm, s.d. 13.85). Basic measurements of the excavated whole flakes for the two sites are provided in Table 5. The angles of striking platforms for EG10 average to 109.30° (range between 80°-135°, s.d. 12.89), and for EG12 average to 107° (range between 80°-130°, s.d. 11.40). Although no average and s.d. available, the range of striking platforms (between 70°-129°) documented for Olduvai FLK N levels 1 & 2 (M.D. Leakey, 1971, p. 83) compare with the values recorded for EG10 and EG12.

The flake type system devised by Toth (1982) was adopted to look at the stages of flaking represented at EG10 and EG12. Following intensive examination of the Koobi Fora assemblages, Toth (1982, 1985, 1987) recognized six major flake types based on the presence/absence of cortex on the platform and on the dorsal surface of the whole flakes. As is shown in Figure 11, a majority of the EG10 and EG12 whole flakes were flake types 2 and 3. The preponderance of these flake types with cortical platforms corresponds with the unifacial mode of flaking prevalent at Kada Gona (Toth, *pers. com.*). Because of their immediate proximity to sources of raw materials, it may be that EG10 and EG12 represent focal points for flaking activities, and usable flakes may have been preferentially selected and removed for utilization at different locations of the ancient Gona landscape (Toth *et al.*, this volume). The

Figure 7

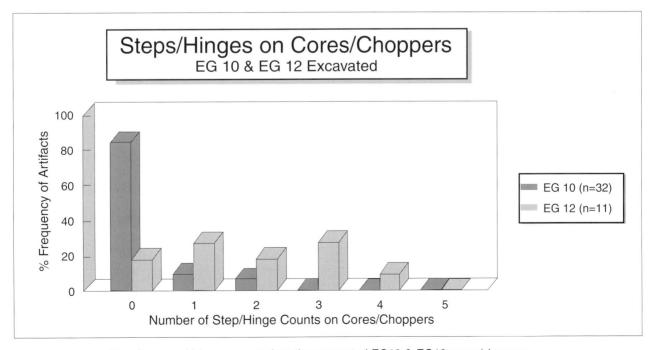

7.   *Steps and hinges counted on the excavated EG10 & EG12 cores/choppers.*

Figure 8

8. *Photo showing the Gona stone artifacts, a) bifacial core/chopper made of trachyte, b) bifacial core/chopper made of chert, c) exhaustively worked chalcedony core/chopper, d) a whole flake and e) cores and whole flakes from EG10 and EG12. (Photo d, © David Brill; Photo e, Tim White)*

Figure 9

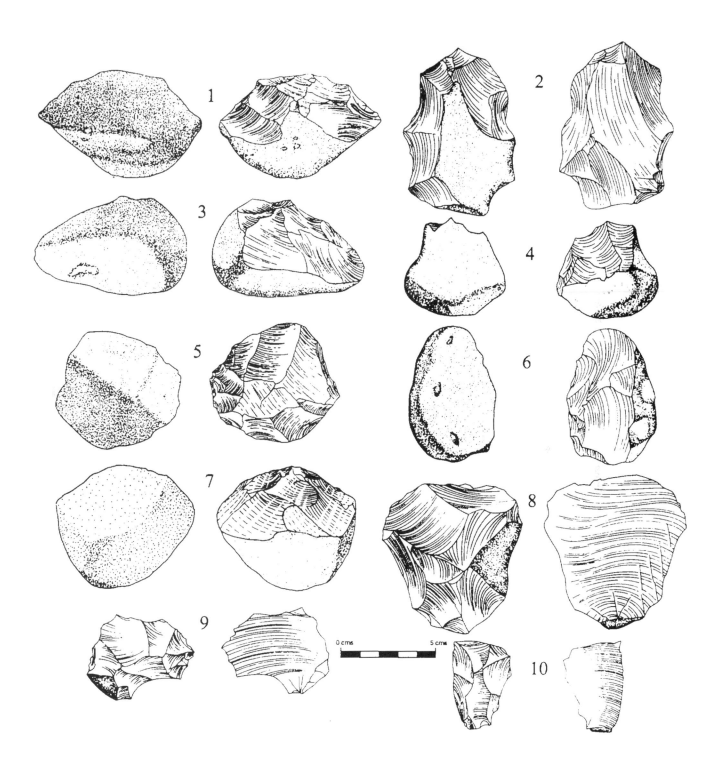

9. *Drawings of the EG10 and EG12 excavated artifacts. 1) unifacial side chopper, 2) discoid, 3) unifacial side chopper, 4) unifacial end chopper, 5) partial discoid, 6) unifacial side chopper, 7) unifacial side chopper, and 8-10) whole flakes.*

Table 5

	EG10	EG12
No. of Artifacts	114	62
Raw Materials		
Trachyte	78.95	66.13
Rhyolite	11.40	17.74
Basalt	7.02	6.45
Other	2.63	9.68
% Total	100.00	100.00
Maximum Dimensions		
Mean	42.18	40.94
Std	15.56	13.85
Range	(20-85)	(20-71)
Length		
Mean	37.38	34.50
Std	15.34	12.84
Range	(14-78)	(15-66)
Breadth		
Mean	34.63	35.55
Std	13.74	13.23
Range	(14-77)	(19-66)
Thickness		
Mean	13.18	12.13
Std	6.26	5.76
Range	(3-33)	(4-30)
Dorsal Scars		
Mean	3.07	3.00
Std	1.72	1.54
Range	(0-10)	(0-8)
Platform Breadth		
Mean	22.86	24.56
Std	11.37	13.27
Range	(4-60)	(5-58)
Platform Thickness		
Mean	10.21	10.10
Std	6.04	5.65
Range	(2-34)	(2-29)
Platform Scars		
Mean	0.32	0.37
Std	0.70	0.80
Range	(0-4)	(0-4)
Bulb Range		
Mean	109.30	107.00
Std	12.89	11.40
Range	(80-135)	(80-130)
Breadth/Length		
Mean	0.96	1.08
Std	0.26	0.30
Thickness/Breadth		
Mean	0.39	0.34
Std	0.11	0.11

5. Summary of the basic attributes of the EG10 and EG12 excavated whole flakes.

hominins were skilled in striking flakes off cores, but why they repeatedly used the cortical surface of only one face of the cores for a platform is unclear. This pattern of flaking is consistent both at EG10, EG12 and the other sites known around Kada Gona. The dorsal scar counts of the EG10 whole flakes average 3.07 mm (range between 0-10 mm, s.d. 1.72) and the EG12 average to 3 mm (range between 0-8 mm, s.d. 1.54) showing moderate to extensive flaking.

Table 6

Angular Fragments (n=405)		
	Mean	20.12
	Std	8.78
	Range	(6-46)
Split Flakes, Left (n=26)		
	Mean	31.81
	Std	16.01
	Range	(13-77)
Split Flakes, Right (n=30)		
	Mean	37.77
	Std	16.68
	Range	(12-78)
Snapped Flakes, Proximal (n=15)		
	Mean	27.33
	Std	10.2
	Range	(12-41)
Snapped Flakes, Distal (n=6)		
	Mean	30.17
	Std	6.82
	Range	(18-38)
Split & Snapped Flakes (n=18)		
	Mean	28.90
	Std	9.34
	Range	(13-47)
Core Fragments/Broken Cobbles (n=5)		
	Mean	55
	Std	12.70
	Range	(35-73)

6. Maximum dimensions of the EG10 excavated débitage.

Angular fragments make up c. 65% of the excavated assemblages identified into the *débitage* category. The remaining, including broken flakes (split and snapped) and core fragments account for nearly 10% of the total excavated *débitage*. Because the majority of the artifacts fall into angular fragments, this class of artifacts is the best indicator of the types of raw materials used at Gona. About 68% of the angular fragments were made of trachyte, 24% made of rhyolite, 6% of basalt and the remaining of other raw materials, thus clearly showing trachyte to have been the most preferred and utilized raw material at EG10. A total of 405 angular fragments were analyzed, and in maximum dimensions average to 20.12 mm (range between 6-46 mm, s.d. 8.78). A total of 56 split flakes (26 left and 30 right), 23 snapped flakes (15 proximal, 2 medial and 6 distal), 18 split and snapped flakes, and 5 core fragments (one broken cobble) were identified. The maximum dimensions of these artifact categories are presented in Table 6.

Figure 10

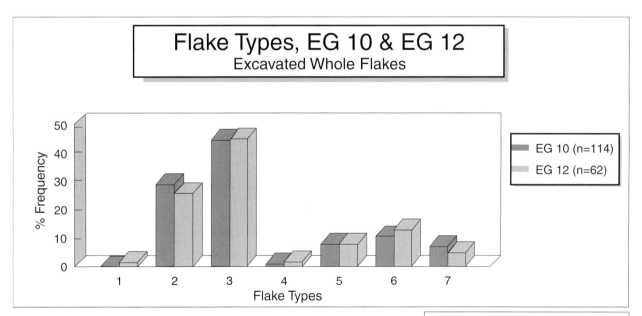

10. *Flake types of EG10 and EG12 excavated whole flakes.*

Key (after Toth, 1985):
1 = All cortex platform all cortex dorsal
2 = All cortex platform, part cortex dorsal
3 = All cortex platform, no cortex dorsal
4 = No cortex platform, all cortex dorsal
5 = No cortex platform, part cortex dorsal
6 = No cortex platform, no cortex dorsal
7 = Indeterminate

Figure 11

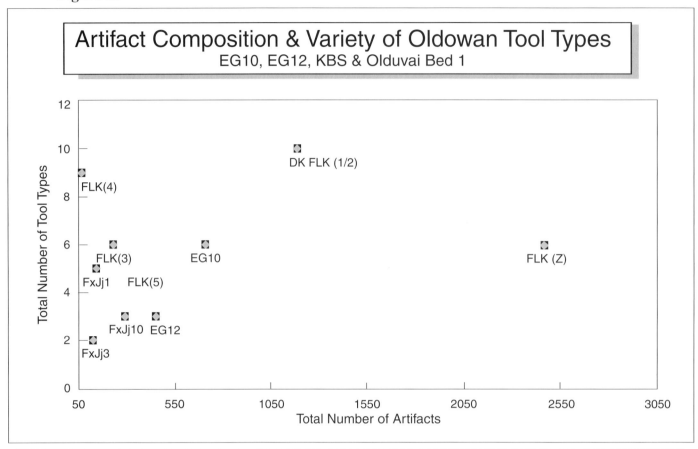

NOTE: Artifact types such as protobifaces, awls, burins, spheroids, laterally trimmed flakes, sundry tools, etc. are excluded. Artifacts identified into these types were very few in numbers at Olduvai, and represented only in one or two sites. Furthermore, none of these artifacts seem to have been part of the tool repertoire of Oldowan hominids, and their exclusion seems justified.

11. *Comparison of the variety of tool types, EG10, EG12, the KBS sites and the Olduvai Bed I and Lower Bed II assemblages. Artifact data for EG10 and EG12 are based on both the surface and excavated cores/choppers. Data for the KBS sites after Isaac & Harris (1997), and for Olduvai after M.D. Leakey (1971).*

DISCUSSION

The Earliest Stone Technology

Despite years of intensive surveys, the older Gona deposits (c. 2.9-2.6 Ma) have not yet yielded any evidence of stones or fossilized bones modified by a hominin agent. Current understanding of the state of Palaeolithic research in East Africa indicates that flaked stone technology made an apparently abrupt entrance into the archaeological record by c. 2.6 Ma. As pointed out by G. Isaac (1976a, p. 491), "...the empirical discovery of the effects of conchoidal fracture was a threshold." The sudden appearance of thousands of flaked stones at Gona and the subsequent widespread manufacture and use documented over broader parts of Plio-Pleistocene Africa confirm his suggestions. Once the initial discovery was made, ancestral hominins engaged in intensive flaking activities making thousands of artifacts, and the practice of stone tool production and use appears to have disseminated quickly, especially in areas in close proximity to raw material sources. The presence of numerous archaeological localities at Gona with abundant artifacts by 2.6 Ma, and their absence in older deposits is intriguing. The mastery and control of stone working shown c. 2.6 Ma points towards the likelihood that the direct forebears of the first toolmakers may not be complete technological novices, and suggests that they probably practiced manipulating tools by regularly using unmodified stones and such perishable items as wooden clubs, tree branches, etc. in their various day-to-day activities (see McGrew, 1993). Unfortunately, these materials do not preserve and leave fossil signatures as well as do flaked stones, and the evidence for Early Pliocene hominin tool use may remain difficult to prove archaeologically.

The Oldowan was initially proposed by Louis Leakey, and later defined by M.D. Leakey (1971) based on her meticulous and detailed analysis of the Upper Bed I & Lower Bed II assemblages of Olduvai Gorge. The Oldowan tradition is characterized mainly by simple cores/choppers and *débitage*, battered hammerstones or spheroids, and occasional retouched pieces. The final shape of the stone artifacts was often dictated by the shape, size and flaking-quality of the original cobbles, and the extent of core reductions accomplished during the reduction of the cores (Toth, 1985, 1987; Schick and Toth, 1993). Currently, there are a large number of Plio-Pleistocene sites with Oldowan assemblages dated between 2.6-1.5 Ma, and all the artifacts of this tradition are non-standardized. The probable exception may be the so called "Karari" (c. 1.5 Ma), mainly differentiated by the preponderance of heavy duty scrapers/ cores, whereby the hominins knocked flakes off sequentially from "prepared platform cores," a technique argued to be effective for maximizing flake outputs and for making optimal use of raw materials (Harris, 1978; Ludwig, 1999). The Karari temporally overlaps with the Acheulean, and its status in terms of artifact tradition is still unclear.

M.D. Leakey defined the Oldowan primarily based on the morphology of the "tools," and some of the terminologies introduced in her descriptions were intended to reflect their function. She inferred that many of the various tool forms, such as choppers, scrapers, heavy duty/light-duty tools etc., were intentionally shaped for specific activities, while others were described and classified based on their morphology as spheroids, polyhedrons, discoids, etc. In particular, the artifacts identified as spheroids/subspheroids and proto-bifaces (which become relatively abundant above Middle Bed II) were considered to be more advanced compared to the core/chopper dominated assemblages of the Lower Beds. M.D. Leakey proposed the "Developed Oldowan" to accommodate these "evolved" forms. However, experimental knapping by Schick and Toth (1994) showed that the spheroids/subspheroids were largely made of quartz, and that intensive use of quartz chunks as hammers over several accumulated hours of flaking could result in these forms taking spherical shapes. They concluded that the abundance of spheroids/subspheroids probably reflects intensive activities with more "habitual" use of tools, and the preponderance of quartz as the main raw material type used at Olduvai during the Early Pleistocene. Thus, overall results from knapping experiments indicate that the final shape of Plio-Pleistocene Oldowan artifacts emerge as byproducts of the core reduction processes instead of resulting from deliberate fashioning by ancestral toolmakers.

In addition, the number of artifacts labeled as protobifaces, awls and burins were very few even within the assemblages of the Lower Beds at Olduvai. Experimental knapping studies have shown that the flaking quality of the raw materials, distances traveled to sources, the original size and morphology of the cobbles available for Plio-Pleistocene hominins, and the extent of core reductions carried out at the sites had a major influence on the final shape and composition of Oldowan artifacts (Toth, 1982, 1985, 1987). Moreover, the primary intent of the toolmakers was the production of simple sharp-edged implements used primarily for cutting up carcasses, and probably for breaking bones for extracting marrows (for e.g., de Heinzelin *et al.*, 1999; Blumenschine & Selvaggio, 1988; Bunn & Kroll, 1986). Until the need by Late Pliocene-Early Pleistocene hominins for such elaborate tools as proto-handaxes, awls and burins and their possible functions are demonstrated through experimental studies, it can be argued that the variety of "tool" types identified in the Lower Beds at Olduvai were quite overstated. As shown in Figure 12, the EG10 and EG12 artifacts, and tool types fall within the range of Olduvai Bed I and the

KBS artifacts after the protobifaces, awls, burins, etc. are excluded.

Because of the excellent knapping skills and control exhibited by the Gona toolmakers, and the similarity illustrated with Early Pleistocene assemblages in artifact making techniques, the EG10 and EG12 assemblages are assigned to the Oldowan Industry. Semaw *et al.* (1997) suggested a "technological stasis" for the Oldowan (2.6-1.5 Ma) based on the following simple observations:

a) The earliest toolmakers had remarkable control and mastery of conchoidal fracture of stone, and the Oldowan persisted with little technological change, the hominins continuing to make simple tools by adopting similar stone working techniques up to c. 1.5 Ma,

b) Plio-Pleistocene Oldowan toolmakers were primarily after sharp-edged cutting implements made with the "least effort" strategy, and the final shape of the artifacts were dictated mainly by the availability and flaking quality of the raw materials, and the extent of the reduction of the cores carried out at the sites (Toth, 1982, 1985, 1987),

c) Current archaeological evidence suggests that the artifacts were used mainly for processing animal carcasses (for e.g., Blumenschine & Selvaggio, 1988; Bunn *et al.*, 1980; Bunn & Kroll, 1986; Beyries, 1993; de Heinzelin *et al.*, 1999), and probably also for processing plant foods (for example, Keeley & Toth, 1981).

d) Compared to the Late Pliocene/earliest Pleistocene, later in the Early Pleistocene relatively higher density concentrations of artifacts and associated fossilized fauna with evidence of cutmarks are observed, along with a marked increase in the intensity of hominin activities (plus more exhaustive core reductions and greater proportions of retouched pieces), but still within a continuum and with no significant departure in the techniques of stone working,

e) Large artifacts made with controlled design, predetermined shape, and symmetry were essentially unknown within the Oldowan, and bifaces with such forms as handaxes and cleavers emerged by c. 1.5 Ma in the archaeological record with the advent of the Acheulean tradition in Africa (Isaac & Curtis, 1974; Gowlett, 1988; Asfaw *et al.*, 1991; Dominguez-Rodrigo *et al.*, 2001).

Therefore, it is appropriate that all the core/flake assemblages dated between c. 2.6-1.5 Ma be subsumed under the Oldowan artifact tradition as defined by M.D. Leakey (1971; Semaw, 2000; Semaw *et al.*, 1997). Because of the similarity in the manufacture techniques utilized, the stone assemblages dated c. 2.6-1.5 Ma group into the Oldowan.

Late Pliocene/Early Pleistocene (2.6-1.5 Ma) Artifact Tradition(s)?

There is an excessive number of Late Pliocene artifact industries/facies introduced into the archaeological literature, including the "Shungura" for the Omo (Chavaillon, 1976), the "Nachikui" for the Lokalalei 1 (Kibunjia, 1994), and the generic labels including the "Pre-Oldowan" and the "Pre-Classic Oldowan," both proposed for accommodating the stone assemblages that are older than 2.0 Ma (Roche, 1989, 1996; Piperno, 1989; Ludwig, 1999). Further, these industries/facies were subsumed under the broader "Omo Industrial Complex" (Kibunjia, 1994) to separate them from the conventional "Oldowan Industry" (*sensu stricto* of M.D. Leakey, 1971) known between 2.0-1.5 Ma. The basic premise underlying the so called "Pre-Oldowan" appears to be the firm conviction by the researchers that Late Pliocene assemblages should look very crude since the earliest toolmakers understood conchoidal fracture on stones only conceptually, but were still novices not fully adept in stone knapping skills (Chavaillon, 1976; Roche, 1989; Piperno, 1989; Kibunjia, 1994; but see also Roche *et al.*, 1999). Additionally, these facies/industries were proposed entirely based on preliminary observations, and detailed descriptions of the respective assemblages are still lacking. Further, no experimental knapping studies were undertaken using raw materials comparable to Plio-Pleistocene assemblages to investigate if the variations in Late Pliocene-Early Pleistocene assemblage characteristics were related to raw material differences. In addition, there are no convincing cases suggested by any of the researchers for supporting the significance of the various facies/industries, and why the 2.0 Ma date was used as a boundary to differentiate the core/flake Oldowan tradition that existed between 2.6-1.5 Ma.

The Omo assemblages reported by Chavaillon (1976) and Merrick (1976) were excavated from artifact horizons stratigraphically placed within Member F of the Shungura Formation (2.4-2.3 Ma). A great majority of the artifacts were made of vein quartz, and most identified as flaking debris resulting from shattered quartz pebbles or chunks made primarily by the 'bipolar' flaking technique. Chavaillon was convinced that the simplicity of the Omo artifacts sets them apart from the conventional Oldowan to merit a separate designation that he named the "Shungura." During Member F times, the closest source of the raw materials, small quartz chunks and pebbles, were ancient stream channels located 20-30 Km east of where the artifacts were made and discarded. "This material definitely does not flake as regularly as lava or other materials might and this may account for the size and the simplicity of the Member F artifacts" (Merrick, 1976, p.480). Thus, for Merrick, the simplicity and small size nature of the artifacts resulted from constraints imposed by the raw materials and distances traveled to sources, and apparently the Omo assemblages did not warrant a new industry other than the Oldowan.

Plio-Pleistocene toolmakers inhabited various parts of the ancient landscape with access to a limited variety of raw materials. Despite the similarities in the techniques of manufacture employed, raw material constraints certainly had major influence on the morphology and composition of contemporary assemblages. Ignoring the role of the various raw materials available for early hominins and their influence on the final form of the artifacts seem to have encouraged researchers to name further industries/facies and perpetuate redundancy for otherwise technologically homogeneous traditions. The following was a cautionary note Isaac forwarded a quarter of a century ago:

> "Distinctive features of stone artifact assemblages can be attributed to differences in the traditions or cultures of the hominins that made them. Clearly before this is done it is desirable to distinguish features which may have been induced largely by *differences in raw materials*, and difference which may reflect *varied activities* by the same people at different times and places. The distinctiveness of the Shungura industries *vis-à-vis* Olduvai and Koobi Fora may be an example of differences induced by contrasting raw materials, which therefore cannot be interpreted as necessarily indicative of other cultural or developmental stage differences."(Isaac, 1976a, p. 496, *original emphasis*).

Two distinct assemblages were recognized at West Turkana, the "less elaborate" from Lokalalei 1 and the "sophisticated" artifacts recently excavated from LA2C (Kibunjia, 1994; Roche *et al.*, 1999). The LA2C "*débitage* scheme" is argued to be exceedingly sophisticated compared to Lokalalei 1 and other Pliocene assemblages (Roche *et al.*, 1999, 2001; Roche & Delanges, 2001). Why such differences in "sophistication" between the Lokalalei 1 and the LA2C assemblages? The major evidence put forward to substantiate this claim was the high incidence of refitting pieces excavated at LA2C. The researchers concluded that, "...the variation observed probably reflects technical solutions to different environments and needs, as well as differences in cognitive and motor skills among early hominin groups characterized by non-synchronous evolutionary processes." (Roche *et al.*, 1999, p. 59). The "non-synchronous evolutionary processes" proposed by Roche *et al.* (1999) entail more than one hominin group as candidates for the two assemblages, and two groups of hominins with different level of technical skills responsible for designing the two "distinct assemblages." The presence of diverse quality of raw materials (coarse to fine-grained) was recognized, and that the heavy reduction of the cores made of the fine-grained raw materials was acknowledged, but the impact of the varying quality of the raw materials available for the hominins and the influence on the reduction stages of the cores represented at the two Lokalalei sites were not adequately explored as possible factors contributing for the assemblage differences.

At Lokalei, the contemporary hominins c. 2.3 Ma and possible candidates for the artifacts were early *Homo* or the immediate descendants of *Australopithecus garhi* and *Australopithecus aethiopicus* (Howell, 1978; Howell *et al.*, 1987; Walker *et al.*, 1986; Suwa *et al.*, 1996; Kimbel *et al.*, 1996; Asfaw *et al.*, 1999). Thus far, no conclusive evidence exists for the exclusion of anyone of these groups as candidates. The regional geographic and temporal overlap of *A. garhi* (at Bouri) and the Gona artifacts c. 2.5 Ma is well documented, and the archaeological evidence at Gona clearly shows that australopithecines had sophisticated understanding of stone flaking techniques (Semaw, 2000; Semaw *et al.*, 1997; de Heinzelin *et al.*, 1999). At Hadar, the 2.3 Ma artifacts are stratigraphically associated with early *Homo*, and the assemblages (although of moderate density) are typical of the Oldowan tradition with no marked sophistication shown compared to the other Pliocene/early Pleistocene assemblages known elsewhere in East Africa (Kimbel *et al.*, 1996). Therefore, attempts to attribute Late Pliocene Oldowan assemblages to distinct hominin groups and proving this archaeologically stands a very slim chance of success.

Therefore, experimental knapping studies of the various raw materials available at Lokalalei appear to be most appropriate and critical at this juncture to explain the reasons for the presence of "the sophisticated" LA2C and "the poorly-made" Lokalalei 1 artifacts at the two contiguous sites separated by only c.1 Km. The refitting data, however, is important evidence for the remarkable preservation and site integrity of LA2C, also clearly showing the intensive flaking activities undertaken at the site, and that the assemblages were not disturbed by post-depositional processes (Schick, 1986, 1987, 1997). The refitting pieces are also significant for understanding of the stages represented during the reduction of the cores, but the fact that the pieces conjoin back together is not necessarily strong supporting evidence for the sophistication of the techniques of flaking employed at LA2C, compared for example to Lokalalei 1, and the evidence may reflect variations in the flaking quality of the raw materials utilized at the two Lokalalei sites.

In a detailed study of the production techniques of Late Pliocene assemblages and investigation of the processes of core reductions based on experimental studies, Ludwig and Harris (1998) observed more stepped flakes with the "less-advanced" Lokalalei assemblages compared to EG10 and EG12. Based on their preliminary observations, Ludwig and Harris (1998) concluded that raw material flaws could have influenced the characteristics of the Lokalalei 1 artifacts. They did not recognize notable variations in the techniques of stone manufacture between Gona and Lokalalei, and questioned the validity of the "Pre-Oldowan." However, the same data was later interpreted showing more steps and hinges at Gona compared to other Pliocene sites, and Ludwig (1999), based on his knapping experiments (experienced vs. novices) on argillite blanks collected from gravels in New Jersey, suggested that the preponderance of steps/hinges at Gona means less-advanced technology by the toolmakers, an interpretation which is inconsistent with his earlier analysis of the same data. Recent knapping experiments on cobbles from the Gona conglomerates suggest that the meaning of steps/hinges has yet to be clearly understood, and it remains uncertain if the evidence can be used as a yardstick for discriminating between the technical skills of an experienced vs. a novice knapper. It appears that further knapping experiments of raw materials comparable to Plio-Pleistocene assemblages are needed to firmly address these questions (Toth *et al.*, this volume; Stout & Semaw, this volume).

Why Flaked Stones by 2.6 Ma?

The sudden behavioral shift in Late Pliocene hominins, the factors that triggered the abrupt appearance of flaked-stones in the archaeological record by c. 2.6 Ma, and the reasons why ancestral humans resorted to a novel means of adaptation by incorporating substantial meat in their diet at this juncture are among the least understood issues in Early Palaeolithic studies. Adequate data are needed to explain and establish the causal factors for these behavioral changes based on the archaeological, faunal and palaeontological record. Investigations are currently underway to understand if the changes c. 2.6 Ma in East Africa were tied to the climatic shifts documented globally (Shackleton, 1995). Some researchers point out that the onset of the build-up of ice sheets in the northern hemisphere beginning c.2.8-2.7 Ma may have had effects on the environments of East Africa and probably contributing to the behavioral changes seen in Late Pliocene hominins (Harris, 1983; de Menocal, 1995; de Menocal & Bloemendal, 1995; Vrba, 1995). Vrba relates the appearance of stone tools with these climatic changes and argues for drastic shifts in the African mammalian fauna (the "turnover pulse") c. 2.5 Ma (Vrba, 1985, 1988, 1990, 1995; but see Hill, 1987; White, 1988, 1995). There is still uncertainty whether or not the changes in East Africa clearly indicate a punctuated event at c. 2.6-2.5 Ma, or if the faunal changes occurred gradually (White, 1995; Behrensmeyer *et al.*, 1997; Bobe *et al.*, 2002). Pollen and carbon isotope studies suggest gradual environmental shifts from humid and wetter regimes to cooler, drier and seasonal situations, and from forest to more open woodland regimes between c. 3.0-2.0 Ma (Bonnefille *et al.*, 1987; Levin *et al.*, 2001). Further, these changes also coincided with the major tectonic and volcanic activities documented in East Africa (Denys *et al.*, 1986; Pickford, 1990; Partridge *et al.*, 1995) and changes in the micromammalian fauna (Denys, *et al.*, 1986; Wesselman, 1995).

Based on the evidence of the proliferation of savanna-adapted animals in East Africa during the Late Pliocene, and the advent of marked seasonality in the geological record, one can surmise that environmental changes may have led to possible gradual disappearance of food items (plant matters?) that ancestral hominins subsisted on for a long time, and at the same time have helped lead to increased abundance of new dietary food items (meat) which had to be accessed by utilizing sharp cutting tools. The archaeological evidence from Gona coupled with the cutmark data from Bouri attest that the first toolmakers used sharp-edged flaked-stones for activities related to animal butchery, resources perhaps available seasonally. Further studies are needed to investigate the possible causal factors for the appearance of flaked stones c. 2.6 Ma and to understand better if the changes were ecologically driven and how Late Pliocene hominins were affected by these. These questions constitute a major focus of the ongoing field research at Gona.

Why are there thousands of flaked stones at Gona by 2.6 Ma? The Afar region was inhabited virtually continuously by ancestral hominins since the Late Miocene/Early Pliocene onward, and the region documents a great number of major biological and behavioral evolutionary events of the last c. 6.0 Ma (White *et al.* 1994; Kimbel *et al.*, 1994; Asfaw *et al.*, 1999; de Heinzelin *et al.*, 1999; Haile-Selassie, 2001 and references therein). The large number of localities with thousands of artifacts documented at Gona by 2.6-2.5 Ma, as well as the fossilized remains of *Australopithecus garhi* and the cutmark evidence from Bouri, clearly demonstrate that Gona and the adjacent areas were consistently inhabited by hominins throughout much of the Pliocene. Gona, and the surrounding areas were among the favored habitats probably due to the presence of ancient streams nearby that carried the cobbles used as raw materials, in conjunction with the availability of fresh drinking water, plant foods, and trees used for shade and refuge from predators. The geological and faunal evidence also show that the area that is currently dry and arid was well-watered and thriving with enormous variety of terrestrial and aquatic animals during the Plio-Pleistocene

The Makers of the Earliest Stone Tools

Early sites at Gona have well-documented toolmaking activities in the Afar Rift by 2.5-2.6 Ma, and the cutmark evidence from Bouri have now extended the known range of early toolmaking hominins at that time at least c. 90 km further south (de Heinzelin *et al.*, 1999). *Australopithecus garhi* from Bouri and *Australopithecus aethiopicus* from the Omo were contemporary with the earliest Gona artifacts. *Australopithecus garhi* was directly associated with the cutmarked animal bones and is the most likely candidate for the earliest stone artifacts (Asfaw *et al.*, 1999;

de Heinzelin *et al.*, 1999), although *Australopithecus aethiopicus* may not be excluded. Slightly later, c. 2.4-2.3 Ma, flaked-stones appear at Omo and in the Turkana basin (Howell, 1978; Howell *et al.*, 1987; Walker *et al.*, 1986). A number of sites in East Africa have produced fossilized specimens of early *Homo* dated to c. 2.3 Ma (Hill *et al.*, 1992; Schrenck *et al.*, 1993; Kimbel *et al.*, 1996; Suwa *et al.*, 1996), and pre-2 Ma *Homo* is considered by many as the most likely candidate for initiating the use of flaked stones. Despite that, stratigraphic association of early *Homo* with Oldowan artifacts is so far shown only at Hadar (Kimbel *et al.*, 1996). Nonetheless, both Late Pliocene australopithecines as well as early *Homo* appear to be possible candidates for the earliest tools, probably the lineage that produced *Homo* becoming a more dependent and "habitual" tool-user through time.

The cranial capacity of the hominins known c. 2.6-2.5 Ma shows no significant departure in size from the ancestral condition. *Australopithecus garhi* and *Australopithecus aethiopicus* had cranial capacities of c. 450 cc and 410 cc respectively, not significantly large compared to any of their immediate predecessors. The evidence seems to suggest that the advent of flaked stones probably acted as a catalyst and played a critical role in increase in hominin brain size through time (Isaac, 1976a; Walker *et al.*, 1986; Asfaw *et al.*, 1999; Schick & Toth, 1993).

Function of the Earliest Stone Tools

The discovery of conchoidal fracture on stones as a behavioral threshold, paved the way for Late Pliocene hominins providing them with the most effective means of access to critical food items, and opening unprecedented venues for control over a wide range of opportunities. As indicated by the cutmark evidence from Bouri, high calorie food sources like meat, which were minimally utilized or unexploited prior to 2.6 Ma became part of the ancestral human diet (de Heinzelin *et al.*, 1999). EG10 and EG12 yielded thousands of flaked stones, but there were no associated fossilized fauna recovered from the excavations. Behavioral information on Late Pliocene hominin diet was inferred for a long time based solely on the evidence of stone artifacts until the discovery of well-preserved fossilized fauna with evidence of cutmarks from the Bouri Peninsula within deposits dated to 2.5 Ma (de Heinzelin *et al.*, 1999). The cutmarked bones belonging to medium-size animals were excavated within fine-grained sediments, but with no associated artifacts. Experimental work and micro-polish studies have also shown that some of the later Koobi Fora Oldowan flakes may have been used for processing plant matters (Keely & Toth, 1981).

Sources of raw materials were lacking near Bouri, and that in part may explain the absence of *in situ* artifacts in association with the excavated fossilized bones there. It is likely that the hominins carried with them stone artifacts preferred for certain tasks (cores or flakes with sharp cutting edges) and "manuports" used as cores for generating sharp-edged flakes as well as missiles used for defense against predators. The geological evidence shows that Late Pliocene hominins at Omo traveled long distances seeking raw materials, and the Bouri hominins appear to have experienced similar situations. Thousands of artifacts were made at Gona, but the absence of any associated *in situ* fossilized bones is intriguing, and investigations have yet to firmly show if this is a result of preservation bias or other factors.

CONCLUSIONS

The paucity and lack of continuity in the Late Pliocene archaeological record in Africa, both temporally and geographically, is puzzling. Small windows are open from time to time providing extraordinary and rare opportunities for a glimpse into hominin artifact manufacture and use, and well-documented cases are very scarce and restricted within the deposits dated between 2.6-2.5 Ma at Gona and Bouri, c. 2.3 Ma at Omo, Lokalalei, and Hadar. The record picks up again later c. 1.89 Ma at Koobi Fora and Fejej. Existing archaeological evidence shows that relatively continuous record and a high density of artifact occurrences in East Africa and elsewhere appeared in the archaeological record after c. 1.8 Ma, implying that the manufacture and use of stone artifacts becomes more "habitual" during the Early Pleistocene. Further archaeological and geological field investigations are crucial for understanding whether or not the paucity of artifact occurrences during the Late Pliocene was the result of preservation bias or other factors. The discovery of the deliberate manufacture of sharp-edged cutting tools on stones c. 2.6 Ma was a behavioral threshold. This major technological breakthrough sparked a novel means of adaptation and at the same time resulted in hominin dependency on technology. The appearance of artifacts in the archaeological record was abrupt and it expanded the breadth of dietary preferences and opportunities to include high calorie food items such as meat. This in turn may have created the stimuli for subsequent brain expansion through the feedback interplay of continued tool use.

The Gona archaeological discoveries have clearly shown that the first toolmakers had excellent mastery and control of the mechanics of conchoidal fracture on stones and produced thousands of artifacts even during the initial stages of artifact manufacture (Semaw, 1997, 2000; Semaw *et al.*, 1997). A majority of the EG10 and EG12 cores were flaked unifacially, but a significant number (over 35%) were also worked bifacially and around much of their circumference, and some were exhaustively reduced. Further, there appears no discernible trend at Gona or elsewhere in East Africa for Late Pliocene-Early Pleistocene core reduction strategies to have gradually evolved from unifacial to bifacial/multifacial stone working. The degree of core-reductions (minimal or exhaustive) conform to the "least effort" strategy of the production of simple sharp-edged cores and flakes used primarily for activities related to animal butchery, and the final shape of the artifacts were often highly influenced by the size, flaking-quality, abundance and distances traveled to sources of raw materials (Toth, 1982, 1985, 1987). The trachyte and rhyolite available from the conglomerates at Gona were of good flaking quality and suitable for making the simple Oldowan artifacts recovered at EG10 and EG12. The preponderance of unifacial flaking at EG10 and EG12 may be a result of the abundance and easy access to these raw materials, or due to other unexplained idiosyncrasies related to the core reduction norms practiced by the hominins around Kada Gona. Further, it is possible that some may be "test cores" resulting from flaking episodes generated while examining the flaking quality of available raw materials (Ludwig, 1999). A number of cases were documented at Gona whereby occasionally encountered fine-grained raw materials such as chalcedony and chert were multifacially and exhaustively reduced probably due to their exotic color and good flaking quality.

The composition and the techniques of manufacture of the Gona, Omo, Hadar and Lokalalei assemblages conform to the simple core/flake Oldowan tradition and no convincing cases have been put forward to justify the multiple Late Pliocene industries/facies existing in the current archaeological literature (Roche, 1989; Piperno, 1989; Kibunjia, 1994). The similarity in the techniques of manufacture and the simplicity of Plio-Pleistocene assemblages suggests a "technological stasis" in the Oldowan Industry (Semaw *et al.*, 1997). Similar conclusions were reached by other researchers following detailed studies of the Gona and other East African Plio-Pleistocene assemblages and referred by Ludwig (1999) as 700,000 years of "methodological stasis" in the Oldowan (2.6-1.7 Ma). His conclusions are agreeable, but the new label proposed seems a bit redundant with the "technological stasis" already proposed earlier by Semaw *et al.* (1997). What is commendable is more experimental work on comparable raw materials to East African Plio-Pleistocene artifacts and investigations of the knapping skills of the hominins to firmly understand the meaning of the variations in Oldowan assemblages.

Extensive areas have been surveyed recently within the Gona Paleoanthropological Research Project (GPRP) study area and new sites identified tens of kms away from the previously documented East Gona localities of EG10 and EG12. New sites recently excavated at Gona show that stone raw materials readily available from nearby sources were ignored by early hominins in favor of better quality materials (such as chert) sought from areas further away (sources still to be investigated), and the evidence seems to point towards a preference for better quality stone raw materials, foresight and planning, and a more sophisticated behavior for the earliest toolmakers than previously known (Semaw *et al.*,

2002). Investigations of the sources of raw materials, distances ancestral tool makers had to travel to acquire suitable stones, and experimental knapping studies of the stone raw materials available to early tool makers are important for understanding the technical skills of the first tool makers. The novel ancestral human adaptation that began with the creation of simple cutting stone tools c. 2.6 Ma underwent continual changes with more advanced artifact traditions emerging in spurts, and worked-stones playing major adaptive roles in the subsistence strategies of humans for the last two-and-half million years.

ACKNOWLEDGMENTS

I would like to thank Professors Kathy Schick and Nick Toth (Co-directors of CRAFT) for inviting me to contribute to this volume, and I am grateful for the overall support at CRAFT Research Center, Indiana University, and essential backing from Friends of CRAFT, Inc., and the Stone Age Institute. I am thankful to the Afar people at Eloha and the administration at Asayta for their camaraderie and field support. I would like to thank the Authority for Research and Conservation of Cultural Heritage (ARCCH) of the Ministry of Sports and Culture of Ethiopia for field permits. The L.S.B. Leakey Foundation, the National Science Foundation, the Wenner Gren Foundation, the National Geographic Society and the Boise Fund supported the fieldwork. Ann and Gordon Getty provided major funding for the 1992-94 field research, and I am grateful for their support. EG10 and EG12 were excavated with my academic advisor Professor J.W.K. Harris (Rutgers University), and his overall support is very much appreciated. Dr. Craig Feibel (Rutgers University) carried out detailed geological study of the Kada Gona deposits in 1993-94 and his contribution is invaluable. Dr. Paul Renne (Berkeley Geochronology Center, CA) dated the tuffs and did the paleomagnetic stratigraphy of the sites. Renne's work was critical for resolving the age of the Gona deposits and I am grateful for his work. The late Professor Desmond Clark loaned a field vehicle for the fieldwork and I am grateful for his support. The overall assistance and support by Dr. Yonas Beyene, Dr. Berhane Asfaw, Professor Tim White, Professor Clark Howell, Dr. Robert Blumenschine, Dr. Michael Rogers, Dr. Manuel Dominguez-Rodrigo, Asahmed Humet and John Cavallo is very much appreciated. Yonas Qená and Paul Jung drew the artifacts.

REFERENCES CITED

Aronson, J.L., Schmitt, T.J., Walter, R.C., Taieb, M., Tiercelin, J.J., Johanson D.C., Naesser, C.W., & Nairn, A.E.M. (1977). New geochronologic and paleomagnetic data for the hominid-bearing Hadar Formation, Ethiopia. *Nature* **267**, 323-27.

Aronson, J.L. & Taieb, M. (1981). Geology and paleogeography of the Hadar hominid site, Ethiopia. In (G.J. Rapp & C.F. Vondra, Eds) *Hominid sites: their geologic settings. American Association for the Advancement of Science Selected Symposium.* Westview Press, Boulder **63**, pp.165-195.

Asfaw, B. Beyene, Y., Semaw, S., Suwa, G., White, T. & WoldeGabriel, G. (1991). Fejej: a new paleontological research area in Ethiopia. *Journal of Human Evolution* **21**, 137-143.

Asfaw, B., Beyene, Y., Suwa, G., Walter, R.C., White, T.D., WoldeGabriel, G. & Yemane, T. (1992). The earliest Acheulean from Konso-Gardula. *Nature* **360**, 732-735.

Asfaw, B., White, T., Lovejoy, O., Latimer, B., Simpson, S., and Suwa, G. (1999). *Australopithecus garhi*: a new species of early hominid from Ethiopia. *Science* **284**, 629-635.

Balout, L. (1955). *Préhistoire de l'Afrique du Nord. Essai de Chronologie.* Paris: Arts et Métiers Graphiques.

Behrensmeyer, A.K., Todd, N, E., Potts, R. & McBrinn, G.E. (1997). Late Pliocene faunal turnover in the Turkana Basin, Kenya and Ethiopia. *Science* **278**, 1589-94.

Beyries, S. (1993). Are we able to determine the function of the earliest palaeolithic tools? In (A. Berthelet & J. Chavaillon, Eds) *The use of tools by non-human primates.* Clarendon Press: Oxford, pp.225-236.

Biberson, P. (1961). *Le Paléolithique Inférieur du Maroc Atlantique.* Rabat: Publications du Service Archéologique du Maroc.

Bishop, W.W. (1959). Kafu stratigraphy and Kafuan artifacts. *South African Journal of Science* **55**, 117-21.

Blumenschine, R.J. & Selvaggio, M., (1988). Percussion marks on bone surfaces as a new diagnostic of hominid behavior. *Nature* **333**, 763-765.

Bobe, R., Behrensmeyer, A.K. & Chapman, R.E. (2002). Faunal change, environmental variability and late Pliocene hominid evolution. *Journal of Human Evolution* **42**, 475-97.

Bonnefille, R., Vincens, A. & Buchet, G. (1987). Palynology, stratigraphy, and palaeoenvironment of a Pliocene hominid site (2.9-3.3 M.Y.) at Hadar, Ethiopia. *Palaeogeography, Palaeoclimatology, Palaeoecology* **60**, 249-281.

Brain, C.K., Churcher, C.S., Clark, J.D., Grine, F.E., Shipman, P., Susman, R.L., Turner, A.& Watson, V. (1988). New evidence of early hominids, their culture and environments from the Swartkrans cave, South Africa. *South African Journal of Science* **84**, 828-35.

Brown, F. H. (1995). The potential of the Turkana Basin for paleoclimatic reconstruction in East Africa. In (E. S. Vrba, G. H. Denton, T. C. Partridge & L. H. Burckle, Eds) *Paleoclimate and Evolution, with Emphasis on Human Origins.* New Haven and London: Yale University Press, pp. 319-330.

Bunn, H.T., Harris, J.W.K.; Kaufulu, Z., Kroll, E., Schick, K., Toth, N. & Behrensmeyer, A.K. (1980). FxJj 50: an early Pleistocene site in Northern Kenya. *World Archeology* **12**, 109-136.

Bunn, H.T. & Kroll, E.M. (1986). Systematic butchery by Plio-Pleistocene hominids at Olduvai Gorge, Tanzania. *Current Anthropology* **27**, 431-445.

Chavaillon, J. (1976). Evidence for the technical practices of early Pleistocene hominids, Shungura Formation, Lower Omo Valley, Ethiopia. In (Y. Coppens, F. C. Howell, G. Isaac & R.E.F. Leakey, Eds) *Earliest Man and Environments in the Lake Rudolf Basin.* Chicago: University of Chicago Press, pp. 565-573.

Chavaillon, J., Chavaillon, N., Hours, F. & Piperno, M. (1979). From the Oldowan to the Middle Stone Age at Melka-Kunture (Ethiopia). Understanding cultural changes. *Quaternaria* **21**, 87-114.

Clark, J.D. (1976). The African origins of man the toolmaker. In (G.Ll. Isaac & E. R. McCown, Eds) *Human Origins Louis Leakey and the East African Evidence.* W.A Benjamin Inc., Philippines, pp.1-53.

Clark, J.D. (1992). The Earlier Stone Age/ Lower Palaeolithic in North Africa and the Sahara. In (F. Klees & R. Kuper, Eds) *New Light on the Northeast African Past.* Koln: Heinrich-Barth-Institut, pp. 17-37.

Clark, J.D. (1994). The Acheulian Industrial Complex in Africa and elsewhere. In (R.S. Corruccini & R.L. Ciochon, Eds) *Integrative Paths to the Past: Palaeoanthropological Advances in Honor of F. Clark Howell.* New Jersey: Prentice Hall Publishers, pp.201-215.

Clark, J.D. & Kurashina, H. (1979). Hominid occupation of the East Central Highlands of Ethiopia in the Plio-Pleistocene. *Nature* **282**, 33-39.

Clark, J. D., Asfaw, B., Assefa, G., Harris, J.W.K., Kurashina, H., Walter, R. C., White, T. D. & Williams, M.A.J. (1984). Palaeoanthropologic discoveries in the Middle Awash Valley, Ethiopia. *Nature* **307**, 423-428.

Clark, J. D., de Heinzelin, J., Schick. K. D., Hart, W. K., White, T. D., WoldeGabriel, G., Walter, R. C., Suwa, G., Asfaw, B., Vrba, E. & H. Selassie, Y. (1994). African *Homo erectus*: old radiometric ages and young Oldowan assemblages in the Middle Awash Valley, Ethiopia. *Science* **264**, 1907-1909.

Cooke, H.B.S. (1976). Suidae from Plio-Pleistocene strata of the Rudolf basin. In (Y. Coppens, F.C. Howell, Gl. Ll. Isaac & R.E.F. Leakey Eds) *Earliest Man and Environments in the Lake Rudolf Basin.* Chicago University Press, Chicago, pp. 251-63.

Corvinus, G. (1976). Prehistoric exploration at Hadar, Ethiopia. *Nature* **261**, 571-572.

Corvinus, G. & Roche, H. (1976). La préhistoire dans la région de Hadar (Bassin de l'Awash, Afar, Ethiopie): premiers résultats. *L'Anthropologie* **80** (2), 315-324.

Corvinus, G. & Roche, H. (1980). Prehistoric exploration at Hadar in the Afar (Ethiopia) in 1973, 1974, and 1976. In (R.E.F. Leakey & B.A. Ogot, Eds) *Proceedings, VIIIth Panafrican Congress of Prehistory and Quaternary Studies*, Nairobi, pp. 186-188.

de Heinzelin, J. (1983). The Omo Group. In (J. de Heinzelin, Ed) *Stratigraphic and Related Earlth Sciences Studies in the Lower Omo Basin, Southern Ethiopia*. Musee Royal de l'Afrique Centrale, Treuvren, Belgique, Annals Sciences Geologiques, **85**, pp. 187-89.

de Heinzeliin, J., Clark, J. D., White, T.W., Hart, W., Renne, P., WoldeGabriel, G., Beyene, Y., & Vrba, E. (1999). Environment and behavior of 2.5-million-year-old Bouri hominids. *Science* **284**, 625-629.

de Menocal, P.B. (1995). Plio-Pleistocene African climate. *Science* **270**, 53-59.

de Menocal, P. B. & Bloemendal (1995). Plio-Pleistocene climatic variability in subtropical Africa and the paleoenvironment of hominid evolution. In (E.S. Vrba, G.H. Denton, T. C. Partridge & L. H. Burckle, Eds) *Paleoclimate and Evolution, with Emphasis on Human Origins*. New Haven and London: Yale University Press, pp. 262-288.

Denys, C., Chorowicz, J. & Tiercelin, J. J. (1986). Tectonic and environmental control on rodent diversity in the Plio-Pleistocene sediments of the African Rift System. In (L.E. Frostick, R.W. Renaut, I. Reid & J. J. Tiercelin, Eds) *Sedimentation in the African Rifts*. Oxford: Alden Press Ltd, pp.362-372.

Dominguez-Rodrigo, M., Serralonga, J., Juan-Tresserras, J., Alcala, L., & Luque, L. (2001). Woodworking activities by early humans: a plant residue analysis on Acheulian stone tools from Peninj (Tanzania). *Journal of Human Evolution* **40**, 289-299.

Feibel, C. S., Brown F. H. & Mc Dougall, I. (1989). Stratigraphic context of hominids from the Omo Group deposits: northern Turkana basin, Kenya and Ethiopia. *American Journal of Physical Anthropology* **78**, 595-622.

Fitch, F.J. & Miler J.A. (1970). Radioisotopic age determinations of Lake Rudolf artifact site. *Nature* **226**, 226-8.

Fitch, F.J. & Miler J.A. (1976). Conventional potassium-argon and argon 40/argon 39 dating of the volcanic rocks from East Rudolf. In (Y. Coppens, F.C. Howell, Gl. Ll. Isaac & R.E.F. Leakey, Eds) *Earliest Man and Environments in the Lake Rudolf Basin*. Chicago University Press, Chicago, pp. 123-47.

Gowlett, J.A.J. (1988). A case of Developed Oldowan in the Acheulean? *World Archaeology* **20**, (1) 13-26.

Gowlett, J.A.J. (1990) Archaeological studies of human origins & early prehistory in Africa. In (P. Robertshaw, Ed) *A History of African Archaeology*. James Currey, London, pp.13-38.

Gowlett, J.A.J., Harris, J.W.K., Walton, D. & Wood, B.A. (1981). Early archaeological sites, hominid remains and traces of fire from Chesowanja, Kenya. *Nature* **294**, 125-29.

Haile-Selassie, Y. (2001). Late Miocene hominids from the Middle Awash, Ethiopia. *Nature* **412**, 178-81.

Hall, C.M., Walter, R.C. & York, D. (1985). Tuff above "Lucy" is over 3 ma old. Eos 66, 257.

Harris, J. W. K. (1978). T*he Karari Industry: its Place in East African Prehistory*. Ph.D. Thesis, University of California, Berkeley, CA.

Harris, J. W. K. (1983). Cultural beginnings: Plio-Pleistocene archaeological occurrences from the Afar, Ethiopia. *African Archaeological Review* **1**, 3-31.

Harris, J. W. K.& Semaw, S. (1989). Pliocene archaeology at the Gona River, Hadar. *Nyame Akuma* **31**, 19-21.

Harris, J.W.K., Williamson, P.G., Verniers, J., Tappen, M.J., Stewart, K., Helgren, D., de Heinzelin, J., Boaz, N.T. & Bellomo, R. (1987). Late Pliocene hominid occupation in Central Africa: the setting, context, and character of the Senga 5A site, Zaire. *Journal of Human Evolution* **16**, 701-728.

Harris, J.W.K., Williamson, P.G., Morris, P.J., de Heinzelin, J., Verniers, J., Helgren, D., Bellomo, R.V., Laden, G., Spang, T.W., Stewart, K. & Tappen, M.J. (1990). In (N. Boaz, Ed) *Archaeology of the Lusso Beds*. Memoir 1. Martinsville: Virginia Museum of Natural History.

Hay, R.L. (1971). Geologic background of Beds I and II stratigraphic summary. In (M.D. Leakey, Ed) *Olduvai Gorge, Vol. III.* London: Cambridge University Press, pp. 9-18.

Hill, A. (1987). Causes of perceived faunal change in the latter Neogene of East Africa. *Journal of Human Evolution* **16**, 583-96.

Hill, A., Ward, S., Deino, A., Curtis, G. & Drake, R. (1992). Earliest Homo. *Nature* **355**, 719-722.

Howell, F.C. (1978). Overview of the Pliocene and the earlier Pleistocene of the lower Omo Basin, southern Ethiopia. In (C.J. Jolly, Ed) *Early Hominids of Africa*. London: Duckworth, pp. 85-130.

Howell, F. C., Haesaerts, P. & de Heinzelin, J. (1987). Depositional environments, archaeological occurrences and hominids from Members E and F of the Shungura Formation (Omo Basin, Ethiopia). *Journal of Human Evolution* **16**, 665-700.

Isaac, G. Ll. (1976a). The activities of early African hominids: a review of archaeological evidence from the time span two and a half to one million years ago. In (G. Ll. Isaac & E.R. McCown, Eds) *Human Origins Louis Leakey and the East African Evidence*. W.A. Benjamin, Inc, Philippines, pp. 483-514.

Isaac, G. Ll. (1976b). Plio-Pleistocene artifact assemblages from East Rudolf, Kenya. In (Y. Coppens, F. C. Howell & G. Ll. Isaac Eds) *Earliest Man and Environments in the Lake Rudolf Basin*. University of Chicago Press, Chicago, pp. 552-564.

Isaac, G. Ll. (1997). Introduction. In (G.Ll. Isaac & B. Isaac, Eds) *Koobi Fora Research Project, Plio-Pleistocene Archaeology Vol. 5.* Oxford: Clarendon Press, pp.1-11.

Isaac, G. Ll. & Curtis, G. H. (1974). Age of the Acheulian industries from the Peninj Group, Tanzania. *Nature* **249**, 624-627.

Isaac, G.Ll. & Harris, J.W.K. (1997). Sites stratified within the KBS Tuff: Reports. In (G. Ll. Isaac & B. Isaac, Eds) *Koobi Fora Research Project, Plio-Pleistocene Archaeology Vol. 5.* Oxford: Clarendon Press, pp. 71-114.

Isaac, G.Ll & Isaac, B., Eds (1997). *Koobi Fora Research Project, Plio-Pleistocene Archaeology Vol. 5.* Oxford: Clarendon Press, Oxford.

Isaac, G.Ll., Harris J.W.K. & Marshall, F. (1981). Small is informative: the application of the study of mini-sites and least effort criteria in the interpretation of the Early Pleistocene archaeological record at Koobi Fora, Kenya. *Proc. Union Internacional de Ciencias Prehistoricas Y Protohistoricas; X Congress, Mexico City,* Mexico, pp. 101-119.

Johanson, D. C., Taieb, M. & Coppens, Y. (1982). Pliocene hominids from the Hadar Formation, Ethiopia (1973-1977): stratigraphic, chronological, and paleoenvironmental contexts, with notes on hominid morphology and systematics. *American Journal of Physical Anthropology* **57**, 373-402.

Jones, P.R. (1994). Results of experimental work in relation to the stone industries of Olduvai Gorge. In (M. D. Leakey, Ed) *Olduvai Gorge- Excavations in Beds III, IV and the Masek Beds (1968-71), Vol. 5.* Cambridge University Press.

Kalb, J. E. (1993). Refined stratigraphy of the hominid-bearing Awash Group, Middle Awash Valley, Afar Depression, Ethiopia. Newsletters on *Stratigraphy* **29** (1), 21-62.

Kalb, J.E., Oswald, E.B., Mebrate, A., Tebedge, S. & Jolly, C.J. (1982a). Stratigraphy of the Awash group, Afar, Ethiopia. *Newsletters on Stratigraphy* **11**, 95-127.

Kalb, J.E., Oswald, E.B., Tebedge, S., Mebrate, A., Tola, E. & Peak, D. (1982b). Geology and stratigraphy of Neogene deposits, Middle Awash Valley, Ethiopia. *Nature* **298**, 17-25.

Kaufulu, Z.M. & Stern, N. (1987). The first stone artifacts to be found *in situ* within the Plio- Pleistocene Chiwondo Beds in northern Malawi. *Journal of Human Evolution* **16**, 729-740.

Keeley, L. H. & Toth, N. P. (1981). Microwear polishes on early stone tools from Koobi Fora, Kenya. *Nature* **293**, 464-465.

Kibunjia, M. (1994). Pliocene archeological occurrences in the Lake Turkana basin. *Journal of Human Evolution* **27**, 159-171.

Kibunjia, M. Roche, H., Brown, F.H. & Leakey, R.E. F. (1992). Pliocene and Pleistocene archeological sites of Lake Turkana, Kenya. *Journal of Human Evolution* **23**, 432-438.

Kimbel, W.H., Johanson, D.C. & Rak, Y. (1994). The first skull and other new discoveries of *Australopithecus afarensis* at Hadar, Ethiopia. *Nature* **368**, 449-451.

Kimbel, W.H., Walter, R. C., Johanson, D. C., Reed, K. E., Aronson, J. L., Assefa, Z., Marean, C. W., Eck, G. G., Bobe, R., Hovers, E., Rak, Y., Vondra, C., Yemane, T., York, D., Chen,Y., Evensen, N. M. & Smith, P. E. (1996). Late Pliocene *Homo* and Oldowan tools from the Hadar Formation (Kada Hadar Member), Ethiopia. *Journal of Human Evolution* **31**, 549-561.

Kuman, K. (1994a). The archaeology of Sterkfontein-past and present. *Journal of Human Evolution* **27**, 471-95.

Kuman, K. (1994b). The archaeology of Sterkfontein: preliminary findings on site formation and cultural change. *South African Journal of Science* **90**, 215-19.

Kuman, K. (1996). The Oldowan industry from Sterkfontein: raw materials and core forms. In (G. Pwiti & R. Soper, Eds) *Aspects of African Archaeology, Papers from the 10th Congress of the Panafrican Association for Prehistory and Related Studies.* Zimbabwe Publications, Harari, pp. 139-148.

Kuman, K. (1998). The earliest South African Industries. In (M.D. Petraglia & R. Korissettar, Eds) *Early Human Behavior in Global Context. The Rise and Diversity of the Lower Paleolithic Record.* London and New York: Routledge, pp. 151-186.

Kuman, K., Field, A.S. & Thackeray, J.G. (1997). Discovery of new artifacts at Kromdraai. *South African Journal of Science* **93**, 187-93.

Leakey, L.S.B. (1936). *Stone Age Africa: an Outline of Prehistory in Africa.* Oxford University Press, London.

Leakey, M.D. (1971). *Olduvai Gorge, Vol. III.* London: Cambridge University Press.

Leakey, R.E. (1970). Fauna and artifacts from a new Plio-Pleistocene locality near Lake Rudolf in Kenya. *Nature* **226**, 223-4.

Levin, N., Quade, J., Semaw, S., Schick, S., Toth, N. & Simpson, S.W. (2001). Plio-Pleistocene environments of Gona, Ethiopia: the isotopic record for pedogenic carbonate and fossil teeth. Abstract, *The Geological Society of America*, Annual Meeting.

Ludwig, B. V. (1999). *A Technological Reassessment of East African Plio-Pleistocene Lithic Artifact Assemblages.* Ph. D. Dissertation, Rutgers University, New Brunswick, NJ.

Ludwig, B. V. & Harris, J.W.K. (1998). Towards a technological reassessment of East African Plio-Pleistocene lithic assemblages. In (M.D. Petraglia & R. Korisettar, Eds) *Early Human Behavior in Global Context. The Rise and Diversity of the Lower Paleolithic Record.* London and New York: Routledge, pp. 84-107.

McDougall, I, Maier, R., Sutherland-Hawkes, P, and Gleadow, A.J.W. (1980). K/Ar age estimate for the KBS Tuff, East Turkana, Kenya. *Nature* **284**, 230-4.

McDougall, I., Brown, F.H., Cerling, T.E. & Hillhouse, J.W. (1992). A reappraisal of the geomagnetic polarity time scale to 4 Ma using data from the Turkana Basin, East Africa. *Geophysical Research Letters* **19** (23), 2349-2352.

McGrew, W.C. (1993). The intelligent use of tools: twenty propositions. In (K.Gibson & T. Ignold, Eds) *Tools, Language and Cognition in Human Evolution.* Cambridge University Press, pp. 151-170.

Merrick, H.V. (1976). Recent archaeological research in the Plio-Pleistocene deposits of the Lower Omo, southwestern Ethiopia. In (G. Ll., Isaac & I. McCown, Eds) *Human Origins Louis Leakey and the East African Evidence.* Menlo Park: W.A. Benjamin, pp.461-81.

Merrick, H.V. & Merrick, J.P.S. (1976). Recent archaeological occurrences of earlier Pleistocene age from the Shungura Formation. In (Y. Coppens, F.C. Howell, G. Ll., Isaac & R.E.F. Leakey, Eds) *Earliest Man and Environments in the Lake Rudolf Basin.* Chicago: Chicago University Press, pp. 574-584.

Partridge, T. C., Wood, B. A. & de Menocal, B. (1995). The influence of global climatic change and regional uplift on large-mammalian evolution in East and South Africa. In (E.S. Vrba, G.H. Denton, T.C. Partridge & L.H. Burckle, Eds) *Paleoclimate and Evolution, with Emphasis on Human Origins.* New Haven and London: Yale University Press, pp. 331-355.

Pickford, M. (1990). Uplift of the roof of Africa and its bearing on the evolution of mankind. *Human Evolution* **5** (1), 1-20.

Piperno, M. (1989). Chronostratigraphic and cultural framework of the *Homo habilis* sites. In (G. Giacobini, Ed) *Hominidae. Proceedings of the 2nd International Congress of Human Paleontology.* Jaca Book, Milan, pp. 189-95.

Roche, H. (1989). Technological evolution in early hominids. *OSSA* **4**, 97-98.

Roche, H. (1996). Remarque sur les plus anciennes industries en Afrique et en Europe. Colloquium VIII *Lithic Industries, Language and Social Behaviour in the First Human Forms.* IUPSS Congress, Forli, Italy, pp.55-68.

Roche, H. (2001). Stone knapping evolution among early hominids, knapping stone a uniquely hominid behavior. *International Workshop, 21-24 November 2001, Abbaye des Prémontrés, Pont-À-Mousson* **54**, pp. 23.

Roche, H. & Tiercelin, J. J. (1977). Découverte d'une industrie lithique ancienne *in situ* dans la formation d'Hadar, Afar central, Éthiopie. *C.R. Acad. Sci.* Paris D **284**, 187-174.

Roche, H. & Tiercelin, J. J. (1980). Industries lithiques de la formation Plio-Pléistocène d'Hadar: campagne 1976. In (R. E. F. Leakey & B. A. Ogot, Eds) *Proceedings, VIIIth Panafrican Congress of Prehistory and Quaternary Studies.* Nairobi, pp. 194-199.

Roche, H., Delagnes, A., Brugal, J.P., Feibel, C., Kibunjia, M., Mourre, V., & Texier, P.J. (1999). Early hominid stone tool production and technical skill 2.34 Myr ago in West Turkana, Kenya. *Nature* **399**, 57-60.

Roche, H. & Delagnes, A. (2001). Evidence of controlled and reasoned stone knapping at 2.3 Myr, West Turkana (Kenya). *International Workshop, 21-24 November 2001, Abbaye des Prémontrés, Pont-À-Mousson* **54**, pp. 24.

Sahnouni, M. (1998). The Lower Palaeolithic of the Maghreb, Excavations and Analyses at Ain Hanech, Algeria. *Oxford: British Archaeological Reports, International Series 689. Cambridge Monographs in African Archaeology* **42**, Archaeopress.

Sahnouni, M. & de Heinzelin, J. (1998). The site of Ain Hanech revisited: new investigations at this Lower Pleistocene site in Northern Algeria. *Journal of Archaeological Science* **25**, 1083-1101.

Schrenk, F., Bromage, T.G., Betzler, C.G., Ring, U. & Juwayeyi, Y. (1993). Oldest Homo and Pliocene biogeography of the Malawi Rift. *Nature* **365**, 833-836.

Schick, K. (1986). *Stone age sites in the making: experiments in the formation and transformation of archaeological occurrences.* Oxford: British Archaeological Reports, International Series, Cambridge Monographs in African Archaeology **319**.

Schick, K. (1987). Modeling the formation of Early Stone Age artifact concentrations. *Journal of Human Evolution* **16**, 789-808.

Schick, K. (1997). Experimental studies of site formation processes. In (G. Ll. Isaac & B. Isaac, Eds) *Koobi Fora Research Project, Plio-Pleistocene Archaeology Vol. 5.* Oxford: Clarendon Press, pp. 244-256.

Schick, K. & Toth, N. (1993). *Making Silent Stones Speak.* Simon & Schuster, New York.

Schick, K. & Toth, N. (1994). Early Stone Age and case study into the nature and function of spheroids and subspheroids. In (R.S. Corruccini & R. L. Ciochon, Eds) *Integrative Paths to the Past: Palaeoanthropological Advances in Honor of F. Clark Howell.* Prentice Hall, Englewood Cliffs, New Jersey, Vol. 2, pp. 429-449.

Semaw, S. (1997). *Late Pliocene Archeology of the Gona River Deposits, Afar, Ethiopia.* Ph.D. Thesis. Rutgers University, New Brunswick, NJ.

Semaw, S. (2000). The world's oldest stone artifacts from Gona, Ethiopia: their implications for understanding stone technology and patterns of human evolution between 2.6-1.5 Million years ago. *Journal of Archaeological Science* **27**, 1197-1214.

Semaw, S., Renne, P., Harris, J.W.K., Feibel, C. S., Bernor, R. L., Fesseha, N. & Mowbray, K. (1997). 2.5-million-year-old stone tools from Gona, Ethiopia. *Nature* **385**, 333-336.

Semaw, S., Schick, K., Toth, N., Simpson, S., Quade, J., Rogers, M., Renne, P., Stout, D. & Dominguez-Rodrigo, M. (2002). Recent discoveries from Gona, Afar, Ethiopia. *Journal of Human Evolution* **42**, A33.

Shackleton, N. J. (1995). New data on the evolution of Pliocene climatic variability. In (E. S. Vrba, G. H. Denton, T.C. Partridge & L.H. Burckle, Eds) *Paleoclimate and Evolution, with Emphasis On Human Origins.* New Haven and London: Yale University Press, pp. 242-248.

Sussman, R.L. (1991). Who made the Oldowan tools? Fossil evidence for tool behavior in Plio-Pleistocene hominids. *Journal of Anthropological Research* **47**, 129-51.

Suwa, G., White, T. & Howell, F. C. (1996). Mandibular postcanine dentition from the Shungura Formation, Ethiopia: crown morphology, taxonomic allocation, and Plio- Pleistocene hominid evolution. *American Journal of Physical Anthropology* **101**, 247-282.

Taieb, M., Coppens, Y., Johanson, D.C. & Kalb, J. (1972). Dépôts sedimentaires et faunes du Plio-Pléistocène de la basse vallée de l'Awash (Afar centrale, Ethiopie). C. R. Acad. Sc., D. **275**, 819-22.

Texier, P. (1995). The Oldowan assemblage from NY 18 site at Nyabusosi (Toro-Uganda). C.R. Acad. Sci. Paris, **320**, 647-53.

Tobias, P.V. (1976). White African: an appreciation and some personal memories of Louis Leakey. In (G.Ll. Isaac & E. R. McCown, Eds) *Human Origins Louis Leakey and the East African evidence.* W.A Benjamin Inc., Philippines, pp. 55-74.

Toth, N. (1982). *The Stone Technologies of Early Hominids at Koobi Fora, Kenya: an Experimental Approach.* Ph.D. Thesis. University of California, Berkeley, CA.

Toth, N. (1985). The Oldowan reassessed: a close look at early stone artifacts. *Journal of Archeological Science* **12**, 101-120.

Toth, N. (1987). Behavioral inferences from early stone artifact assemblages: an experimental model. *Journal of Human Evolution* **16**, 763-787.

Van Riet Lowe, C. (1957). The Kafuan Culture. In (J.D. Clark, Ed) *Third Pan-African Congress on Prehistory.* Livingstone 1955, London Chatto & Windus, pp.207-209.

Vrba, E. S. (1985). Environment and Evolution: alternative causes of the temporal distribution of evolutionary events. *South African Journal of Science* **81**, 229-236.

Vrba, E. S. (1988). Late Pliocene climatic events and hominid evolution. In (F. Grine, Ed) *The Evolutionary History of the "Robust" Australopithecine.* New York: Aldine de Gruyter, pp. 405-426.

Vrba, E. S. (1990). The environmental context of the evolution of early hominids and their culture. In (R. Bonnichsen & M. Sorg, Eds) *Bone Modification.* Center for the study of the 1st Americans. Orono, Maine, pp. 27-42.

Vrba, E. S. (1995). On the connections between paleoclimate and evolution. In (E.S. Vrba, G.H. Denton, T.C. Partridge & L.H. Burckle, Eds) *Paleoclimate and Evolution, with Emphasis on Human Origins.* New Haven and London: Yale University Press, pp. 24-45.

Walker, A., Leakey, R. E., Harris, J. M. & Brown, F. H. (1986). 2.5-Myr *Australopithecus boisei* from west of Lake Turkana, Kenya. *Nature* **322**, 517-522.

Walter, R. C. (1980). *The Volcanic History of the Hadar Early Man Site and the Surrounding Afar Region of Ethiopia.* Ph.D. Thesis, Case Western Reserve University, OH.

Walter, R. C. (1994). Age of Lucy and the First Family: single-crystal ^{40}Ar/^{39}Ar dating of the Denen Dora and lower Kada Hadar Members of the Hadar Formation, Ethiopia. *Geology* **22**, 6-10.

Walter, R.C. & Aronson, J.L. (1982). Revisions of K/Ar ages for the Hadar hominid site, Ethiopia. *Nature* **296**, 122-127.

Wesselman, H.B. (1995). Of mice and almost-men: regional paleoecology and human evolution in the Turkana Basin. In (E.S. Vrba, G.H. Denton, T.C. Partridge & L.H. Burckle, Eds) *Paleoclimate and Evolution, with Emphasis on Human Origins.* New Haven and London: Yale University Press, pp. 356-368.

White, T.D. (1988). The comparative biology of "Robust" Australopithecus: clues from context. In (F. Grine, Ed). *Evolutionary History of the "Robust" Australopithecines.* Aldine de Gruyter, New York, pp. 449-78.

White, T.D. (1995). African omnivores: global climatic change and Plio-Pleistocene hominids and suids. In (E. S. Vrba, G. H. Denton, T. C. Partridge & L.H. Burckle, Eds) *Paleoclimate and Evolution, with Emphasis on Human Origins.* New Haven and London: Yale University Press, pp. 369-384.

White, T.D; Suwa, G. & Asfaw, B. (1994). *Australopithecus ramidus,* a new species of early hominid from Aramis, Ethiopia. *Nature* **371**, 306-12.

WoldeGabriel, G., White, T.D., Suwa, G., Renne, P.R., de Heinzelin, J., Hart, W.K. & Heiken, G. (1994). Ecological and temporal placement of early Pliocene hominids at Aramis, Ethiopia. *Nature* **371**, 330-33.

WoldeGabriel, G., Haile-Selassie, Y., Renne, P.R., Hart, W.K., Ambrose, S.H., Asfaw, B., Heiken, G. & White, T. (2001). Geology and palaeontology of the Late Miocene Middle Awash Valley, Afar Rift, Ethiopia. *Nature* **412**, 175-178.

CHAPTER 3

THE NORTH AFRICAN EARLY STONE AGE AND THE SITES AT AIN HANECH, ALGERIA

By Mohamed Sahnouni

ABSTRACT

Palaeolithic archaeologists usually view the North African Early Stone Age of little archaeological value other than providing scanty information on a scarce human presence in this region. While this may be true for many reported sites due to various problems surrounding them, Ain Hanech in northeastern Algeria is a key locality for documenting North African hominin behavior and adaptation. The sites at Ain Hanech are estimated to date to 1.78 Ma based upon biostratigraphic and paleomagnetic evidence. The recent excavations uncovered a savanna-like faunal assemblage associated with Oldowan artifacts contained in a silty matrix. Taphonomic evidence suggests that the sites were minimally disturbed, and therefore preserving hominin behavioral information. The lithic artifacts, excavated from three distinct deposits at the original site of Ain Hanech and the newly discovered site of El-Kherba nearby, are fresh and represent coherent assemblages, including small debitage. The assemblages may be considered as a North African variant of the Oldowan Industrial Complex. Meat was likely a significant part of North African hominin diet.

INTRODUCTION

North Africa is likely the area which hominins inhabited before their spread out of the African continent. A number of early archaeological sites reported during the 1950's and 1960's are considered to be evidence of early human presence in this region. Located primarily in Morocco and Algeria, the most famous of these are the Casablanca sequence and the locality of Ain Hanech on the Algerian High Plateau. Until relatively recently, however, modern investigations had not been carried out at these archaeological localities, and there has been a lack of precise information regarding their stratigraphy and chronological framework, depositional context, and paleoenvironmental reconstruction. In addition, aspects of hominin behavior and ecology accessible by modern methods had not been addressed, including selection, production, use and transport of stone artifacts; food acquisition and processing; and land-use patterns.

To update our knowledge on the Northwest African Early Palaeolithic and document hominin dispersal and adaptation into this region, new investigations were carried out in two major Lower Palaeolithic areas during the last decade. The first area is the Casablanca sequence in Atlantic Morocco where a French team (Raynal & Texier, 1989; Raynal et al., 1990; Raynal et al., 1995; Raynal et al., 2001) revised the work done by P. Biberson in the 1950's and 1960's. The second is the Ain Hanech area in northeastern Algeria where the author has initiated new research since 1992 (Sahnouni, 1998; Sahnouni & de Heinzelin, 1998; Sahnouni et al., 1996). The research undertaken at Ain Hanech consisted primarily of systematic survey of the area, excavation of areas adjacent to the classical site and at newly discovered localities nearby, study of the stratigraphy and chronology, exploration of site formation processes, and analysis of the lithic artifact assemblages.

Many archaeologists consider that Early Palaeolithic assemblages throughout North Africa are of little archaeological value other than documenting a profusion of sites in the region and providing basic information on the spread of hominins into this region.

However, as it will be seen, the new investigations at Ain Hanech reveal that it is feasible to address early hominin behavior and adaptation in North Africa. Presently, Ain Hanech allows us to tackle various aspects of hominin behavior such as the manufacture, use, and discard of artifacts; acquisition and processing of animal subsistence; and vertical and horizontal distribution patterns of Oldowan occurrences. This chapter presents a synthesis on the current status of the Early Stone Age in North Africa, emphasizing the recent studies of Ain Hanech sites. Tentative conclusions are drawn on hominin occupation and adaptation in North Africa, hoping that the new interpretation of Ain Hanech sheds light on the time and nature of their dispersal into this region.

STATUS OF THE EARLIEST NORTH AFRICAN ARCHAEOLOGICAL SITES

Figure 1 shows the reported earliest Northwest African archaeological sites, assigned to the Pre-Acheulean civilization formerly designated "the Pebble Culture". These are mostly located in Morocco and in Algeria. A single find of bifacially flaked core/choppers encountered within a sandy-clay deposit has been reported in Tunisia (Gragueb & Oueslati, 1990). In Morocco, a series of sites located on the Atlantic coast in the vicinity of the town of Casablanca has been investigated by Biberson (1961). These sites are: Arbaoua, Oued Mda, Douar Doum, Terguiet el-Rahla, Carriere Deprez, Carriere Shneider (lower and upper), Chellah, Souk Arba-Rhab, and Sidi Abderrahman (niveau G). A number of these do not constitute well-defined archaeological sites but are localities from which mainly pebble tools were picked up from the surface. In Algeria, the archaeological sites assigned to the Pre-Acheulean industrial tradition are the following: Ain Hanech (Arambourg, 1949; 1970; Sahnouni, 1987; 1998), Mansourah (Laplace-Jauretche, 1956), Djebel Meksem (Roubet, 1967), and Monts Tessala (Thomas, 1973) (these sites are located in the north), Aoulef (Hugot, 1955), Reggan (Ramendo, 1963), Saoura (Alimen & Chavaillon, 1960), and Bordj Tan Kena (Heddouche, 1980) (these sites are located in the Sahara).

A recent review of the earliest Northwest archaeological sites (Sahnouni, 1998) indicates that the current available information on most of these sites, upon which the human antiquity in the Maghreb is based, is not suitable for current standards for Lower Palaeolithic studies. They provide outdated data and unreliable information due to problems related to the circumstances of their discovery, their stratigraphic and sedimentological contexts, the physical condition of the stone artifacts, and their dating and faunal associations. The status of knowledge on these sites is briefly outlined:

1. Most of all the Maghrebian Pre-Acheulean sites have been discovered without systematic archaeological survey. They have been located either casu-

ally or coincidentally in the course of urban development. Only a few of them have been encountered following geological or paleontological expeditions, and these have been investigated without a real archaeological perspective or appreciation. As a consequence, most of the archaeological materials were casually collected, with only the "pebble tools" being systematically selected and examined. Furthermore, these selected artifacts constituted the basis for a proliferation of typological and classificatory analyses. For each site there is almost always an invention of a specific type list and classification system, in which new "pebble-tool" types were created and added (Alimen & Chavaillon, 1962; Biberson, 1967; Heddouche, 1981, Hugot, 1955; Ramendo, 1963).

2. Very few Pebble Culture assemblages are reported to have been found in stratigraphic context or with depositional information. Such pebble tool assemblages have been considered as evidence of a very old human presence in Morocco (Biberson, 1961). Recently, workers who have been revising Biberson's stratigraphic sequence have thrown serious doubts on the antiquity of the earlier Pebble Culture assemblages (Raynal & Texier, 1989; Raynal *et al.*, 1995).

3. With the exception of few sites, all the pebble artifacts encountered are abraded to varying degrees, making them not only inappropriate for hominin behavioral inferences, but also doubtful as proof of earlier human antiquity in the Maghreb. They were often collected from the surface or from eroded conglomerates. Such occurrences lack necessary contextual information, and some of them are merely pseudo-artifacts like those collected from Carriere Deprez cave site (currently Ahl Oughlam) (Raynal *et al.*, 1990).

4. With regard to dating, one of the crucial problems of North African palaeolithic archaeology is the lack of suitable material for dating, where there are no volcanic rocks to provide sound radiometric ages. Uranium-series dating is applicable only to the end of the Lower Palaeolithic sequence. Therefore, dating of the archaeological remains from earliest occupation sites relies on associated fauna. Unfortunately, not all the Maghrebian Pre-Acheulean sites include fossil animal bones of biostratigraphic interest.

In summary, archaeological occurrences identified throughout much of the 20th century were found without systematic survey, and were collected from either the surface or dismantled high-energy deposits. Only the "pebble tools" have been considered for description and analysis. These are usually abraded and rolled, indicating their secondary context, and of doubtful authenticity. The association of fauna is limited, and when it was available it was restricted to taxonomic study.

Figure 1

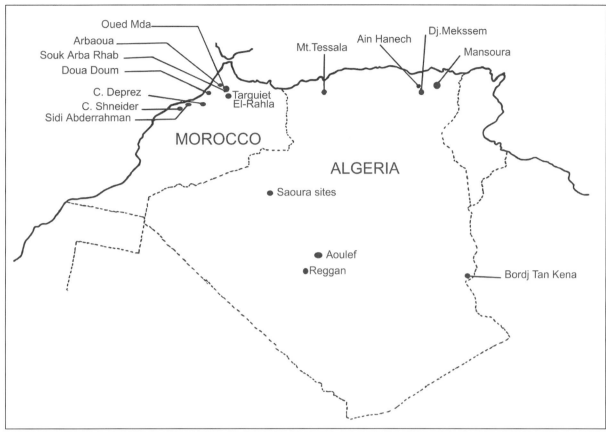

1. *North African Early Stone Age sites. The sites are mostly located in Algeria and Morocco, the most famous of which are Ain Hanech and the Casablanca sequence.*

Lastly, radiometric dating is impossible due to the lack of suitable materials such as volcanic rocks. There has clearly been a tremendous need for renewed investigations of the North African Palaeolithic record that incorporates systematic survey and excavation of in situ occurrences, firmer chronological control, analysis of complete artifact and faunal assemblages for possible behavioral information, and better understanding of the geological and environmental context. Such investigations will be profiled here looking at recent investigations in Morocco and especially a new round of investigations in Algeria at the site of Ain Hanech.

THE EARLY STONE AGE IN ATLANTIC MOROCCO

Perhaps the most important Lower Palaeolithic cultural sequence in North Africa remains the one established by Biberson (1961). Large quarries on the Moroccan Atlantic coast, opened up for modern building materials, have exposed a series of marine deposits interbedded with terrestrial sediments, allowing Biberson to construct a cultural-historical sequence showing the evolution of the Lower Palaeolithic industries through time. He divided the "Pebble-Culture" into

4 successive stages. Stage I includes the oldest artifacts obtained using simple technological gestures (unidirectional). The site of Targuiet-el-Rahla illustrates this stage. His stage II incorporates "pebble tools" characterized by bidirectional flaking. The site of Carriere Deprez in Casablanca represents this stage. In stage III the multidirectional technique appeared where the artifacts are considered to be more evolved. This stage is represented by the site of Souk-el-Arba du Rhab. The last stage (IV) is represented by level G of the Sidi Abderrahman sequence, and is characterized by the emergence of the first Acheulean artifacts. Chronologically, Biberson correlated stages I and II with the marine climatic cycle Mesaoudien dated to Late Pliocene, and stages III and IV with the cycle Maarifien dated to Early Pleistocene. Biberson (1976) revised this cultural historical nomenclature by replacing the term "Pebble Culture" with Pre-Acheulean and condensing the four stages into two major phases. Based on the new classification, the "Pebble Culture" stages I and II constitute the Ancient Pre-Acheulean while the stages III and IV form its evolved phase: Evolved Pre-Acheulean.

However, recent systematic investigations of the Plio-Pleistocene and Pleistocene littoral deposits in the

Casablanca area have substantially modified Biberson's earlier interpretations on the earliest human antiquity in Morocco and the evolution of the Palaeolithic industries (Raynal & Texier, 1989; Raynal *et al.*, 1995; Raynal *et al.*, 2001). The revised investigations emphasize the total absence of evidence of a very early human presence in Atlantic Morocco, demonstrating that assemblages of Pebble Culture Stage I are either surface finds or reworked materials. Artifacts assigned to Pebble Culture stage II were extracted from high-energy deposits. Materials of Pebble Culture Stage III are from polycyclic colluviums. Stage IV of the Pebble Culture is reconsidered as Acheulean by these authors instead of a "Pebble Culture" tradition. In addition, new investigations at the site of Ahl Al Oughlam (formerly Carriere Deprez) in Casablanca (Morocco), indicate that Pebble Culture assemblages there, which had been assigned to the Stage II by Biberson, appear to be pseudo-artifacts generated by high-energy deposits (Raynal *et al.*, 1990).

Therefore, the new evidence suggests that the earliest human presence in Atlantic Morocco is later than Biberson assumed. The oldest occupation dates to late Lower Pleistocene, and appears to be Acheulean as illustrated by the level L of Thomas-1 quarry cave site. This site yielded an early Acheulean assemblage made of quartzite and flint, comprising bifacial choppers, polyhedrons, cleavers, bifaces, trihedrons, and flakes. The associated fauna, probably slightly older than that of Tighenif (ex Ternifine) in Algeria, includes hippo, zebra, gazelles, suid, and micromammals species. Based on fauna and paleomagnetic data, an age of 1Ma is suggested for the level L (Raynal *et al.*, 2001).

THE EARLY SONE AGE AT AIN HANECH, ALGERIA

Historical Sketch

The Ain Hanech site is situated in northeastern Algeria on the High Plateau between the Tellian Atlas and the Saharan Atlas Mountains (Figure 1). The site was discovered by C. Arambourg (1947) in the course of his paleontological survey of the region around the city of Setif. First, Arambourg rediscovered the Pliocene site of Ain Boucherit, which had previously yielded some fossil remains to Pomel (1893-1897). He excavated this locality and retrieved a mammalian fauna comprising mastodon, elephant, suid, equids, and bovids. Then he discovered and excavated the nearby site of Ain Hanech, uncovering a Lower Pleistocene mammal fauna associated with Mode I technology artifacts. The fauna included elephant, hippo, rhino, bovids, and horse. The artifacts incorporated primarily polyhedrons, subspheroids, and few spheroids similar to those known at Olduvai Gorge (Upper Bed I/Lower Bed II) (Leakey, 1971). Although the core tools assemblage comprise very few spheroids, they have been termed as

an industry of "spheroids à facettes" (Arambourg & Balout, 1952; Balout, 1955). The analysis of the industry reconstructed a sequence reduction showing the technological manufacture of core-forms (Sahnouni, 1987), and suggested the Oldowan character of the industry (Sahnouni, 1993).

On the occasion of the II Pan African Congress of Prehistory held in Algiers in 1952, a field trip to northeastern Algeria included a visit to Ain Hanech. Some participants picked up Acheulean bifaces from the surface in the vicinity of the excavation cleaned for the event (Arambourg & Balout, 1952). This led to the erroneous idea that Ain Hanech was an Acheulean site. However, subsequent research demonstrated the entire absence of Acheulean artifacts within the Oldowan occurrences (Arambourg, 1953). In fact, as it will be shown later in this chapter, these artifacts represent another human occupation occurring higher up in the stratigraphic sequence.

New Investigations

The paleontological investigations carried out by Arambourg had definitely established the occurrences of Plio-Pleistocene fauna in northeastern Algeria. Moreover, an Oldowan industry was discovered for the first time in North Africa. However, Arambourg excavated Ain Hanech without the methodological rigor used these days in palaeolithic archaeology. Thus, several questions remained unresolved, including: 1) accurate stratigraphical information concerning the site and its surroundings, 2) the age of the sediments and associated materials, 3) the nature of the association between the fauna and artifacts, and 4) the behavioral implications of the archaeological occurrences.

To address these pertinent issues, new investigations were initiated at Ain Hanech in 1992-93 and 1998-1999 (Sahnouni *et al.*, 1996; Sahnouni & de Heinzelin, 1998; Sahnouni *et al.*, in press). These consisted of surveying the area, excavating areas adjacent to the original site and newly discovered localities nearby, studying the stratigraphy and chronology, investigating site formation history, and analyzing the lithic artifacts. These renewed investigations are still in progress. A synthesis of the major results thus far is presented in this chapter.

The Sites at Ain Hanech

The new explorations show that Ain Hanech is not a single site but rather it is an area with a complex of Plio-Pleistocene sites expanding over an area of approximately one km² (Figure 2). The sites include Ain Boucherit, Ain Hanech, El-Kherba, and El-Beidha. Ain Boucherit is a paleontological locality situated approximately 200m to the southeast of Ain Hanech on the west side of the Ain Boucherit stream. It had yielded a Late Pliocene fauna including the following taxa: *Anancus osiris, Mammuthus africanavus, Hipparion libycum, Equus numidicus, Kolpochoerus phacocheroides,*

Figure 2

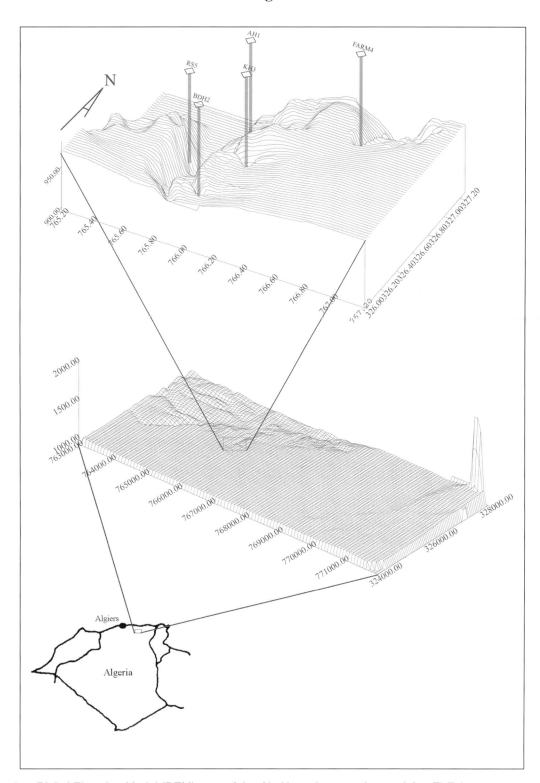

2. *Digital Elevation Model (DEM) map of the Ain Hanech research area (after El-Eulma map #
    94, 1/50.000, scale: UTM). Middle: General view of the area surrounding Ain Hanech. The
    Ain Hanech area is depicted in the rectangle. Top: Detailed view of Ain Hanech site, including
    Ain Hanech Farm (FARM4), Ain Hanech classical site (AH1), the newly discovered localities
    of El-Kherba (KH3) and El-Beidha (BDH2). RS5 refers to the reference stratigraphic section
    where the Ain Boucherit fossil-bearing stratum is located towards the bottom, and the
    Acheulean finds are contained in the calcrete deposit sealing the stratigraphic sequence.*

Figure 3

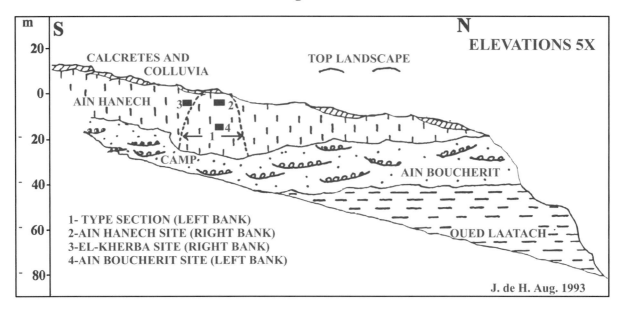

3. *Synthetic view of the regional stratigraphy. Note that Ain Hanech (2) and El-Kherba (3) sites are altimetrically equal levels. Ain Boucherit (4) is located 13m below Ain Hanech (by J. de Heinzelin).*

Sivatherium maurusium, Hippopotamus, Bos palaethiopicus, Parantidorcas latifrons, Damaliscus cuiculi, Orenagor tournoueri, and *Canis anthus primaevus* (Arambourg, 1970; 1979). The site of Ain Hanech is located near a small local cemetery on the private property of the Thabet family in a sedimentary outcrop cut by the deep ravine of the seasonal Ain Boucherit stream. The newly discovered archaeological localities of El-Kherba and El-Beidha are situated in the immediate vicinity south of the classical site and at distance from it of about 300m and 800m, respectively. While archaeological investigations are well underway at Ain Hanech and El-Kherba, the site of El-Beidha has not been explored so far. However, elephant fossil bones associated with a few lithic artifacts were recovered at El-Beidha in the course of digging a test trench (Sahnouni, 1998).

STRATIGRAPHIC AND CHRONOLOGICAL BACKGROUND

Basin deposits occur in pockets throughout the Eastern Algerian Plateau, consisting of a very thick sequence of fluvial and lacustrine sediments sometimes reaching a depth of several hundred meters (Vila, 1980). The Ain Hanech region is formed within the Beni Fouda sedimentary basin with deposits ranging from the Miocene through the Pleistocene and Holocene. In the Ain Hanech vicinity the deposits comprise three main formations, from oldest to youngest, Oued El-Attach Formation, the Ain Boucherit Formation, and Ain Hanech Formation (Figure 3). The Ain Hanech Formation (Figure 4) is a 30 m-thick sequence of

cyclothemic layers, from O to T, of primarily fluvial origin. The site is correlated with Unit T Upper (Sahnouni & de Heinzelin, 1998).

Dating

Paleomagnetic analysis of the Ain Hanech Formation indicates that Unit T and underlying Unit S have normal polarity, and that Units P, Q and R below show reversed magnetic polarity (Figure 4). Taking into account the Plio-Pleistocene affinities of the vertebrate fauna and archaeological context, the normal polarity would fit best within the Olduvai (N) subchron, occurring between 1.95 and 1.77 Ma, rather than the Jaramillo (N) subchron (Sahnouni *et al.*, 1996). Paleontologically, Ain Hanech fauna incorporates biochronologically relevant taxa that went extinct before the Jaramillo Event (Figure 5), including *Kolpochoerus, Equus numidicus, Equus tabeti,* and *Mammuthus meridionalis.* The Ain Hanech *Kolpochoerus* is within the range of *K. heseloni* from the *Notochoerus scotti* zone at Koobi Fora (Sahnouni *et al.*, in press). At Koobi Fora, *K. heseloni* is below KBS tuff dated to 1.88 Ma. *E. numidicus,* found at the Late Pliocene site of Ain Boucherit, persisted at Ain Hanech (Sahnouni *et al.* in press). *E. numidicus* is close to *Equus* from Shungura Member G (G4-13) (Eisenmann, 1985) dated to 2.32-1.88 Ma (Brown, 1994; Brown *et al.*, 1985). *E. tabeti* is recorded in other Early Pleistocene sites, including Koobi Fora (Kenya) from the *Metridiochoerus andrewsi* and *Metridiochoerus compactus* zones of the Koobi Fora Formation (Eisenmann, 1983), and at Ubeidiya in Israel (Eisenmann, 1986), which is estimated to date to 1.4 Ma

Figure 4

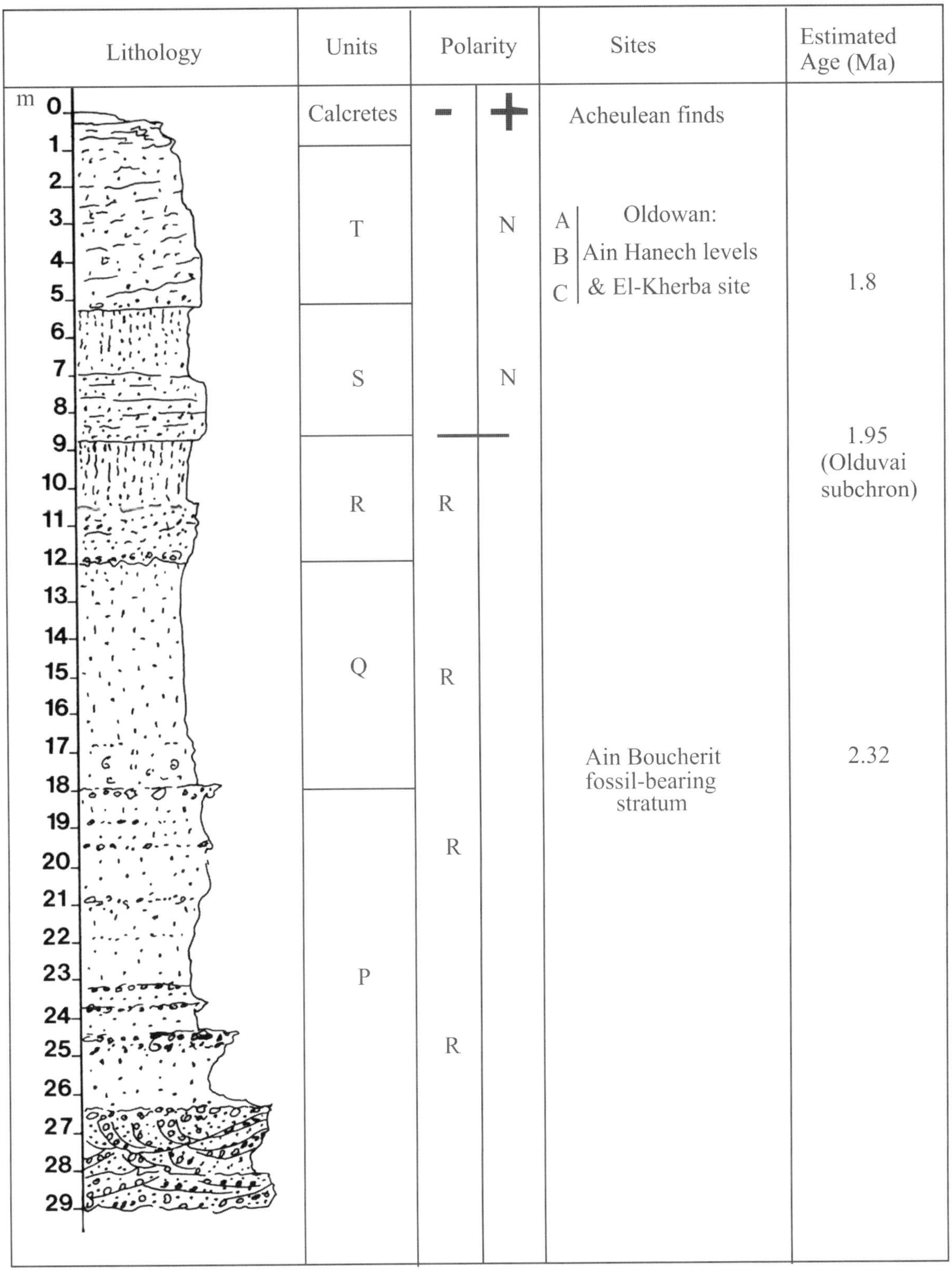

Lithology	Units	Polarity	Sites	Estimated Age (Ma)
m 0–5	Calcretes	− +	Acheulean finds	
	T	N	A Oldowan: B Ain Hanech levels C & El-Kherba site	1.8
	S	N		
	R	R		1.95 (Olduvai subchron)
	Q	R	Ain Boucherit fossil-bearing stratum	2.32
		R		
	P	R		

4. Reference profiles of Ain Hanech Formation, showing its component sedimentary units, their geomagnetic polarity, and the associated sites. The Ain Boucherit fossil-bearing stratum with Plio-Pleistocene fauna is contained in the Unit Q with reverse polarity. Ain Hanech Oldowan levels and the newly discovered locality of El-Kherba are contained within the Unit T with normal polarity. The Acheulean finds derive from the calcrete deposit sealing the stratigraphic sequence.

Figure 5

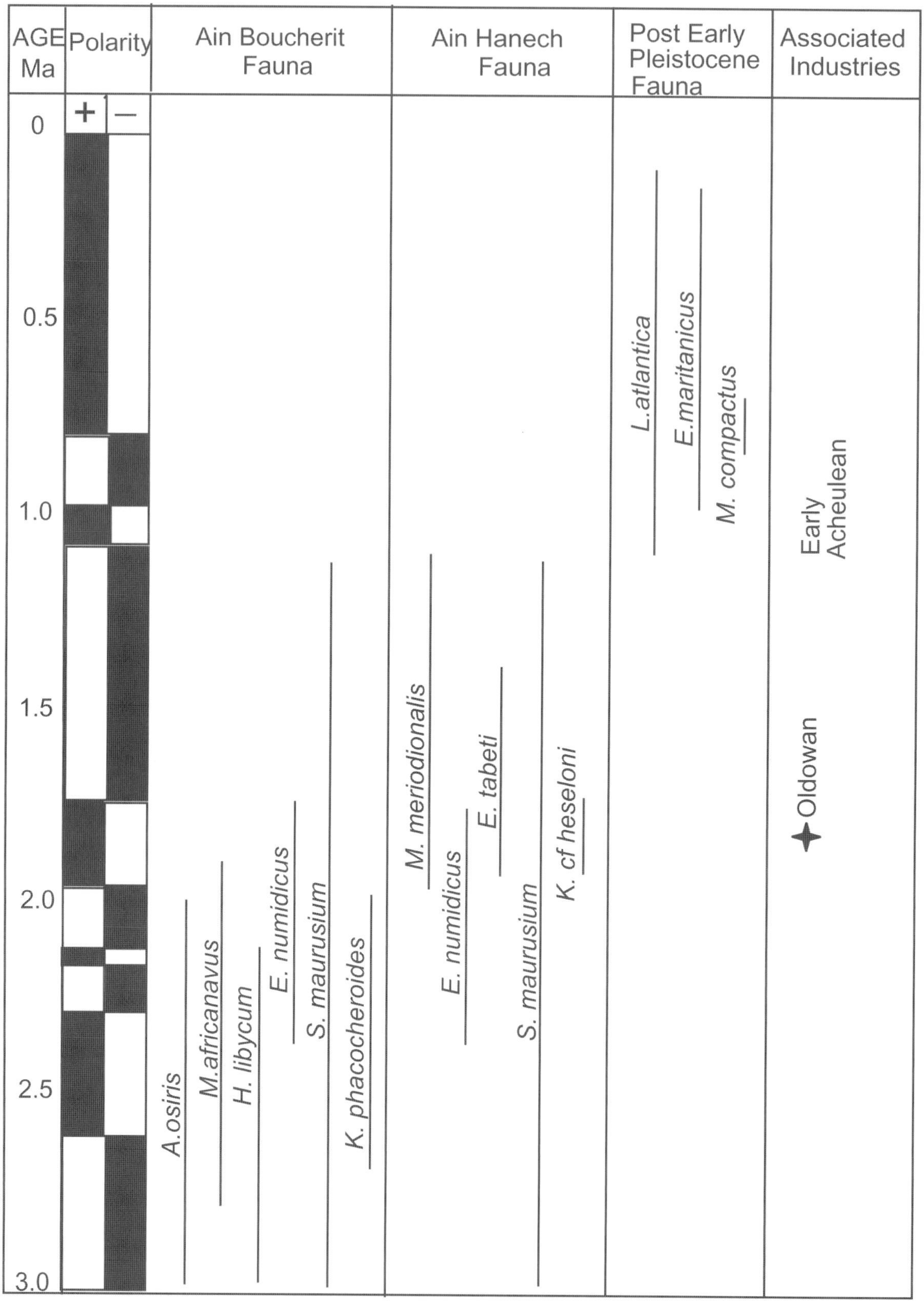

5. *Chronological range of mammalian taxa of biochronological significance from Ain Boucherit, Ain Hanech, and post Early Pleistocene in Northwest Africa. Note that taxa found at Ain Hanech went extinct before the Jaramillo (N) paleomagnetic subchron, suggesting that Ain Hanech normal polarity is most likely Olduvai (N) subchron dated to 1.95-1.78 Ma.*

Figure 6

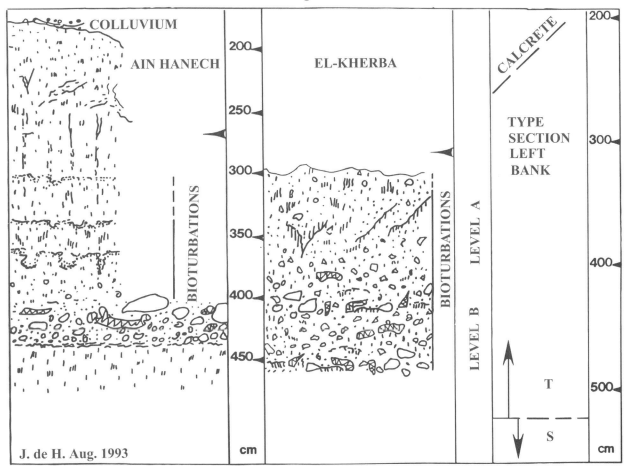

6. *Stratigraphical profiles of the archaeological sites, and their correlation with the Ain Hanech Formation type section.*

(Bar-Yosef & Goren-Inbar, 1993). *M. meridionalis* did not persist beyond the Early Pleistocene in North Africa (Coppens *et al.*, 1978), and by 1 Ma it was replaced by *Loxodonta atlantica* in Thomas Quarry (Unit L) (Raynal *et al.*, 2001). Archaeologically, the Ain Hanech artifact assemblage is very similar to that of the East African Oldowan (Sahnouni, 1993; 1998).

Stratigraphic Position of the Acheulean Occurrences

Bifaces were previously thought to be associated with the Oldowan (Arambourg & Balout, 1952; Arambourg, 1953). Presently, based upon stratigraphic and archaeological evidence in current investigations, it is certain that Acheulean artifacts are totally independent from the Ain Hanech Oldowan industry proper. These are contained in the calcretes deposit located six meters higher up in the stratigraphic sequence (Figure 4). I collected several Acheulean bifaces from eroded calcrete areas, and most of them still bear calcrete concretions. Therefore, the Acheulean occurrences represent a later phase of hominin occupation. No Acheulean artifacts have been found in 166 m² of excavation of the

underlying Oldowan-bearing deposits, which have yielded 1502 of fresh unabraded Oldowan artifacts.

Stratigraphic Profiles of Ain Hanech and El-Kherba Sites

Two archaeological localities have been recently excavated. The first is Ain Hanech adjacent to Arambourg's trench, and the second is the newly discovered locality of El-Kherba. Based mainly upon altimetric and stratigraphic evidence, the two stratigraphic profiles are correlated with Unit T of the Ain Hanech Formation type section on the left bank (Sahnouni & Heinzelin, 1998) (Figure 6).

Stratigraphic Profile of Ain Hanech Site

The stratigraphic profile of the Ain Hanech locality consists of a 3.5m thickness and includes the following features from bottom to top (Figure 6): 1) at the base, a 50 cm thick layer consisting of dark sandy clay with limestone cobbles of medium size dimensions (ca. 80x60x40mm) and black flint pebbles, containing vertebrate fossil animal bones and Mode I artifacts; 2) a layer of finely mottled white silt with few calcic grains;

3) heterogeneous and heterometric gravels in a mixture of sand, calcic granules, and silt lenses; fossil bones contained in the upper part; 4) mottled silt overlain by three successive sandy layers interbedded with mottled silt; traces of bioturbation diminishing from base to top; 5) the upper part, 1.20m thick consisting of mottled white and light reddish brown silts, with vertical traces of roots and vague reduced traces of tree stumps.

Stratigraphic Profile of El-Kherba Site

The deposits at El-Kherba are about 2 m thick (Figure 6). Artifacts and bones are in various positions and are incorporated in a mottled silty matrix. There are mottles of carbonate lumps, more or less decalcified, and traces of decayed organisms and trampling. In the upper part, larger mottled structures simulate sliding or mud flow. The color near 0.60m depth above the base of excavation trench is yellow to very pale brown, reddish yellow, and pure white. The color near 1.10m depth is light gray and a more reduced pure white.

Archaeological Levels at Ain Hanech

Stratigraphic studies and, especially, the expansion of the archaeological excavations at Ain Hanech allowed us to identify three Oldowan deposits (Sahnouni *et al.*, in press). They are from the youngest to the oldest, A, B, and C (Figure 7). Level C was recognized in the course of digging a stratigraphic test trench below the conglomerate supporting the base of level B. This layer is easily discernible, as it is separated from level B by 1m of sterile deposits. However, layers B and A are hard to delineate because they are contained in a rather homogeneous sedimentary matrix. Stratigraphically, levels A and B include, at the bottom, a gravel layer abruptly overlain by a silty stratum. These sedimentary deposits suggest an alluvial floodplain cut by a meandering river channel. By the time of deposition of level A, the river had probably created an oxbow lake. It may be inferred that the hominid activities took place during level B on the riverbank, and during level A on the floodplain proper.

Figure 7

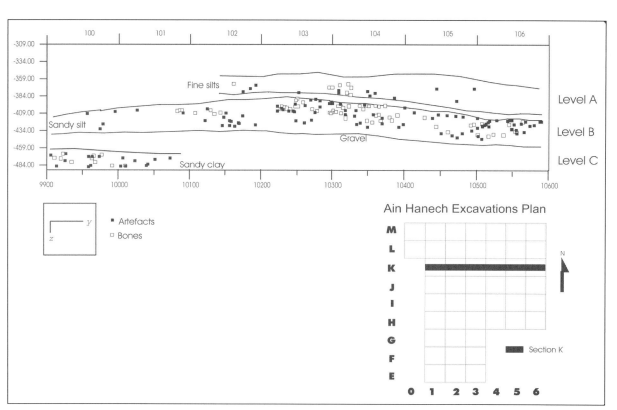

7. The discerned archaeological levels at Ain Hanech site, namely A, B, and C. Level C is still unexplored, as it was exposed in the course of a limited test trench. The position of the section (referred to K) relative to the grid system is shown in bold below.

Fauna and its Paleoecological Implications

The faunal remains retrieved from the new excavations are currently under analysis, and therefore, a complete faunal list is not provided in this study. What can be said presently, however, is that in addition to previously published taxa, preliminary data point to the presence of species that were not known previously at Ain Hanech, such as *Kolpochoerus* cf *heseloni* and *Equus numidicus* (Sahnouni *et al.*, in press). There is also a new species of *Pelorovis* (Sahnouni *et al.*, 2002).

Table 1 presents the faunal species list published by Arambourg (1970; 1979). Several elements suggest the antiquity of the site, such as the elephant, equid, suid, giraffid, and bovids. They also suggest an open savanna-like environment as inferred from the presence of *Gazella pomeli*, *Numidocapra*, and probably *Connochaetes* (Geraads, 1981). Carbon isotope analysis carried out by T. Cerling (in Sahnouni, 1998) on two equid molar teeth suggests a C_3 plant dominated diet for Ain Hanech equids, and corroborates the faunal environmental inferences. C_3 plants, with δ^{13} C values ranging between -33 and 22 per mil, grow in temperate regions with winter rainfall and cool temperatures (Vogel, 1978).

The new excavated faunal assemblage also incorporates modified bones bearing hominin and carnivore signatures. For example, the upper specimen (cat. #: AH98-M102-50) is a femur diaphysis fragment of a medium-sized animal, likely of an antelope, which appears to have been broken intentionally by hominins (Figure 8, upper). The maximum dimensions are: 90mm, 23mm, and 14mm, respectively length, breadth, and thickness. The longitudinally-fractured bone bears typical human-made breakage patterns (Lyman, 1994), such as an impact point of percussion, oval scar, opposite rebound point, and concave fracture surface.

Another bone specimen (Figure 8, lower) (cat. #: AH98-L105-35) that bears tooth marks is a rib proximal end fragment of a large-sized animal, probably an adult rhinoceros. Its dimensions are: length 85mm, breadth 50mm, and thickness 31mm. The tooth marks are located on both surfaces of the bone. On the medial surface a series of four linear marks or grooves are present. On the lateral surface there are two pairs of circular-shaped punctures. The size of these punctures suggests that a canid probably made them.

Context of the Ain Hanech Sites

Many archaeologists had thought that Ain Hanech occurrences were unsuitable for behavioral inquiry because they might have been disturbed by hydraulic agencies (Clark, 1992). To shed light on the issue, the new investigations undertaken at Ain Hanech also focused on assessing the site formation processes of the occurrences. The question was tackled by inspecting the

Table 1

Animal species	Common names
Mammuthus meridionalis	Elephant
Ceratotherium mauritanicum	Rhinoceros
Equus tabeti	Horse
Hippopotamus amphibius	Hippo
Kolpochoerus phacochoeroides	Suid
Sivatherium maurusium	Giraffe
Crocuta crocuta	Hyena
Oryx el eulmensis	Antelope
Girrafa (?) pomeli	Giraffe
Bos bubaloides	Cattle
Bos praeafricanus	Cattle
Numidocapra crassicornis	Ovicaprid
Gazella pomeli	Gazelle
Taurotragus gaudri	Eland
Canis cfr. atrox	Canid
Alcelaphus sp	Alcelaphine
Connochaetes	Wildebeest

1. Animal species recovered previously at Ain Hanech according to Arambourg (1970 & 1979) and subsequent changes made by other authors (Geraads, 1981; Eisenmann, 1985).

excavated archaeological material in terms of sedimentary matrix in which they were accumulated, taphonomic conditions of bones, and the concentration of artifacts. The detailed results of that study have recently been published (Sahnouni, 1998; Sahnouni & de Heinzelin, 1998). They are briefly summarized here.

All the inspected criteria converge to the same conclusion: that Ain Hanech and El-Kherba cultural remains were minimally disturbed and likely preserved behavioral information. The remains were contained in a fine sedimentary matrix, indicating burial in a floodplain deposit as a result of a low velocity regime. However, minimal site reworking and rearrangement of small remains might have occurred.

Except for a few that underwent some post-depositional alteration, fossil animal bones are well-preserved. When observable, they do not exhibit the types of cracking or flaking that imply long periods of exposure (Behrensmeyer, 1978; Sahnouni & de Heinzelin, 1998) (Figure 9). Anatomically, the bone assemblages include all categories of skeletal elements, eliminating the possibility of substantial hydraulic sorting. In addition, they show neither a strongly preferred orientation nor high

Figure 8

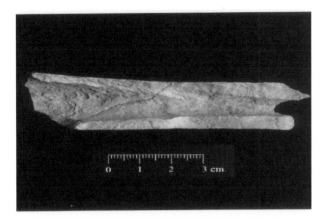

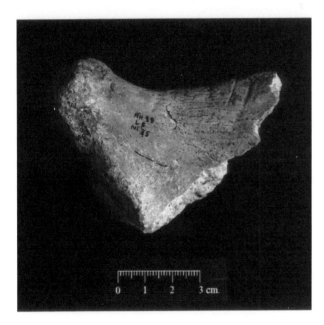

8. *Modified bone surfaces bearing hominid and carnivore signatures. Top: femur diaphysis fragment (cat. #: AH98-M102-50) of a medium-sized antelope, bearing human-made breakage patterns, such as an impact point of percussion, oval scar, opposite rebound point, and concave fracture surface. Bottom: rib proximal end fragment of an adult rhinoceros (cat. #: AH98-L105-35). Four grooves are located on the medial surface, and two pairs of circular-shaped punctures are present on the lateral surface.*

inclination. Detailed taphonomic analysis of the fauna is underway.

Likewise, stone artifacts are fresh, displaying a coherent assemblage composition, including cores and debitage (Figure 10), with debitage overwhelmingly represented. In addition, small flakes (<2cm of maximum dimension) exhibit similar patterns (Sahnouni & Heinzelin, 1998: 1093) to that produced by Schick (1986) in her experimental work, suggesting that flaking occurred in the site. The lithics do not show any significant dip or inclination, but they do display slight preferred orientation patterns. Microscopic analysis of well-preserved artifacts made of flint, reveal microwear polishes preserved on some specimens, making it unlikely they underwent significant hydraulic disturbance.

EXCAVATIONS AT AIN HANECH AND EL-KHERBA

Two main archaeological localities were excavated; the original site of Ain Hanech and that of El-Kherba. At Ain Hanech an area of 118m² by 1.50m depth was excavated, and a volume of 177m³ of sediments was removed. At El-Kherba an area of 33m² by 1m depth was dug and 33m³ of sediments were removed. At both archaeological sites a rich assemblage was recovered, though in low density. On average, at Ain Hanech the density is 25 finds per m³ in Level B and 4 finds per m³ in Level A. At El-Kherba the density is 5 finds per m³. A total of 2475 archaeological remains was recovered at Ain Hanech, including 1243 fossil bones and 1232 stone artifacts (these totals exclude small fragments <2cm). At El-Kherba the excavations yielded 631 archaeological finds, including 361 bones and 270 stone artifacts.

There is no doubt that lithic artifacts and fossil animal bones form a spatio-temporal association, likely reflecting behavioral activities. For instance, the excavations at Ain Hanech in Level B exposed a partial skeleton of a rhino in an articulated position and associated with several stone artifacts (Figure 11: square J2). The skeleton includes the scapula, vertebra, and pelvis, with a humerus located slightly further away. At El-Kherba the excavations uncovered the association of several animal remains (equid, giraffid, and small and large bovids) surrounded with stone artifacts (Figure 12). The behavioral implications of these archaeological associations are currently under study.

THE ARTIFACT ASSEMBLAGES

Overall Presentation

The new excavations at Ain Hanech yielded a total of 1502 lithic artifacts (excluding elements <2cm in maximum dimension, which total 2405). This analysis will only consider artifacts greater or equal to 2cm in maximum dimension. This general total breaks down as follows: Level C: n=31 (2%); Level B: n=947 (63%); Level A: n=254 (16.9%); and El-Kherba: n=270 (17.9%). The artifacts from Level C are not included in the analysis since they represent a small assemblage excavated from a test trench. As can be seen from table 2, the assemblage from Level B is the most abundant. The assemblages from Level A and El-Kherba are nearly evenly represented. They are all coherent assem-

Figure 9

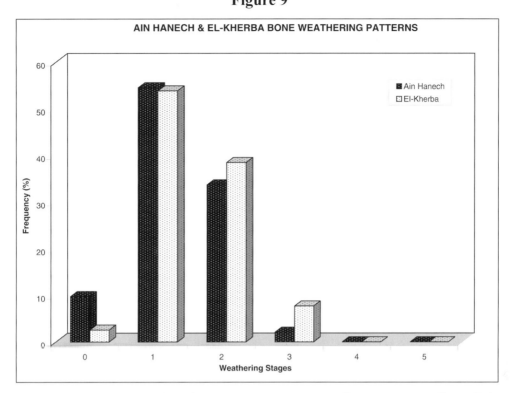

9. *Ain Hanech (n=288) and El-Kherba (n=39) bone weathering patterns. Classes are according to Behrensmeyer, 1978.*

Figure 10

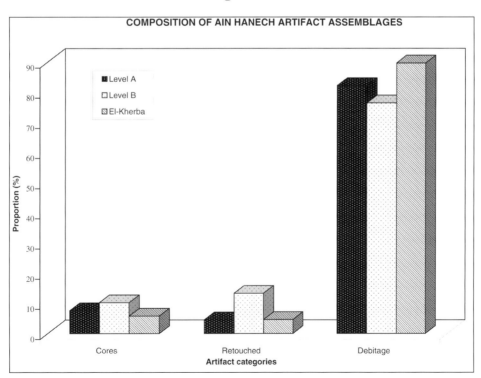

10. *General artifact assemblage composition of the Ain Hanech sites, including elements equal or below 2cm in maximum dimension. Ain Hanech Level A (n=827), Level B (n=2097), and El-Kherba (n=952). Note the high frequencies of debitage in all the assemblages. Small debitage totals: n=573 (69.2%) for Level A, n=1150 (54.8%) for Level B, and n=682 (71.6%) for El-Kherba.*

Figure 11

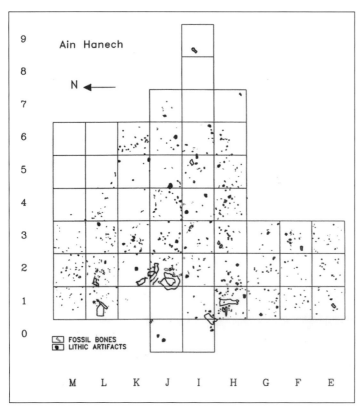

11. *Horizontal distribution of the archaeological material at Ain Hanech (Level B). Note the rhino remains in squares J2 and H1.*

blages and include small debitage. Although their frequency in Level B is relatively low, overall the small elements <2cm are fairly well-represented in all the assemblages. This suggests a minimal disturbance of the occurrences and that the lithic production occurred at the sites. In terms of artifact composition, the assemblages comprise the following categories (Table 2): Cores (22.3%), whole flakes (22.3%), retouched pieces (27.3%), and fragments (25.4%). The frequency of the retouched pieces is slightly inflated, especially in Level B, because a portion of the blanks constitutes pebbles that were directly modified into tools (see section on retouched pieces below). Identifiable percussors (hammerstones) are only present in Level A and El-Kherba assemblages.

Raw Materials

With the exception of a very few pieces made of quartzite and sandstone, all artifacts from Ain Hanech and El-Kherba were made of limestone and flint. Based on the geological map (Map of El-Eulma # 94, 1/50.000, 1977), the limestone component includes several varieties formed during the Cretaceous (Campanian-Maestrichtian, Cenomanian, and Santonian stages). Although the flint component displays homogenous textural patterns, it includes a variety of colors: black, dark brown, light brown, gray, and green.

The black flint is the most predominant and represents a rock formed in the Ypresian-Lower Lutecian stage. The primary geological sources of limestone and flint were the rocky hills surrounding the Beni Fouda basin. Pieces of limestone and flint would break up over time and would have been transported by streams draining into the alluvial plain. They became water-smoothed cobbles and pebbles mostly characterized by a polyhedral shape because of their short transport, and were deposited in beds of river and stream courses. Thus, these two rocks were readily available to hominins as a source of raw material.

In terms of artifact number, flint (56.3%) is relatively more common than limestone (43%). The use of quartzite and sandstone, recorded only at Ain Hanech locality, was negligible. If frequencies of rock types are compared between the two archaeological localities, flint is found in somewhat higher frequencies at El-Kherba (61.2%) than at Ain Hanech (54.7%). Inversely, limestone was slightly more frequent at Ain Hanech (44.8%) than at El-Kherba (38.7%).

Figure 13 shows frequencies of categories of artifacts based upon rock types. As can be seen, all categories of artifacts were manufactured by both limestone and flint. Whole flakes were nearly equally in limestone and flint. Fragments are proportionally somewhat more common in limestone compared to flint. Interestingly,

Figure 12

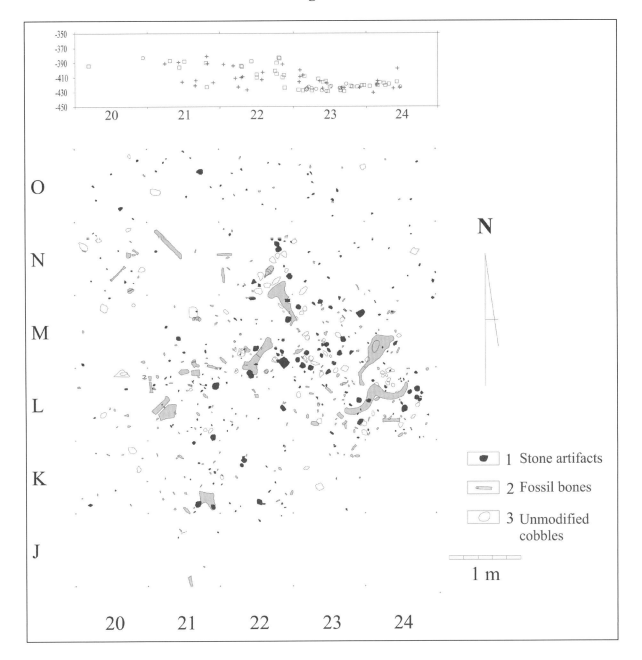

12. *Horizontal distribution of the archaeological material at El-Kherba. Note the concentration of fossil animal bones with artifacts. The animal remains include a new species of Pelorovis (horn cores in square L24), hippo (pelvis bones in squares M24 and N22), and Sivatherium (femur in square M22). The graph in the top shows the vertical dispersion (in mm) of the remains from squares L20 to L24. (□ = fossil bone; + = stone artifact; ○ = unmodified stone).*

Table 2

Artifact categories	Level B		Level C		Level A		El-Kherba		All assemblages	
	N	%	N	%	N	%	N	%	N	%
Cores	6	0.3	212	22.3	62	24.4	55	20.3	335	22.3
Retouched Pieces	15	0.9	276	29.1	37	14.5	83	30.7	411	27.3
Whole flakes	6	0.3	194	20.4	72	28.3	63	23.3	335	22.3
Fragments	4	0.2	239	25.2	71	27.9	68	25.1	382	25.4
Percussors	0	0	0	0	2	0.7	1	0.3	3	0.1
Split cobbles	0	0	26	2.7	10	3.9	0	0	36	2.3
Total	31	2	947	100	254	100	270	100	1502	100

2. *General breakdown of the lithic assemblages greater or equal 2cm in maximum dimension. The artifacts from Level C represent a small sample as they have been excavated in a test trench. Fragments include snapped and split flakes, and other fragments.*

the retouched pieces are much more frequent in flint than in limestone. This trait can be explained by the fact that a significant proportion of retouched pieces were manufactured on pebbles.

When the frequencies of rock types within each site are considered, we find that both limestone and flint were used but somewhat differently. For instance, limestone is relatively more frequent than flint at Ain Hanech level A (9.7% versus 6.3%). In contrast, flint is more heavily flaked than limestone in both Ain Hanech level B (36.2% versus 28%) and El-Kherba site (11.3% versus 7.5%).

Cores

The cores (Figures 14 and 15) total 329 specimens and are fairly variable with respect to raw materials, dimensions, technological patterns, and overall morphology. They were likely flaked primarily for flake production. In terms of raw materials, the cores are primarily made of limestone (30.8%) followed by flint (12.9%).

With regard to dimensions, the cores are extremely variable (Table 3). The reason is that the limestone and flint are different in size clasts. The limestone is primarily available as cobbles while flint is particularly small and abundant as pebbles (Figure 16). The mean for the maximum dimension for all cores is 69.97mm. However, the maximum dimension mean varies between cores made of limestone and those made of flint, respectively 83.04mm and 47.23mm.

Technologically, the cores are considerably variable and are characterized by an absence of stereotypic forms. On the whole, they show a whole range of flaking intensity, preserving residual cortical areas though in variable degrees. The specimens preserving no cortex are very few totaling near 3%. The number of scar

counts varies between 2 and 28, suggesting a lack of standardization and variability in exploitation of cores. In terms of flaking methods, the employed modes are unidirectional, and to a lesser extent bidirectional. The edge angle varies as well, ranging between 59° and 130°, with a mean of 94.11°. The variability is likely due to the type of rocks flaked. For example the mean angle for cores made of limestone is 97.26° while the mean angle for those made of flint is 85°.

Typologically, the limestone cores may also comprise specimens that were flaked for a desired shape or for useful edges. They are usually unifacially, bifacially, and polyfacially flaked. Using Mary Leakey's typological system, several morphological types may be recognized (Figure 14) including unifacial and bifacial choppers, polyhedrons, subspheroids, and facetted spheroids. Overall, the polyhedrons are the most dominant category (67%) and are moderately to heavily trimmed on at least three different faces. The spheroids represent only 1.89% (4 specimens).

Whole Flakes

A total of 329 flakes greater or equal to 2cm was recovered at Ain Hanech sites (Figure 17), including 72 in Level A, 194 in Level B, and 63 in El-Kherba. The flakes are made nearly equally of flint (50.45%) and limestone (48.93%). Except for Level A where flakes in limestone prevail (12.76% versus 8.81%), those made of flint predominate in both Level B (30.09% versus 28.57%) and El-Kherba (11.55% versus 7.59%). There is a single flake made of quartzite and occurs in Level B.

Figure 13

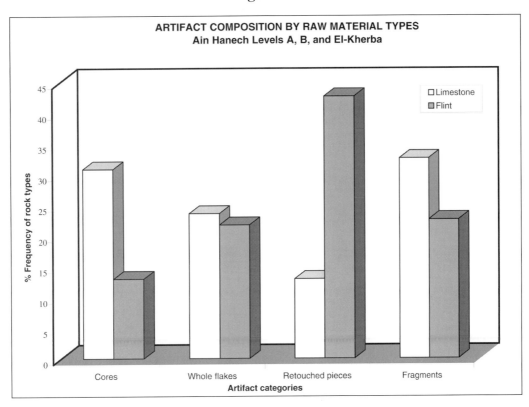

13. *Artifact composition by raw material types. Note that retouched pieces are numerous in flint.*

Table 3

Sites	Variable	Mean	Std. Dev.	Range
	Length	76.33	24.08	25-145
Ain Hanech A (n=51)	Breadth	66.76	21.49	20-122
	Thickness	53.82	20.28	10-93
	Length	72.44	31.63	21-208
Ain Hanech B (n=212)	Breadth	60.72	27.29	15-183
	Thickness	48.56	23.55	25-111
	Length	64.57	20.68	31-90
Ain Hanech C (n=7)	Breadth	53.14	17.32	23-74
	Thickness	43.28	21.78	9-72
	Length	63.22	27.53	21-128
El-Kherba (n=53)	Breadth	52.35	25.30	10-117
	Thickness	39.77	21.86	9-91

3. *Comparison of core dimensions from Levels A, B, and C, and from El-Kherba site.*

Figure 14

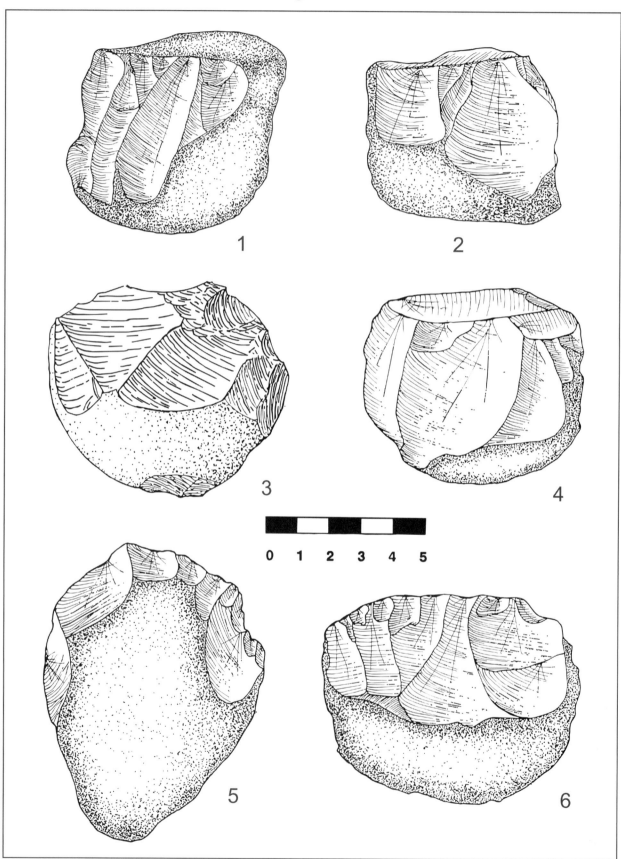

14. *Examples of cores made of limestone from Ain Hanech sites, including 1, 2: unifacial and bifacial cores from Level A; 3, 4: unifacial core and polyfacial core from Level B; 5, 6: unifacial core and bifacial core from El-Kherba.*

Figure 15

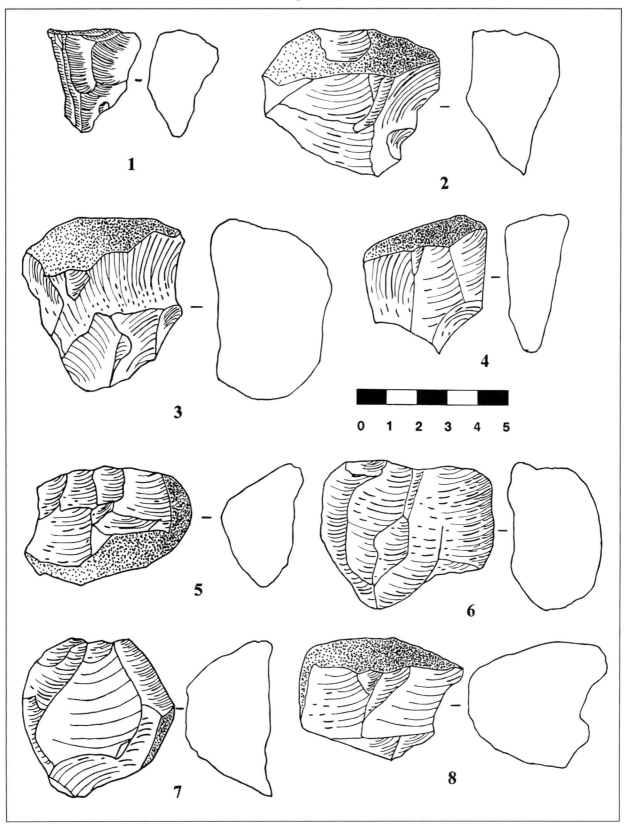

15. *Examples of cores made of flint from Ain Hanech sites, including 1-3 from Level A; 4-5 from Level B; 7 from Level C; 6 and 8 from El-Kherba.*

Figure 16

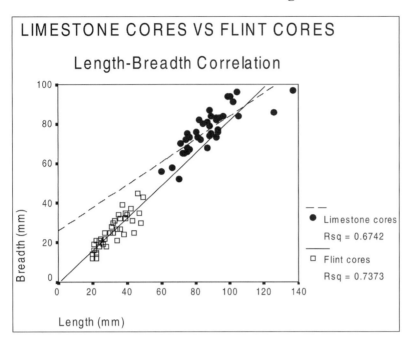

16. *Limestone cores versus flint cores based on length-breadth correlation. Note that limestone cores (black dots) are significantly bigger than cores made of flint (squares).*

Table 4

Sites	Variable	Mean	Std. Dev.	Range
	Length	38.16	16.57	13-106
Ain Hanech A (n=71)	Breadth	27.10	13.72	11-84
	Thickness	11.06	6.11	4-45
	Length	37.86	16.73	13-106
Ain Hanech B (n=194)	Breadth	27.91	14.03	8-84
	Thickness	10.91	6.21	3-45
	Length	34.00	15.54	17-96
El-Kherba (n=63)	Breadth	23.29	10.76	8-64
	Thickness	9.94	4.79	4-30

4. *Dimensions of flakes from Ain Hanech Levels A and B, and from El-Kherba site.*

Table 5

Statistics	Level A	Level B	El-Kherba	All
Mean	95.03	98.34	90.08	94.48
Std. Deviation	17.17	16.32	18.22	17.23
Minimum	58	59	53	53
Maximum	127	140	123	140

5. *Interior platform angle statistics of flakes from Ain Hanech Levels A and B, and from El-Kherba.*

Whole Flake Dimensions and Shape

Overall, the flakes are small. The maximum dimension ranges between 13mm and 106mm, with a mean of 36.67mm. However, the maximum dimension varies slightly from assemblage to the other (Table 4). The mean of length is: 38.16mm for Level A, 37.86mm for Level B, and 34mm for El-Kherba. El-Kherba flakes are particularly smaller (Figure 18) because they are primarily made of flint, which occurred in small clasts. Flakes from Level A and B show a greater size range. Based upon experimental studies (Toth, 1982), smaller flakes tend to be preferentially produced by the manufacture of flake scrapers, and flakes, whose length ranges between 27.8mm and 35.1mm were produced by manufacturing larger artifacts such as choppers and polyhedrons. Both trends are observed in Ain Hanech flakes.

The variability in flake dimensions is due to differences in raw materials. As can be seen in (Figure 19), flakes made of limestone are larger than those made of flint. Raw material shape and size account for these differences: limestone flakes were removed from large and thick cobbles, whereas those made of flint were detached from smaller pebbles.

To have an indication of flake shapes, two ratios were calculated: breadth-length (B/L) and thickness-breadth (T/B). The plot of these two ratios (Figure 20) suggests that Ain Hanech sites flakes are primarily moderately long and relatively thick. Although rare, there are also flakes with extreme shapes, such as short and thin and slightly long and very thick. According to experimental flaking (Toth, 1982), flakes resulting from manufacturing Acheulean artifacts such as bifaces, picks, and flake scrapers tend to be shorter and thinner, while flakes produced by flaking choppers, polyhedrons and core scrapers are relatively longer and thicker. Thus, the majority of the Ain Hanech flakes correspond to those produced in the production of Oldowan-type cores.

Flake Types

Figure 21 shows the overall breakdown of flakes types for each assemblage based on presence/absence of cortex on the dorsal surface and platform of the flake (Toth, 1985). These include the following types:

Type I: Cortical platform with all cortical dorsal surface;

Type II: Cortical platform with partially cortical dorsal surface;

Type III: Cortical platform with non-cortical dorsal surface;

Type IV: Non-cortical platform with all cortical dorsal surface;

Type V: Non-cortical platform with partially cortical dorsal surface;

Type VI: Non-cortical platform with non-cortical dorsal surface.

In general, all the types are represented in the flake population. Nevertheless, the first three flake types, those with cortical platforms (types I, II, and III), are the most abundant amounting to 69.1%, 57.3%, and 60.4%, respectively in Level A, Level B, and El-Kherba. Differences in flake types are also observed within assemblages, especially Level A and El-Kherba (Figure 22). In both assemblages, limestone flakes type I are highly numerous, and flakes type IV are relatively uncommon in Level A. The proportion of flakes type VI in flint is high at El-Kherba. The flake types study shows that flaking occurred at the sites because all flake stages are depicted, and the assemblages were minimally disturbed by water action. The discrepancies in flake types suggest an overall light reduction of limestone cores, and a relative heavy reduction of flint cores.

Flake Platform Types and Interior Platform Angle

Figure 23, displaying flake platform types for flakes greater or equal to 2cm, shows that near 60% of platforms are cortical represented by flake types I, II, and III. These are followed by platforms with one scar, amounting 35.9%. The platforms with more than one scar constitute only 4.23% of flakes. These platform patterns do not significantly change when assemblages are compared, except that cortical platforms are relatively numerous at Ain Hanech Level A and El-Kherba and non-cortical platforms are frequent at El-Kherba. Similar patterns are also observed between platforms made of limestone and those made of flint. However, it should be pointed out that platforms with more than one scar are largely made of flint, suggesting that flint cores were more exhaustively flaked than the limestone ones.

Flake platform trends observed in Ain Hanech and El-Kherba assemblages are compatible with Oldowan technologies. The cortical platforms and those with one scar are the byproduct of unifacial and bifacial cores, respectively. The platforms with more than one scar are rare, likely due to the absence of core preparation for producing sophisticated artifacts known in more advanced traditions, such as bifaces, cleavers, etc.

The interior platform angle formed by the platform and flake ventral surface is variable, ranging between 53° and 132° (Table 5). The mean is 94.48°. The interior platform angle variability is likely due to rock type. Overall, flakes made of flint have an obtuse flaking angle, while those of limestone have an acute flaking angle.

Flake Dorsal Scars and Scar Patterning

The number of scars counted on flake dorsal surfaces varies between 0 (all cortex) and 11 (Figure 24). Overall, flakes show few dorsal scars (45.2% with no or one scar, 38% with 2-4 scars, and 17% with 5 scars and above). Overall, these patterns indicate that extremely heavy flaking was not common. In terms of raw materials, however, significant differences appear between

Figure 17

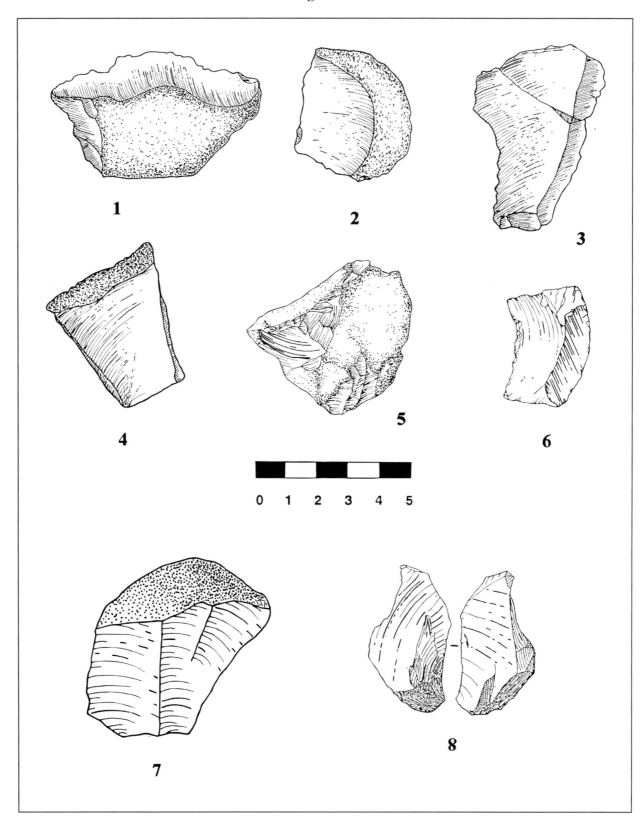

17. *Example of whole flakes from Ain Hanech sites, including 1-3 from Level A (limestone); 4-6 from Level B: 4 is made of limestone, and 5 and 6 are made of flint; 7 is made of limestone and is from Level C; and 8 is made of flint and is from El-Kherba. Flakes 6 and 8 were interpreted as meat cutting tools as evidenced by usewear analysis.*

Figure 18

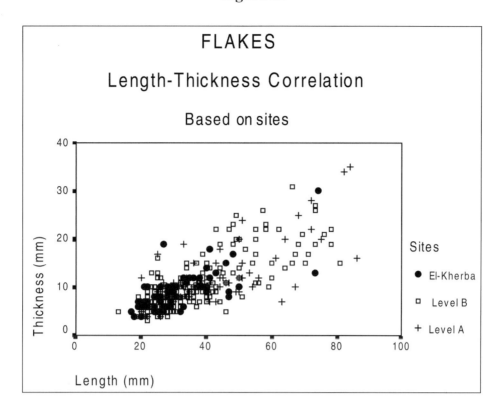

18. *Scatter plot showing whole flakes dimensions (length by thickness) based on sites. Note that El-Kherba flakes are primarily small because the majority are made of flint.*

Figure 19

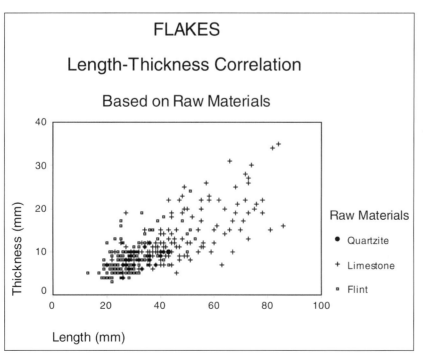

19. *Scatter plot showing whole flake dimensions (length by thickness) based on raw materials. Note that flakes made of flint are smaller overall due to the use of small-sized flint cores.*

Figure 20

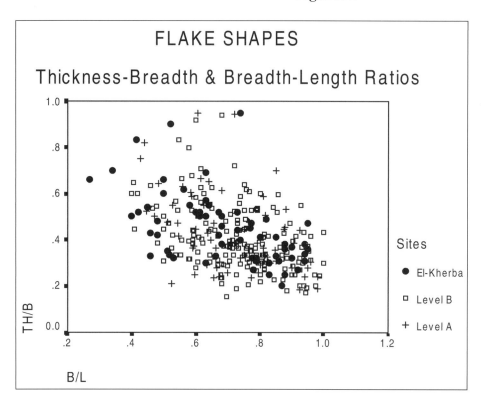

20. *Scatter plot showing flake shape for each site using thickness-breadth and breadth-length ratios. Overall, the flakes are moderately long and relatively thick in all assemblages. These flake shapes are compatible with the categories of artifacts producing flakes found at Ain Hanech sites, e.g. choppers, polyhedrons, and cores.*

Figure 21

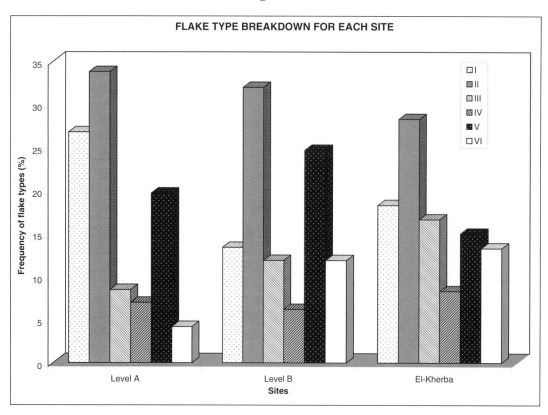

21. *Overall flake type breakdown for the three sites. All the types are present, suggesting that all stages of flaking are represented and minimal disturbance of the occurrences.*

Figure 22

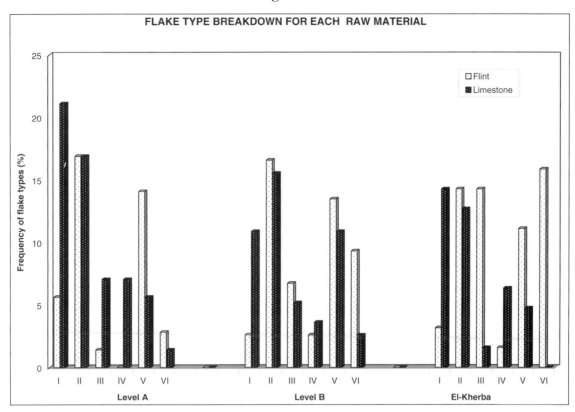

22. *Flake type breakdown based on raw materials. Note the relative variability between sites with Level A being some-what similar to El-Kherba. In these two assemblages, flakes type I, made of limestone, are numerous, suggesting a primary and minimal flaking of limestone cores. Flakes type VI, made of flint, are more frequent at El-Kherba, indicating heavy flaking of flint cores.*

Figure 23

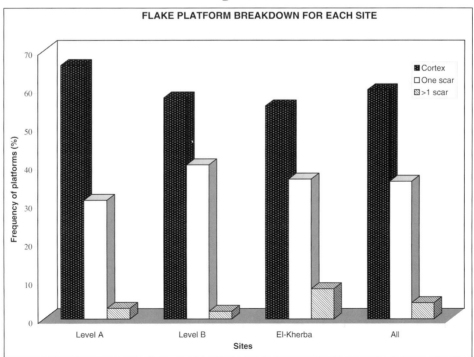

23. *Flake platform breakdown for each site. Cortical and one-scar platforms are the most common, suggesting the absence of extensive platform preparation and, perhaps, minimal flaking of cores.*

Figure 24

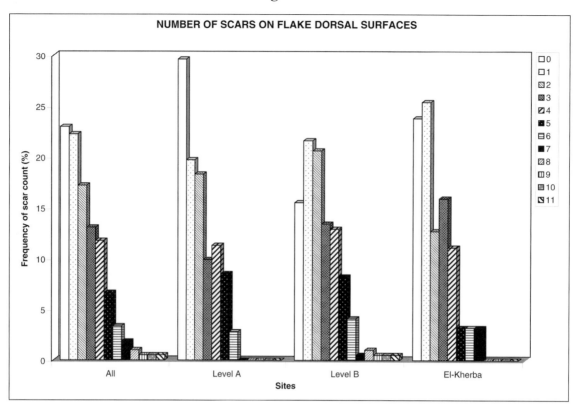

24. *Diagram showing number of scars on flake dorsal surfaces. Overall, scars are not numerous on dorsal surfaces, suggesting minimal core preparation. Note that there is a slight variability between sites.*

Figure 25

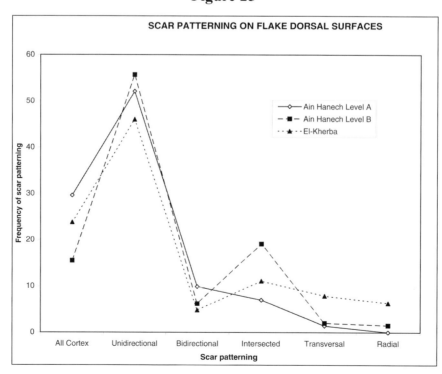

25. *Diagram showing scar patterning on flake dorsal surfaces. Note that unidirectional flaking patterns predominate.*

scar counts on dorsal surfaces of flakes made of limestone and those made of flint. In limestone flakes, dorsal surfaces with total cortex and those with one scar predominate. On the contrary, in flint flakes, dorsal surfaces with total cortex are not common at all, while those with 1, 2, 3, 4, and 5 represent the highest frequencies, suggesting heavy flaking of flint cores.

There is a relative variability in number of scars on flake dorsal surfaces between assemblages. Dorsal surfaces with total cortex are the most common in Ain Hanech Level A, while those with one scar are numerous at both Ain Hanech Level B and El-Kherba. This is due to the substantial presence of flakes made of flint in the two later assemblages.

In terms of scar patterning, the most common flaking direction encountered is the unidirectional, averaging 51.26% for all assemblages (Figure 25). This is followed by the intersected scar patterning (a scar crossed over by a perpendicular scar) (12.4%), indicating the removal of flakes at right angles. Lastly the bidirectional and radial scar patterning represent the lowest frequencies. The bulk of limestone and flint flakes are characterized by unidirectional scar patterning, indicating that this flaking mode prevailed for both types of rocks.

Retouched Pieces

The recent investigations yielded a total of 411 retouched pieces at Ain Hanech (Figure 26), including 37 (2.4% of the general total) in Level A, 276 (18.37% of the general total) in Level B, and 83 (5.52% of the general total) in El-Kherba. The test trench in Level C yielded 15 specimens. Figure 27 displays the breakdown of retouched types by site and overall. Six categories are recognized: scrapers, denticulates, notches, endscrapers, awls, and burins. The most abundant types are scrapers and denticulates, totaling 50% and 32% respectively. Endscrapers (8.5%) and notches (7%) are relatively frequent. Lastly, awls and burins are very rare. Similar frequencies of retouched pieces categories are observed in the other assemblages.

Raw Materials and Blanks

The retouched pieces were primarily made of flint (77% of the total of retouched pieces), while those made of limestone total just 21%. Within each assemblage, specimens made of flint still predominate although those made of limestone are relatively numerous in Level B (16.5%) (Figure 28).

There are three types of blanks that were used to shape retouched pieces at Ain Hanech sites: flakes, fragments, and small pebbles. The fragments comprise split and snapped flakes, and angular fragments. These three types of blanks are more or less evenly represented. However, the pebbles were slightly more retouched than the other blanks, especially in Level B (Figure 29). In Level A the blanks are primarily flakes, and at El-Kherba fragments represent the most commonly retouched blanks. The modified blanks are of medium size, ranging from 2cm to 8.5 cm with a mean of 3.4cm. Those from El-Kherba are slightly smaller (Table 6).

Retouch Characteristics

In all the assemblages the retouch is chiefly marginal, extending a few millimeters from the edge. Very few specimens are characterized by a slightly extended retouch on the surface. The retouch is commonly located on the lateral and distal sides of the blanks. The types of retouch inclination characterizing the tools are primarily abrupt (>90°), semi-abrupt (between 60° and

Table 6

Sites	Variable	Mean	Std. Dev.	Range
Ain Hanech A (n=37)	Length	37.38	14.34	20-86
	Breadth	26.32	11.29	12-58
	Thickness	12.41	4.78	5-25
Ain Hanech B (n=276)	Length	35.88	12.88	20-83
	Breadth	25.43	10.06	8-67
	Thickness	11.66	4.97	4-37
El-Kherba (n=83)	Length	29.75	12.23	19-83
	Breadth	20.35	9.21	4-73
	Thickness	9.58	3.76	2-23

6. Dimensions of retouched pieces from Ain Hanech Levels A and B, and from El-Kherba site.

Figure 26

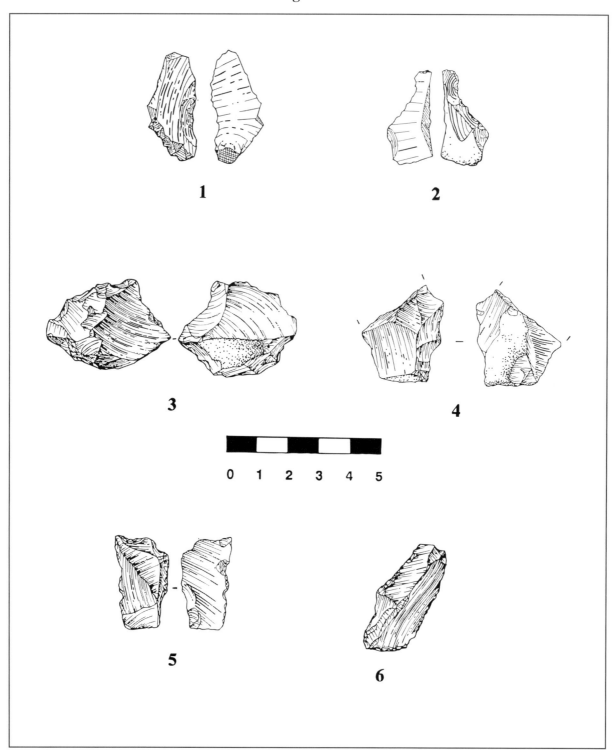

26. *Examples of retouched pieces from Ain Hanech sites, including 1-2: scrapers from Level A; 3-5: denticulates-like from Level B; and 6: scraper from El-Kherba. All the specimens are made of flint.*

90°), and simple (between 30° and 60°). Of these, simple retouch is the least common. The mean retouch angle is 77.6°. It is hard to infer the precise implications of these retouch inclination patterns without an in-depth functional analysis. However, they may signify that retouched pieces were employed for various tasks, seemingly not only those necessitating very sharp edges.

Summary of the Lithic Assemblages

To summarize the lithic assemblages in this study, the following points may be highlighted:

1. In all assemblages, two types of rock were primarily used to manufacture the artifacts: limestone and flint. Flint was used more commonly in Ain Hanech Level B. These rocks were available in the general vicinity of the site in riverbeds in the form of cobbles and pebbles. There is no evidence indicating long-distance transport of raw materials, but selection of gravel suitable shapes for flaking did occur.

2. Although debitage of both limestone and flint were modified into retouched pieces, Ain Hanech toolmakers apparently had a preference for blanks made of flint in manufacturing retouched tools. Moreover, small pebbles were also directly transformed into tools, suggesting a probable expedient component of the industry.

3. A relative variability is noted between the assemblage of Level B and those of Level A and El-Kherba, especially in terms of artifact density, core categories and flaking extent, and flake type proportions. This variability supports the stratigraphic distinction of the archaeological levels (Levels A, B, C, and El-Kherba site), and may reflect diachronic occupations with different activities. The nature of these activities and their implications still requires detailed studies.

USE OF ARTIFACTS

The Ain Hanech assemblages yielded a considerable quantity of artifacts made of fresh high quality flint. Flint is an ideal material for usewear studies (Keeley, 1980). Lawrence Keeley carried out a preliminary microwear study on a sample of artifacts made of flint (in Sahnouni & Heinzelin, 1998). Evidence of microwear was found on three specimens from Ain Hanech Level B. Both meat and bone polishes were identified on two whole flakes and on a denticulate-like retouched fragment. The evidence shows that Ain Hanech stone artifacts were used for food processing, indicating that meat was a component of hominin diet in this area.

Figure 27

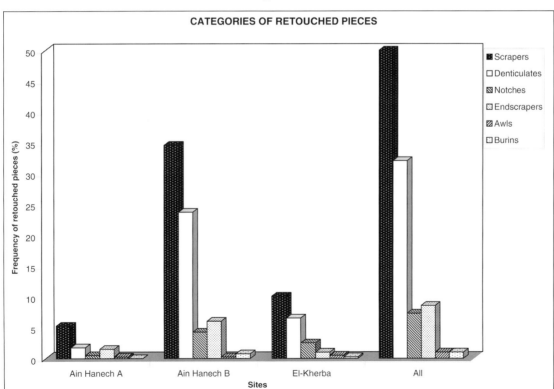

27. *Diagram showing categories of retouched pieces identified at Ain Hanech sites. Note the most common categories are scrapers and denticulates.*

Figure 28

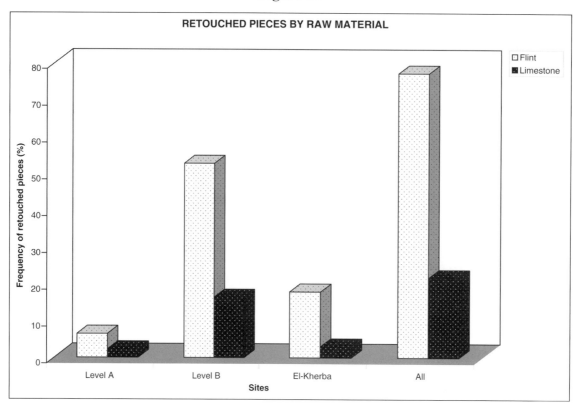

28. *Frequency of retouched pieces by raw materials. Note retouched pieces are primarily made of flint.*

Figure 29

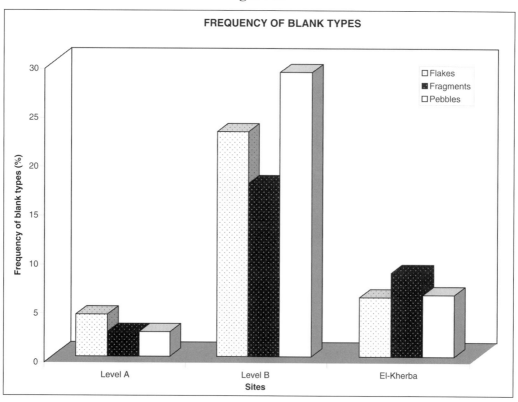

29. *Frequency of blank types. Note several kinds of blanks were transformed into retouched at Ain Hanech sites, especially small pebbles that were directly modified into tools, suggesting the expedient component of the artifacts.*

THE PLACE OF AIN HANECH INDUSTRY WITHIN THE OLDOWAN INDUSTRIAL COMPLEX

The lithic study shows that Ain Hanech assemblages are formed by two fundamental groups of artifacts. The first group represents cores, and the second group includes flakes, fragments, and retouched pieces. It must be strongly emphasized that not a single biface or even protobiface was recovered from the extensively excavated archaeological horizons. Moreover, the 2002 field season yielded hundred of Oldowan artifacts but yet not a single Acheulean artifact. Acheulean artifacts occur 6 meters higher in the stratigraphy of Ain Boucherit/Ain Hanech sequence. In this respect, the Ain Hanech assemblages clearly belong to the Mode I Technology tradition (Clark, 1969) characterizing the oldest known lithic technologies, e.g. Oldowan Industrial Complex. It may be also seen as a variant of Oldowan technological entity in North Africa.

However, it is worthwhile to ascertain as much as possible the place of Ain Hanech industry relative to other Oldowan assemblages defined elsewhere in Africa. This will help document the regional diversity of the African Early Pleistocene Paleo-cultural entities. Based on the presence/absence of categories of artifacts, the Ain Hanech assemblages are briefly compared with those from Olduvai in Tanzania (Oldowan and Developed Oldowan-A-) and Koobi Fora in Kenya (KBS Industry and Karari Industry). Unfortunately there is no other integral mode I technology assemblage in North Africa with which it can be compared. Although the difference of classification systems employed by each author and raw materials that might have influenced the final product, the comparison emphasizes what are the most common Oldowan artifacts depicted in Ain Hanech assemblages compared to other African Oldowan assemblages.

The Oldowan tradition at Olduvai Gorge is dated to 1.85-1.6 Ma and occurred through the Bed I and lower Bed II (Leakey, 1971). The Developed Oldowan-A- is restricted to the lower Bed II and lower part of Middle Bed II. The KBS Industry and Karari Industry, ranging from approximately 1.9 to 1.3 Ma, are defined as variants of the Oldowan Industrial Complex based upon a number of archaeological occurrences located at Koobi Fora in Kenya. The KBS Industry is older and restricted to the Lower Member of the Koobi Fora Formation (Isaac, 1976), and is correlated with the Oldowan. The Karari Industry is younger, dated to the Upper Member of the Koobi Fora Formation, and is considered as a variant of Oldowan (Harris, 1978).

Table 7 compares artifact frequencies between Ain Hanech and Olduvai, and Ain Hanech and KBS Industry and Karari Industry. With few exceptions, all Oldowan artifacts known at Olduvai and Koobi Fora are found at Ain Hanech. The Ain Hanech assemblages lack only discoids and protobifaces. The other categories are present although with slight differences in their frequencies. Despite the absence of discoids and protobifaces, the Ain Hanech assemblages composition is closer to Olduvai assemblage than to KBS and Karari ones. Both Ain Hanech and Olduvai include subspheroids and spheroids, although Ain Hanech spheroids are faceted. These two artifact categories are entirely absent from both KBS Industry and Karari Industry. They also incor-

Table 7

Artifacts	Ain Hanech[1] (n=1471)	Oldowan[2] (n=537)	Dv.Oldowan[2] (n=681)	KBS Industry[3] (n=24)	Karari Industry[4] (n=511)
Choppers	4.4	51.0	26.0	45.8	21.7
Proto-bifaces	0	1.3	2.3	—	2.54
Polyhedrons	7.2	10.0	4.8	12.5	17.22
Discoids	0	9.1	2.9	8.3	17.40
Subspheroids & Spheroids	2.65	6.0	23.5	0	0
Heavy duty scrapers	0.3	8.6	2.6	—	17.02
Light duty scrapers	26.3	10.2	9.4	8.3	21.13
Burins	0.3	1.7	—	—	0
Awls	0.3	—	2.8	—	0
Sundry tools	—	2.0	6.2	8.3	2.93

7. *Comparison of Ain Hanech assemblages with Oldowan, Developed Oldowan, KBS Industry, and Karari Industry (1Ain Hanech industry includes assemblages from Level A, Level B, and El-Kherba; 2after M. Leakey, 1975; 3after Isaac, 1976; 4after Harris, 1978).*

porate other artifacts, such as choppers, polyhedrons, heavy-duty scrapers, light-duty scrapers, burins, and awls. Furthermore, technological similarities between Ain Hanech and Olduvai core assemblages were demonstrated (Sahnouni, 1993).

CONCLUSIONS

This chapter has presented current knowledge on the North African Early Stone Age, focusing on the new study of the sites at Ain Hanech in northeastern Algeria. As shown in this study, North Africa yields archaeological sites with Oldowan industries. Furthermore, new investigations show that Ain Hanech provides valuable information on North African hominins behavior and adaptation. The new investigations have included study of the stratigraphy and chronology, assessment of site formation processes, analysis of the lithic assemblages, and exploration of the overall behavioral implication of the Oldowan occurrences. Tentative conclusions include:

1. There is a claim tending to promote a sort of "short chronology" scenario for the earliest human presence in North Africa (Raynal *et al.*, 2001). This scenario proposes that the earliest occupation in this region is Acheulean and is dated to 1 Ma. While this may apply to Atlantic Morocco, the Ain Hanech evidence reaffirms and strengthens the remote human antiquity in this part of the African continent. The site of Ain Hanech is the oldest archaeological occurrence in North Africa. The human presence in this region is Oldowan and may go back to 1.78 Ma (Figure 30), suggesting an earlier dispersal of hominins into North African than commonly assumed.

2. The Ain Hanech industry may be considered as a North African variant of the Oldowan Industrial Complex. The lithic assemblages incorporate a full range of artifact categories from cores to microdebitage, and are similar to those recovered in East African Plio-Pleistocene sites, e.g. Olduvai Bed I and Lower Bed II, and Koobi Fora sites. Like other Oldowan assemblages, the Ain Hanech industry is characterized by a simple technology and a low degree of standardization reflected by the variability of flaking on cores and by the retouch on flakes and fragments. Moreover, the assemblages comprise categories of artifacts similar to those found in East African Oldowan assemblages.

3. Although it is still premature to discuss subsistence patterns because detailed results of the faunal analysis are not available yet, preliminary evidence indicates that animal tissues likely constituted a significant part of Ain Hanech hominin diet. Remains of various animal species were recovered, including equid, large and small bovids, rhino, hippo, and elephant. Manufactured artifacts were used for processing the animal carcasses, such as meat cutting and bone scraping. Bones bearing carnivore signatures were also recovered, suggesting that there was likely acquisition of meat resources at these sites. A detailed study will be carried out to document subsistence patterns, including strategies employed for meat acquisition, and assessment of consumption patterns.

4. In terms of site interpretation, it can be visualized that the Ain Hanech area witnessed repeated occupations by hominins at a shallow river embankment, perhaps directed by the availability of good quality raw materials and game for meat acquisition. In ongoing archaeological explorations, material from the three Oldowan deposits is being analyzed and compared/contrasted to investigate behavioral patterns, including artifact manufacture, transport and discard, subsistence acquisition, and land use.

ACKNOWLEDGMENTS

The author thanks several institutions and people for making further research at Ain Hanech possible, including the Algerian Ministry of Culture and Communications for the research permit; the University of Algiers, The L.S.B. Leakey Foundation, The Wenner-Gren Foundation, CRAFT Research Center at Indiana University and Friends of CRAFT for their support; the local authorities for their interest in our research project; the Thabet family for their warm hospitality; and finally the archaeology students of the University of Algiers who actively participated in the excavations despite the difficult field conditions.

Figure 30

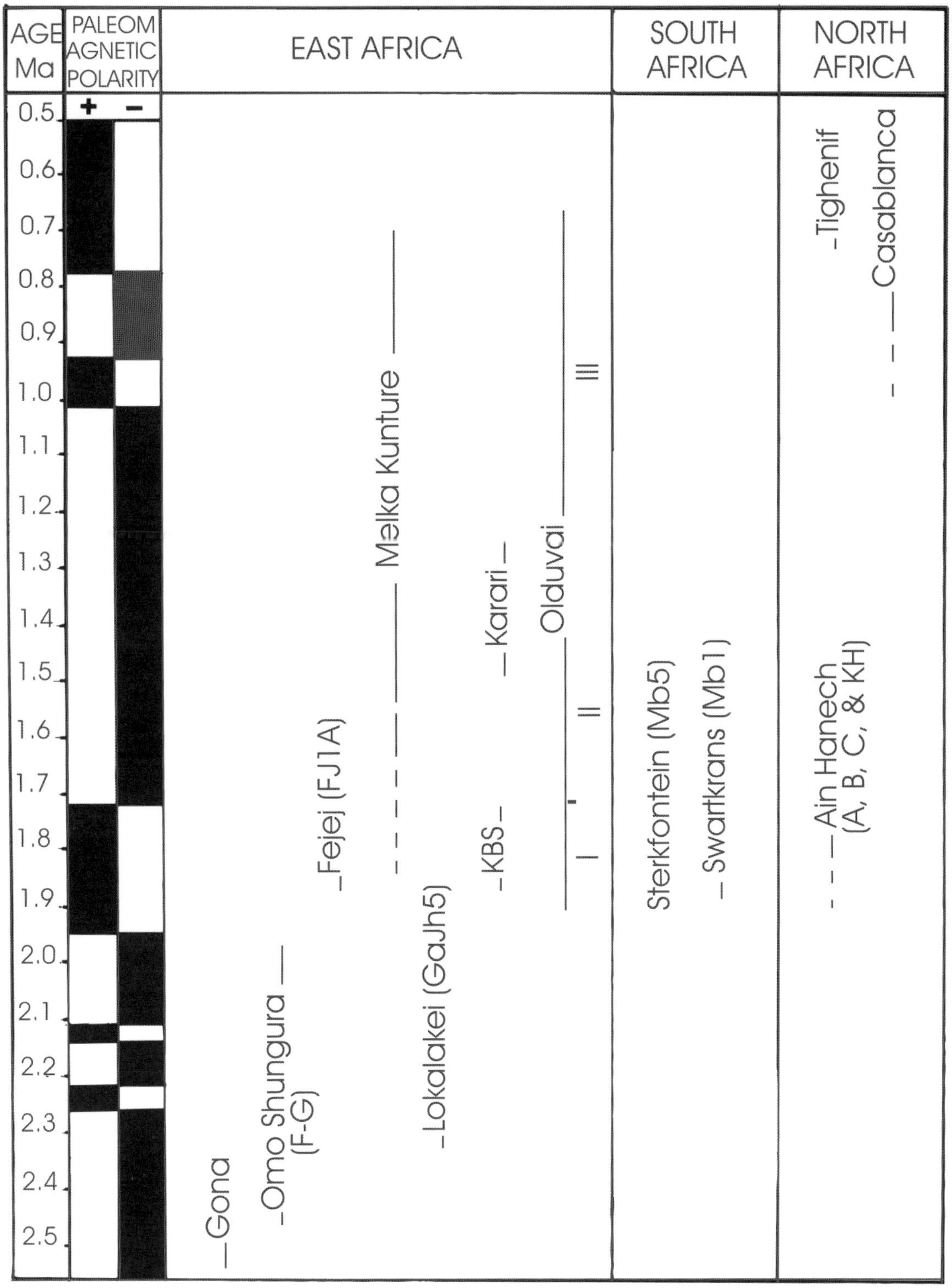

30. *Chronological position of the Ain Hanech sites relative to other early East African earliest archaeological sites.*

REFERENCES CITED

Alimen, M.H. & Chavaillon, J. (1960). Présentation de "galets aménagés" des niveaux successifs du Quaternaire ancien de la Saoura. *Bulletin Société Préhistorique Française* **57**, 373-374.

Alimen, M.H. & Chavaillon J. (1962): Position strati-graphique et évolution de la Pebble Culture au Sahara Nord-Occidental. *IV Congrès Panafricain de Préhistoire. Annales du Musée Royal de l'Afrique Centrale*, pp. 3-24.

Arambourg, C. (1947). Les Vertébrés fossiles des formations continentales des Plateaux Constantinois. *Bulletin de la Société d'Histoire Naturelle de l'Afrique du Nord* **38**, 45-48.

Arambourg, C. (1949). Présentation d'objets énigmatiques provenant du Villafranchien d'Algérie. *Société Géologique de France (c.r. sommaires des Séances)* **7**, 120-122.

Arambourg, C. (1953). Nouvelles observations sur le gise-ment de l'Ain Hanech près de Saint-Arnaud (Constantine). *Comptes Rendus Académie des Sciences* **236**, 2419-2420.

Arambourg, C. (1970). *Les Vertébrés du Pleistocène de l'Afrique du Nord.* Archives du Museum National d'Histoire Naturelle, 1-127.

Arambourg, C. (1979). *Les Vertébrés villafranchiens d'Afrique du Nord.* Paris: Singer-Polignac.

Arambourg, C. & Balout, L. (1952). Du nouveau à l'Ain Hanech. *Bulletin de la Société d'Histoire Naturelle de l'Afrique du Nord* **43**, 152-159.

Balout, L. (1955). *Préhistoire de l'Afrique du Nord.* Paris: Arts & Métiers Graphiques.

Bar-Yosef, O. & Goren-Inbar, N. (1993). *The lithic Assemblages of Ubeidiya: A Lower Paleolithic Site in the Jordan Valley.* Jerusalem: Institute of Archaeology.

Behrensmeyer, A. K. (1978). Taphonomic and Ecologic Information from Bone Weathering. *Paleobiology* **4**, 150-62.

Biberson, P. (1961). *Le Paléolithique inférieur du Maroc Atlantique.* Rabat: Publications du Service des Antiquités du Maroc.

Biberson, P. (1967). Galets aménagés du Maghreb et du Sahara. *Collection Fiches Typologiques Africaines.* Paris : Museum National Histoire Naturelle.

Biberson, P. (1976). Les plus anciennes industries du Maroc. *Proceedings of the IX U.I.S.P.P. Congress, Nice (France). Colloque V: Les plus anciennes industries en Afrique.*, pp. 118-139.

Brown, F. H. (1994). Development of Pliocene and Pleistocene chronology of the Turkana Basin, East Africa, and its relation to other sites. In (R. S. Corruccion & R. L. Ciochon Eds): *Integrative paths to the past. Paleoanthropological advances in honor of F. Clark Howell.* New Jersey: Prentice Hall, pp. 285-312.

Brown, F. H., McDougall I., Davies T. & Maier, R. (1985). An integrated Plio-Pleistocene chronology for the Turkana Basin. In (E. Delson Ed) *Ancestors: The hard evidence.* New York: Alan R. Liss, pp. 82-90

Clark, J.G.D. (1969). *World Prehistory: A new outline.* Cambridge: Cambridge University Press.

Clark, J.D. (1992). Earlier Stone Age/Lower Palaeolithic Northwest Africa in North Africa and Sahara. In (F. Klees & R. Kuper, Eds) *New light on the Northeast African Past.* Koln: Heinrich-Barth-Institut, pp. 17-37.

Coppens, Y., Maglio, V.J., Madden, C.T. & Beden, M. (1978). *Proboscidea.* In (V. J. Maglio & H. B. S. Cooke, Eds) *Evolution of African Mammals.* Cambridge, Massachusetts: Harvard University Press, pp. 336-367.

Eisenmann, V. (1983). Family *Equidae.* In (J. M. Harris, Ed.) *Koobi Fora Research Project.* Oxford: Clarendon Press, pp. 156-214.

Eisenmann, V. (1985). Les équidés des gisements de la Vallée de l'Omo en Ethiopie. In (Y. Coppens & F. Howell, Eds) *Les faunes Plio-Pléistocènes de la basse Vallée de l'Omo (Ethiopie). Tome 1: Les Périssodactyles, les Artiodactyles (Bovidae), expédition internationale 1967-1976.* Paris: CNRS, pp. 13-65.

Eisenmann, V. (1986). Les équidés du Pleistocène d'Oubeidiyeh (Israel). *Mémoire et travaux du Centre de Recherches Français de Jerusalem* **5**, 191-212.

Geraads, D. (1981). *Bovidae* et *Giraffidae* (Artiodactyla, Mammalia) du Pleistocène de Ternifine (Algérie). *Bulletin Museum National d'Histoire Naturelle, Paris* **3**, **4 série, section C**, 47-86.

Gragueb, A. & Oueslati, A. (1990). Les formations quater-naires des Côtes nord-est de la Tunisie et les industries préhistoriques associées. *L'Anthropologie* **91**, 259-292.

Harris, J.W.K. (1978). *The Karari Industry. Its place in East African Prehistory.* Ph.D. Thesis, University of California, Berkeley.

Heddouche, A.E.K. (1980). Découverte d'une industrie à galets aménagés au Sahara Nord-Oriental. *Libyca* **28**, 105-112.

Heddouche, A.E.K. (1981). Les galets aménagés de Bordj Tan Kena, Illizi (Algérie). *Libyca*, **30-31**, 9-18.

Hugot, H. (1955). Un gisement à pebble tools à Aoulef. *Travaux de l'Institut de Recherche Saharienne* **8**, 131-153.

Isaac, G. (1976). Plio-Pleistocene artifact assemblages from East Rudolf, Kenya. In (F. C. Howell, Y. Copens, Gl. Isaac & R.E.F. Leakey, Eds) *Earliest man and environment in the Lake Rudolf Basin.* Chicago: University of Chicago Press, pp. 552-564.

Keeley, L.H. (1980). *Experimental determination of stone tool uses.* Chicago: University of Chicago Press.

Laplace-Jauretche, G. (1956). Découverte d'un gisement à galets taillés (Pebble Culture) dans le Quaternaire ancien du Plateau de Mansourah (Constantine). *Bulletin Société Préhistorique Française* **53**, 215-216.

Leakey, M.D. (1971). *Olduvai Gorge, Volume 3. Excavations in Beds I and II, 1960-1963.* Cambridge: Cambridge University Press.

Leakey, M.D. (1975). Cultural Patterns in the Olduvai sequence. In (K.W. Butzer and G. Isaac, Eds) *After the Australopithecines.* The Hague: Mouton, pp. 477-493.

Lyman, R.L. (1994). *Vertebrate Taphonomy.* Cambridge: Cambridge University Press.

Pomel, A. (1893-1897). *Monographies des Vertébrés fossiles de l'Algérie.* Alger: Service de la Carte Géologique de l'Algérie.

Ramendo, L. (1963). Les galets aménagés de Reggan (Sahara). *Libyca* **11**, 42-73.

Raynal, J.P. & Texier, J. P. (1989). Découverte d'Acheuléen ancien dans la carrière Thomas 1 à Casablanca et problème de l'ancienneté de la présence humaine au Maroc. *Comptes Rendus Académie des Sciences* **308, série II**, 1743-1749.

Raynal, J.P., Texier, J. P., Geraads, D. & Sbihi-Alaoui, F. Z. (1990). Un nouveau gisement paléontologique plio-pleistocène en Afrique du Nord: Ahl Al Oughlam (ancienne carrière Deprez) à Casablanca (Maroc). *Comptes Rendus Académie des Sciences* **310, série II**, 315-320.

Raynal, J.P., Magoga, L., Sbihi-Alaoui, F. Z. & Geraads, G. (1995). The Earliest occupation of Atlantic Morocco: the Casablanca evidence. The earliest occupation of Europe. In (W. Roebroeks & T. van Kolfschoten, Eds) *The earliest occupation of Europe.* Leiden : University of Leinden, pp.255-262.

Raynal, J.P., Sbihi Alaoui, F. Z., Geraads, D., Magoga, L. & Mohi, A. (2001). The earliest occupation of North Africa: the Moroccan perspective. *Quaternary International* **75**, 65-75.

Roubet, C. (1967). Découverte de nouveaux galets aménagés dans la région sétifienne. *Libyca* **15**, 9 14.

Sahnouni, M. (1987). *L'industrie sur galets du gisement villafranchien supérieur d'Ain Hanech.* Alger: Office des Publications Universitaires.

Sahnouni, M. (1993). Étude comparative des galets taillés polyédriques, subsphériques et sphériques des gisements d'Ain Hanech (Algérie Orientale) et d'Olduvai (Tanzanie). *L'Anthropologie, (Paris)* **97**, 51-68.

Sahnouni, M. (1998). *The Lower Palaeolithic of the Maghreb: Excavations and analyses at Ain Hanech, Algeria.* Oxford: Archaeopress.

Sahnouni, M., Heinzelin, J. de, Brown, F. & Saoudi, Y. (1996). Récentes recherches dans le gisement oldowayen d'Ain Hanech, Algérie. *Comptes Rendus Académie des Sciences 323*, **série II a**, 639-644.

Sahnouni, M. & Heinzelin, J.de (1998). The site of Ain Hanech revisited: New Investigations at this Lower Pleistocene site in Northern Algeria. *Journal of Archaeological Science* **25**, 1083-1101.

Sahnouni, M., Hadjouis, D., Abdesselam, S., Ollé, A., Verges, J.M., Derradji, A., Belahrech, H. & Medig, M. (2002). El-Kherba: a Lower Pleistocene butchery site in Northeastern Algeria. Abstract for the Paleoanthropology Society Meetings. *Journal of Human Evolution* **42**, A31.

Sahnouni, M., Hadjouis, D., Made, J. van der, Derradji, A., Canals, A., Medig, M., Belahrech, H., Harichane, Z. & Rabhi, M. (In press). Further research at the Oldowan site of Ain Hanech, northeastern Algeria. *Journal of Human Evolution.*

Schick, K. D. (1986). *Stone age sites in the making.* Oxford: BAR International Series 319.

Thomas, G. (1973). Découverte d'industrie du groupe de la "Pebble Culture" sur le versant nord des monts du Tessala (Algérie). Sa place dans la stratigraphie du Pleistocène inférieur et moyen de l'Oranie. *Comptes Rendus Académie des Sciences* **276, série D**, 921-924.

Toth, N. (1982). *The stone technologies of Early Hominids at Koobi Fora, Kenya. An experimental approach.* Ph.D. Thesis. University of California, Berkeley.

Toth, N. (1985). The Oldowan reassessed: A close look at Early Stone artefacts. *Journal of Archaeological Science* **12**, 101-120.

Vila, J.M. (1980). *La chaîne alpine d'Algérie Orientale et des confins algéro-tunisiens.* Doctoral Thesis. Université Pierre & Marie Curie, Paris.

Vogel, J. C. (1978). Isotopic assessment of the dietary habits of ungulates. *South African Journal of Science* **74**, 298-301.

CHAPTER 4

THE ACQUISITION AND USE OF LARGE MAMMAL CARCASSES BY OLDOWAN HOMININS IN EASTERN AND SOUTHERN AFRICA:
A SELECTED REVIEW AND ASSESSMENT

BY TRAVIS RAYNE PICKERING AND MANUEL DOMÍNGUEZ-RODRIGO

ABSTRACT

Damage to fossil bone surfaces, in the form of cutmarks and percussion marks, establishes causal links between early hominin stone tool technology and spatially associated large mammal remains from archaeological sites located throughout eastern and southern Africa and dated *c.* 2.5–1.0 million years old. The presence of abundant tooth marks in faunal assemblages from most of these sites also implicates carnivores as significant actors in the formation of the bone accumulations. We review arguments based on this taphonomic evidence from selected, well-excavated sites and conclude that although Oldowan hominins likely engaged in a full range of carcass-acquiring behaviors, depending on immediate circumstances, they were not relegated solely to the role of passive scavengers, as some influential models of early hominin behavior imply.

INTRODUCTION

The 1990s proved to be a productive and exciting decade for paleoanthropology. In addition to the announcement of several new early hominin species from various African sites (i.e., *Ardipithecus ramidus, Australopithecus anamensis, A. bahrelghazali, A. garhi*), work at Gona, in Ethiopia's Awash River Valley, yielded the world's oldest known stone tools, dated radioisotopically between 2.6–2.5 million years old (Ma) (Semaw *et al.*, 1997). Further, cutmarked and hammerstone damaged animal bones were recovered from Gona-aged deposits in the nearby Middle Awash paleoanthropological study area (at two localities in the Hata Member of the Bouri Formation) (de Heinzelin *et*

al., 1999)–providing the earliest-occurring, indisputable causal links between hominin stone tool technology and the exploitation of large mammal carcasses[1] (Figure 1).

Interestingly, early *Homo* is not represented in the fossil record of the Awash River Valley during this time period. Instead, a species of *Australopithecus*, *A. garhi*, is present (Asfaw *et al.*, 1999). It is not possible to assert definitively that *A. garhi* was responsible for the Gona stone tools and evidence of their use in the Middle Awash as carcass reducing implements, but circumstantial evidence (i.e., the absence of early *Homo*) makes a fairly compelling case for *A. garhi* as the author of the earliest known stone tool assemblages.

These earliest assemblages are classified by most experts as belonging to the Oldowan Industrial Complex (e.g., Isaac, 1984; Leakey, 1966, 1971; Semaw, 2000; Semaw *et al.*, 1997; Schick & Toth, 1993; contra, Kibunjia, 1994; Piperno, 1989; Roche, 1989, 1996), and various studies in the past 20 years have demonstrated causal links between Oldowan tools and large mammal bones–similar to that evidenced in the Awash River Valley–throughout fossil localities in eastern and southern Africa. Here we review and evaluate a selected sample of these studies with the goal of assessing early hominin carcass foraging capabilities. We restrict our discussion to well-studied faunal assemblages, most of which are associated spatially with Oldowan and Developed Oldowan stone tool assemblages and are in good to excellent archaeological contexts. This means that assemblages such as those from Senga (Democratic Republic of Congo) are not included because of problems with re-deposition (Boaz *et al.*, 1992; de Heinzelin, 1994). Finally, there are yet to be convincing inferences of hominin influence on prehis-

Figure 1

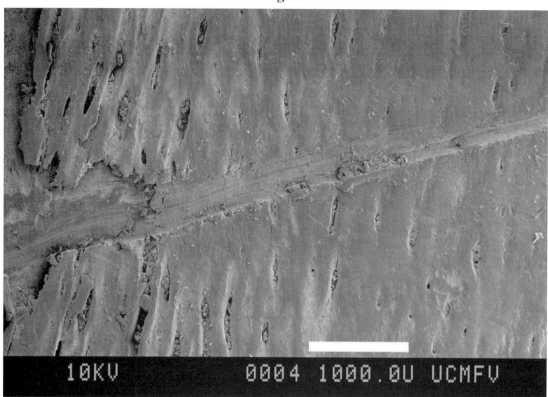

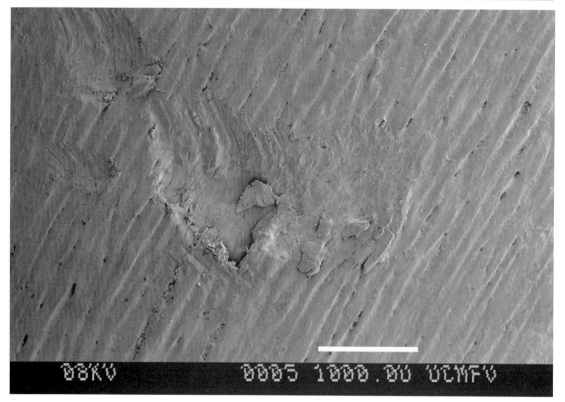

1. *Scanning Electron Microscope micrographs showing representative examples of two major classes of bone surface damage imparted by hominins–stone tool cutmarks (top) and hammerstone percussion damage (bottom). Note the internal microstriations within the main groove of the cutmark and the patches of striations emanating from the percussion pit (see Potts & Shipman, 1981; Blumenschine & Selvaggio, 1988). Identification of these microscopic features associated with incidences of bone surface damage, an understanding of the anatomical placement and patterning of such damage, and secure knowledge of the geomorphological context of the faunal assemblage in which the damage occurs are classes of evidence used by faunal analysts to infer hominin behavior in archaeofaunas.*

toric faunal assemblages *in the absence of stone tool marks*, even though various researchers are currently using primate models to establish criteria for identifying the involvement of pre-stone-tool-using hominins in bone assemblage formation (e.g., Pickering & Wallis, 1997; Plummer & Stanford, 2000; Tappen & Wrangham, 2000; Domínguez-Rodrigo, 1999a). Thus, we do not discuss this topic further, even though we are of the opinion–based on the observations that modern humans and many of our living primate relatives (e.g., chimpanzees, baboons) are avid meat-eaters (reviewed in Stanford & Bunn, 2001)–that hominins likely engaged in significant meat eating before the advent of stone tool technology.

FLK 22 (*ZINJANTHROPUS*): BED I, OLDUVAI GORGE, TANZANIA

The formational history of the Earlier Stone Age faunal assemblage from FLK 22 (*Zinjanthropus*) (more commonly referred to as FLK *Zinj*) has been the subject of more heated debate than the formation of any other archaeofauna of comparable age. We believe that there are several factors that contribute to this intense focus on the *c.* 1.75 Ma FLK *Zinj* fauna, including its meticulous excavation by Mary Leakey, its large size (~60,000 macro- and micromammalian specimens combined), its exquisite preservation of bone surfaces and its spatial association with fossil hominin remains and Oldowan stone tools–rendering the site *the* classic example of a so-called "home base" or Type C site (e.g., Isaac, 1978, 1984).

However, the interpretation of FLK *Zinj* as a hominin home base is not universally accepted. Lewis Binford (e.g., 1981, 1985, 1986, 1988) was the first archaeologist to question this interpretation of the site, arguing that it was instead a locale on the ancient landscape where early hominins scavenged for marginal scraps of flesh and marrow from ungulate carcasses killed and consumed primarily by carnivores. Binford's (1981) conclusions about the relatively minor involvement of hominins in the formation of the FLK *Zinj* fauna had a major impact on the course of actualistically-driven, zooarchaeological studies in Africa for the next 20 years[2].

It is important to note, however, that Binford was not the only early researcher to challenge previous assertions about the modern human behavioral capabilities of Plio-Pleistocene hominins. Based on his painstaking taphonomic analyses of faunal assemblages recovered from Swartkrans Cave (South Africa), C.K. Brain argued as early as 1970 that rather than being competent, bone-accumulating hunters (as traditionally proposed; e.g., Etkin, 1954; Washburn, 1959; Washburn & Howell, 1960; Washburn & Lancaster, 1968), many early hominins were actually the victims of predation. Although Brain's conclusions dealt specifically with the South African australopithecines, a group of species that

presumably lacked stone tool technology–his general notion of early hominins as "the hunted" rather than "the hunters" must still be viewed as a major impetus (along with Binford's arguments about the FLK *Zinj* fauna) in the re-evaluation of early hominin foraging capabilities.

This re-evaluation eventually reached its zenith during the mid-1980s through mid–1990s, and was spearheaded by zooarchaeologist Rob Blumenschine. Returning to the FLK *Zinj* fauna, Blumenschine (1995) proposed a complex series of events resulting in its formation. Hominins first entered this sequence as scavengers, when they transported the marrow-bearing portions of ungulate carcasses–already largely defleshed by felid predators–away from acquisition sites to FLK *Zinj*, where these bones were then broken by the hominins for marrow. Subsequently, tertiary level scavengers (i.e., hyenas) impacted the assemblage by removing bone portions, which contained grease left unexploited by the hominins.

This model of bone assemblage formation at FLK *Zinj* was based on Blumenschine's study of bone surface damage in the ungulate limb bone subassemblage. First, midshaft sections of long bones at FLK *Zinj* preserve carnivore tooth marks in frequencies comparable to tooth mark frequencies on midshafts in experimentally-created bone assemblages in which carnivores had primary access to ungulate limb bones (Blumenschine, 1988; Capaldo, 1995, 1997; Selvaggio, 1994a). Second, limb bone fragments from FLK *Zinj* also preserve numerous hammerstone percussion marks. This suggests that the primary carnivore consumers did not regularly breech the bones for marrow, and that this resource was thus available to the hammerstone-wielding hominins. Finally, the abundant tooth marks on epiphyseal and metaphyseal ("near-epiphyseal", in the terminology employed by Blumenschine and his co-workers) specimens at FLK *Zinj*, suggested that these bone portions were ignored by hominins, but subsequently exploited by hyenas–carnivores that possess the masticatory and digestive apparatuses capable of rendering grease from the trabaculae of long bone ends.

In many aspects, this three-stage model of assemblage formation at FLK *Zinj* is powerful, but its major weakness is that it does not fully appreciate the abundant cutmark evidence preserved in the assemblage, first reported by Henry Bunn (1981, 1982; Bunn & Kroll, 1986) and Rick Potts and Pat Shipman (1981). While Potts & Shipman's (1981) interpretation of the FLK *Zinj* cutmark evidence (they argue that the cutmarks resulted not from hominins cutting meat from carcasses, but from the removal of tendons and skin) is not incompatible with the notion of early hominins as marginal scavengers, Bunn's (1981, 1982; Bunn & Kroll, 1986) interpretation is decidedly contrary to this view of early hominin behavior. Bunn argues that the abundance and placement of the FLK *Zinj* cutmarks suggest "an efficient [hominin] strategy of carcass skinning, joint dismemberment, and meat removal and for a significant amount of meat-eating by [hominins] nearly two million years ago" (Bunn & Kroll, 1986: 432)–behaviors that are consistent with primary or, at least, very early access to fully-fleshed carcasses.

Some researchers, however, remained unconvinced that the cutmark evidence at FLK *Zinj* is relevant to inferences about the quantity of muscle tissue removed from carcasses by early hominins. Binford (1986) retorted that the cutmark patterns at FLK *Zinj* are consistent with removal of desiccated meat from bone, while Blumenschine (1986a, 1988, 1995) countered that these patterns could have resulted from the removal of marginal meat scraps left after felids consumed the bulk of muscle masses; both responses imply that hominins were marginal (at least secondary or tertiary level) scavengers.

The general ambiguity surrounding the usefulness of cutmark data for helping to resolve the question of assemblage formation at FLK *Zinj* prompted one of us (MD-R) to conduct a series of experimental studies, in which the interface of carcass flesh availability and cut-

mark patterns was examined. A major finding of this research program calls into question the notion that felid carcass-consumers would regularly provide hominin scavengers with edible scraps of adhering tissue. Only negligible amounts of scavengeable flesh scraps were documented on 28 medium-sized (i.e., 150-350 kg) ungulate carcasses after ravaging by lions in the Maasai Mara National Reserve (Kenya) (Domínguez-Rodrigo, 1999b) (Figure 2). More specifically, upper limb bones (i.e., the humerus and femur) and intermediate limb bones (i.e., the radioulna and tibia) displayed a paucity of adhering flesh scraps after lion ravaging (Figure 3). Even more importantly, midshaft sections of upper limbs displayed a *complete* lack of flesh scraps and, similarly, flesh scraps on the midshaft portions of intermediate limb bones were poorly represented (Figure 4). These results suggest that cutmarks on upper and intermediate limb bone midshafts most likely indicate early access to fully fleshed carcass parts by hominins[3]; hominins would have no reason to put a cutting edge to a long bone midshaft previously defleshed in its entirety by a felid consumer[4].

Bunn (1981, 1982, 2001) conducted the most com-

Figure 2

2. *Lion and remnants of a lion-ravaged wildebeest carcass in the Maasai Mara National Reserve (Kenya). Note that the skeleton has been defleshed completely by lions, leaving marrow and brains as the only soft tissue resources available to potential secondary and tertiary level scavengers.*

Figure 3

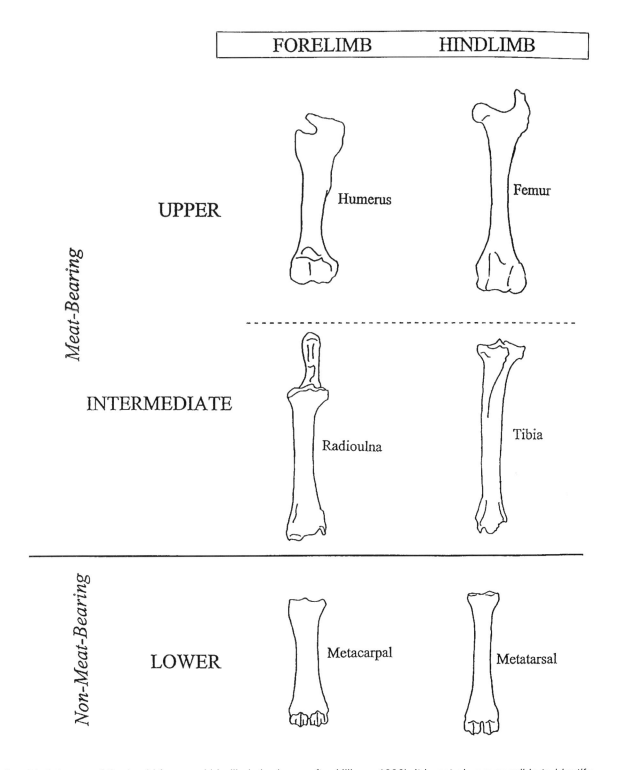

3. *Limb bones of the bovid fore- and hindlimb (redrawn after Hillson, 1996). It is not always possible to identify an archaeological bone fragment to a specific element. In most cases, however, an experienced analyst can confidently categorize a diaphyseal fragment of bovid limb bone as deriving from an upper limb element (i.e., the humerus or femur), an intermediate limb element (i.e., the radioulna or tibia) or a lower limb element (i.e., the metacarpal or metatarsal), based on an assessment of the fragment's cortical thickness, apparent or projected cross-sectional shape and other features such as nutrient foraminae (see Domínguez-Rodrigo, 1999a). This generalized categorization is useful because of the differential distribution of flesh scraps across these element categories in felid-ravaged carcasses, and because that differential distribution influences subsequent cutmark patterns inflicted by hominin scavengers (see text for discussion).*

Figure 4

4.   *Close-up of the remnants of a wildebeest carcass ravaged by lions in the Masaai Mara National Park (Kenya). Note that the femur–an upper limb bone in the classification scheme employed in this chapter–and especially its midshaft, has been completely defleshed by lions. The lack of edible flesh remaining on this bone portion after lion ravaging would discourage a potential hominin scavenger from using a stone tool to cut in this region of the femur; thus, cutmarks observed on upper limb bone midshaft fragments usually indicate early access (i.e., before ravaging by carnivores) to carcasses by hominins.*

prehensive study of the FLK *Zinj* cutmarks, and concluded:

> "Cut marks are concentrated on the meaty limbs [i.e., upper and intermediate limb bones] of both smaller and larger animal carcasses. Sixty-two percent of all cut-marked bone specimens from smaller animals and 39.2% of all cut-marked specimens from larger animals are from the meaty limb bones. A large majority of the cut marks on meaty limb bones occur on mid-shaft specimens rather than on or immediately adjacent to epiphyses. At least 61.3% of the cut marks on meaty limbs of smaller animals and at least 68.9% of the cut marks on meaty limbs of larger animals occur on shaft specimens, and these values are generous underestimates" (Bunn & Kroll, 1986: 436-437).

Oliver's (1994) reanalysis of the FLK *Zinj* fauna corroborates the findings of Bunn and Kroll–there are more cutmarked upper and intermediate limb bones than cutmarked lower limb bones (i.e., metapodials). And both studies support the notion that Bed I hominins at Olduvai Gorge gained access to ungulate carcasses possessing large amounts of edible flesh, contradicting the three-stage model of assemblage formation, favored by Blumenschine and colleagues.

However, Blumenschine's (1995) finding that tooth mark frequencies in the FLK *Zinj* midshaft subassemblage are comparable to those observed in experimental studies in which carnivores had primary access to ungulate limb bones, still demands explanation. First, in response to the midshaft tooth mark evidence at FLK *Zinj*, zooarchaeologists need to recognize the fact that *animals other than large carnivores* can and do impart tooth marks on bones. Using modern chimpanzees as models for early hominins, Pickering & Wallis (1997) demonstrated that large-bodied hominoids are capable of producing tooth marks identical to those created by carnivores, in terms of the types of damage produced, in the morphology of damage marks and in overall degree and veracity of damage (see also, Brain, 1976; Maguire *et al.*, 1980; Plummer & Stanford, 2000; Tappen & Wrangham, 2000). Similarly, even baboons (in both captive and wild-ranging situations) leave tooth marks on bones comparable in frequency and distribution to those made by carnivores on human-produced bone assemblages (Domínguez-Rodrigo, 1999a). Thus, it is possible that some, if not many, of the tooth marks on FLK *Zinj* fossils were imparted by non-carnivores–including possibly hominins.

Assuming, though, that all the FLK *Zinj* tooth marks were inflicted by carnivores, it is important to note that the percentage of tooth marks on the midshaft subassemblage is similar not only to experimental situations in which carnivores were the primary defleshers and demarrowers of bones (Blumenschine, 1995). Domínguez-Rodrigo (1999a) has also shown experimentally that a similar percentage of tooth marked midshafts results from cases in which hominins are the primary carcass defleshers, but leave some marrow-bearing bones unbroken, available secondarily to scavenging carnivores. Based on analysis of long bone completeness at FLK *Zinj*, there is some reason to believe that this might be a realistic scenario of hominin behavior at the site. There is significant number of complete bones (9–10% of the total minimum number of elements) in the FLK *Zinj* faunal assemblage (Potts, 1988). This observation means that quite a few marrow-bearing bones were left unbroken at the site.

Further, this observation can lead to at least two conclusions. First, if hominins (who *all participants* in the debate–e.g., Binford, 1981; Blumenschine, 1988, 1995; Bunn & Kroll, 1986; Domínguez-Rodrigo, 1999a–agree had some role in the formation of the FLK *Zinj* fauna) were relegated, as passive scavengers, solely to marrow exploitation, one would *not* expect high numbers of unbroken limb bones at the site (Domínguez-Rodrigo, 2002). In other words, why would limb bones already defleshed by felid predators be transported by hominins to the site and then left unbroken? Second, Capaldo (1995) has demonstrated that hyenas destroy nearly all grease- and marrow-bearing elements available in small accumulations of bone. Further, most researchers (e.g., Bunn & Kroll, 1986; Blumenschine, 1988; Blumenschine & Marean, 1993; Marean *et al.*, 1992; Domínguez-Rodrigo, 1999a) agree that hyenas ravaged the FLK *Zinj* fauna *after* hominin involvement in the assemblage. Thus, the presence of whole marrow-bearing bones in the fossil sample could suggest a glut of these elements available after hominin use (Capaldo, 1995), but not exploited completely by hyena scavengers, because of their overabundance[5]. Together, these conclusions argue for early (abundant cutmarks on flesh bearing elements) and frequent

(abandonment of limb bones without full exploitation of their marrow reserves) access to fully fleshed carcasses by Bed I hominins–a decidedly different idea of early hominin foraging capabilities than that proposed by advocates of marginal scavenging models.

BEYOND FLK *ZINJ* IN EAST AFRICA

While many other Plio-Pleistocene archaeological sites preserve large stone tool and faunal assemblages, few display the exquisite bone surface preservation observed at FLK *Zinj*. This is true for many of the important Koobi Fora (Kenya) sites excavated in the 1970s (Bunn, 1982, 1997). The best of these relatively poorly preserved Koobi Fora faunal assemblages is the 2,000+ piece sample from FxJj 50 (Okote Member, Koobi Fora Formation, dated *c.* 1.6 Ma). This fauna consists of the remains of at least 22 individual mammals and was formed on the floodplain of a stream channel, where hominins may have had access to the river cobbles that form the basis of the large stone tool assemblage found in association with the bones (Bunn, *et al.*, 1980).

As with FLK *Zinj*, FxJj 50 conforms to the expectations of a hominin "home base" or Type-C site, because there is a spatial association of abundant stone tools and fauna at the site–in addition to demonstrated causal links between these classes evidence, in the form of cutmarks and percussion marks on many bone surfaces (Bunn, 1981, 1982, 1997; Bunn *et al.*, 1980). Recently, a more detailed examination was made of a sample of these bone surface modifications, and it was concluded that most cutmarks in the FxJj 50 assemblage occur on the midshafts of upper and intermediate limb bone specimens (Domínguez-Rodrigo, 2002). This pattern is similar to that observed in the FLK *Zinj* fauna, and again, suggests early access by hominins to ungulate carcasses. In addition, tooth mark patterns at FxJj 50 argue against passive scavenging by hominins, with fewer midshafts preserving tooth marks than in experiments in which carcasses were ravaged solely by carnivores. Further, the epiphyseal and metaphyseal samples from FxJj 50 preserve tooth mark frequencies nearly identical to those observed in experimental settings in which hyenas ravaged carcasses *after* their use by hominins (e.g., Blumenschine, 1988, 1995; Capaldo, 1995).

Beyond these observations, Koobi Fora also preserves an interesting dichotomy in the material composition of penecontemporaneous archaeological sites (all occurring in deposits bracketed by the KBS Tuff [1.88 Ma] and the Okote Tuff [1.6 Ma]) between regions within the study area. Henry Bunn (1994) has documented a new kind of archaeological occurrence in the ephemeral lake margin contexts of the Koobi Fora Ridge and Ileret region (located in the western portion of the study area, along the shore of modern Lake Turkana). This new kind of occurrence is defined by cutmarked bone specimens *in the absence of associated stone tools*, and thus, is unlike sites in the fluvial contexts of the Karari Ridge (located in the northeastern portion of the study area), where these two classes of evidence (cutmarked bones and stone tools) co-occur within the same sites (e.g., FxJj 50).

Bunn interprets this disparity with reference to the paleogeography of the Turkana Basin *c.* 1.6 Ma. Unlike today, the prehistoric basin was dominated not by a large, permanent lake, but rather by the perennial ancestral Omo River, which flowed north to south through the region (Brown & Feibel, 1986; 1988; Feibel, 1988). The Karari Ridge area is nearby the proposed confluence of the axial drainage system of this ancient river. Thus, hominins carrying out activities in Karari Ridge *c.* 1.6 Ma had ready access to stone cobbles, carried in hydraulically from the eastern margins of the basin. One can therefore infer that there was little pressure for curation of artifacts at the Karari Ridge, and that this may explain the abundance of discarded tools at archaeological sites throughout the area. In contrast, the nearest source of stone tool raw materials available to early hominins in the ephemeral lake margin of the Koobi Fora Ridge was 15 km away, in gravel channels to the east and northeast. Bunn (1994) has thus suggested that hominins (sensibly) did not discard stone tools while foraging at the Koobi Fora Ridge, and that this probably explains the lack of artifacts in this area *c.* 1.6 Ma. The abundance of cutmarked bones from the Koobi Fora Ridge, however, does attest to the fact that hominins did possess and use stone tools in the region.

There is also a striking difference in the taxonomic composition of the fauna from the Koobi Fora and Karari Ridges. Most of the bones from the Koobi Fora Ridge are hippopotamus, while most of those from sites in the Karari are bovid. Bunn (1994: 261) has suggested that the near-lakeshore paleohabitat of the Koobi Fora Ridge may simply have supported more hippopotamuses than the Karari Ridge paleohabitat, and that hominins were thus "utiliz[ing] carcasses in proportion to their availability in the different areas." Alternatively, archaeological occurrences at the Koobi Fora Ridge "may document some single-event locations of the consumption of meat and/or marrow that did not involve significant transport of carcass parts. In contrast, the pattern at the Karari Ridge, based as it is on home base/central place sites, may document a later stage in the process of carcass utilization in which [hominins] transported more bovid bones and fewer hippopotamus bones to the known sites" (Bunn, 1994: 262).

A disparate pattern of hammerstone damage in the two sub-regions may support Bunn's latter suggestion. Nearly all limb bones from sites at the Karari Ridge were broken by hominins, while only just over half of the hippopotamus limb bones from the Koobi Fora Ridge sites have been breeched by hammerstone percussion. Bunn posits that the enormous amounts of skeletal muscle available on hippopotamus limbs might have offset the need for hominins to break open these bones for marrow. In addition, it is likely that the paucity of stone raw material at the Koobi Fora Ridge resulted in a short supply of suitable hammerstones available for hominin use in that region (N. Toth, personal communication). Finally, we believe that it is also possible that a "meat-stripping and marrow-abandonment" strategy could have been a response to the relatively open, exposed situations (i.e., potential carnivore pressure) in which hominins found themselves, if the Koobi Fora Ridge sites do, indeed, represent happened-upon carcasses[6]. In contrast, assuming the Karari Ridge sites represent hominin central place locales, it makes sense that there would be additional time in these *relatively* secure areas for more thorough exploitation of the transported parts of smaller (e.g., bovid) carcasses.

Finally, some of the best evidence of large mammal carcass exploitation by Oldowan hominins in East Africa, other than that from FLK *Zinj* and Koobi Fora, comes from the site of BK (*c.* 1.2 Ma), Bed II, Olduvai Gorge (Tanzania) (Monahan, 1996) and the ST site complex (*c.* 1.5 Ma) at Peninj (Tanzania) (Domínguez-Rodrigo *et al.*, 2002). Analyses of bone surface damage (see discussion above about tooth marks and cutmarks) in these assemblages indicate that hominins had primary access to fleshed (not ravaged) carcasses, and that carnivores subsequently scavenged the remnants of these carcasses. Even though published studies of the BK and ST site complex faunas are fairly recent (and, thus, have not yet received the same level of scrutiny as publications about FLK *Zinj*), they are already important simply because they expand the application of actualistic research on faunal assemblage formation beyond better known sites such as FLK *Zinj* and FxJj 50.

FARTHER AFIELD:
HOMININS IN THE STERKFONTEIN
VALLEY (SOUTH AFRICA)

The Sterkfontein Valley is home to a multitude of paleontological sites, which preserve the abundant remains of large terrestrial mammals, including hominins. Only three of these sites, however, also preserve Oldowan stone tool assemblages, Kromdraai A (Kuman *et al.*, 1997), Swartkrans (Members 1–3) (Clark, 1993) and Sterkfontein Member 5 (Oldowan Infill) (Kuman, 1994a, 1994b, 1998; Kuman & Clarke, 2000)[7]. It is not yet possible to date these breccia infills absolutely, although faunal indicators suggest the following broad age estimates: Kromdraai A, *c.* 2.0–1.0 Ma; Swartkrans Members 1–3, *c.* 1.8–1.0 Ma; Sterkfontein Member 5 (Oldowan Infill), *c.* 2.0–1.7 Ma (Brain, 1993a; Cooke, 1994; Kuman, 1994a, 1994b; McKee *et al.*, 1995; Vrba, 1981).

Researchers have established causal links (i.e., the observation of stone tool cutmarks) between the abundant faunas and stone tools in only two of the South

African Oldowan assemblages, Swartkrans Member 3 and Sterkfontein Member 5 (Oldowan Infill) (Brain, 1993b; Pickering, 1999)[8]. Interestingly, the only taxonomically identifiable hominin species recovered from both of these assemblages is *Australopithecus (Paranthropus) robustus* (e.g., Grine, 1989, 1993; Kuman & Clarke, 2000; Pickering, 1999). Further, studies of fossil hominin hand bones from Swartkrans suggest that individuals from whom these bones derived possessed all the requisite morphology to produce Oldowan stone tools (e.g., Susman, 1988a, 1988b, 1989).

This finding has led some researchers to conclude that *A. (P.) robustus* was the stone tool maker and user at Swartkrans. It is important to note, however, that early *Homo* was contemporaneous with *A. (P.) robustus* in the Sterkfontein Valley and at nearby sites, such as Drimolen, throughout the Plio-Pleistocene (e.g., Clarke, 1977a, 1977b; Clarke *et al.*, 1970; Grine, 1989, 1993; Keyser *et al.*, 2000). With this knowledge in mind, there are taphonomic reasons to question the assumption that an abundance of (taxonomically diagnostic) *A. (P.) robustus* craniodental material necessarily means that all the (taxonomically un-diagnostic) hominin postcranial material is also attributable to *A. (P.) robustus* (e.g., Trinkaus & Long, 1990; Pickering, 2001).

Regardless of the taxonomic identity of the stone tool maker/user at Swartkrans, the Member 3 cutmark evidence preserved there is dramatic because, although Members 1 and 2 at the site also preserve stone tools, no faunal remains in these earlier members have been found to display tool damage (Brain, 1981, 1993b). While the tool-modified bone sample from Member 3 is small (14 cutmarked pieces and two chopmarked pieces), a majority of the damaged bones (N=13) are long bone midshaft specimens.

Unfortunately, we are not able to assess the relative timing of hominin access to carcasses using this information in combination with models developed by one of us (Domínguez-Rodrigo, 1999a, 1999b; see discussion above). This is because the types (i.e., upper, intermediate or lower) of cutmarked long bone fragments in Member 3 have not yet been specified. However, efficient carcass foraging by Swartkrans hominins can be inferred by examination of another class of evidence preserved in the Member 3 assemblage.

Two hundred seventy burned bone pieces have been recovered from this deposit (Brain, 1993c; Brain & Sillen, 1988; Sillen & Hoering, 1993). Chemical and histological studies of these pieces suggest that they were heated in fires that reached temperatures of modern, human-tended *Celtis* wood fires. Further, these burned bone specimens were recovered from a six-meter-deep profile, throughout a laterally restricted area, which was a gully during Member 3 times. Referring specifically to excavation grid square W3/S3, Brain (1993c: 240) states:

"[B]urnt bones occur in 23 excavation spits, each

10 cm deep, indicating that the bones were heated in frequently recurring fires during the deposition period of this stratigraphic unit, which may have spanned several thousand years. The spatial distribution of the burnt bones suggests strongly that the fires occurred within the confines of the [gully], which was beneath a dolomite roof and had an inclined entrance towards the southeast."

In addition, the horizontal distribution of the burned bones *within* the gully might be viewed as supporting Brain's interpretation of these specimens (as evidence of hominin control of fire) as correct. Had the gully been wood-choked and naturally ignited in prehistory one might expect that bone across *every* excavation grid square would be burned for any particular horizon. This is not the case; twelve of the 30 gully excavation grid squares preserve no burned bone. Finally, several of the burned bone specimens are also cutmarked, further bolstering the image of technologically competent hominins, consuming the meat of large mammals, around tended fires during Member 3 times at Swartkrans (Brain, 1993b).

SUMMARY AND CONCLUSION

This brief but critical review of arguments about large mammal carcass acquisition and use by Oldowan hominins in eastern and southern Africa (c. 2.5–1.0 Ma), highlights the exciting actualistic research conducted by various zooarchaeologists over the past 20 years. These innovative researchers have taken taphonomic studies beyond the level of mere "cautionary tales", and have provided colleagues not only with hypotheses about early hominin foraging capabilities, but also with the methods and techniques to test these hypotheses. It is our view that the most important of these methods are utilized during naturalistic and experimental studies with modern carnivores and humans, and are predicated on the precise recording of intraskeletal and intra-elemental placement of taxonomically indicative bone surface modifications (i.e., carnivore toothmarks versus hominin tool damage)[9].

Rob Blumenschine and several of his colleagues took the early lead in much of this type of work, which resulted in models used to explain the anatomical distribution and frequency of carnivore tooth marks and hominin-induced percussion damage in the FLK *Zinj* archaeofauna. This important body of work, however, did not deal as specifically with another, major class of bone surface modification preserved in FLK *Zinj* faunal assemblage–stone tool cutmarks, damage that often indicates butchering activities by hominins.

Analyses of bone surface damage in the FLK *Zinj* and other important Oldowan faunal assemblages from East and South Africa (e.g., FxJj 50, Koobi Fora; BK, Olduvai; ST site complex, Peninj; and possibly Swartkrans Member 3) demonstrate that hominin-inflicted cutmarks occur predominately on the midshaft sections of upper and intermediate limb bone specimens at these prehistoric sites (e.g., Brain, 1993b; Bunn, 1981, 1982, 1997, 2001; Bunn & Kroll, 1986; Domínguez-Rodrigo *et al.*, 2002; Oliver, 1994; Monahan, 1996). It has been demonstrated, under naturalistic conditions, that large carnivores (such as modern lions) with primary access to dead ungulates seldom abandon such carcasses with substantial portions of meat still adhering to upper and intermediate limb bone midshafts (e.g., Domínguez-Rodrigo, 1999b). Thus, an abundance of cutmarked upper and intermediate limb bone midshaft specimens at FLK *Zinj* and other Oldowan archaeological sites, seems to indicate that hominins often gained early access to fully fleshed ungulate carcasses during the Early Stone Age in eastern and southern Africa. Such early access to carcasses by hominins implies active hunting and/or power scavenging, a term coined by Bunn (1996: 322) for "the aggressive, confrontational driving away of primary predators or primary scavengers at kills" (see also, Bunn, 2001; Bunn & Ezzo, 1993).

This does not mean that we accept the notion that Oldowan hominins never acquired substantial carcass resources from passive scavenging opportunities. Countless researchers have emphasized the inherent behavioral flexibility of various mammalian taxa, in general, and of large-brained primates (early hominins included among this group), more specifically.

This said, however, does not diminish the importance of the conclusion that early hominins possessed the capability to acquire large mammal carcasses soon after these animals died. It thus follows that Oldowan hominins also possessed fairly sophisticated cognitive capabilities, because early access to such carcasses by these relatively diminutive, clawless and dull-toothed creatures would have required the skilled use of tools as weaponry and/or group coordination during hunting and power scavenging episodes.

ACKNOWLEDGMENTS

We thank Nick Toth and Kathy Schick for the invitation to contribute to this volume. We recognize the great debt owed to all the paleoanthropologists mentioned herein; their work is admired and appreciated. TRP thanks Henry Bunn and Bob Brain, in particular, for all the valuable lessons and opportunities through the years. TRP thanks the Pickering family for unwavering support and understanding. MD-R thanks all the African archaeologists, especially those conducting zooarchaeological research, from whom he has learned a great deal.

ENDNOTES

1. In 1999, we recovered several cutmarked bone specimens (associated spatially with stone tools) as surface finds throughout the Gona study area, from archaeological localities at the same stratigraphic interval as the 2.5–2.6 Ma sites, based on preliminary outcrop tracing in the field. To date, however, no cutmarked or hammerstone percussed bone specimens have been recovered *in situ* from Pliocene deposits in the Gona study area. Archaeological sites slightly younger (all dated radioisotopically *c.* 2.3 Ma) than those from Gona occur in the Omo (Ethiopia), Hadar (Ethiopia) and West Turkana (Kenya) study areas. Of these, only the Hadar site, A.L. 666, has produced a bone specimen possibly modified by hominins (a bovid scapula with a suspected cutmark) (Kimbel *et al.*, 1996), even though other early stone tool sites such GaJh 5 (Lokalalei, West Turkana) and FtJj 1 and FtJj 5 (Omo) did preserve associated fauna (e.g., Kibunjia, 1994; Merrick, 1976; Merrick & Merrick, 1976; Howell *et al.*, 1987).

2. We note here that the debate about the formation of the FLK *Zinj* archaeofauna, sparked by Binford, was based on two separate classes of evidence— ungulate skeletal part profiles and bone surface modifications. Perceptively, all participants in this long-running discussion have made reference to *both* of these classes of evidence. To place too much weight on the interpretation of skeletal part profiles, at the expense of an emphasis on bone surface damage, severely restricts any conclusion about the formation of a fossil assemblage (see, e.g., White, 1992). We make this assertion for several reason, including, most importantly: (1) the growing realization, based on ethnoarchaeological studies, that there is no (reliable) static model of a "human pattern" of carcass and carcass part transport (e.g., Bartram, 1993; Domínguez-Rodrigo & Marti Lezana, 1996; contra, White, 1952; Perkins & Daly, 1968); and (2) the well-established fact that skeletal part ratios in faunal assemblages of disparate origins can resemble one another simply because many different taphonomic processes often remove the same less dense bones and bone portions from an assemblage, regardless of accumulating agent(s) (summarized in Lyman, 1984, 1994; Bartram & Marean, 1999). Thus, our discussion here, although often referring to specifics of skeletal part representation at Oldowan sites, maintains a focus on interpretations based primarily on patterns of bone surface damage.

3. This finding is contrary to what we view as the "traditional expectation" of hunting (a proxy term for early access to carcasses by hominins) being inferred based on a concentration of cutmarks on limb epiphyses—which is usually thought to indicate disjointing of limb units (e.g., Shipman, 1986).

4. The same will hold true for partially, but largely, defleshed limb bone midshafts. Bunn (2001: 206-207) articulates the idea well:

   "It is, of course, possible to start with a largely defleshed bone and then experimentally slice away at the visible muscle attachment areas, as Selvaggio (1994b) has shown…I would suggest, however, that butchers with any interest in preserving the sharpness of their knife blades are not going to repeatedly hack into the visible bone surfaces when the adhering meat can be shaved free without hitting the bone directly enough to produce cutmarks…Cutmarks are mistakes; they are accidental miscalculations of the precise location of the bone surface when muscle masses obscured it. As soon as a butcher can see the bone surface, few if any cutmarks will be inflicted thereafter in that area."

5. We realize that animals could have died naturally in places where hominins had previously or would eventually accumulate archaeofaunas, and that this phenomenon (rather than abandonment by hominins after flesh-stripping) could account for unbroken marrow bones at FLK *Zinj*. One way to provide support for the natural-death-source-of-whole-bones alternative is with the discovery of whole bones offsite in Bed I deposits at Olduvai. Such discoveries would broaden the environmental context of FLK *Zinj* and render whole bone representation at the site more coincident than it now seems. As the situation now stands (i.e., the absence of offsite discoveries of whole limb bone specimens), accepting this alternative will lead the skeptic to ask if animals were only dying at localities where hominins were accumulating carcass remains?

6. We do note, however, that predator risk was likely to have been minimal in prehistoric alluvial environments, based on analogy with modern lakeshore situations in East Africa—which support a low degree of carnivore overlap in use of space (e.g., Blumenschine, 1986b; Domínguez-Rodrigo, 2001). It is also important to acknowledge Monahan's (1998) reminder that hominins would have predicated their carcass and carcass part transport decisions based on a complex interplay of other variables *in addition to potential predation pressure*—including the number of hominin carcass-carriers and carcass size.

7. C.K. Brain (1958) recovered several stone tools— two of which are definitive (a chert flake and quartzite pebble) and a few others of which are more ambiguous—in the Kromdraai B deposit. We

also note that it is difficult to categorize definitively the stone tool assemblages from Kromdraai A and Swartkrans Member 3, which could be classified as Acheulean, rather than as Developed Oldowan (Kuman *et al.*, 1997; Clark, 1993).

8. Disappointingly, the 28,274-piece macromammalian faunal assemblage from the Sterkfontein Member 5 Oldowan Infill preserves only one definitively cutmarked bone specimen, a bovid scapula fragment (Pickering, 1999). An additional specimen, a bovid rib fragment, displays probable (but not definite) cutmarks. The re-deposited context of the Oldowan Infill stone tool assemblage, the inferred geomorphological setting of the cave during that time period (Kuman, 1994a, 1994b; Kuman & Clarke, 2000) and other taphonomic indicators (Pickering, 1999) all suggest hominins did not dwell in the cave during the deposition of the recovered materials; this may account for the paucity of cutmarked bone specimens in the Oldowan Infill assemblage when compared to the relative abundance of cutmarked specimens in Swartkrans Member 3 (see discussion in text), an assemblage inferred to have been created by hominin occupants of that cave, during Member 3 times (Brain, 1993b).

9. Many of the researchers mentioned in this paper are pioneers in this approach to questions in Early Stone Age archaeology, and it is encouraging to know that others are further refining recording processes and techniques (e.g., Abe *et al.*, 2000).

REFERENCES CITED

Abe, Y., Marean, C.W., Nilssen, P. & Stone, E. (2000). Taphonomy and zooarchaeology of the Die Kelders Cave 1 Middle Stone Age large mammal remains. *Journal of Human Evolution* **38**, A2.

Asfaw, B., White, T., Lovejoy, O., Latimer, B., Simpson, S. & Suwa, G. (1999). *Australopithecus garhi*: A new hominid from Ethiopia. *Science* **284**, 629-635.

Bartram, L.E. (1993). *An Ethnoarchaeological Analysis of Kua San (Botswana) Bone Food Refuse*. Ph.D. dissertation, University of Wisconsin-Madison.

Bartram, L.E. & Marean, C.W. (1999). Explaining the "Klasies Pattern": Kua ethnoarchaeology, the Die Kelders Middle Stone Age archaeofauna, long bone fragmentation and carnivore ravaging. *Journal of Archaeological Science* **26**, 9-29.

Binford, L.R. (1981). *Bones: Ancient Men and Modern Myths*. New York: Academic Press.

Binford, L.R. (1985). Human ancestors: Changing views of their behavior. *Journal of Anthropological Archaeology* **4**, 292-327.

Binford, L.R. (1986). Response to Bunn and Kroll. *Current Anthropology* **27**, 444-446.

Binford, L.R. (1988). The hunting hypothesis, archaeological methods and the past. *Yearbook of Physical Anthropology* **30**, 1-9.

Blumenschine, R.J. (1986a). Response to Bunn and Kroll. *Current Anthropology* **27**, 446.

Blumenschine, R.J. (1986b). *Early Hominid Scavenging Opportunities: Implications of Carcass Availability in the Serengeti and Ngorongoro Ecosystems*. Oxford: British Archaeological International Series, **283**.

Blumenschine, R.J. (1988). An experimental model of the timing of hominid and carnivore influence on archaeological bone assemblages. *Journal of Archaeological Science* **15**, 483-502.

Blumenschine, R.J. (1995). Percussion marks, tooth marks, and experimental determinations of the timing of hominid and carnivore access to long bones at FLK *Zinjanthropus*, Olduvai Gorge, Tanzania. *Journal of Human Evolution* **29**, 21-51.

Blumenschine, R.J. & Marean, C.W. (1993). A carnivore's view of archaeological bone assemblages. In (J. Hudson, Ed) *From Bones to Behavior: Ethnoarchaeological and Experimental Contributions to the Interpretation of Faunal Remains*. Carbondale (IL): Southern Illinois University Press, pp. 273-300.

Blumenschine, R.J. & Selvaggio, M.M. (1988). Percussion marks on bone surfaces as a new diagnostic of hominid behavior. *Nature* **333**, 763-765.

Boaz, N.T., Bernor, R.L., Brooks, A.S., Cooke, H.B.S., de Heinzelin, J., Dechamps, R., Delson, E., Gentry, A.W., Harris, J.W.K., Meylan, P., Pavlakis, P.P., Sanders, W.J., Stewart, K.M., Verniers, J., Williamson, P.G. & Winkler, A.J. (1992). A new evaluation of the significance of the Late Neogene Lusso Beds, Upper Semliki Valley, Zaire. *Journal of Human Evolution* **22**, 505-517.

Brain, C.K. (1958). *The Transvaal Ape-Man-Bearing Deposits*. Pretoria: Transvaal Museum.

Brain, C.K. (1970). New finds at the Swartkrans australopithecine site. *Nature* **225**, 1112-1119.

Brain, C.K. (1976). Some principles in the interpretation of bone accumulations associated with man. In (G.Ll. Isaac & E.R. McCown, Eds) *Human Origins: Louis Leakey and the East African Evidence*. Menlo Park (CA): Benjamin Cummings, pp. 96-116.

Brain, C.K. (1981). *The Hunters or the Hunted? An Introduction to African Cave Taphonomy*. Chicago: University of Chicago Press.

Brain, C.K. (1993a). Structure and stratigraphy of the Swartkrans Cave in the light of the new excavations. In (C.K. Brain, Ed) *Swartkrans: A Cave's Chronicle of Early Man*. Pretoria: Transvaal Museum, pp. 22-33.

Brain, C.K. (1993b). Taphonomic overview of the Swartkrans fossil assemblages. In (C.K. Brain, Ed) *Swartkrans: A Cave's Chronicle of Early Man*. Pretoria: Transvaal Museum, pp 255-264.

Brain, C.K. (1993c). The occurrence of burnt bones at Swartkrans and their implications for the control of fire by early hominids. In (C.K. Brain, Ed) *Swartkrans: A Cave's Chronicle of Early Man*. Pretoria: Transvaal Museum, pp. 229-242.

Brain, C.K. & Sillen, A. (1988). Evidence from the Swartkrans Cave for the earliest use of fire. *Nature* **336**, 464-466.

Brown, F.H. & Feibel, C.S. (1986). Revision of lithostratigraphic nomenclature in the Koobi Fora region, Kenya. *Journal of the Geological Society* **143**, 297-310.

Brown, F.H. & Feibel, C.S. (1988). "Robust" hominids and Plio-Pleistocene paleogeography of the Turkana Basin, Kenya and Ethiopia. In (F.E. Grine, Ed) *Evolutionary History of the "Robust" Australopithecines*. New York : Aldin de Gruyter, pp. 325-341.

Bunn, H.T. (1981). Archaeological evidence for meat-eating by Plio-Pleistocene hominids from Koobi Fora and Olduvai Gorge. *Nature* **291**, 574-577.

Bunn, H.T. (1982). *Meat-Eating and Human Evolution: Studies on the Diet and Subsistence Patterns of Plio-Pleistocene Hominids in East Africa*. Ph.D. dissertation. University of California, Berkeley.

Bunn, H.T. (1994). Early Pleistocene hominid foraging strategies along the ancestral Omo River at Koobi Fora, Kenya. *Journal of Human Evolution* **27**, 247-266.

Bunn, H.T. (1996). Response to Rose and Marshall. *Current Anthropology* **37**, 321-323.

Bunn, H.T. (1997). The bone assemblages from the excavated sites. In (G.Ll. Isaac & B. Isaac, Eds) *Koobi Fora Research Project: Volume 5, Plio-Pleistocene Archaeology*. Oxford: Claredon Press, pp. 402-444.

Bunn, H.T. (2001). Hunting, power scavenging, and butchering by Hadza foragers and by Plio-Pleistocene *Homo*. In (C.B. Stanford & H.T. Bunn, Eds) *Meat-Eating and Human Evolution*. New York: Oxford University Press, pp. 199-218.

Bunn, H.T. & Ezzo, J.A. (1993). Hunting and scavenging by Plio-Pleistocene hominids: Nutritional constraints, archaeological patterns, and behavioural implications. *Journal of Archaeological Science* **20**, 431-452.

Bunn, H.T., Harris, J.W.K., Isaac, G.Ll., Kaufulu, Z., Kroll, E.M., Schick, K.A., Toth, N. & Behrensmeyer, K. (1980). FxJj 50: An early Pleistocene site in northern Kenya. *World Archaeology* **12**, 109-136.

Bunn, H.T. & Kroll, E.M. (1986). Systematic butchery by Plio/Pleistocene hominids at Olduvai Gorge, Tanzania. *Current Anthropology* **27**, 431-452.

Capaldo, S.D. (1995). *Inferring Hominid and Carnivore Behavior from Dual-Patterned Archaeological Assemblages.* Ph.D. dissertation. Rutgers University.

Capaldo, S.D. (1997). Experimental determinations of carcass processing by Plio-Pleistocene hominids and carnivores at FLK 22 (*Zinjanthropus*), Olduvai Gorge, Tanzania. *Journal of Human Evolution* **33**, 555-597.

Clark, J.D. (1993). Stone artefact assemblages from Members 1 -3, Swartkrans Cave. In (C.K. Brain, Ed) *Swartkrans: A Cave's Chronicle of Early Man.* Pretoria: Transvaal Museum, pp. 167-194.

Clarke, R.J. (1977a). *The Cranium of the Swartkrans Hominid, SK 847, and Its Relevance to Human Origins.* Ph.D. dissertation. University of the Witwatersrand.

Clarke, R.J. (1977b). A juvenile cranium and some adult teeth of early *Homo* from Swartkrans, Transvaal. *South African Journal of Science* **73**, 46-49.

Clarke, R.J., Howell, F.C. & Brain, C.K. (1970). More evidence of an advanced hominid at Swartkrans. *Nature* **225**, 1217-1220.

Cooke, H.B.S. (1994). *Phacochoerus modestus* from Sterkfontein Member 5. *South African Journal of Science* **90**, 99-100.

de Heinzelin, J. (1994). Rifting, a long-term African story, with considerations on early hominid habitats. In (R.S. Corruccini and R.L. Ciochon, Eds) *Integrative Paths to the Past: Paleoanthropological Advances in Honor of F. Clark Howell.* Englewood Cliffs (NJ): Prentice Hall, pp. 313-320.

de Heinzelin, J., Clark, J.D., White, T., Hart, W., Renne, P., WoldeGabriel, G., Beyene, Y. & Vrba, E. (1999). Environment and behavior of 2.5-million-year-old Bouri hominids. *Science* **284**, 625-629.

Domínguez-Rodrigo, M. (1999a). Meat-eating and carcass procurement at the FLK *Zinj* 22 site, Olduvai Gorge (Tanzania): A new experimental approach to the old hunting-versus-scavenging debate. In (H. Ullrich, Ed) *Lifestyles and Survival Strategies in Pliocene and Pleistocene Hominids.* Schwelm: Edition Archaea, pp. 89-111.

Domínguez-Rodrigo, M. (1999b). Flesh availability and bone modifications in carcasses consumed by lions: Palaeoecological relevance in hominid foraging patterns. *Palaeogeography, Palaeoclimatology, Palaeoecology* **149**, 373-388.

Domínguez-Rodrigo, M. (2001). A study of carnivore competition in riparian and open habitats of modern savannas and its implications for hominid behavioral modeling. *Journal of Human Evolution* **40**, 77-98.

Domínguez-Rodrigo, M. (2002). Hunting and scavenging in early humans: The state of the debate. *Journal of World Prehistory* (in press).

Domínguez-Rodrigo, M., de Luque, L., Alcalá, L., de la Torre Sainz, I., Mora, R., Serrallonga, J. & Medina, V. (2002). The ST site complex at Peninj, West Lake Natron, Tanzania: Implications for early hominid behavioral models. *Journal of Archaeological Science* **29**, 639-665.

Domínguez-Rodrigo, M & Marti Lezana, R. (1996). Estudio etnoarqueológico de un campamento temporal Ndorobo (Maasai) en Kulalu (Kenia). *Trabajos de Prehistoria* **53**, 131-143.

Etkin, W. (1954). Social behavior and the evolution of man's mental faculties. *The American Naturalist* **88**, 129-142.

Feibel, C.S. (1988). *Paleoenvironments of the Koobi Fora Formation, Turkana Basin, Northern Kenya.* Ph.D. dissertation. University of Utah.

Grine, F.E. (1989). New hominid fossils from the Swartkrans Formation (1979-1986 excavations): Craniodental specimens. *American Journal of Physical Anthropology* **79**, 409-449.

Grine, F.E. (1993). Description and preliminary analysis of new hominid craniodental fossils from the Swartkrans Formation. In (C.K. Brain, Ed) *Swartkrans: A Cave's Chronicle of Early Man.* Pretoria: Transvaal Museum, pp. 75-116.

Hillson, S. (1996). *Mammal Bones and Teeth: An Introductory Guide to Methods of Identification.* Dorchester: Dorset Press.

Howell, F.C., Haesaerts, P. & de Heinzelin, J. (1987). Depositional environments, archaeological occurrences and hominids from Members E and F of the Shungura Formation (Omo Basin, Ethiopia). *Journal of Human Evolution* **16**, 665-700.

Isaac, G.Ll. (1978). The food-sharing behavior of proto-human hominids. *Scientific American* **238**, 90-108.

Isaac, G.Ll. (1984). The archaeology of human origins: Studies of the Lower Pleistocene in East Africa 1971 - 1981. *Advances in Archaeology* **3**, 1-87.

Keyser, A.W., Menter, C.G., Moggi-Cecchi, J., Pickering, T.R. & Berger, L.R. (2000). Drimolen: A new hominid-bearing site in Gauteng, South Africa. *South African Journal of Science* **96**, 193-197.

Kibunjia, M. (1994). Pliocene archaeological occurrences in the Lake Turkana basin. *Journal of Human Evolution* **27**, 159-171.

Kimbel, W.H., Walter, R.C., Joahnson, D.C., Reed, K.E., Aronson, J.L., Assefa, Z., Marean, C.W., Eck, G.G., Bobe, R., Hovers, E., Rak, Y., Vondra, C., Yemane, T., York, D., Chen, Y., Evenson, N.M. & Smith, P.E. (1996). Late Pliocene *Homo* and Oldowan tools from the Hadar Formation (Kada Hadar Member), Ethiopia. *Journal of Human Evolution* **31**, 549-561.

Kuman, K. (1994a). The archaeology of Sterkfontein: Preliminary findings on site formation and cultural change. *South African Journal of Science* **90**, 215-220.

Kuman, K. (1994b). The archaeology of Sterkfontein–past and present. *Journal of Human Evolution* **27**, 471-495.

Kuman, K. (1998). The earliest South African industries. In (M.D. Petraglia & R. Korisettar, Eds) *Early Hominid Behavior in Global Context: The Rise and Diversity of the Lower Palaeolithic Record.* London: Routledge Press, pp. 151-186.

Kuman, K. & Clarke, R.J. (2000). Stratigraphy, artefact industries and hominid associations for Sterkfontein, Member 5. *Journal of Human Evolution* **39**, 827-847.

Kuman, K., Field, A.S. & Thackeray, J.F. (1997). Discovery of new artefacts at Kromdraai. *South African Journal of Science* **93**, 187-193.

Leakey, M.D. (1966). A review of the Oldowan Culture from Olduvai Gorge, Tanzania. *Nature* **210**, 462-466.

Leakey, M.D. (1971). *Olduvai Gorge: Excavations in Beds I and II, 1960 - 1963.* Cambridge: Cambridge University Press.

Lyman, R.L. (1984). Bone density and differential survivorship of fossil classes. *Journal of Anthropological Archaeological* **3**, 259-299.

Lyman, R.L. (1994). *Vertebrate Taphonomy.* Cambridge: Cambridge University Press.

Maguire, J.M., Pemberton, D. & Collett, M.H. (1980). The Makapansgat Limeworks Grey Breccia: Hominids, hyaenas, hystricids or hillwash? *Palaeontologia Africana* **23**, 75-98.

Marean, C.W., Spencer, L.M., Blumenschine, R.J. & Calpado, S.D. (1992). Captive hyaena bone choice and destruction, the schlepp effect and Olduvai archaeofaunas. *Journal of Archaeological Science* **19**, 101-121.

McKee, J., Thackeray, J.F. & Berger, L.R. (1995). Faunal assemblage seriation of southern African Pliocene and Pleistocene fossil deposits. *American Journal of Physical Anthropology* **96**, 235-250.

Merrick, H.V. (1976). Recent archaeological research in the Plio-Pleistocene deposits of the Lower Omo, southwestern Ethiopia. In (G.Ll. Isaac and I. McCown, Eds) *Human Origins: Louis Leakey and the East African Evidence.* Menlo Park: W.A. Benjamin, pp. 461-482.

Merrick, H.V. & Merrick, J.P.S. (1976). Archaeological occurrences of earlier Pleistocene age from the Shungura Formation. In (Y. Coppens, F.C. Howell, G.Ll. Isaac and R.E.F. Leakey, Eds) *Earliest Man and Environments in the Lake Rudolf Basin.* Chicago: University of Chicago Press, pp. 574-584.

Monahan, C.M. (1996). New zooarchaeological data from Bed II, Olduvai Gorge, Tanzania: Implications for hominid behavior in the Early Pleistocene. *Journal of Human Evolution* **31**, 93-128.

Monahan, C.M. (1998). The Hadza carcass transport debate revisited and its archaeological implications. *Journal of Archaeological Science* **25**, 405-424.

Oliver, J.S. (1994). Estimates of hominid and carnivore involvement in the FLK *Zinjanthropus* fossil assemblage: Some socioecological implications. *Journal of Human Evolution* **27**, 267-294.

Perkins, D. & Daly, P. (1968). A hunters' village in Neolithic Turkey. *Scientific American* **219**, 97-106.

Pickering, T.R. (1999). *Taphonomic Interpretations of the Sterkfontein Early Hominid Site (Gauteng, South Africa) Reconsidered in Light of Recent Evidence.* Ph.D. dissertation. University of Wisconsin-Madsion.

Pickering, T.R. (2001). Taphonomy of the Swartkrans hominid postcrania and its bearing on issues of meat-eating and fire management. In (C.B. Stanford and H.T. Bunn, Eds) *Meat-Eating and Human Evolution.* New York: Oxford University Press, pp. 33-51.

Pickering, T.R. & Wallis, J. (1997). Bone modifications resulting from captive chimpanzee mastication: Implications for the interpretation of Pliocene archaeological faunas. *Journal of Archaeological Science* **24**, 1115-1127.

Piperno, M. (1989). Chronostratigraphic and cultural framework of the *Homo habilis* sites. In (G. Giacobini, Ed) *Hominidae: Proceedings of the 2ⁿᵈ International Congress of Human Paleontology.* Milan: Jaca Book, pp. 189-195.

Plummer, T.W. & Stanford, C.B. (2000). Analysis of a bone assemblage made by chimpanzees at Gombe National Park, Tanzania. *Journal of Human Evolution* **39**, 345-365.

Potts, R.B. (1988). *Early Hominid Activities at Olduvai.* New York: Aldine de Gruyter.

Potts, R.B. & Shipman, P. (1981). Cutmarks made by stone tools on bones from Olduvai Gorge, Tanzania. *Nature* **291**, 577-580.

Roche, H. (1989). Technological evolution in early hominids. *Ossa* **4**, 97-98.

Roche, H. (1996). Remarque sur les plus anciennes industries en Afrique et en Europe. In *Colloquium VIII: Lithic Industries, Language and Social Behaviour in the First Human Forms.* Forli: IUPSS Congress, pp. 55-68.

Schick, K.D. & Toth, N. (1993). *Making Silent Stones Speak: Human Evolution and the Dawn of Technology.* New York: Touchstone.

Selvaggio, M.M. (1994a). *Identifying the Timing and Sequence of Hominid and Carnivore Involvement with Plio-Pleistocene Bone Assemblages from Carnivore Tooth Marks and Stone Tool Butchery Marks on Bone Surfaces.* Ph.D. dissertation. Rutgers University.

Selvaggio, M.M. (1994b). Carnivore tooth marks and stone tool butchery marks on scavenged bone: Archaeological implications. *Journal of Human Evolution* **27**, 215-227.

Semaw, S. (2000). The world's oldest stone artefacts from Gona, Ethiopia: Their implications for understanding stone technology and patterns of human evolution between 2.6 - 1.5 million years ago. *Journal of Archaeological Science* **27**, 1197-1214.

Semaw, S., Renne, P., Harris, J.W.K., Feibel, C.S., Bernor, R.L., Fesseha, N. & Mowbray, K. (1997). 2.5-million-year-old stone tools from Gona, Ethiopia. *Nature* **385**, 333-336.

Shipman, P. (1986). Scavenging or hunting in early hominids: Theoretical framework and tests. *American Anthropologist* **88**, 27-43.

Sillen, A. & Hoering, T. (1993). Chemical characterization of burnt bones from Swartkrans. In (C.K. Brain, Ed) *Swartkrans: A Cave's Chronicle of Early Man.* Pretoria: Transvaal Museum, pp 243-249.

Stanford, C.B. & Bunn, H.T. (Eds). (2001). *Meat-Eating and Human Evolution.* New York: Oxford University Press.

Susman, R.L. (1988a). Hand of *Paranthropus robustus* from Member 1, Swartkrans: Fossil evidence for tool behavior. *Science* **240**, 781-784.

Susman, R.L. (1988b). New postcranial remains from Swartkrans and their bearing on the functional morphology and behavior of *Paranthropus robustus*. In (F.E. Grine, Ed) *Evolutionary History of the "Robust" Australopithecines*. New York: Aldine de Gruyter, pp. 149-172.

Susman, R.L. (1989). New hominid fossils from the Swartkrans Formation (1979-1986 excavations): Postcranial specimens. *American Journal of Physical Anthropology* **79**, 451-474.

Tappen, M. & Wrangham, R. (2000). Recognizing hominoid-modified bones: The taphonomy of colobus bones partially digested by free-ranging chimpanzees in the Kibale Forest, Uganda. *American Journal of Physical Anthropology* **113**, 214-234.

Trinkaus, E. & Long, J.C. (1990). Species attribution of the Swartkrans Member 1 first metacarpals: SK 84 and SK 5020. *American Journal of Physical Anthropology* **83**, 419-424.

Vrba, E.S. (1981). The Kromdraai australopithecine site revisited in 1980: Recent excavations and results. *Annals of the Transvaal Museum* **33**, 17-60.

Washburn, S.L. (1959). Speculations on the interrelations of the history of tools and biological Evolution. In (J.N. Spuhler, Ed) *The Evolution of Man's Capacity for Culture.* Detroit: Wayne State University Press, pp. 21-31.

Washburn, S.L. & Howell, F.C. (1960). Human evolution and culture. In (S. Tax, Ed) *Evolution after Darwin: Vol. II, The Evolution of Man.* Chicago: University of Chicago Press, pp. 33-56.

Washburn, S.L. & Lancaster, C. (1968). The evolution of hunting. In (R. Lee R & I. DeVore, Eds). *Man the Hunter.* New York: Aldine Press, pp. 293-303.

White, T.D. (1992) *Prehistoric Cannibalism at Mancos 5MTUMR-2346.* Princeton: Princeton University Press.

White, T.E. (1952). Observations on the butchering techniques of some aboriginal peoples, Part 1. *American Antiquity* 337-338.

CHAPTER 5

AFTER THE AFRICAN OLDOWAN: THE EARLIEST TECHNOLOGIES OF EUROPE

BY FERNANDO DÍEZ-MARTÍN

ABSTRACT

The starting point of this paper is the issue of the first occupation of Europe, which has been subject of an intense debate for the past few decades and has been polarized by two antagonistic perspectives: the 'long chronology' and the 'short chronology'. To present the state of our knowledge on this topic, a brief summary of some recent key publications is considered. Then, the geological and chronological framework of the earliest European sites with Mode 1 features is provided along with a general overview of the technological characteristics observed in them. Finally, some of the possible explanatory hypotheses of these pre-500 Ka. European Mode 1 industries are presented and briefly discussed.

KEYWORDS:

Europe, Lower and Middle Pleistocene, Mode 1 technologies, Lower Palaeolithic, variability, occupation.

INTRODUCTION: THE EARLIEST OCCUPATION OF EUROPE, A MATTER OF CONTROVERSY

At the beginning of the 1990's, the idea of an early human colonization of Europe still was a strong paradigm among some scholars. This situation was possible in part due to a tradition that provided a long list of Lower Pleistocene sites and isolated finds scattered around the continent. The meeting on *The First Europeans*, held in 1989 (Bonifay & Vandermeersch, eds., 1991), is a significant example of how the archae-ological record then supported this interpretation. The French Massif Central and Southeastern regions seemed to provide a wealth of early evidence, some of which could be taken back to the Upper Pliocene (among them, Saint-Elbe, Soleilhac and Chilhac are the best known cases). All over Europe there were sites that, considering their absolute dating, faunal remains, geostratigraphy or, simply, their technological features, seemed to date to the Lower and Early Middle Pleistocene, such as the cases of Vallonet (southeastern France), El Aculadero and Orce (southern Spain), Isernia and Monte Poggiolo (Italy), Kärlich (Germany) or Stránská Skála (Czech Republic). Finally, Lower Pleistocene fluvial terraces completed this picture with a high density of sites, including those within river deposits in Spain, Italy and France.

All these data seemed to lend consistent support to the human occupation of Europe almost immediately after the African origin of the genus *Homo*. Although such expansion eventually spread throughout Europe, the main accumulation of discoveries was located in the Mediterranean regions. Every single archaeological case exhibited a technology similar to that seen at the early African sites: Oldowan-like choppers, polyhe-drons, discoids, débitage and casually retouched flakes. From this perspective, this 'simple' or 'Mode 1' technology (industries characterized by Oldowan-like cores such as 'choppers,' 'discoids,' 'scrapers,' etc., and an assortment of flakes and flake fragments, as well as a lack of more formal tools such as handaxes, cleavers, or picks) had been present in Europe for a long period of time. Moreover, these had been replaced by the Acheulean or 'Mode 2' industries containing such formal tools only at the beginning of the Middle Pleistocene (Bonifay & Vandermeersch, 1991).

This scenario, the **'long chronology' hypothesis** (also called the 'mature Europe' or 'old Europe' hypothesis), was never unanimously accepted (Dennell, 1983) and came into serious dispute in 1993. That year witnessed a scientific meeting at Tautavel (France) that brought together an important and heterogeneous group of European prehistorians who critically reviewed the issue of the first occupation of Europe (Roebroeks & van Kolfschoten, eds., 1995). By reviewing the archaeological assemblages related to its earliest occupation, almost all of these scholars concluded that the long chronology scenario was untenable. Roebroeks and van Kolfschoten (1994) used this new perspective to introduce their **'short chronology' hypothesis**, in which they stated that the first European occupation took place at about 500 Ka. ago and that before this date Europe had been almost empty. The new paradigm was based on the assumption that there was no irrefutable evidence of such an ancient human presence before 0.5 Ma. No human fossils had been found and the archaeological

sites referred as "very old" were considerably problematic due to a number of reasons: in some cases they came from disturbed high-energy contexts or from sediments not securely dated; in others, lithic assemblages were apparently natural and not humanly manufactured; and finally, many finds from fluvial terraces and other areas were isolated pieces and therefore non-diagnostic.

These arguments had a "ripple effect" that seemed rapidly to weaken the Old Europe view. Certainly, a very consistent point in favor of the new scenario was the absence of human remains before the Middle Pleistocene period. The **Orce** (Spain) skull fragment, found in sediments dated to about 1.07 Ma., had generated an intense controversy between those scientists who defended the human status of the fossil (Gibert *et al.*, 1994) and those who preferred to see it as a juvenile equid (Agustí & Moyà-Solà, 1987). These days, most researchers are inclined to reject this evidence as human (Moyà-Solà & Köhler, 1997; Gibert *et al.* 1998a). At the same time, the phalanx recovered in the Spanish Lower Pleistocene karstic site of **Cueva Victoria** (Palmqvist *et al.*, 1996a) is also debatable.

Table 1

site	country	chronology proposed	dating method	lithic assemblage (no.)	problematic	references
Prezletice	Czech Republic	890-600 Ka.	BS/PM	Mode 1 (870)	doubtful industry	Valoch, 1995
Stránská Skála	Czech Republic	>780 Ka.	BS/PM	Mode 1	doubtful industry	Valoch, 1995
Chilhac III	France	1.9-1.8 Ma.	BS	Mode 1 (48)	secondary site	Chavaillon, 1991
Soleilhac	France	900 Ka.	BS/ME/PM	? (400)	unclear sequence, biostratigraphy	Raynal *et al.*, 1995
Vallonet	France	900 Ka.	BS/PM/ESR	Mode 1 (59)	doubtful industry secondary site?	Lumley *et al.*, 1988
Kärlich A	Germany	>780 Ka.	BS/PM	? (3)	doubtful and scarce industry	Bosinski, 1995
Colle Marino	Italy	>700 Ka.	ME	—	reliability of regional correlations	Mussi, 1995
Monte Peglia	Italy	>780 Ka.	BS	? (5)	doubtful industry	Mussi, 1995
El Aculadero	Spain	780-500 Ka.	ME	Mode 1 (2769)	stratigraphic sequence revision	Raposo & Santonja, 1995
Cúllar-Baza I	Spain	780-500 Ka.	BS/ME	? (6)	scarce assemblage	Santonja & Villa, 1990
Duero/Guadalquivir (upper terraces)	Spain	780-500 Ka.	ME	-	isolated pieces	Santonja & Villa, 1990
Korolevo VII/VIII	Ukraine	>780 Ka.	PM/TL	Mode 1 (1900)	ambiguous dating	Gladiline & Sitlivy, 1990

Table 1: Some problematic early European sites in stratigraphic context (dating method: BS = biostratigraphy; ESR = electro spin resonance; ME = morphostratigraphy; PM = paleomagnetism; TL = thermoluminiscence).

Figure 1

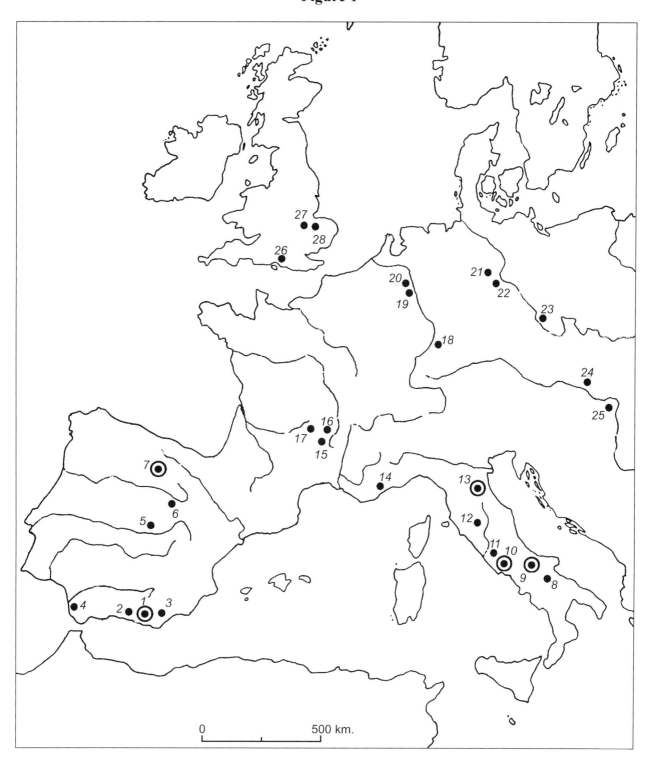

1. Geographic location of the European sites mentioned in the text:
    SPAIN:1. Barranco León and Fuente Nueva 3; 2. Orce; 3. Cúllar-Baza I; 4. El Aculadero; 5. Áridos; 6. Ambrona;
    7. Atapuerca TD6. ITALY: 8. Notarchirico; 9. Isernia la Pineta; 10. Ceprano; 11. Colle Marino; 12. Monte Peglia;
    13. Monte Poggiolo. FRANCE: 14. Vallonet; 15. Soleilhac; 16. Chilhac; 17. Saint-Elbe. GERMANY: 18. Mauer;
    19. Miesenheim; 20. Kärlich; 21. Schöningen; 22. Bilzingsleben. CZECH REPUBLIC: 23. Prezletice; 24. Stránská
    Skála. HUNGARY: 25. Vértesszölös. GREAT BRITAIN: 26. Boxgrove; 27. High Lodge; 28. Barnham.

Some of the major site occurrences that, after further scrutiny, totally or partially confirm Roebroeks and van Kolfschoten's 'short chronology' hypothesis will be discussed here (a more extensive list of sites and their characteristics is presented in table 1; also see figure 1 for the location of the archaeological sites mentioned). The French Massif Central has currently lost its status as a source for early Palaeolithic sites. Many of the reported localities were indeed very small collections or even isolated pieces that, on occasion, might be explained as geofacts, naturally produced by volcanism in the region (Raynal *et al.*, 1995). Some of the better-known sites have important problems and, so far, do not provide consistent evidence. For example, **Chilhac III**, a site that has provided a small collection of doubtless artifacts (Chavaillon, 1991: 82), has had significant stratigraphic disturbance. The association of lithics and Villafranchian fauna (Late Pliocene and Early Pleistocene) supported a date of 1.8 Ma. for this site based on biochronology. However, the deposits in which lithic implements and fauna have been found seem to have been affected by solifluxion and transport that could well have mixed materials of different ages (Villa, 1991: 211; Raynal *et al.*, 1995: 138).

Soleilhac, for its part, is poorly known. The lithic collection has not been published in detail and, even though an age of 0.9 Ma. has been proposed (Bonifay, 1991: 70), this date is inconclusive. The stone tools are associated with a post-Villafranchian fauna, a not very precise biochronological marker that could easily be of Middle Pleistocene age (Roebroeks & van Kolfschoten, 1994: 498). Paleomagnetic and morphostratigraphic data, which supports the older date, can only be taken as tentative because the geological history of the site is poorly known (Raynal *et al.*, 1995: 139-140).

Vallonet Cave, in southwest France, is a site frequently brought to this debate. The fertile deposits are located in a small 5 m long area and have been dated consistently at 0.9 Ma. based on biostratigraphy, palaeomagnetism and ESR dates (de Lumley *et al.*, 1988; Yokoyama *et al.*, 1988). However, the question here does not pertain to the chronology but rather to the nature of the lithic assemblage. The 59 supposedly artifactual objects (76% made from limestone), come from deposits that contain sands and limestone cobbles. Natural sedimentary processes such as wave action could have produced natural fractures in the pebbles in this matrix. This suggests that the lithic assemblage is in fact a collection of non-artifactual objects (Roebroeks & van Kolfschoten, 1994). Other criticism of this site has focused on the fact that the supposed archaeological accumulation might be to some extent a secondary context and mixed assemblage (Villa, 1996: 71).

El Aculadero, in southern Spain, had been considered by many to be the most plausible candidate for an ancient human settlement in Iberia (Querol & Santonja, 1983), having been assigned to stage III of the now outdated *galet aménagé* culture of North Africa (Biberson, 1961). Although this site did not provide absolute dating or faunal remains, its chronological framework was estimated by considering the typological characteristics observed in the large lithic collection (2769 artifacts including choppers and flakes), and by the data provided by the morphostratigraphic sequence. Thus, a pre-Acheulean and pre-Middle Pleistocene framework was proposed (Santonja & Villa, 1990: 53). More recently, the geomorphologial sequence has been revised and, as result, the site is now considered to be significantly younger, dating to the end of the Pleistocene (Raposo & Santonja, 1995: 18).

It seemed that, taking into account all this evidence (or better, the lack of it), the short chronology hypothesis was a very strong and accurate scenario to explain the first human occupation of Europe. According to this perspective, at about 0.5 Ma. the species *Homo heidelbergensis* extensively occupied the continent, producing a technology related to the fully developed Acheulean complex. The main archaeological evidence confirming this would be the British site of **Boxgrove** (Roberts *et al.*, 1995: 171-172) — a primary context site in which handaxes, associated with a *H. heidelbergensis* tibia, were produced — and the **Mauer** (Germany) mandible, both correlated with oxygen isotope stage 13, between 524 and 478 Ka[1]. In addition, after this date, Acheulean sites are very common all over Europe (see various contributions in Roebroeks & van Kolfschoten, eds., 1995).

Unfortunately, the pristine consistency of this proposal did not last long. In 1995 the research team working at the Spanish site complex of **Atapuerca** published their new discoveries from a 6-m² test excavation at **Gran Dolina** site, level **TD6** (Carbonell *et al.*, 1995a; Parés & Pérez-González, 1995). According to the new information available, stone tools, fauna and human fossils, ascribed to the new species *Homo antecessor* (Bermúdez de Castro *et al.*, 1997), were found at Aurora stratum, a layer bracketed dating to the Lower Pleistocene period, before the Matuyama/Brunhes transition at 780 Ka. This new information had considerable impact in the European paleoanthropological community. Apparently, this was the first time that a site could provide all the elements to refute the short chronology and to please the most skeptic prehistorians. Immediately, the TD6 occurrences supported and revitalized an idea that had been circulating before (Villa, 1983: 12-14): the first occupation of Europe took place at about 1 Ma. ago. The Atapuerca team defended this perspective in their 'mature Europe' hypothesis (Carbonell *et al.*, 1995b) (figure 2).

The 'mature Europe' or 'long chronology hypothesis' has been supported by other evidence that seems to

---

[1] The Italian site of **Notarchirico** could well fit within this chronological framework. Although radiometric analyses have provided some contradict dates, the microfaunal assemblage suggests a post-500 Ka. chronology for the entire series (Mussi, 1995:32). The lowermost stratigraphic levels include handaxes and they are interstratified with other levels in which only cores and flakes occur (Piperno *et al.*, 1999).

Figure 2

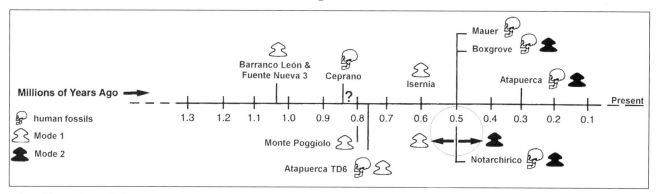

2.   *Chronological distribution of Mode 1 and Mode 2 sites in the European record.*

be more consistent than that previously dismissed by Roebroeks and van Koflschoten. On the one hand, the human cranium discovered at the Italian locality of **Ceprano** (Ascenci *et al.*, 2000) has an estimated age of 900-800 Ka., based on regional geological correlations. The sediments bearing the Ceprano calvaria are considered to be slightly younger than layers containing volcaniclasts in the Priverno basin and dated by Ar/Ar analysis to around 1.0 Ma. The morphological traits observed in the fossil are in agreement with this age. Recently, it has been remarked that the Ceprano specimen represents a 'bridge' between *H. ergaster/erectus* and *H. heidelbergensis* (Manzi *et al.*, 2001) and, therefore, the hypothesis that the Italian fossil might be the first adult cranial specimen of *H. antecessor* is possible (it must be remembered that a similar phylogenetic bridge pattern was suggested for *H. antecessor* morphology). On the other hand, further and more intensive investigation at sites like **Fuente Nueva 3** or **Monte Poggiolo** is of importance here and will be discussed below.

Although, at present, the debate on this issue is quite polarized between those who still support the 'short chronology' scenario (Roebroeks & van Kolfschoten, 1998; Gamble, 1999: 115-123) and those who back the 'mature Europe' hypothesis, it seems evident that Roebroek's and van Kolfschoten's perspective (at least in its original formulation) does not constitute the most useful model. Unexpectedly, the debate on the first human occupation of Europe has become more lively and the evidence more complex than ever. According to the archaeological record at hand, discussion of the *first* occupation of Europe primarily refers to Southern/Mediterranean Europe. It is important to note that this region may be more exposed to the influences of the first human settlement in North Africa (Sahnouni & Heinzelin, 1998; Raynal *et al.*, 2001) or the Caucasus (Gabunia *et al.*, 2000) and is certainly ecologically more hospitable for habitation by early human groups than are northern latitudes (Turner, 1992; 1999a).

THE EARLIEST ARCHAEOLOGICAL SITES AND THEIR CONTEXTS

This chapter will focus on the archaeological sites that provide the most consistent evidence supporting the mature Europe hypothesis. Although not everyone agrees with this perspective (see especially Roebroeks & van Kolfschoten, 1998), the available information firmly brackets these sites within a period of time that precedes 500 Ka. The archaeological sites to be considered within this time span (1.0 Ma. to 0.5 Ma.) are, from older to younger: **Barranco León, Fuente Nueva 3, Atapuerca TD6, Monte Poggiolo**, and **Isernia la Pineta.**

Fuente Nueva 3 and Barranco León, Spain

Both of these Mode 1 sites are located in the northeastern area of the Guadix-Baza Basin (Spain), well known for its rich and complete Pleistocene macromammal sequence (Martínez-Navarro *et al.*, 1997). The sites form part of the lacustrine sediments of the Baza Formation, which contain limestones, carbonate silts and dark mudstones. The stratigraphic section at Fuente Nueva 3 has three fossiliferous strata deposited in a low-energy situation: level 1 is a 5 cm thick layer rich in faunal remains; level 2 is 2-5 cm thick and is an archaeological stratum with lithic objects; level 3 contains both fauna and artifacts. The absence of tooth marks and the low percentage of carnivores indicate that the accumulation of the fauna was not due to predator activity. At Barranco León there is a 35 cm thick level of fine-grained sand, containing archaeological remains. The correlation between both sites has been impossible so far due to lateral facies changes, but it has been suggested that Barranco León might be stratigraphically below Fuente Nueva 3 (Oms *et al.*, 2000).

The chronological framework of these two sites has been estimated by a number of different methods. First, the macromammal fauna at Fuente Nueva 3 shows a Lower Pleistocene context (Martínez-Navarro *et al.*, 1997: 615-617) related to the much better known assemblages of Fuente Nueva 2 and Venta Micena (Palmqvist *et al.*, 1996b). In terms of rodents, the presence of *Allophaiomys bourgondiae* in the micromammal assemblage, more primitive than the *Microtus nivaloides* found at Vallonet and Atapuerca TD6, confirms that the biostratigraphic position of Fuente Nueva 3 (and subsequently Barranco León) should be located in an earlier Lower Pleistocene period. The first palaeomagnetic studies at FN3 showed a succession of reversed strata in the sequence, interpreted as part of the Matuyama chron (Martínez-Navarro *et al.*, 1997: 613). However, as the first analysis consisted of only 24 samples, new palaeomagnetic studies have been carried out recently, in which a total of 110 samples were analyzed at both sites (Oms *et al.*, 2000). This new study confirmed the reversed magnetization throughout the two sections and a Matuyama placement for both sites. These results, along with the biostratigraphic correlation, suggest an age that could be older than the Jaramillo normal subchron (0.99-1.07 Ma).

Gran Dolina TD4 and TD6, Spain

Gran Dolina is one of the sites located in the Sierra de Atapuerca archaeological complex (Spain). It is an 18 m thick sediment-filled gallery whose stratigraphic section has been exposed by a railway trench that cuts through the southwest area of the Sierra, made of Cretaceous limestone along with its associated cave breccias (Carbonell *et al.*, 1999a: 316). The cave infill is divided into 11 levels from, ascending TD1 to TD11, and contains both interior (TD1 and TD2) and exterior sediments (the rest of the sequence) (Parés & Pérez-González, 1999: 330-332). The levels with archaeological remains are TD4, TD6, TD10 and TD11. Only the first two levels fall in the scope of this paper. TD3-4 is a 2m thick unit consisting of sandy lutite that contains limestone clasts and has provided a small collection of five, doubtless human-made, lithic artifacts (Carbonell & Rodríguez, 1994). TD6 is a clastic unit that includes the Aurora stratum, a 20 cm thick lutite (fine-grained sedimentary) layer with limestone clasts that contains the important association of *Homo antecessor* fossils (Bermúdez de Castro *et al.*, 1997) and Mode 1 stone tools.

The chronology of the sequence has been assessed by three different methods. Paleomagnetic data show that at level TD7 there is a well-defined normal to reverse polarity switch that has been interpreted as the boundary between Matuyama and Brunhes chrons, that is to say, the transition between Lower to Middle Pleistocene at 780 Ma (Parés & Pérez-González, 1995; 1999). According to this information, the archaeological strata at TD4 and the overlying TD6, then, are older than this date. Micromammal biostratigraphy supports this chronological framework as well (Cuenca-Bescós *et al.*, 1999). The TD3-TD8 sedimentary section includes the rodent *Mimomys savini*, an important stratigraphic marker replaced by *Arvicola cantiana* during

the Middle Pleistocene (Cromerian complex). No traces of the latter have been found in the Gran Dolina sequence, but the concurrence of a primitive *Mimomys* and an evolutionary stage of *Microtus* at TD5 and TD6 is related to the Late Biharian biochron at the end of the Lower Pleistocene (780-857 Ka.). Finally, a combination of U-series and ESR analyses has been carried out on fossil teeth at different stratigraphic units. The results for TD6 provided a mean age of 731± 63 Ka., which is in agreement with the paleomagnetic and biostratigraphic data (Falguères *et al.*, 1999). Taking all this information into account, TD6 level should be located at the end of the Early Pleistocene (i.e. >780 Ka.).

Monte Poggiolo, Italy

This Mode 1 site is found in northeast Italy, on the southeastern margin of the Po river valley. The archaeological locality lies on the Monte Poggiolo hill, 180 m. above sea level, in 5 m thick sandy coastal gravels. The sedimentary sequence of the area belongs to the Early Pleistocene: marine blue clays are overlain, to the southeast, by the Monte Poggiolo fluviatile sediments and, to the northwest, by coastal yellow sands. Pedogenic processes, tectonic activity and faulting have affected the sequence from the Middle Pleistocene onwards (Antoniazzi *et al.*, 1996).

The lowermost sediments, the infracoastal blue clays, have been correlated to a Matuyama, pre-Jaramillo age. ESR analysis carried out in the Monte Poggiolo area (where the transition from the blue clays to the fluviatile sediments is gradual) has furnished an age of 1.5 Ma. (Yokoyama *et al.*, 1992). A quarry opened in the yellow sands, 20 km. away from the archaeological site, has provided a macromammal collection that has been dated to the early Galerian biochron, 1.0-0.9 Ma. (Azzaroli *et al.*, 1988). This interpretation is in agreement with the paleomagnetic and ESR analyses, which locate these sediments in the Brunhes-Jaramillo interval, at about 1 Ma. (Mussi, 1995). Although the fluviatile deposits bearing the archaeological accumulation have no faunal remains, they have been correlated stratigraphically with the yellow sands and interpreted as a lateral facies of gravel-beach deltaic deposits that, due to a marine transgression, cut into the yellow sandy coastal sediments. Paleomagnetism and ESR analysis on quartz grains carried out on the detrital sediments of the archaeological site confirm this hypothesis, as they have provided an upper Matuyama age, around 800 Ka. (Peretto *et al.*, 1998). The geological and chronological information provided seems consistent with the idea that the whole sequence belongs to the Lower Pleistocene and that the fluviatile sediments, in which the important archaeological accumulation has been found, fall late in the Matuyama Chron.

The lithic assemblage is in fresh condition and shows no traces of significant fluvial or marine post depositional transport. This is supported by the 76 refittings recorded at the site. Some of them amazingly reconstruct the complete original core and each refitted group is found in the same stratigraphic level and in a narrowly defined area (Peretto *et al.*, 1998).

Isernia La Pineta, Italy

Isernia is a Mode 1 site located in central Italy. The site lies in lacustrine and fluviatile sediments that belong to the Pleistocene deposits that fill a Tertiary tectonic basin (Cremeschi & Peretto, 1988). The archaeological and faunal associations are present in four horizons separated in two different sectors (50 m. from each other). The stratigraphic series is as follows: in Sector I, level 3c is the oldest in the sequence and lies on a paleosurface, on top of travertine deposits which belong to the last episode of a lacustrine event; level 3a lies over fluviatile silty deposits that covered the travertine layers and contains an extremely dense faunal accumulation (mainly elephant, rhino and bison remains) associated with stone tools. After a phase of tectonic uplift and volcanic events, the fluvial deposits become dominant, and it is in this new regime that the other two horizons are found: level 3S10 and, in Sector II, level 3a which contains in a very thin layer, a very important accumulation of stone artifacts and almost no bones. The upper part of the series contains tuff sediments that indicate new volcanic activity. The dense concentrations of lithics and faunal remains, interpreted as living floors, have a good level of integrity and seem to indicate low energy deposition pattern (Villa, 1996).

The chronology of the site has been a matter of discussion for several years. K-Ar analysis on sanidine crystals from level 3a furnished an age of 736±40 Ka that seemed in accordance with a 550±50 Ka. date obtained in a later stratum at the top of the sequence. Paleomagnetic analysis showed, in addition, a reverse polarity that was interpreted as the Matuyama chron. (Coltorti *et al.*, 1982). However, the controversy rests on the fact that the biostratigraphic data do *not* support the absolute and paleomagnetic dating and, indeed, provide some contradictory measurements (Cremaschi & Peretto, 1988). The presence in the site of the rodent *Arvicola cantiana*, which marks the replacement of *Mimomys savini* at about 500 Ka., suggests that the site

is younger (Roebroeks & van Kolfschoten, 1994). Recently, analysis of the macrofaunal assemblage, has led some paleontologists to suggest an age of about 0.6 Ka (Petronio & Sardella, 1999), which seems plausible and in agreement with recent Ar/Ar dates (Coltorti *et al.*, 2000).

THE FIRST EUROPEAN TECHNOLOGIES: A GENERAL OVERVIEW

In this section, the general technological trends observed in the lithic assemblages reviewed in this chapter are presented. For this purpose, the most recent information available for each archaeological collection will be used (Gibert *et al.*, 1998b; Oms *et al.*, 2000; Carbonell *et al.*, 1999a; Peretto *et al.*, 1998; Peretto, ed., 1994). However, to provide a detailed comparison between sites is a difficult task. It is important to note that it is problematic to compare the quantitative data offered by the different researchers. Some publications are comprehensive and show clear counts for all categories considered (i.e., artifact raw materials, dimensions, representative types and their proportions, etc.). Unfortunately, other publications are less complete. To provide a synthetic summary of the oldest European technologies, the emphasis here will be on general comparisons of quantitative information among the different sites, though in some instances more specific, qualitative information must be referred to.

Table 2 shows the number of lithic implements from each site and the general composition of assemblages by artifact classes or categories. At a glance, we can see that 50% of all artifacts are flakes, an important element in all the collections. On the other hand, the core category is quantitatively small in all the assemblages. To cite the extreme cases as examples, there is a core/flake ratio of 1:7 at Monte Poggiolo and of 1:18 at Barranco León. There is a mean ratio of 1:11 when all the assemblages are considered. Only the Spanish sites have been reported to contain hammerstones and manuports. In the case of Fuente Nueva and Barranco León a number of Jurassic dolomite cobbles have been considered manuports (Gibert *et al.*, 1998b: 21). At Atapuerca TD6, 19 quartzite, sandstone and limestone cobbles

Table 2

site	total lithics	cobbles/ hammerstones (%)	cores	flakes	retouched/ trimmed flakes	indet./ others
Barranco León	112	5 (4.46)	4 (3.57)	71 (63.39)	13 (11.60)	19 (16.96)
Fuente Nueva 3	120	20 (16. 66)	6 (5)	70 (58.33)	10 (8.33)	14 (11.66)
Atapuerca TD6	268	19 (7.08)	19 (7.08)	145 (54.10)	27 (10.7)	58 (21.64)
Monte Poggiolo	1319	—	153 (11.59)	1154 (87.49)	12 (0.90)	—
Isernia la Pineta	2567	—	160 (6.23)	1113 (43.35)	1294 (50.40)	—

Table 2: General composition of the assemblages under study.

have been recorded. These pieces have rounded or cubic shapes. The first shape type has been interpreted as hammerstones and the second as blanks chosen to be flaked (Carbonell *et al.*, 1999a: 672).

Raw Material Selection

The raw materials at all the sites are considered to be local in origin. Although this statement can apply as a general rule, it is interesting to evaluate each particular case in some depth, considering both the location of the procurement areas and the variety of rocks used at each spot (figure 3).

At Barranco León (located in the distal area of an alluvial system) and Fuente Nueva (in a lacustrine environment), almost all the transformed implements (excluding, then, the referred manuports) are made of high quality flint. At the first site, the grayish flint recovered has a Jurassic origin and its outcrop source has been identified two km. to the northwest. The Fuente Nueva siliceous variety has a white color and occurs four km. from the archaeological site (Gibert *et al.*, 1998b: 23). The two quartzite pieces (one at each site) probably have a more remote origin, in the gravel deposits to the north (six to eight km. away).

Atapuerca TD6 exhibits the highest variety of rock types selected, all of them available from no more than three km. away from the site (Mallol, 1999). Flint and quartzite, in this order, are the best-represented materials. Flint appears in two different forms: a Neogene flint from the marls located at the south of the Sierra and a Cretaceous flint related to the karstic system. All varieties of metamorphic rocks, including the quartzite cobbles used, come from the terraces of the Arlanzón River that borders the Sierra to the south. Flint and quartzite play an important role in the assemblage composition and are the main materials selected not only in the

reduction sequence, sometimes flaked in an exhaustive manner, but also in the retouch activities. The relatively important presence of limestone, a ubiquitous material in the Sierra, is the best example of a casual and *ad hoc* selection of blanks for flaking purposes.

The Monte Poggiolo lithic collection is produced entirely from flint. Siliceous rocks naturally occur in the archaeological area, which, as already mentioned, is located in fluviatile gravels rich in flint and limestone cobbles (Antoniazzi & Piani, 1992: 241). However, petrographic studies suggest that the siliceous materials selected belong to at least two heterogeneous rock types that could have different origins. Therefore, more varied procurement sources than the very local river delta channels may be possible (Peretto *et al.*, 1998: 361). In addition, there is not a clear relation between the two different flint types and specific reduction strategies. It seems that the flaking processes have been equally intense regardless of cobble quality.

The main rock type used at Isernia is flint. Its origin is local and comes from the outcrop of an eroding formation less than two km. from the site. The siliceous material appears as small blocks (less than 10 cm.) that can be related to two different varieties (Sozzi *et al.*, 1994: 51): the first is fine-grained and has an homogeneous texture; the second is a brecciated, coarse-grained and lower quality type, often showing fissures and weakness planes in its structure. The second important raw material is limestone. This rock has also a local origin and comes from the alluvial outcrops located in the vicinity of the archaeological site. Generally the angular limestone cobbles selected exhibit larger dimensions than the flint utilized (the length of the limestone specimens can reach 20 cm.). Difference in shape, dimension and quality between flint and limestone correlates with different selection and use (and therefore, in their contribution to the specific categories). While the flint

Figure 3

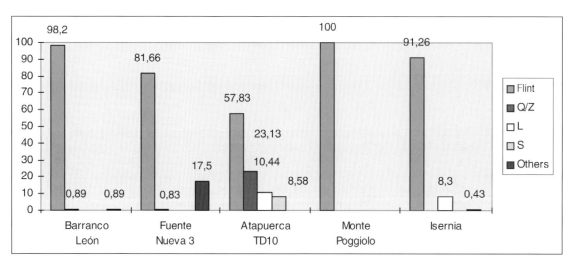

3. *Percentage of raw materials used in the assemblages under study (Flint; Q/Z, quartzite and quartz; L, limestone; S, sandstone).*

shows a low, non-intensive level of reduction (i.e. cobbles with few or isolated scars and pieces that fit in the range of the traditional "chopper" typological group), the limestone is intensively flaked and much more involved in the reduction processes.

Knapping Processes

Cores and Reduction Strategies

Although in every assemblage the core class tends to be of a Mode 1 nature, there are significant differences among sites in terms of the qualitative information available (table 2, figure 4). For instance, at Fuente Nueva and Barranco León cores are scarce and exhibit homogeneity in the reduction processes performed (Gibert *et al.*, 1998b: 21). In both localities, almost all the cores have been exploited in a centripetal manner, although one of these pieces, from Fuente Nueva, shows a laminar tendency (at least two laminar scars can be recognized). The only exception to this discoid tendency is a small quartzite "chopper-core" from Fuente Nueva, which shows simple unifacial exploitation. These flint cores are of small size, particularly in the case of Barranco León. Taking into consideration the fact that most of the flakes recovered at both sites tend to be larger than the cores, it seems clear that these cores can be considered exhausted nuclei.

At Atapuerca several reduction patterns are present, the unifacial and bifacial being the most representative. Most cases (68% of the cores) exhibit the orthogonal technique (by which the flakes are detached using an angle close to 90°). Depending on the raw material selected, the orthogonal method involves one or two faces (on fluvial materials) or multiple faces (on flint). It has been suggested (Carbonell *et al.*, 1999a: 664) that the Neogene flint blanks were introduced in the cave

partially reduced, as no cortical flakes of this material have been found in the locality. This hypothesis seems plausible, taking into consideration that the big cobbles found outside the cave should have been flaked with anvil technique to detach smaller pieces. Only one discoid core has been recorded, a quartzite piece with centripetal scars, but some morphological features observed in a number of flakes in Cretaceous flint (ie. radial scars on dorsal surfaces along with faceted platforms) suggest that centripetal exploitation was widely used on flint.

Cores at Monte Poggiolo have the highest value of all sites considered. While in the past all cobble cores at this site tended to be classified as chopping tools, today these objects are seen as examples of an opportunistic reduction sequence (Peretto *et al.*, 1998: 362). Most of the nuclei exhibit simple reduction patterns (mainly unifacial and bifacial) and a small number of scars, which has been related to a fairly low degree of reduction. In fact, among the unifacilly flaked cobbles, 20% show only one scar. The other reduction patterns (ie. multifacial or discoidal) appear in much lower numbers and, unlike the simple reduction cases already cited, represent exhausted residues of a more intense reduction.

At Isernia, the use of both bipolar and direct percussion techniques has been reported. Bipolar technique was suitable to reduce the small flint cobbles, whereas direct percussion was employed to reduce the limestone blocks. Due to the use of bipolar technique here (with hard-hammer percussion more dominant at other sites), the cores present an important variety of morphological types that, nevertheless typologically fall within a range of unifacial and bifacial cores with discoid pieces being rare (Crovetto, 1994: 247-248).

Figure 4

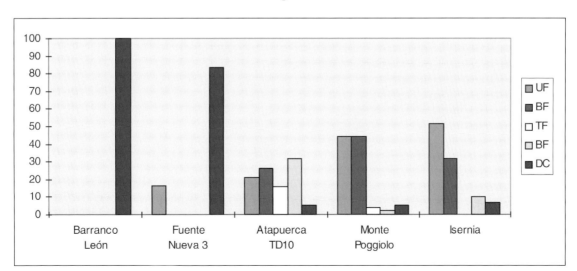

4.   *Percentage of core types by facial patterns in the assemblages under study (UF, unifacial; BF, bifacial; TF, trifacial; MF, multifacial; DC, discoid).*

Débitage

Débitage (flakes and debris) is the most abundant category in all the assemblages. The technical features observed in débitage are related to the reduction strategies carried out in the different sites, with very little preparation of cores evident. The casual knapping patterns reported at Monte Poggiolo are associated with flakes produced in the early stages of reduction (cortical and semi-cortical dorsal surfaces are abundant) and with simple platforms. Consequently, most of the platforms represented are cortical (30%) or plain, i.e. single scar (45%). Whereas the core assemblages of Fuente Nueva and Barranco León, as already mentioned, must be considered exhausted, the morphology of the flakes recovered at both sites show that they preferentially come from a non-prepared core technology (cortical and plain platforms are predominant in flakes). This indicates that a more intense reduction is observed at these sites than at Monte Poggiolo, as most flakes show non-cortical dorsal surfaces. This pattern of exhausted cores at these two sites seems to be a function of the higher quality flint being worked. One of the most remarkable aspects related to the flake assemblage from Atapuerca is its homogeneity in size regardless of raw material type (Carbonell *et al.*, 1999a: figs. 11-16). However, some other morphological characteristics do vary according to the type of rock selected. Flint flakes, primarily made from Cretaceous cobbles, do not show traces of cortex in their dorsal face or platforms, which is consistent with the interpretation that the early stages of knapping were performed outside the cave. Quartz and quartzite flakes, in contrast, are more representative of all different stages of the reduction sequence.

At Isernia, as the experimental work shows (Crovetto *et al.*, 1994), the knapping technique employed has played a very significant role in the flake morphologies. First, the large number of small flakes and angular fragments appear to be the result of bipolar technique. In addition, the flakes produced tend to repeat the same morphological pattern: they exhibit non-cortical dorsal faces (only 5% of the flakes are totally cortical, while 58% are completely non-cortical) and non-cortical platforms, which sometimes are shattered (Crovetto, 1994: 193-194).

Retouched Pieces

All the assemblages considered here include a number of retouched specimens. It is worth pointing out, however, that, although from a typological perspective and for classificatory purposes, all these objects are included in the same 'retouched' category, at some sites they do not appear to represent an intentional modification of natural edges.

Barranco León and Fuente Nueva have provided a considerable collection of specimens included in the retouched class. Most of these flakes have been defined as trimmed pieces (Martínez-Navarro *et al.*, 1997: 618;

Oms *et al.*, 2000), in which the modification of the edges is shallow and marginal and probably due to utilization (although no use-wear analysis has yet been carried out to confirm this). More traditional retouched tools (classified as notches, denticulates and side scrapers) are present in small numbers.

Retouched tools are very rare at Monte Poggiolo. Twelve pieces exhibit traces of simple retouch of their edges. Although these objects have been characterized as denticulates (7) and side scrapers (5), their placement within such formal retouch types is problematic because of the crude and coarse character of the retouch (Peretto *et al.*, 1998: 361).

At Atapuerca, 60% of the retouched pieces are made on flint. The type of retouch observed is mainly simple (along one edge), fairly invasive, and unifacial (Carbonell *et al.*, 1999a: table 5) and does not show the more formal morphologies. Thus, many of the retouched pieces exhibit a sporadic/casual and discontinuous retouch (38%) that cannot be formally typed. The remaining retouched pieces have been classified as denticulates (34%) and scrapers (28%).

The category of retouched pieces described at Isernia is interesting in being far more varied than at the rest of the sites considered here. It includes the traditional retouched flakes, as well as retouch on fragments, small cobbles, cores and plaquettes (Crovetto, 1994: 228-229). Most of these pieces exhibit edges modified into denticulates (up to 90% among the flint retouched pieces), which tend to be coarse and non-systematic. Only in rare instances is it possible to find continuous (although sinuous) retouch, that would generally be typed in a scraper class (1.5%). Considering the nature of most of the pieces and that most of the retouch is denticulated, most of the retouched pieces at Isernia are generally placed within classical types of denticulate scrapers, end scrapers, *becs* and Tayac points.

However, experimental studies (Crovetto *et al.*, 1994) have been able to replicate most of the technical and morphological features observed in the archaeological assemblage and explain the way in which "retouched pieces" were produced. When flint blocks were placed on an anvil and repeatedly struck forcefully with a hammerstone until the core is exhausted, a recurrent pattern emerged: many of the artifacts produced (flakes, chunks, and cores) had the appearance of coarse and non-systematic denticulates, thick carinated scrapers, *becs* and end-scrapers. These results, which replicate very closely the features observed in the archaeological collection, led to the conclusion that the Isernia retouched pieces are for the most part not intentionally shaped but rather the accidental by-products of the bipolar technique employed (Crovetto *et al.*, 1944: 151). Thus, the category of retouched objects at Isernia is likely very over-represented. Only flat scrapers with a continuous retouch could be regarded as retouched tools *sensu stricto*, which means 0.6% of the total assemblage would fall into a retouched tool category (a percentage substantially lower than that obtained by typological means).

Tool Use: Use-wear Analysis

Samples of artifacts have been studied at Atapuerca, Monte Poggiolo and at Isernia in order to search for microwear evidence. At Atapuerca a sample of 24 out of 43 artifacts that were examined for microwear exhibit traces of use. However, functional interpretation has been possible for only 17 specimens, including 9 pieces inferred to have been used on meat, 5 on wood, 2 on bone and 1, possibly, on hide (Carbonell *et al.*, 1999a: table 3). Although flakes were mainly used to work on soft tissues, and denticulate pieces performed tasks on hard matter, this is admittedly a small sample and there is no clear-cut evidence of functional specialization among different typological groups.

Monte Poggiolo has provided a sample of 27 utilized flakes. Working edge angles are predominantly acute (42%) and simple (34%). It has been possible to identify microwear patterns on 21 artifacts with regard to material worked. Ten pieces exhibit traces of use on soft animal biomass and have been related to butchery activities. Traces on the other pieces have been interpreted as produced by working on vegetal tissues (wood in six cases and an unidentified grass in the other four). Microwear on the remaining artifact has been related to scraping activities on an undeterminate hard material (Peretto *et al.*, 1998: 454).

At Isernia a sample of 218 flakes from sector II have been analyzed. It includes 134 unmodified flakes and 84 denticulate pieces (some of them, from a typological perspective, classified as *becs*). The result of the investigation shows that those carinated objects with a denticulate retouch do not present traces of microwear, which seems to be in agreement with the interpretation provided by the experimental studies (i.e., they simply are the by-products of the bipolar reduction strategies). Only the non-retouched flake group has provided traces of use. In these cases, the working edge angle is very acute. Micro-retouches and micro-fractures affecting natural edges have been found in several pieces, a pattern that has been interpreted as damage produced by use. In addition, traces of polish have occasionally been found in some specimens, which suggest that the flakes were used only briefly (Longo, 1994).

Summary of Technological Traits Observed in the Earliest European Lithic Assemblages

Available data about the lithic assemblages of Barranco León, Fuente Nueva, Atapuerca TD6, Monte Poggiolo and Isernia can be summarized as follows (figure 5):

1. In general, most raw materials selected always have a local origin. All the rocks are easily found in the vicinity of the archaeological sites, in a range of less than 4 km. At Barranco León and Fuente Nueva it is possible, however, that quartzite cobbles used have a more distant origin. To confirm this second pattern, it would be necessary to improve knowledge about the surrounding catchment areas.

2. Furthermore, in most cases, the types of rock selected are not diverse and tend to have been used according to their overall relative frequency in the site area, which seems to have been the most important factor in procurement strategies. Atapuerca is an exception, although no specific selection of rocks according to texture quality is observed.

3. Although there is no particular selection of raw materials, it is worthwhile to remember that reduction strategies, blank dimensions, and raw material quality have been major factors determining the final morphology of the cores and flakes. For instance, flint at Fuente Nueva and Barranco León is of good quality while the small flint cobbles from Isernia often show fissures and weakness planes.

4. Cobbles and blocks have been reduced using hard-hammer, bipolar or throwing percussion techniques. Knapping was carried out at the sites, although at Atapuerca, some of the early stages of the flint reduction sequence were carried out at the quarry, outside the cave system.

5. Cores show patterns of non-systematic and opportunistic reduction. This appears to be the case particularly at Monte Poggiolo and Isernia, where *ad hoc* percussion techniques were predominant. The non-systematic character of these objects is supported by the lack of prepared platforms. At most sites, unifacial and bifacial reduction models, generally showing a low degree of reduction, predominate. Multifacial and discoid methods are present in low numbers. However, Barranco León and Fuente Nueva show a different pattern. In both cases, the core assemblage consists of exhausted discoid specimens and, considering the fact that the core category is poorly represented, it should be pointed out that only the final stages of the reduction sequence are present.

6. Flake production seems to have been the main goal of the craftsmen. Flake morphologies indicate non-systematic reduction systems and tend to exhibit the early stages of the reduction sequence (cortical and sub-cortical dorsal faces, cortical and plain platforms). Again, specific raw materials introduce variations to this pattern: in Atapuerca, unlike fluvial rocks, flint flakes tend to be non-cortical, while non-cortical flakes are predominant at Barranco León and Fuente Nueva.

Figure 5

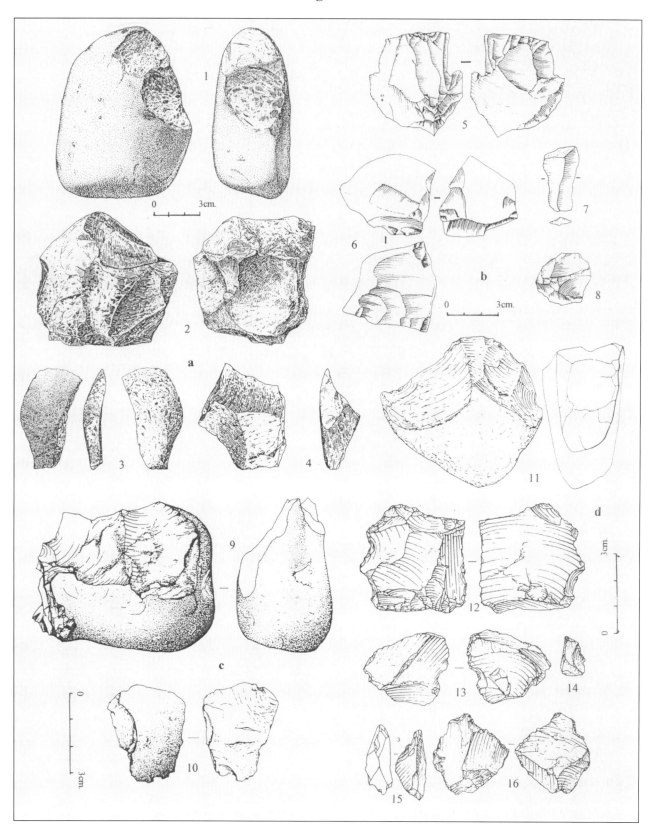

5. *Lithic industry. a. Atapuerca TD6: 1. limestone core-chopper; 2. Neogene flint core; 3. quartzite flake; 4. Neogene flint flake; b. Fuente Nueva 3: 5 and 6. cores; 7. flake; 8. trimmed flake; c. Monte Poggiolo: 9. core; 10. flake; d. Isernia: 11. limestone core-chopper; 12, 14 and 16. denticulates; 13. notch; 15. burin. (adapted from Carbonell et al., 1999a; Martínez-Navarro et al., 1997; Peretto et al., 1998; Crovetto, 1994).*

7. The retouched tools are relatively few. They include casually trimmed and discontinuously retouched pieces and other flakes that can be assigned (though not unproblematically) to more classical types. Most of these types are denticulates and side scrapers. The retouch tends to be coarse and crude, sometimes invasive, and simple. Thus, retouched tools seem to be part of the same opportunistic strategy performed in the block reduction processes. Small format tools with specific and recurrent edge morphologies do not seem to have been an objective.

8. Taking into consideration all these technical and morphological characteristics, the lithic assemblages considered fall within the range of Oldowan-like Mode 1 technologies, as described by Clark (1977: 23).

ON THE MEANING OF THE FIRST EUROPEAN MODE 1 TECHNOLOGIES

While there is a broad consensus that the technological traits observed in the oldest archaeological occurrences of Europe *formally* can be related to the characteristics first seen in the African Oldowan technocomplex (Ludwig & Harris, 1998; Schick & Toth, 2001) or to Mode 1 (using a generic terminology, perhaps more appropriate for these non-African cases), less agreement exists at the *interpretative* level. Nowadays there are two main different positions addressing this question, which, not surprisingly, tend to follow the two theoretical spheres represented by the short chronology and the mature Europe hypotheses.

The first perspective (closer to the mature Europe scenario) interprets these occurrences as, in a literal sense, pre-Acheulean technologies. Mode 1 assemblages at these first sites could be considered to have some sort of discrete *cultural* background and might be linked, in some ways, with the Oldowan "tradition". This view, which assumes that similar techno-typological traits indicate cultural relationships of some kind, is deeply rooted in the European Paleolithic (Otte, 1996: fig. 116). The European archaeological record for many decades has yielded Lower and Middle Pleistocene assemblages consisting of crude cores and flakes without handaxes or cleavers. Recurrently, these sites have been interpreted as part of a pebble tool culture with an African Oldowan origin that developed separately from and contemporaneously with the Acheulean (a good example of this view can be found in Tieu, 1991). A more recent perspective could indirectly support this cultural-historical interpretation. Some authors (Foley, 1987; Foley & Lahr, 1997) have suggested that human technology can be read from a phylogenetic point of view and that technological patterns reflect patterns of human biological evolution and dispersal rather than cultural connections. This view could be summarized in the assertion "one species, one technology."

Unfortunately, the relation between specific human groups and certain technologies is an idea that does not seem to match the archaeological evidence (ie. Cosgrove, 1999; Bar-Yosef & Kuhn, 1999). Most scholars agree that Paleolithic technological variability is a much more complex phenomenon and that a number of different factors can be involved in conditioning the traits observed in any assemblage (i.e. raw materials, ecological constraints, type of occupation). The influence of those factors should serve as a caution for those who infer cultural affinities of this sort, especially considering that early European Mode 1-like assemblages occur well after the origin of Mode 2 in East Africa at about 1.6 Ma.

From the short chronology perspective, this framework is used to point out that early European Mode 1 industries should be interpreted from a para-Acheulean rather than a pre-Acheulean perspective and that the lack of large flake bifacial tools, is due to several factors, some of them simply adaptive, not related to any specific cultural background or connection. This discussion shares some similarities with the debate about the meaning of Mode 1 industries in East Asia (Schick, 1994: 586-591). In the European case, workers have paid attention mainly to the following aspects (Rolland, 1992: 88): 1) taphonomic effects combined with small sample sizes, 2) raw material constraints, and 3) specific site function. However, although all these factors must be regarded as relevant in any lithic collection, none of them provides conclusive explanations for the absence of Mode 2 features in the earliest European occurrences.

With regard to the first of these factors, it should be stressed here that the available early European lithic collections are sparse and may not adequately represent the full range of prehistoric technological features at these sites. Due to intense geomorphologic dynamics in European Quaternary deposits (i.e. periglacial and fluvial forces involved in sorting and reworking materials) it might have been difficult to preserve most of the archaeological remains, which inevitably would lead to sample bias (by which bifacial tools might be lost from the ancient sites), although it is important to note that preservation biases affect the entire Pleistocene sequence and they do not appear to be an insurmountable obstacle to recover Middle Pleistocene large tools. This particular problem is well exemplified at the South African site of Sterkfontein, as has been recently reported (Kuman, 1998: 177).

While this concern does not apply to the Italian sites studied here (for both of which the archaeological collections are abundant), it certainly must be kept in mind in regards to the Spanish record. Orce Basin assemblages are regrettably meager. For instance, the 120 artifacts recovered at Fuente Nueva actually belong to four different aggregates, the largest in stratigraphic context being a collection of only 49 artifacts found in level 2 (Oms *et al.*, 2001: tab. 1). Barranco León has provided the largest sample so far, a homogeneous archaeological sample of 107 pieces. Atapuerca TD6 is, as already noted, a 6 m² test excavation and therefore, the 268 artifacts recovered represent only a preliminary indication of its potential richness (which will be fully uncovered when the extensive excavation over an area of 80 m², now in the lower part of TD10 level, reaches Aurora stratum). However, assuming that the final lithic definition still has to be completed, the relatively high lithic density (45 lithic objects/m² and, considering the entire archaeological and paleontological assemblages, more than 743 items/m²) provides a more reliable sample.

The second factor, the influence of raw material, has been repeatedly cited as an important factor of variability and a possible influence on the final techno-typological appearance. It has been pointed out, for instance, that the crudeness seen at Isernia is due to the poor quality and small dimensions of the raw material available in the surrounding area (Mussi, 1995: 40). While rock quality may account for the technological pattern in this case, it is more difficult to consider its relevance in the other sites. High-quality flint was selected at the Orce basin occurrences, while at Monte Poggiolo the size or quality of the raw material does not seem to be related to the opportunistic lithic production. At Atapuerca the same variety of raw materials (and in similar proportions) is selected and used to produce the Mode 1 industry from TD6, the Mode 2 industry from Galería and the Mode 3 industry from TD10-11 (Carbonell *et al.*, 1999a: fig. 19). It should be kept in mind, additionally, that raw material influence in reduction processes has been generally overstated. As several studies demonstrate (Moloney, 1996; Brantingham *et al.*, 2000), almost every lithic feature can be obtained using any type and quality of raw material. It could be expected, then, that some Mode 2 "progressive signs" might have emerged and been noticed in these lithic assemblages.

With regard to the third factor, ecological and functional aspects might also provide some clues relative to the composition of the early assemblages where handaxes are not found. As has been pointed out, Mode 1 tool-kits "represent opportunistic, least effort solutions to the problem of obtaining sharp edges from stone" (Isaac, 1981: 184). Using this perspective as a reference background, the occurrence of Mode 1 collections after the origin of the Acheulean techno-complex could be interpreted as efficient solutions serving multiple tasks when a mere *ad hoc* technology was sufficient. The interpretation of the lithic assemblage of the German site of Miesenheim I illustrates the use of the "least effort functional hypothesis" to explain European Mode 1 collections. Dated to around 450-400 Ka, when biface occurrences appear to be present in the continent, it exhibits a production of simple cores and flakes. This opportunistic tool-kit has been interpreted as an effective response to a human activity dominated by a secondary scavenging of carcasses obtained by carnivores (Turner, 1999b: 380).

Whether or not this case can be used to explain the lack of handaxes in the earliest European lithic assemblages remains uncertain, because we do not understand the functional meaning of handaxes well enough to explain the dichotomy "handaxes *versus* cores and flakes" in terms of site function. For instance, handaxes have been related to large mammal meat access (Jones, 1980; Toth, 1982), as several archaeological cases show, such as the elephant butchery sites of Áridos and Ambrona in Spain (Santonja *et al.*, 1980; Santonja *et al.*, 1997) or Boxgrove in Britain (Ashton & McNabb, 1994). On the other hand, a similar association of large mammal remains and non-bifacial tool-kits covering a broad chronological and geographical range has been reported, for instance at the Ethiopian site of Gadeb (Clark & Kurashina, 1979: 38) and the Italian localities of Isernia (Giusberti & Peretto, 1991) and Notarchirico A-B (Piperno *et al.*, 1999: 86-106). (A comprehensive summary of African and European Pleistocene archaeological sites related to elephant access can be found in Martos-Romero, 1999). This evidence is in agreement with experimental data showing the functional versatility and success of Mode 1 repertoires (Toth, 1987).

Another interesting approach has stressed possible differences between both technologies in terms of specific responses to landscape mobility and use. Recently, two cases have been reported in Africa showing this pattern within the Acheulean complex time span. In the Ethiopian Eastern Middle Awash, Mode 1 and 2 industries co-occur occupying different environmental settings. Mode 1 assemblages are recovered in stable floodplains, whereas Mode 2 assemblages are related to high-energy tributary streams (Clark & Schick, 2000: 67). At Peninj (Tanzania), Mode 1-like assemblages are located in the floodplains near the ancient lake margin while Mode 2 assemblages are located away from the lake and close to fluvial contexts (Domínguez-Rodrigo *et al.*, 2005). In addition, at Peninj, Mode 1 assemblages are located in patches whereas Mode 2 occurrences seem to be more scattered. Both of these cases may represent the patterning of distinct technological behaviors in response to different geographic and ecological locations, already observed in the earliest Acheulean of Olduvai Bed II (Hay, 1976: 181).

The European example of this "ecological hypothesis" for the Mode 1/Mode 2 co-occurrences (after Domínguez-Rodrigo *et al.*, in press) is represented by Keeley's work (1980). Studying the geographical distribution of handaxes in Britain, he concluded that the sites with bifaces, multitask tools he suggests, could be related to economic activities (hunting and gathering) carried out away from the home bases, where a higher proportion of débitage and small tools should be expected (*ibid.*: 161). From this perspective, Ashton (1998) has recently proposed his "static resource model" to explain the way in which lithic assemblages could vary in terms of biface/non-biface presence. He argues that this dichotomy should be explained as a response to environmental differences, mainly local resource availability.

If this view is correct, differences between Mode 1 and Mode 2 assemblages would simply reflect specific technological responses to the variety of locations and specialized activities that occur within the group's home range. However, in the European context, this hypothesis has not yet been demonstrated to be consistent enough to suggest that the opportunistic technology seen at the earliest European sites is due mainly different activities at different locations. (It would, of course, be necessary to explain why early European settlements show such a homogeneous technological response in different locations and possibly with different activities). Although there are some well-documented Middle Pleistocene Acheulean sites related to task-specific activities, such as the cave site of Galería in the Atapuerca complex (Carbonell *et al.*, 1999b), regional-scale studies available do not confirm a clear pattern of technologies related to specific tasks.

One of the most extensive and complete Middle Pleistocene landscape approaches in Europe is the long-term survey and excavation analyses carried out in the Duero River basin, in the Spanish Northern Meseta (Santonja & Pérez-González, 1984, 1997). Until recently, all the Acheulean occurrences reported in this area were found in the bank terraces of the fluvial system, mostly in lower sections of tributary rivers. This persistent fluvial location pattern led to the conclusion that both movements and settlements took place preferentially along fluvial environments (*ibid.*, 1984: 328). This perspective assumes that hominid home ranges (and presumably all the different site-types and activities related to them) were mostly restricted to these areas, in which handaxes occur homogeneously and no specific Mode 1-like/Mode 2 dichotomy is observed.

However, in recent years our understanding of landscape occupation patterns in this region has broadened, and we now know that the Tertiary high plateaus adjacent to the river valleys were also intensively occupied (Diez-Martín, 2000). Although this area, more than 100 m above the river valley level, is a distinctive ecosystem, no substantial difference has been found in the composition of the lithic assemblages in this highland area versus the fluvial setting, other than minor formal aspects related to constrains imposed by the access to raw material sources. In addition, the scatter and patches analysis developed in these geomorphologically homogeneous plateaus does not exhibit major differences in the composition of tool-kits between denser site locations apparently visited repeatedly and apparently visited sporadically.

Perhaps one of the most solid arguments supporting the idea that European Lower Pleistocene technology does not represent a distinctive entity is the fact that Mode 1-like assemblages and Mode 2 assemblages clearly co-occur during the Middle Pleistocene. If, as we know, Mode 1 technology is "time-transgressive" (Clark & Schick, 2001: 55) and overlaps with the Acheulean during a long period of time, then we could speculate that in fact the opportunistic pattern seen in the oldest European lithic assemblages must be due to some sort of ecological, functional/locational, or social (age or sex division) influences not yet defined (even though European archaeological evidence supporting this view is not conclusive, as discussed above). However, our large-scale knowledge of this particular issue is rather deficient.

Middle Pleistocene (post-500 Ka.) lithic assemblages without handaxes and cleavers are well known in Europe, in sites such as High Lodge (Roberts *et al.*, 1995), Bilzingsleben (Mania, 1995), Schöningen (Thieme, 1999), Vertesszölös (Dobosi, 1990; Kretzoi & Vértes, 1965) or Notarchirico E1, E, C and Alpha (Piperno, 1999; Piperno *et al.*, 1999). The nature of these sites and other related industries, commonly known as Clactonian, Tayacian or Taubachian (Otte, 1996), and their relationship with synchronous Acheulean has been the subject of a long debate and diverse interpretations (for instance and mainly referring to the Clactonian debate, Breuil, 1932; Ohel, 1979; Svoboda, 1987; Mithen, 1994). Recently some of these Middle Pleistocene flake industries have been included by a number of authors as mere variants within the range of the Acheulean . For instance, recent excavations at the British site of Barnham (Ashton *et al.*, 1994) have demonstrated that the Clactonian is contemporaneous with the Acheulean and that differences between them should be due to raw material availability variations across the landscape or over time (Wenban-Smith, 1998: 96).

Leaving aside the particular Clactonian case, in which archaeological work has provided a new set of explanatory options to consider (see the comprehensive critical review by White, 2000), the main problem concerning Middle Pleistocene occurrences lacking handaxes is related to their technological characteristics. Unfortunately, as has already been pointed out (Wenban-Smith, 1998: 93), the debate over these assemblages has focused on the presence/absence of bifacial tools, while other technological patterns related to the nature of the flake tool component have been ignored. More attention should be given to this specific question, in order to conclude whether or not, beyond the lack of bifaces, these Middle Pleistocene retouch strategies are comparable to the non-systematic and casual retouch reported at Monte Poggiolo or Atapuerca TD6, for instance. For this reason, it is difficult to fully accept the general statement that the only difference between Mode 1 and Mode 2 technologies is the absence or presence of bifacial artifacts (Villa, 2001: 119) and that, therefore, non-bifacial Lower and Middle Pleistocene occurrences share the same technological traits.

Reality may be far more complex, and it is suggested here (as a working hypothesis needing comparative analytic studies in order to be tested) that, some of the standardization patterns observed in shaping large bifacial, Mode 2 tools should have correlates in small flake tools in those assemblages. Analyzing Lower and Middle Pleistocene flake assemblages can provide broader information about their respective technological characteristics. Some data that could be used as a starting point and that might support this idea (or at the very least indicate that it is a research avenue worth exploring) would include the high quality of scraper retouch at High Lodge (Ashton & McNabb, 1992: 166), as well as the qualitative variety and diversity of retouched tools reported at sites such as Bilzingsleben (Mania *et al.*, 1999: 303), Vértesszölös (Krétzoi & Vértes, 1965: 80), Schöningen (Thieme, 1999: 181), Notarchirico Alpha (Piperno, 1999: 309) or Visogliano 13-39 (Abbazzi *et al.*, 2000: 1182), technological traits not seen in the earliest retouched tool assemblages.

The adaptive/functional explanations discussed above envisage a scenario in which the earliest morphological traits originated as responses to particular challenges produced *within* Europe. Another set of explanations relies on the possibility that those early Mode 1 assemblages could actually be showing, due to different reasons (not necessarily cultural), a distinctive technological behavior, brought from *outside* the continent by its first migrant populations. These perspectives are diverse and tend to explain this phenomenon in a variety of ways, not necessarily in agreement with each other. However, all of them stress the important role played by migratory factors and their influence in the technological behavior of the first Europeans. One of these perspectives, suggested by Toth and Schick (1993: 352), pointed out that populations exiting Africa and reaching Eurasia could have lost the Acheulean component of their technologiös and could have returned to the use of Mode 1 patterns. Following their argument, while moving out of Africa, human populations could have had problems in finding suitable rocks to create large bifacial stone tools and, due to weak social and communicative networks or cognitive skills, they might not have been able to maintain their original Acheulean repertoire. This phenomenon, possibly along with other factors such as the adaptive modification of the Mode 2 tool-kit due to newly available materials such as bamboo (Schick, 1994: 587), would have been responsible for the spread of "Mode 1-like" assemblages in extensive areas of eastern Asia.

Taking into consideration that most consistent data for the oldest Asian occupation reaches back to about 1 Ma Schick & Zhuan(1993), Rolland, (1992) suggested that the first humans would have migrated to Europe via East Asia, moving along with animal dispersals and carrying with them their technological mutation, from a Mode 2 back to a Mode 1 industry. This Eastern Asian-European link seemed to be the only one possible, taking into account that the other areas surrounding Europe showed to have been first occupied with a Mode 2 technology: Ubeidiya in the Near East, dated between 1.5 and 1 Ma. (Bar-Yosef, 1994) and the Acheulean in the Maghreb, dated at about 1 Ma. (Raynal *et al.*, 1995).

The two viewpoints presented above assume that all archaeological sites found beyond East Africa, whether or not they look Acheulean and whatever the reason responsible for this fact, derive from a fully established Mode 2 industry. These perspectives are based on a scenario in which the first human traces out of Africa post-date the earliest Acheulean occurrences (ie. <1.6 Ma) ; (Schick & Toth, 2001: 70). The last years have witnessed a number of significant archaeological findings that have challenged (and in some cases been able to change) the established Out of Africa paradigm and have refreshed the debate on this issue. If we accept alternative data reported in Central Asia (Dennell *et al.*, 1988) and the Far East (Swisher *et al.*, 1994; Wanpo *et al.*, 1995), the first hominin radiation eastwards would be pushed back in time to around 2 Ma. (Turner, 1999a: 568), doubling the traditionally accepted chronology of first human expansion into this region. However, this evidence is controversial in the nature of the findings (the case of Longgupo cave, Etler *et al.*, 1997; Swartz & Tattersall, 1996) or disputed in the chronological results (Pakistan and Java cases, Hemingway, 1989; Klein, 1999: 271). Although a very early sortie out of Africa would be the perfect explanation for Mode 1 industries in large parts of Asia and the isolated *Homo erectus* evolution in this geographic *cul-de-sac* (Tattersall, 1997: 48), doubts concerning the validity of these very old chronologies necessitates putting this possibility on hold at this time.

With regard to the European case, more consistent examples supporting an early sortie Out of Africa and a pre-Acheulean presence at the gates of Europe have been found in North Africa and the Caucasus regions. The Algerian site of Ain Hanech (Sahnouni & Heinzelin, 1998; Sahnouni, this volume), bracketed by means of paleomagnetism and biostratigraphy within the Olduvai subchron (1.95-1.78 Ma.), has provided a rich collection of fauna associated with numerous Oldowan stone tools. At the Georgian site of Dmanisi (Gabunia *et al.*, 2000; 2001) an impressive collection of human fossils (three crania and 2 mandibles assigned to *Homo ergaster or Homo georgicus*) (Vekua *et al.*, 2002) has been associated with faunal remains and more than one thousand Mode 1 stone tools have suggested an age of about 1.85 Ma. for the occurrences. Although these data support the idea that humans reached the areas surrounding Europe before the origin of Mode 2 industries, it does not imply that the earliest European Mode 1 industries are directly descendant from these migration events. First, there is a significant chronological gap of almost one million years between the Algerian and Georgian sites the earliest European occurrences, and the European evidence does not seem to support a direct link with these early sites in Georgia and North Africa.

Some scholars have pointed out that early human dispersals during the Plio-Pleistocene transition into Eurasia may well have been sporadic, considering the lack of chronological and geographical continuity observable in the evidence currently at hand (Bar-Yosef & Belfer-Cohen, 2001). The present data seems consistent with a scenario in which hominins reached northern latitudes in diverse waves of intermittent trial occupations that possibly did not succeed. The same problem of an intermittent record may also be the case in North Africa. Such incursions could possibly have involved periodic incursions into Europe via Gibraltar or Sicily. Despite having been questioned for decades, such migration routes continue to be considered by a number of authors (Rolland, 1998; Bar-Yosef, 1998; Arribas & Palmqvist, 1999) and could be backed by other Early Pleistocene open-water crossing cases, such as the colonization of Flores Island by *H. erectus* (Morwood *et al.*, 1998). (A good critique to the Sicilian alternative, however, is provided by Villa, 2001).

More in agreement with the scenario of intermittent, non-continuous dispersals would be the hypothesis recently presented by Carbonell and colleagues (1999c). The authors argue that after the origin of Mode 2 industries in East Africa, some groups persisted in using Mode 1 patterns. Eventually, due to technological competition (Mode 2 would have been a more efficient response to environmental changes, adaptive requirements, or population pressure), Mode 1 makers would have been pushed out of African core areas and, shortly before 1 Ma, would have arrived in Europe bearing their non-handaxe tradition. Other authors have already pointed out the weakness of this proposal (Villa, 2001), although it seems that, leaving aside the absence of empirical archaeological evidence supporting this hypothesis, its main problem is related with the technological competition scenario. While it is obvious that Mode 2 exhibits a more complex technological and operative behavior and as a result, probably, more complex social networks (Kohn & Mithen, 1999), its functional and adaptive advantages need to be clarified (as already mentioned in this paper) in order to explain such a powerful displacement of supposedly contemporaneous Mode 1 producers.

Nevertheless, this hypothesis deals with an interesting idea that has been largely ignored by most scholars. It is generally assumed that the invention of Mode 2 technology marks a powerful *terminus post quem* boundary (or the starting point of a distinct period), beyond which handaxe production would have rapidly generalized. However, it seems more likely, as already pointed out (Bar-Yosef, 1998: 227), that such an innovation process might have been slower. Factors like conservationism, geographical distance/isolation (Clark, 1961: 23-24) or cultural identity (the idea of social boundaries and resistance based on cultural identity is a pertinent — although largely unexplored — issue in this

discussion, as is shown in modern human groups by Barth, 1969) might have been responsible for a more diverse technological picture after 1.6 Ma. Finally, we might also add to these factors, a possible intermittent radiation through North Africa, taking into account the affinities reported between the ancestral traits observed in the earliest European human fossils of Atapuerca and Ceprano and Middle Pleistocene African specimens (Aguirre & Carbonell, 2001; Manzi *et al.*, 2001).

DISCUSSION AND CONCLUSIONS

The scope of this chapter has been to present in some detail the technological traits observed in the oldest archaeological occurrences of Europe. At this point, however, to talk about the *earliest European technology* also requires addressing the question of what constitutes the first settlement of the continent. The issue of the first settlement of Europe has been a matter of debate for decades.

I have discussed above the different positions taken by European scholars on this issue since the 1990's, using some key publications that can properly exemplify the two major competing hypotheses in recent years: the short chronology and the long chronology perspectives.

Although in the last few years research has provided fresh data about the oldest sites and their technological patterns, the debate is still open. For instance, while the paper by Roebroeks and van Kolfschoten (1994) has been very important in creating a push for a more critical approach to the purported 'very old' European sites, it does not completely resolve ambiguities still existing in the record they dismissed. It is important to note that, although evidence from sites such as Vallonet, Soleihac or the Upper Duero river terraces have problems in their interpretation (e.g., in terms of chronology, taphonomy, or small lithic samples), it is still impossible to firmly reject them. The lack of better data at a number of sites does not necessarily render them useless in this debate. It only speaks to the fact that, we will need additional information in order to state whether or not these cases are relevant to the pre-500 Ka. occupation pattern.

In order to present the earliest European technological traits, the archaeological occurrences that in our current state of our knowledge seem more reliable have been presented above. These are Barranco León, Fuente Nueva 3, Atapuerca TD6, Monte Poggiolo, and Isernia la Pineta. I consider that current interdisciplinary investigations at these occurrences provide reasonably good geological contexts, reliable chronology, and a wealth of archaeological data that identifies these as bona fide prehistoric sites within the covering the 1.0-0.5 Ma time interval. Isernia should be considered the youngest example within this time range. A definite chronology for this site is still a subject of study and discussion, but its archaeological significance is inescapable as per the data cited above.

All five archaeological assemblages considered in this survey exhibit opportunistic and non-systematic reduction patterns that *formally* fit within the Mode 1 technological complex. Much can be said about the meaning of these features, and I would agree that some assemblages, especially Fuente Nueva and Barranco León, are too small to make a conclusive technological diagnosis. Bearing in mind this problem and awaiting further information, a significant portion of this paper has been devoted to discussing the meaning of the technological features observed in these occurrences.

Some interpretations for their Mode 1 appearance could lead to the conclusion that the patterns observed might be due to specific adaptive reasons rather than to a *sensu stricto* pre-Acheulean technology. Much attention has been paid in the last years to the functional/adaptive influence in lithic technological variability. This perspective has proven to be highly valuable in our understanding of some formal or stylistic patterns (i.e. the Lower-Middle-Upper European Acheulean technological divisions do not represent a linear evolutionary progression, as was previously thought). However, as has been suggested for the "Clactonian question" (White, 2000: 54), none of the cases cited in this paper seem to clearly demonstrate that Mode 1 traits observed in pre-500 Ka. sites are basically due to functionality, ecological constraints, or raw material availability. In order to support an explanation based on variability within a Mode 2 technology, more information needs to be compiled.

At present, different timings of human incursions into North and South Europe appear to be the most plausible scenario, and these should have played a key role in the characterization of the dichotomy observed before and after 500 Ka. To state that a *proper* occupation of the continent would not have occurred until human populations reached northern areas runs the risk of taking an inappropriate north-Eurocentric perspective and underestimates both the European geographical and ecological diversity and our present-day archaeological knowledge.

It is very important not to mistake data quantity for data quality. From the quantitative perspective, it is obvious that the wealth of archaeological sites after 500 Ka. marks a substantial shift in the information available. Nowadays, this is an unchallengeable archaeological fact. But there are some important, further implications that archaeologists can infer from this pattern. First, the dramatic shift towards an increase in the number of sites implies that the 500 Ka chronological boundary marks the time by which human populations had acquired the adaptive skills needed to survive in these northern latitudes of Europe on a more continual basis for the first time. Literally, this migration event can be referred as to the first *conquest* or *acquisition* of the European landmass by human settlers and their descendants: *H. heidelbergensis* and *H. neanderthalensis*. Furthermore, the sparse archaeological evidence

prior to the 500 Ka boundary might inform us in important ways about skills and capabilities (e.g. adaptive skills or complexity of social networks) that the very first migrants might *not* have yet acquired to allow them to succeed in developing long-term settlements in the region.

If our intention is to reconstruct the culture history of the first human occupation of Europe at a continental scale and not just at specific regional scales, and also to fit this history within the long-term trajectory of hominid adaptations and colonization in the Old World (Roebroeks, 2001: 452), *both* the earliest incursions and the later, longer-term occupations necessarily complement each other, and both need to be taken into account. In this holistic framework, a number of connected questions can be posed: Within Europe, why did the Mediterranean region witness the first sparse traces of human occupation? Is it just a matter of proximity with other regions already occupied, i.e. North Africa, the Levant, or the Caucasus? Is it related to specific ecological reasons and/or geographical barriers? Why did the species involved in this occupation not succeed in terms of chronological continuity? What explains the ability of *Homo heidelbergensis* to settle and, literally, conquer the continent?

With the empirical data we do have, it is impossible to answer most of these questions and therefore provide a reasonable picture of the European migration events before the arrival of *Homo sapiens*. For obvious reasons, we are able to deal more comfortably with the post-500 Ka. occupation of Europe, and to provide a more complete set of explanations for the questions surrounding this phenomenon. A recent paper by Roebroeks (*ibid.*) summarizes the key reasons that might have allowed the permanent settlement of *H. heidelbergensis* in northern Europe, although his conclusions may also apply to the continent as a whole. It is important to keep in mind, however, that in omitting discussion of the earliest migratory pulse and its implications, we miss the opportunity to present a complete picture of this issue.

In sum, in our current state of knowledge available data (including the earliest hominin fossils from Atapuerca and Ceprano) suggest that the first human incursions into Europe, presumably sporadic and not completely successful, took place in the Mediterranean area before Mode 2 hominins spread throughout the continent. Although at present a clear correlation between the first arrival (or set of arrivals) and non-hand-axe occurrences can be established, the nature of this correlation is not properly clarified. It might be that we are seeing degeneration or drift away from original Mode 2 patterns due to mobility or social/cognitive constraints affecting earliest migrant groups, or some sort of technological continuity related to the first human presence in North Africa and the Caucasus and associated Mode 1 technologies. It is possible that we are just seeing a residual migratory example of the technological diversity present in Africa even after the origin of Acheulean technologies (with the co-occurrence of Mode 1). Whatever the case, it would seem reasonable to predict that, for the foreseeable future at least, questions surrounding the early migrations Out of Africa — their timing, their routes, and their associated technologies — will continue to play a major role in our understanding of the first European technologies.

ACKNOWLEDGMENTS

This paper has been made possible thanks to the financial support provided by the Basque Government through its post-doctoral fellowship program. I sincerely thank K. Schick and N. Toth for accepting my stay and facilitating my work at the CRAFT Research Center, Indiana University and Friends of CRAFT, Inc. for critical logistical support and access to resources that enabled this study.

REFERENCES CITED

Abbazzi, L., Fanani, F., Ferretti, M. P. & Rook, L. (2000). New human remains of archaic *Homo sapiens* and Lower Palaeolithic industries from Visogliano (Duino Ausina, Trieste, Italy). *Journal of Archaeological Science* **27**, 1173-1186.

Aguirre, E. & Carbonell, E. (2001). Early human expansions into Eurasia: The Atapuerca evidence. *Quaternary International* **75**, 11-18.

Agustí, J. & Moyà-Solà, S. (1987). Sobre la identidad del fragmento craneal atribuído a Homo sp, en Venta Micena (Orce, Granada). *Estudios Geológicos* **43**, 535-538.

Antoniazzi, A. & Piani, G. (1992). Il sito di Monte Poggiolo nell'ambito delle conoscenze geologiche regionali. In (C. Peretto, Ed) *I primi abitanti della Valle Padana: Monte Poggiolo nel quadro delle conoscenze europee.* Milano: Jaca Book, pp. 237-254.

Antoniazzi, A., Antoniazzi, A., Failla, A., Peretto, C. & Piani, G. (1996). The stratigraphy of the site of Ca'Belvedere di Monte Poggiolo. *Proceedings of the XIII International Congress of Prehistoric and Protohistoric Sciences. Forlì (Italy)* **6**, 2: 853-861.

Arribas, A. & Palmqvist, P. (1999). On the ecological connection between sabre-tooth and hominids: faunal dispersal events in the Lower Pleistocene and a review of the evidence for the first human arrival in Europe. *Journal of Archaeological Science* **26**, 571-585.

Ashton, N. & McNabb, J. (1992). The interpretation and context of the High Lodge industries. In (N. Ashton, N., J. Cook, S. Lewis, & J. Roe Eds) *High Lodge. Excavations by G. de Sieveking, 1962-8, and J. Cook, 1988.* London: British Museum Press, pp. 164-168.

Ashton, N. & McNabb, J. (1994): Bifaces in perspective. In (N. Ashton & A. David Eds) *Stories in stone.* London: Lithic Studies Society, occasional papers 4, pp. 182-191.

Ashton, N., McNabb, J., Irving, B., Lewis, S. & Parfitt, S. (1994). Contemporaneity of Clactonian and Acheulean flint industries at Barnham, Suffolk. *Antiquity* **68**, 585-589.

Ashton, N., Healy, F., & Pettit, P., Eds, (1998) Stone Age archaeology. *Essays in honour of John Wymer.* Oxford: Oxbow Monograph.

Ascenci, A., Mallegni, F., Manzi, G., Segre, A. & Segre Naldini, E. (2000). A re-appraisal of Ceprano calvaria affinities with *Homo erectus*, after the new reconstruction. *Journal of Human Evolution* **39**, 443-450.

Azzaroli, A., de Giuli, C., Ficcarelli, C. & Torre, G. (1988). Late Pliocene to Early mid-Pleistocene mammals in Eurasia: faunal succession and dispersal events. *Palaeogeography, palaeoclimatology, palaeoecology* **66**, 77-100.

Barth, F. (1969). Introduction. In (F. Barth Ed) *Ethnic groups and boundaries. The social organization of culture difference.* Boston: Little, Brown and Company, pp. 9-153.

Bar-Yosef, O. (1994). The Lower Palaeolithic of the Near East. *Journal of World Prehistory* **8**, 211-265.

Bar-Yosef, O. (1998) Early colonizations and cultural continuities in the Lower Paleolithic of western Asia. In (M. Petralia & R. Korisettar Eds) *Early human behavior in global context.* London: Routledge, pp. 231-279.

Bar-Yosef, O. & Kuhn, S. (1999). The big deal about blades: laminar technology and human evolution. *American Anthropologist* **101**, 322-338.

Bar-Yosef, O. & Belfer-Cohen, A. (2001). From Africa to Eurasia. Early dispersals. *Quaternary International* **75**, 19-28.

Bermúdez de Castro, J. M., Arsuaga, J. L., Carbonell, E., Rosas, A., Martínez, I. & Mosquera, M. (1997). A hominid from the Lower Pleistocene of Atapuerca, Spain: possible ancestor to Neanderthals and modern humans. *Science* **276**, 1392-1395.

Biberson, P. (1961). *Le Paléolithique inférieur du Maroc atlantique.* Rabat: Publications des Services des Antiquités du Maroc, 17.

Bonifay, E. (1991). Les premières industries du Sud-Est de la France et du Massif-Central. In (E. Bonifay & B. Vandermeersch Eds) *Les premiers européens.* Paris: Éditions du C.T.H.S., pp. 63-80.

Bonifay, E. & Vandermeersch, B. Eds (1991) *Les premiers européens.* Paris: Éditions du C.T.H.S.

Bonifay, E. & Vandermeersch, B. (1991). Vue d'ensemble sur le très ancien Paléolithique de l'Europe. In (E. Bonifay & B. Vandermeersch Eds) *Les premiers européens.* Paris: Éditions du C.T.H.S., pp. 309-318.

Brantingham, P., Olsen, J., Rech, J. & Krivoshapkin, A. (2000). Raw material quality and prepared core technologies in Northeast Asia. *Journal of Archaeological Science* **27**, 255-271.

Breuil, H. (1932). Les industries à éclat du Paléolithique ancien. *Préhistoire* **I-2**, 16-190.

Carbonell, E. & Rodríguez, X. P. (1994). Early Middle Pleistocene deposits and artefacts in the Gran Dolina site (TD4) of the Sierra de Atapuerca (Burgos, Spain). *Journal of Human Evolution* **26**, 291-311.

Carbonell, E., Bermúdez de Castro, J. M., Arsuaga, J. L., Díez, J. C., Rosas, A., Cuenca-Bescós, G., Sala, R.; Mosquera, M. & Rodríguez, X. P. (1995a). Lower Pleistocene hominids and artifacts from Atapuerca-TD6 (Spain). *Science* **269**, 826-830.

Carbonell, E., Mosquera, M., Rodríguez, X. P. & Sala, R. (1995b). The first human settlement of Europe. *Journal of Anthropological Research* **51**, 107-114.

Carbonell, E., García-Antón, M. D., Mallol, C., Mosquera, M., Ollé, A., Rodríguez, X. P., Sahnouni, M., Sala, R. & Vergès, J. M. (1999a). The TD6 level lithic industry from Gran Dolina, Atapuerca (Burgos, Spain): production and use. *Journal of Human Evolution* **37**, 653-693.

Carbonell, E., Márquez, B., Mosquera, M., Ollé, A., Rodríguez, X. P., Sala, R. & Vergès, J. M. (1999b). El Modo 2 en Galería. Análisis de la industria lítica y sus procesos técnicos. In (E. Carbonell, A. Rosas & J. C. Díez Eds) *Atapuerca: ocupaciones humanas y paleoecología del yacimiento de Galería.* Valladolid: Junta de Castilla y León, pp. 299-352.

Carbonell, E., Mosquera, M., Rodríguez, X. P., Sala, R. & Van der Made, J. (1999c). Out of Africa: the dispersal of the earliest technical systems reconsidered. *Journal of Anthropological Archaeology* **18**, 119-136.

Chavaillon, J. (1991). Les ensembles lithiques de Chilhac III (Haute-Loire). Typologie, situation stratigraphique et analyse critique et comparative. In (E. Bonifay & B. Vandermeersch Eds) *Les premiers européens.* Paris: Éditions du C.T.H.S., pp. 81-91.

Clark, G. (1977) *World Prehistory in new perspective* (third edition). Cambridge: Cambridge University Press.

Clark, J. D. & Kurshina (1979). Hominid occupation of the East-Central highlands of Ethiopia in the Plio-Pleistocene. *Nature* **282**, 33-39.

Clark, J. D. & Schick, K. (2000). Acheulean archaeology of the Eastern Middle Awash. In (J. Heinzelin, J. D. Clark, K. Schick & H. Gilbert Eds) *The Acheulean and the Plio-Pleistocene deposits of the Middle Awash Valley, Ethiopia.* Tervuren: Annales du Musée Royale de l'Afrique Centrale, pp. 51-121.

Coltorti, M., Cremaschi, M., Delitala, M., Esu, D., Fornaseri, M., McPherron, A., Nicoletti, M., Van Otterloo, R., Peretto, C., Sala, B., Schimidt, V. & Sevink, J. (1982). Reversed magnetic polarity at an early Lower Palaeolithic site in Central Italy. *Nature* **300**, 173-176.

Coltorti, M., Corrado, S., di Bucci, D., Marzoli, A., Saso, G., Peretto, C., Ton-That, T. & Villa, I. (2000). New chronostratigraphical and palaeoclimatic data from the Isernia la Pineta site. *The Plio-Pleistocene boundary and the Lower-Middle Pleistocene transition: tupe areas and sections. Abstracts.* (SEQS meetings, Bari, Italy, 25-29 September 2000).

Cosgrove, R.(1999). Forty-two degrees south: the archaeology of Late Pleistocene Tasmania Palaeoecology and Pleistocene occupation in south central Tasmania. *Journal of World Prehistory* **13-4**, 357-402.

Cremeschi, M. & Peretto, C. (1988). Les sols d'habitat du site Paléolithique d'Isernia la Pineta (Molise, Italie centrale). *L'Anthropologie* **92-4**, 1017-1040.

Crovetto, C. (1994). Le industrie litiche. Analisi tecnico-tipologica dei reperti di scavo. In (C. Peretto Ed) *Le industrie litiche del giacimento paleolitico di Isernia la Pineta.* Isernia: Cosmo Iannone, pp. 183-267.

Crovetto, C., Ferrari, M., Peretto, C. & Vianello, F. (1994). Le industrie litiche. La sperimentazione litica. In (C. Peretto Ed) *Le industrie litiche del giacimento paleolitico di Isernia la Pineta.* Isernia: Cosmo Iannone, pp. 119-182.

Cuenca-Bescós, G., Laplana, C. & Canudo, J. I. (1999). Biochronological implications of the *Arvicolidae* (Rodentia, Mammalia) from the Lower Pleistocene hominid-bearing level of Trinchera Dolina 6 (TD6, Atapuerca, Spain). *Journal of Human Evolution* **37**, 353-374.

de Lumley, H., Fournier, A., Krzepkowska, J. & Echasoux, A. (1988). L'industrie du Pleistocene inférieur de la grotte du Vallonet, Roquebrune-Cap-Martin, Alpes Maritimes. *L'Anthropologie* **92**, 501-614.

Dennell, R. (1983). *European economic Prehistory. A new approach.* London: Academic Press.

Dennell, R., Rendell, H. & Hailwoold, E. (1988) Early tool-making in Asia: two-million-year-old artefacts in Pakistan. *Antiquity* **62**, 98-106.

Dobosi, V. (1990). Description of the archaeological material. In (M. Kretzoi & V. Dobosi Eds) *Vértesszölös. Man, site and culture.* Budapest: Akadémiai Kiadó, pp. 311-395.

Diez-Martín, F. (2000). *El poblamiento paleolítico en los páramos del Duero.* Valladolid: Secretariado de Publicaciones de la Universidad de Valladolid.

Domínguez-Rodrigo, M., Alcalá, L., Luque, L. & Serrallonga, J. (2005). Some insights into the paleoecology and behavioral meaning of the early Oldowan and Acheulean sites at Peninj (West Lake Natron, Tanzania) during the Upper Humbu Formation. In (M. Sahnouni Ed) *Les cultures paléolithiques d'Afrique.* Paris: Éditions ArtCom, 129-156.

Falguères, C., Bahain, J-J., Yokoyama, Y., Arsuaga, J. L., Bermúdez de Castro, J. M., Carbonell, E., Bischoff, J. & Dolo, J-M. (1999). Earliest humans in Europe: the age of TD6 Gran Dolina. Atapuerca, Spain. *Journal of Human Evolution* **37**, 343-352.

Foley, R. (1987). Hominid species and stone-tool assemblages: how are they related?. *Antiquity* **61**, 380-392.

Foley, R. & Lahr, M. (1997). Mode 3 technologies and the evolution of modern humans. *Cambridge Journal of Archaeology* **7**, 3-36.

Gabunia, L., Vekua, A., Lordkipanidze, D., Swisher, C., Ferring, R., Justus, A., Nioradze, M., Tvalchrelidze, M., Antón, S., Bosinski, G., Jöris, O., de Lumley, M-A., Masjuradze, G. & Mouskhelishvili, A. (2000). Earliest Pleistocene hominid cranial remains from Dmanisi, Republic of Georgia: taxonomy, geological setting and age. *Science* **288**, 1019-1025.

Gabunia, L., Antón, S., Lordkipanidze, D., Vekua, A., Justus, A. & Swisher, C. (2001). Dmanisi and dispersal. *Evolutionary Anthropology* **10-5**, 158-170.

Gamble, C. (1999). *The Palaeolithic societies of Europe.* Cambridge: Cambridge University Press.

Gibert, J., Sánchez, F, Malgosa, A. & Martínez, B. (1994). Découvertes de restes humanis dans les gisements d'Orce (Granada, Espagne). *C.R..A..C.P.* **319-II**, 963-968.

Gibert, J., Campillo, D., Arques, J. M., García Olivares, E., Borja, C. & Lowenstein, J. (1998a). Hominid status of the Orce cranial fragment reasserted. *Journal of Human Evolution* **203**, 17.

Gibert, J., Gibert, L., Iglesias, A. & Maestro, E. (1998b). Two Oldowan assemblages in the Plio-Pleistocene deposits of the Orce region, southern Spain. *Antiquity* **72**, 17-25.

Giusberti, G. & Peretto, C. (1991). Evidences de la fracturation intentionnelle d'ossements animaux avec moëlle dans le gisement de La Pineta de Isernia (Molise), Italie. *L'Anthropologie* **95**, 765-778.

Hay, R. (1976). *The geology of Olduvai Gorge.* Oxford: Clarendon Press.

Hemingway, M. (1989). Early artefacts from Pakistan? Some questions for the excavators. *Current Anthropology* **30-3**, 317-318.

Isaac, G. (1981). Archaeological tests of alternative models of Early Hominid behaviour: excavation and experiments. *Philosophical transactions of the Royal Society of London* **292**, 177-188.

Jones, P. (1980). Experimental butchery with modern stone tools and its relevance for Palaeolithic archaeology. *World Prehistory* **12**, 153-165.

Kohn, M. & Mithen, S. (1999). Handaxes: products of sexual selection? *Antiquity* **73**, 518-526.

Kretzoi, M. & Vértes, L. (1965). Upper Biharian (Intermindel) pebble-industry occupation site on Western Hungary. *Current Anthropology* **6-1**, 74-87.

Klein, R. (1999). *The human career.* Chicago: University of Chicago Press.

Kuman, K. (1998). The earliest South African industries. In (M. Petralia & R. Korisettar Eds) *Early human behavior in global context.* London: Routledge, pp. 151-186.

Ludwig, B. & Harris, J. (1998). Towards a technological reassessment of East African Plio-Pleiocene lithic assemblages. In (M. Petralia & R. Korisettar Eds) *Early human behavior in global context.* London: Routledge, pp. 84-107.

Mallol, C. (1999). The selection of lithic raw materials in the Lower and Middle Pleistocene levels TD6 and TD10A of Gran Dolina (Sierra de Atapuerca, Burgos, Spain). *Journal of Anthropological Research* **55**, 385-407.

Mania, D. (1995). The earliest occupation of Europe: the Elbe-Saale region. In (W. Roebroeks & T. Van Kolfschoten Eds) *The earliest occupation of Europe.* Leiden: Analecta Praehistorica Leidensia. Leiden, pp. 85-101.

Mania, D., Mania, U., & Vlcek, E. (1999). The Bilzingsleben site. *Homo erectus*, his culture and his ecosphere. In (H. Ullrich Ed) *Hominid evolution. Lifestyle and survival strategies.* Berlin: Archaea, pp. 293-311.

Manzi, G., Mallegni, F. & Ascenci, A. (2001). A cranium for the earliest Europeans: phylogenetic position of the hominid form Ceprano, Italy. *PNAS* **98**, 10011-10016.

Martínez-Navarro, B., Turq, A., Agustí, J. & Oms, O. (1997). Fuente Nueva 3 (Orce, Granada, Spain) and the first human occupation of Europe. *Journal of Human Evolution* **33**, 611-620.

Martos-Romero, J. A. (1999). Elefantes e intervención humana en los yacimientos del Pleistoceno inferior y medio de África y Europa. *Trabajos de Prehistoria* **55-1**, 19-38.

Mithen, S. (1994). Technology and society during the Middle Pleistocene: hominid group size, social learning and industrial variability. *Cambridge Archaeological Journal* **4**, 3-32.

Moloney, N. (1996). The effects of quartzite pebbles on the technology and typology of Middle Pleistocene assemblages in the Iberian Peninsula. In (N. Moloney, L. Raposo & M. Santonja Eds) *Non-flint stone tools and the Palaeolithic occupation of the Iberian Peninsula.* BAR, International Series 649, pp. 107-124.

Morwood, M., O'Sullivan, P., Aziz, F. & Raza, A. (1998). Fission-track ages of stone tools and fossils on the east Indonesian island of Flores. *Nature* **392**, 173-176.

Moyà-Solà, A. & Köhler, M. (1997). The Orce skull: anatomy of a mistake. *Journal of Human Evolution* **33**, 91-97.

Mussi, M. (1995). The earliest occupation of Europe: Italy. In (W. Roebroeks & T. Van Kolfschoten Eds) *The earliest occupation of Europe.* Leiden: Analecta Praehistorica Leidensia. Leiden, pp. 27-51.

Ohel, M. (1979). The Clactonian: an independent complex or an integral part of the Acheulean? *Current Anthropology* **20-4**, 685-713.

Oms, O., Parés, J. M., Martínez-Navarro, B., Agustí, J., Toro, I., Martínez-Fernández, G. & Turq, A. (2000). Early human occupation of Western Europe: Paleomagnetic data for two Paleolithic sites in Spain. *PNAS* **19**, 10666-10670.

Otte, M. (1996). *Le Paléolithique inférieur et moyen en Europe.* Paris: Armand Colin.

Palmqvist, P., Pérez-Claros, J.A., Gibert, J. & Santamaría, J.L. (1996a). Comparative morphometric study of a human phalanx from the Lower Pleistocene site at Cueva Victoria (Murcia, Spain), by means of Fourier analysis, shape coordinates of landmarks, principal and relative warps. *Journal of Archaeological Science* **23**, 95-107.

Palmqvist, P., Martínez-Navarro, B. & Arribas, A. (1996b). Prey selection by terrestrial carnivores in Lower Pleistocene paleocommunity. *Paleobiology* **22-4**, 514-534.

Parés, J. M. & Pérez-González, A. (1995). Paleomagnetic age for hominid fossils at Atapuerca archaeological site, Spain. *Science* **269**, 830-832.

Parés, J. M. & Pérez-González, A. (1999). Magnetochronology and stratigraphy at Gran Dolina section, Atapuerca (Burgos, Spain). *Journal of Human Evolution* **37**, 325-342.

Peretto, C., Ed (1994). *Le industrie litiche del giacimento paleolitico di Isernia la Pineta.* Isernia: Cosmo Iannone.

Peretto, C., Ornella Amore, F., Antoniazzi, A., Antoniazzi, A., Bajain, J-J., Cattani, L., Cavallini, E., Esposito, P., Falguères, C., Gagnepain, J., Hedley, I., Laurent, M., Lebreton, V., Longo, L., Milliken, S., Monegatti, P., Ollé, A., Pugliese, N., Miskovsky-Renault, J., Sozzi, M., Ungaro, S, Vannuci, S., Vergès, J. M., J-J. Wagner & Yokoyama, Y. (1998). L'industrie lithique de Ca'Belvedere di Monte Poggiolo: stratigraphie, matière première, typologie, remontages et traces d'utilization. *L'Anthropologie* **102-4**, 243-365.

Petronio, C. & Sardella, R. (1999). Biochronology of the Pleistocene mammal fauna from Ponte Galeria (Rome) and remarks of the Middle Galerian faunas. *Rivista Italiana di Paleontologia e Stratigrafia* **105**, 155-164.

Piperno, M. (1999). Studio tecnico-tipologico e tipometrico delle schegge e degli strumenti su scheggia del livello Alfa. In (M. Piperno Ed) *Notarchirico. Un sito del Pleistocene medio iniziale nel bacino di Venosa.* Venosa: Ossana, pp. 309-337.

Piperno, M., Cassoli, P., Tagliacozzo, A. & Fiore, I. (1999). I livelli della serie di Notarchirico. In (M. Piperno Ed) *Notarchirico. Un sito del Pleistocene medio iniziale nel bacino di Venosa.* Venosa: Ossana, pp. 75-135.

Querol, M. A. & Santonja, M., Eds (1983) *El yacimiento de cantos trabajados de El Aculadero (Puerto de Santa María, Cádiz).* Excavaciones Arqueológicas en España, 106. Madrid: Ministerio de Cultura.

Raynal, J-P., Magoga, L. & Bindon, P. (1995). Tephrofacts and the first human occupation of the French Massif Central. In (W. Roebroeks & T. Van Kolfschoten Eds) *The earliest occupation of Europe.* Leiden: Analecta Praehistorica Leidensia. Leiden, pp. 129-146.

Raynal, J-P., Sbihi Alaoui, F., Geraads, D., Magoga, L. & Mohi, A. (2001). The earliest occupation of North-Africa. The Moroccan perspective. *Quaternary International* **75**, 65-75.

Raposo, L. & Santonja, M. (1995). The earliest occupation of Europe: the Iberian peninsula. In (W. Roebroeks & T. Van Kolfschoten Eds) *The earliest occupation of Europe.* Leiden: Analecta Praehistorica Leidensia, pp. 7-21.

Roberts, M., Gamble, C. & Bridgland, D. (1995). The earliest occupation of Europe: the British Isles. In (W. Roebroeks & T. Van Kolfschoten Eds) *The earliest occupation of Europe.* Leinden: Analecta Praehistorica Leidensia. Leiden, pp. 165-191.

Roebroeks, W. (2001). Hominid behaviour and the earliest occupation of Europe: an exploration. *Journal of Human Evolution* **41**, 437-461.

Roebroeks, W. & Van Kolfschoten, T. (1994). The earliest occupation of Europe: a short chronology. *Antiquity* **68**, 489-503.

Roebroeks, W. & Van Kolfschoten, T. Eds (1995). *The earliest occupation of Europe.* Leiden: Analecta Praehistorica Leidensia.

Roebroeks, W. & Van Kolfschoten, T. (1998). The earliest occupation of Europe: a view from the north. In (E. Aguirre Ed) *Atapuerca y la evolución humana.* Madrid: Fundación Areces, pp. 153-168.

Rolland, N. (1992). The Palaeolithic colonization of Europe: an archaeological and biogeographical perspective. *Trabajos de Prehistoria* **49**, 69-111.

Rolland, N. (1998). The Lower Palaeolithic settlement of Eurasia, with special reference to Europe. In (M. Petralia & R. Korisettar Eds) *Early human behavior in global context.* London: Routledge, pp. 187-219.

Sahnouni, M. & de Heinzelin, J. (1998). The site of Ain Hanech revisited: new investigations at this Lower Pleistocene site in Northern Algeria. *Journal of Archaeological Science* **25**, 1083-1101.

Santonja, M. (1995). El Paleolítico inferior en la Submeseta norte y en el entorno de Atapuerca. Balance de los conocimientos en 1992. *Evolución humana en Europa y los yacimientos de la Sierra de Atapuerca.* Valladolid: Junta de Castilla y León, pp. 421-444.

Santonja, M. & Villa, P. (1990). The Lower Paleolithic of Spain and Portugal. *Journal of World Prehistory* **4-1**, 45-94.

Santonja, M., López-Martínez, N. & Pérez-González, A. Eds (1980). *Ocupaciones achelenses en el Valle del Jarama (Arganda, Madrid).* Madrid: Diputación de Madrid.

Santonja, M., Pérez-González, A., Mora, R., Villa, P., Soto, E. & Sesé, C. (1997). Estado actual de la investigación en Ambrona y Torralba (Soria). *II Congreso de Arqueología Peninsular*, Zamora, pp. 51-56.

Schick, K. (1994). The Movius line reconsidered. Perspectives on the Earlier Paleolithic of Eastern Asia. In (R. Corruccini & R. Ciocon, R. Eds) *Paleoanthropological advances in honor of F. Clark Howell.* New Jersey: Prentice Hall, pp. 569-595.

Schick, K. & Toth, N. (2001). Paleoanthropology at the Millennium. In (G. Feinman & D. Price Eds.) *Archaeology at the Millennium: a sourcebook.* New York: Kluwer/Plenum, pp. 39-108.

Schick, K. & Zhuan, D. (1993). Early Paleolithic of China and Eastern Asia. *Evolutionary Anthropology* **2**: 22-34.

Svoboda, J. (1987). Lithic industries of the Arago, Vértesszöllös and Bilzingsleben hominids: comparison and evolutionary interpretation. *Current Anthropology* **28**, 219-227.

Schwartz, J. & Tattersall, I. (1996). Whose teeth? *Nature* **381**, 201-202.

Sozzi, M., Vannucci, S. & Vaselli, O. (1994). Le industrie litiche. La materia prima impiegata nella scheggiatura. In (C. Peretto Ed) *Le industrie litiche del giacimento paleolitico di Isernia la Pineta.* Isernia: Cosmo Iannone, pp. 45-86.

Swisher, C., Curtis, G., Jacob, G., Getty, T. & Suprijo, A. (1994). Age of the earliest known hominids in Java, Indonesia. *Science* **263**, 1118-1121.

Tattersall, I. (1997). Out of Africa again and again. *Scientific American* **276-4**, 46-53.

Thieme, H. (1999). The oldest spears in the world: Lower Palaeolithic hunting weapons from Schöningen, Germany. In (E. Carbonell *et al.* Eds) *The first Europeans: recent discoveries and current debate.* Burgos: Aldecoa, pp. 169-189.

Tieu, L. T. (1991). *Palaeolithic pebble industries in Europe.* Budapest: Akadémiai Kiadó.

Toth, N. (1982). The Stone Technologies of Early Hominids at Koobi Fora: An Experimental Approach. Ph. D. Dissertation, University of California at Berkeley.

Toth, N. (1987). Behavioral inferences from Early Stone artifact assemblages: an experimental model. *Journal of Human Evolution* **16**, 763-787.

Toth, N. & Schick, K. (1993) Early stone industries and inferences regarding language and cognition. In (K. Gibson & T. Ingold Eds) *Tools, language and cognition.* Cambridge: Cambridge University Press, pp. 346-362.

Turner, A. (1992). Large carnivores and earliest European hominids: changing determinants of resources availability during the Lower and Middle Pleistocene. *Journal of Human Evolution* **22**, 109-126.

Turner, A. (1999a). Assessing earliest human settlement of Eurasia: Late Pliocene dispersions from Africa. *Antiquity* **73**, 563-570.

Turner, E. (1999b). The problems of interpreting hominid subsistence strategies at Lower Palaeolithic sites: Miesenheim I. A case study from the Central Rhineland of Germany. In (H. Ullrich Ed) *Hominid evolution. Lifestyle and survival strategies.* Berlin: Archaea, pp. 365-382.

Vekua, A., Lordkipanidze, D., Rightmire, G.P., Agusti, J., Ferring, R., Maisuradze, G., Mouskhelishvili, A., Nioradze, M., Ponce de Leon, M., Tappen, M., Tvalchrelidze, M., & Zollikofer, C. (2002) A New Skull of Early *Homo* from Dmanisi, Georgia. *Science* 297:85-89.

Villa, P. (1983). *Terra Amata and the Middle Pleistocene archaeological record of Southern France.* Berkeley: University of California Press.

Villa, P. (1991). Middle Pleistocene Prehistory in southwestern Europe: the state of our knowledge and ignorance. *Journal of Anthropological Research* **47-2**, 193-217.

Villa, P. (1996). The First Italians. *Lithic Technology* **21**, 71-79.

Villa P. (2001). Early Italy and the colonization of Western Europe. *Quaternary International* **75**, 113-130.

Wanpo, H., Ciocon, R., Yumin, G., Larick, R., Quiren, F., Schwarcz, H., Yonge, C., Vos, J. & Rink, W. (1995). Early Homo and associated artefacts from Asia. *Nature* **378**, 275-278.

Wenban-Smith, F. (1998). Clactonian and Acheulean industries in Britain: their chronology and significance reconsidered. In (N. Ashton, F. Healy & P. Pettitt Eds) *Stone Age archaeology. Essays in honour of John Wymer.* Oxford: Oxbow Monograph, pp. 90-104.

White, M. (2000). The Clactonian question: on the interpretation of core-and-flake assemblages in the British Lower Paleolithic. *Journal of World Prehistory* **14-1**, 1-63.

Yokoyama, Y., Bibron, R. & Falguères, C. (1988). Datation absolue des planchers stalagmitiques de la grotte du Vallonet, Roquebrune-Cap-Martin (Alpes-Maritimes, France) par la résonance de spin électronique (ESR). *L'Anthropologie* **92**, 429-436.

Yokoyama, Y., Bahain, J-J., Falgueres, C. & Gagnepain, J. (1992). Tentative datation par la méthode de la résonance de spin électronique (ESR) de sédiments quaternaires de la région de Forli. In (C. Peretto Ed) *Il più antico popolamento della Valle Padana nel quadro delle conoscenze europee. Monte Poggiolo.* Milano: Jaca Book, pp. 337-346.

PART II:
ACTUALISTIC APPROACHES TO THE EARLIEST STONE AGE

CHAPTER 6

A COMPARATIVE STUDY OF THE STONE TOOL-MAKING SKILLS OF *PAN*, *AUSTRALOPITHECUS*, AND *HOMO SAPIENS*

By Nicholas Toth, Kathy Schick, and Sileshi Semaw

ABSTRACT

An experimental program was designed to compare and contrast the stone tool-making skills of modern African apes (bonobos or *Pan paniscus*), of prehistoric tool-making hominins from the earliest known Palaeolithic sites at Gona, Ethiopia (sites EG 10 and EG 12) dating to approximately 2.6 million years ago (possibly *Australopithecus garhi*), and of modern humans or *Homo sapiens*. All three species used the same range of raw materials, unmodified water-rounded cobbles of volcanic rock from the 2.6 Ma river conglomerates in the Gona study area. A detailed attribute analysis of the three samples was conducted, examining flaking patterns and artifact products and, from these products, inferring relative levels of stone tool-making skill in the three species.

Results of this comparative analysis indicate that, in the majority of artifact attributes that appear to be linked with skill, the Gona hominin tool-makers grouped either with the modern human sample or were intermediate between the bonobos and modern humans. This indicates that the biomechanical and cognitive skills required for efficient stone tool-making were already present at Gona by 2.6 Ma. Although some of the individual stone artifacts generated by the bonobos [with cranial capacities and probable EQ (encephalization quotient) values similar to estimates for prehistoric hominins contemporary with the Gona sites] would clearly be recognized as artifactual by palaeolithic archaeologists, the label "Pre-Oldowan" might be more appropriate for the overall assemblage of artifacts they produce. The level of flaking skill seen in the bonobo assemblage may represent an earlier stage of lithic technology not yet discovered in the prehistoric record. Or, alternatively, if the Gona sites indeed represent the earliest phases of stone tool-making with no precursors to be found, it may be that by the time of the Gona sites early hominins were already "pre-adapted" to more skilled flaking of stone.

This study also highlights interesting aspects of the stone tool-making behaviors of the early tool-makers at Gona by 2.6 Ma: 1) the Gona hominins were selective in choosing raw materials, sometimes selecting excellent quality raw materials from the river gravels available within the Gona region; 2) the Gona hominins conducted earlier stages of the reduction of cobbles off-site, prior to the transport of cores to the floodplain sites; and 3) the Gona hominins likely transported numerous larger, more usable flakes away from the floodplain sites. This suggests a higher level of early tool-making complexity, and presumably subsistence complexity, than many prehistorians have appreciated.

INTRODUCTION

The evolution of technology has occurred in tandem with human biological evolution during at least the past 2.6 million years. This unique co-evolution ultimately has led to the modern human condition, and it is likely that major cognitive and biomechanical changes occurred during this time through selection for more efficient tool-related activities. Different stages or grades of human evolution tend to be associated with different levels of technology, with a general trend of increasing technological complexity and sophistication through time.

A persistent challenge in paleoanthropology is how to determine levels of cognitive and biomechanical skill based upon the archaeological record, which consists primarily of flaked stone artifacts and, sometimes, associated faunal remains. It would be interesting to be able to observe and compare different levels of flaked stone technology between prehistoric hominins and extant apes and humans. Unfortunately, Early Stone Age hominins are extinct, and modern apes are not known to flake stone intentionally in the wild. However, the lasting products of early tool-making persists in the form of stone artifacts at early archaeological sites at Gona in Ethiopia, and, beginning in 1990, captive bonobos have been producing flaked stone artifacts in an experimental setting (Toth *et al.*, 1993; Schick *et al.*, 1999). This provides a unique opportunity to conduct a three-way comparison of tool-making patterns evident in the earliest known tool-makers, in stone tool-making apes, and in modern human knappers, in order to investigate technological patterns and abilities evident in each group.

This study, part of a long-term investigation of stone tool-making and tool-using abilities in captive African apes, is an attempt to approach this problem through rigorous comparisons of the artifacts produced by the earliest stone tool-makers and those produced by bonobos and modern humans in controlled experiments. In this study, the bonobos and modern humans used volcanic cobbles from the same river gravels that had served as the source of raw materials for the early Gona stone tool-makers. Thus, all three samples, the prehistoric tool-makers and the two groups of experimental tool-makers, were effectively using the same raw material source.

This study provides a valuable three-species comparison (probably in three different genera) of stone tool-making and spanning a time period of 2.6 million years. This makes it possible to make detailed comparisons of the stone technologies of the earliest known stone tool-makers, those produced by modern humans, and those made by modern apes. Inferences can then be made regarding discrete attributes, and combinations of attributes, that emerge as sensitive indicators of relative levels of skill. Further insights can thereby be gained regarding behavioral implications of the early archaeological sites at Gona, as well as regarding ape stone tool-making abilities and a possible evolutionary 'substrate' for the development of technological skills in human evolution.

THE COMPARATIVE SAMPLES

In effect, then, this study is a comparison of technological skill between three species over some 2.6 million years, with good control over raw materials. These three different samples whose artifacts will be compared and contrasted are: 1) an experimental sample produced by African apes who are practiced in stone tool manufacture (bonobos); 2) an archaeological sample produced by early hominins (at Gona, Ethiopia); and 3) an experimental sample produced by modern humans who are experienced stone tool-makers.

African Apes

The African ape sample consisted of two bonobos ("pygmy chimpanzees"), Kanzi and his half-sister Panbanisha, both born and raised in captivity with daily

human contact. At the time of these experiments Kanzi was twenty years old and Panbanisha fifteen (Figures 1 through 12). Kanzi had been knapping stone for ten years by this time and Panbanisha for four years. The average cranial capacity of *Pan* about is 380 cc, and the Homocentric EQ (a human-centered encephalization quotient, or ratio of brain size to body size, in which human EQ=1.0) is 0.38 (Holloway, 2000). The bonobos have learned to flake stone to produce large, usable flakes as cutting tools (cutting through a rope or membrane to access a food resource). They were encouraged to reduce cobbles as far as possible with a stone hammer to produce a range of

Figure 1

1.   *Portrait of Kanzi.*

Figure 2

2.   *Portrait of Panbanisha.*

Figure 3

3. *Kanzi flaking a cobble core.*

Figure 4

4. *Kanzi flaking another cobble core.*

Figure 5

5. *Kanzi using a flake to cut through rope to access a food resource.*

flakes and fragments from which they could select a tool to use for the cutting activity. Since this study, two of Panbanisha's sons (Nyota, now eight years old and Nathan, now six years old) have started flaking stone, primarily from observing Kanzi and Panbanisha. In effect, we now have set forth a flaked lithic cultural tradition in this bonobo group which is now long-term and transgenerational.

Biomechanical studies by Harlacker (2006), studying the kinesiology of Oldowan tool-making in bonobos and in unskilled and skilled modern humans have demonstrated that the bonobos are only accelerating their hammerstones to about one-half of the impact velocities of the expert human sample (3.67 meters/second versus 7.12 meters/second, respectively). This lower impact velocity of the bonobos will almost certainly affect their lithic assemblage, which will be discussed below.

The history of the bonobo acquisition of stone tool-making is presented in Savage-Rumbaugh and Fields (this volume). (For more details on bonobo acquisition and development of stone tool-making skills, see Toth *et al.*, 1993; Schick *et al.*, 1999; and Savage-Rumbaugh *et al.*, 2006). The study of bonobo stone tool-making is part of a long-term, ongoing research program that will soon also include chimpanzees (*Pan troglodytes*) and orangutans (*Pongo pongo*) as subjects.

Gona Hominins

The archaeological sample was from two contemporaneous sites in the Afar Rift at Gona, Ethiopia named EG (East Gona) 10 and EG 12, located a few hundred meters from each other. These two sites are almost identical in their lithic technology (Semaw, 1997), so for this study the two sites were combined for a statistically larger sample size. These sites have been dated to ap-

Figure 6

6. *Sample of bonobo cores. Note the predominance of end choppers. (Small squares on scale represent one cm).*

Figure 7

7. *Bonobo unifacial end chopper.*

Figure 8

8. *Large bonobo end chopper with some bifacial flaking.*

Figure 9

9. Large bonobo unifacial side chopper.

Figure 10

10. Bonobo unifacial chopper.

Figure 11

11. *Bonobo end chopper exhibiting very heavy hammerstone battering.*

Figure 12

12. *Sample of bonobo flakes. Note the abundance of side-struck flakes. Flakes oriented with striking platforms at top.*

proximately 2.6 Ma. Although the species of tool-maker is not known, the only species known in the Afar Rift during this time period is *Australopithecus garhi*, known from the Middle Awash some 60 km from Gona (Asfaw *et al.*, 1999; White *et al.*, 2005). The cranial capacity of this taxon, based on one skull from Bouri, Middle Awash (the holotype, BOU-VP-12/130), is 450 cc, about 70cc larger than the *Pan* mean. The EQ value for *A. garhi* is not known. Fossils representing a non-robust *Australopithecus* that may also be attributable to *A. garhi* include a mandible fragment (GAM-VP-1/1) and parietal fragments (GAM-VP-1/2) from Gamedah, Middle Awash (White *et al.*, 2005) and isolated teeth from the Omo Shungura Formation (Suwa *et al.*, 1996). (See Figures 13, *A. garhi*, and 14 through 23, setting of and artifacts from the Gona EG sites).

Geoarchaeological evidence suggests that there has not been significant hydrological action at theses sites: they are found in fine-grained slickenside clays and contain not only heavier cores but also a debitage sample including very small flakes and fragments (Semaw, 1997, 2000; Semaw *et al.*, 1997, 2003). Thus far, fossil bone has been found only on the surface at these archaeological localities and may have derived from higher deposits, but cut-marked bones have been found in the same stratigraphic level elsewhere in the study area (Dominguez-Rodrigo *et al.*, 2005).

Modern Humans

The modern human sample consisted of two experienced stone tool-makers (NT and KS) that had, at the time of these experiments, flaked stone for over two decades. The mean cranial capacity for modern humans is about 1350 cc, with a Homocentric EQ value of 1.0. Gona cobbles were flaked unifacially and bifacially to reduce them by roughly half of their original cobble mass. As the aim was to produce a control sample of cores and resultant debitage representing approximately 50% cobble reduction, rather than to replicate precisely the Gona assemblage, there was somewhat more bifacial flaking in the human sample than was represented at the Gona sites (roughly 68% of the human flakes are from unifacial flaking, versus ~79% of Gona flakes). Direct, freehand, hard-hammer percussion was employed using Gona cobble percussors, and no special attempt was made to prepare platforms or remove especially long or thin flakes: the goal was to produce serviceable flakes that could be used for cutting activities. The human experimental sample thus provided an important baseline that could be compared to the bonobo

and archaeological samples (Figures 21 and 22) to examine for similarities and differences with the bonobo and Gona stone technologies.

As analysis proceeded, it became clear that preferentially later stages of cobble reduction were typical of the Gona EG sites, whereas the experimental samples of bonobo and human reduction differed in two major aspects: they contained all stages of flaking, from initiation of cobble reduction to cessation of flaking, and, overall, their cores were not as extensively reduced. In the case of the human experiments, this had been a deliberate design to produce a control sample of cores and debitage with reduction of the cobble mass held at approximately

Figure 13

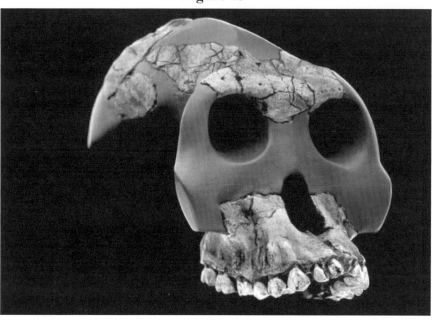

13. *Reconstructed cranium of Australopithecus garhi. Original housed in National Museum of Ethiopia, Addis Ababa. © 1999 David L. Brill*

50%, in order to effectively compare the products to the other two samples. Comparisons with the experimental human sample highlighted the much heavier reduction of the Gona cores. Thus, in an attempt to more closely match the Gona pattern, ten of the experimental human cores (five unifacial choppers, five bifacial choppers) were subsequently reduced further in order to examine assemblage characteristics in such later stages of core reduction. These "later stage" experiments provided a database that were then used in more direct comparisons with the Gona sites, and especially salient results of these comparisons are highlighted in special sections in this analysis.

A major aim of the human experimental sample was to produce data to generate models to understand how the Gona archaeological assemblages could have formed. As will be discussed below, this experimentation has suggested that the Gona sites represent preferentially later stages of core reduction with subsequent selection of certain artifacts that were transported off-site. Another

Figure 14

14. *Gona site EG 10, dated to approximately 2.6 mya.*

Figure 15

15. *The fossil river gravel conglomerate at East Gona: the source of raw material for the Gona hominins and the experimental sample.*

Figure 16

16. *Gona artifacts from EG 10: cores (below) and flakes (above).*

Figure 17

17. *Gona unifacial side choppers from EG 10.*

Figure 18

18. *A set of six Gona EG cores. (Photo courtesy of Tim White).*

Figure 19

19. *The other opposite face of the six Gona cores in Figure 18. (Photo courtesy of Tim White)*

Figure 20

20. *More flakes from Gona EG 10.*

important aim of this study was to compare the human sample with the bonobo sample (each having all stages of reduction represented) to evaluate levels of skill and gain insight into possible stages of development of stone technology in the course of hominin evolution.

METHODOLOGY

This study was designed as a detailed examination of the lithic assemblages produced by three samples: modern African apes (*Pan paniscus*, called bonobos or "pygmy chimpanzees"), prehistoric Gona hominins (possibly *Australopithecus garhi*), and modern humans or *Homo sapiens*. This analysis would also address basic epistemological questions in Palaeolithic archaeology: why do we measure and record the attributes that we do? Beyond pure description, what do the attributes that we study tell us about levels of cognitive and/or biomechanical skills, and what stages of core reduction (e.g. early, late) are represented in an assemblage, and what does this imply regarding transport and land-use patterning?

Analysis and Statistics

It was decided that this analysis would include the detailed statistical descriptions and testing that was employed. For more qualitative attributes (e.g. flake scar types, flake shapes) chi-square tests were employed, while for metrical, quantitative attributes (e.g. linear measurements, weights, and angles) the Mann-Whitney/Wilcoxon test (and sometimes additionally the Kolmogorov-Smirnov test) was used. In both cases, the

threshold for assessing statistical significance was at the .05 confidence level. A summary of statistical tests and overview of results are presented in appendices at the end of this chapter.

Raw Materials

The raw materials for the experiments were selected from the 2.6 Ma river gravels at East Gona, the source for the Gona hominins as well. With the permission of the Ethiopian government, unmodified cobbles (i.e. geological samples) were collected from the surface scatter and in situ gravels. In practice, only one of perhaps every fifty cobbles in the Gona conglomerates was considered suitable for flaking. Cobbles were chosen for 1) their smooth cortex, suggesting a fine-grained rock type; 2) shapes that would be suitable for flaking: not too spherical and not too thin; 3) a range of sizes (from about 10 to 20 cm); and 4) absence of heavy cortical pitting or hairline fractures, which would suggest a poor-quality rock with unwanted vesicles or weathering flaws. All raw materials in the experimental sample (and it appears almost all from the archaeological sites) started as water-worn river pebbles composed of a range of volcanic raw materials.

The experimental sample of raw material was randomly divided into two groups, one for the bonobo subjects and one for the human sample. As will be discussed, the experimental raw material sample was very similar to that exploited by the Gona hominins, so that there was an excellent control of raw materials between the experimental and archaeological samples. The hammer-

Figure 21

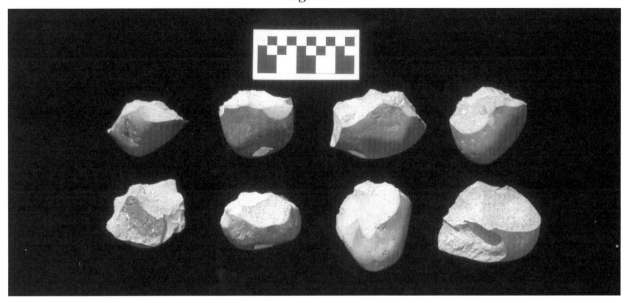

21. *A selection of cores from the human experimental sample. Note the predominance of side choppers.*

Figure 22

22. *A selection of flakes from the human experimental sample. Note the abundance of end-struck flakes.*

stones used to flake cores were also Gona cobbles. Several Gona cores show battering on their cortical surfaces showing that they were also used as percussors, and it is likely that other hammerstones were transported off-site. For this analysis, only stone artifacts greater or equal to 20 mm in maximum dimension were examined.

ATTRIBUTES ANALYZED

The following lithic attributes were used in this study:

1. Assemblage composition (lithic class: cores/flakes/fragments)
2a. Debitage breakdown [whole flakes/broken flakes (splits & snaps)/angular fragments & chunks]
2b. Ratio of split flakes to whole flakes
3. Raw material breakdown (cores)

Figure 23

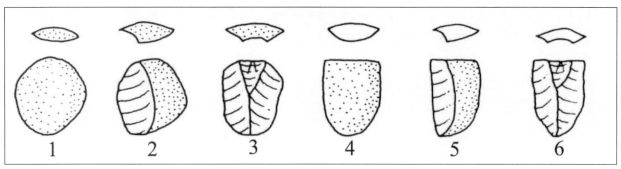

23. *Flake types. Flake types 1-3: cortical platforms with, in order, all dorsal cortex, partial dorsal cortex, and no dorsal cortex. Flake types 4-6: non-cortical platforms with, in order, all dorsal cortex, partial dorsal cortex, and no dorsal cortex. Examination of these flake types in an artifact assemblage can yield critical technological information regarding core reduction.*

4.   Quality of raw material (cores)
5.   Cores: Original form (blank)
6.   Cores: Flaking mode (Unifacial, bifacial, etc.)
7.   Cores: Invasiveness of flaking
8.   Cores: Percentage of circumference flaked
9.   Cores: Quality of flaking
10. Cores: Platform edge battering/microstepping/ crushing
11. Cores: Modified Leakey typology
12. Cores: Weight
13. Cores: Maximum dimension (length)
14. Cores: Breadth
15. Cores: Thickness
16. Cores: Ratio of breadth to length
17. Cores: Modified ratio of breadth to length
18. Cores: Ratio of thickness to breadth
19. Cores: Maximum dimension of largest scar
20. Cores: Ratio of largest scar to core maximum dimension
21. Cores: Percentage of original cobble remaining (estimate)
22. Cores: Number of flake scars
23. Cores: Ratio of step & hinge scars to total scars
24. Cores: Edge angle
25. Cores: Percentage of surface cortex remaining
26a. Flakes: Flake types
26b. Flakes: Simulated flake type population after selection and the removal of the best flakes
26c. Flakes: Simulated flake type population: later stages of flaking
26d. Flakes: Simulated flake type population: later stages after selection and removal
27. Flakes: Location of Cortex
28. Flakes: Dorsal scar pattern
29. Flakes: Flake shape
30. Flakes: Platform battering/microstepping, crushing
31. Flakes: Weight
32. Flakes: Maximum dimension
33. Flakes: Thickness
34. Flakes: Ratio of flake breadth to length
35. Flakes: Ratio of flake thickness to breadth
36. Flakes: Ratio of platform thickness to platform breadth
37. Flakes: Number of platform scars
38. Flakes: Number of dorsal scars
39. Flakes: Ratio of step & hinge scars to all scars
40. Flakes: Exterior platform angle ("core angle")
41. Flakes: Interior platform angle ("bulb angle")
42. Flakes: Percentage of dorsal cortex ("cortex index")

RESULTS

In this study, these 42 separate attributes, both qualitative and quantitative, were examined and compared/ contrasted among the three lithic assemblages, and then interpreted with regard to similarities and differences. When appropriate, statistical tests were conducted to examine for significant differences among the three toolmaker samples in their attribute characteristics.

1. Assemblage: Composition (Lithic Class)

Rationale

The major classes of flaked stone artifacts consist of cores, flakes, and fragments. Fragments comprise a variety of broken portions of flakes, including split flakes (broken more or less perpendicular to the platform), snapped flakes (broken more or less parallel to the platform, into proximal, distal, and sometimes also mid-section snaps), angular fragments (fairly flat fragments, apparent portions of conchoidally flaked pieces but without clear indication of which part is represented), and chunks (miscellaneous, usually thicker, nondescript fragments of conchoidally flaked materials). Here we examine whether there are significant differences in the major flaked artifact categories among the three toolmaker samples (bonobos, Gona knappers, humans).

Results

The overall breakdown of the assemblages into the major lithic classes (cores, flakes, and fragments) by tool-maker is shown in the table and graph below:

Assemblage Composition by Tool-maker

			Core flake frag			
			core	flake	frag	Total
Tool-maker	bonobo	Count	33	158	130	321
		% within Tool-maker	10.3%	49.2%	40.5%	100.0%
	gona	Count	23	178	294	495
		% within Tool-maker	4.6%	36.0%	59.4%	100.0%
	human	Count	31	244	370	645
		% within Tool-maker	4.8%	37.8%	57.4%	100.0%
Total		Count	87	580	794	1461
		% within Tool-maker	6.0%	39.7%	54.3%	100.0%

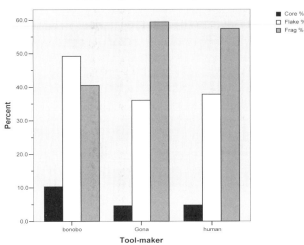

Assemblage Composition by Tool-maker

Chi-Square tests revealed that the Gona and human tool-makers group together in terms of their general assemblage composition (.788 level of significance), while the artifact assemblages produced by the bonobo tool-makers were significantly different from both the Gona and the human samples (at the .000 level of significance). The bonobo assemblage differed from both the Gona and human samples in its much higher core proportion, higher flake proportion, and much lower fragment proportion.

Chi-Square Test: Assemblage Composition, Bonobo v. Gona

	Value	df	Asymp. Sig. (2-sided)
Pearson Chi-Square	30.703[a]	2	.000
Likelihood Ratio	30.716	2	.000
N of Valid Cases	816		

a. 0 cells (.0%) have expected count less than 5. The minimum expected count is 22.03.

Chi-Square Test: Assemblage Composition, Bonobo v. Human

	Value	df	Asymp. Sig. (2-sided)
Pearson Chi-Square	28.157[a]	2	.000
Likelihood Ratio	27.888	2	.000
N of Valid Cases	966		

a. 0 cells (.0%) have expected count less than 5. The minimum expected count is 21.27.

Chi-Square Test: Assemblage Composition, Gona v. Human

	Value	df	Asymp. Sig. (2-sided)
Pearson Chi-Square	.478[a]	2	.788
Likelihood Ratio	.478	2	.787
N of Valid Cases	1140		

a. 0 cells (.0%) have expected count less than 5. The minimum expected count is 23.45.

Discussion

While the Gona and human proportions of cores, flakes, and fragments are similar, the bonobo sample has a higher core percentage (more than double the assemblage percent in the Gona and human samples), indicating the bonobos are removing less debitage per core. Interestingly, they are producing more whole flakes than fragments, while in both the Gona and human samples there are proportionally more fragments than whole flakes. While some lithic analysts might argue that a higher flake-to-fragment ratio may be an indication of greater skill, it is unlikely in the case of the bonobos. Their hammerstone velocities were much lower than the modern human subjects, and the bonobos would often repeatedly strike the core at a location until fracture finally occurred and proceed to "chew" down the cobble with further flaking. The majority of spalls so produced by the bonobos tend to be whole flakes. In contrast, modern humans (and probably Gona hominins) appear to have exploited cores more efficiently, reducing raw material in flake production more readily and with much higher hammerstone velocities, in the process producing more shatter (fragments) relative to whole flakes.

2a. Assemblage: Debitage Breakdown

Rationale

The more subtle differences between debitage categories (whole flakes, split and snapped flakes, and angular fragments/chunks) can be a useful means of comparing different lithic assemblages. Although we had no clearly-defined expectations, we were interested to see if there were any major differences between the tool-making groups in the types of debitage they produced.

Results

The overall breakdown of the debitage type by tool-maker is shown in the table and graph below:

Debitage Breakdown by Tool-maker
(Whole Flakes, Broken Flakes, Angular Fragments)

			Deb type			Total
			Flake	Flk brok	Frag-ment	
Tool-maker	bonobo	Count	158	34	96	288
		% within Tool-maker	54.9%	11.8%	33.3%	100.0%
	gona	Count	178	95	199	472
		% within Tool-maker	37.7%	20.1%	42.2%	100.0%
	human	Count	244	226	144	614
		% within Tool-maker	39.7%	36.8%	23.5%	100.0%
Total		Count	580	355	439	1374
		% within Tool-maker	42.2%	25.8%	32.0%	100.0%

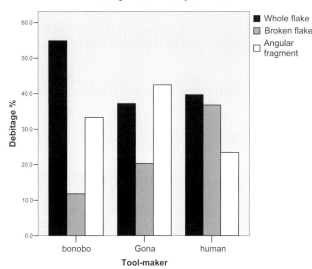

Debitage Breakdown by Tool-maker

Chi-Square tests revealed that each of the three tool-making groups produced debitage breakdowns that were statistically different from the other two groups (at the .000 level of significance).

Chi-Square Test: Debitage Breakdown, Bonobo v. Gona

	Value	df	Asymp. Sig. (2-sided)
Pearson Chi-Square	22.786[a]	2	.000
Likelihood Ratio	22.959	2	.000
N of Valid Cases	760		

a. 0 cells (.0%) have expected count less than 5. The minimum expected count is 48.88.

Chi-Square Test: Debitage Breakdown, Bonobo v. Human

	Value	df	Asymp. Sig. (2-sided)
Pearson Chi-Square	59.767[a]	2	.000
Likelihood Ratio	66.431	2	.000
N of Valid Cases	902		

a. 0 cells (.0%) have expected count less than 5. The minimum expected count is 76.63.

Chi-Square Test: Debitage Breakdown, Gona v. Human

	Value	df	Asymp. Sig. (2-sided)
Pearson Chi-Square	54.975[a]	2	.000
Likelihood Ratio	55.655	2	.000
N of Valid Cases	1086		

a. 0 cells (.0%) have expected count less than 5. The minimum expected count is 139.51.

Discussion

Bonobos produced higher percentages of whole flakes and lower percentages of broken flakes than the other two samples. As mentioned above, it is likely that this pattern results from knapping with lower hammerstone velocities, so that less spontaneous flake fragmentation occurs during the manufacturing process. To test this, as a follow-up experiment, we reduced cobbles with high-velocity/high-impact hammerstone percussion as well as low-velocity/low-impact hammerstone percussion. The low-velocity/low-impact flaking yielded debitage with an appreciably a greater proportion of whole flakes (31% higher) than did high-velocity/high-impact flaking. This supports our contention that the higher percentage of whole flakes in the bonobo sample and the smaller mean flake size are primarily due to their lower-velocity impacts.

In addition, the Gona sample has appreciably more angular fragments and fewer broken flakes than the human sample. It is possible that the elevated proportion of angular fragments may partially be a result of post-depositional breakage of debitage (Hovers, 2003). Also, some subtle signs of conchoidal fracture are more difficult to identify on volcanic archaeological specimens, as their fractured surfaces tend to be slightly weathered; therefore it is likely that some pieces that would be assigned to splits or snaps on fresh specimens would be demoted to angular fragments on archaeological specimens. Finally, as will be discussed below, it appears that a portion of the larger, sharper whole flakes and broken flakes at Gona may have been subsequently transported off-site, which would increase the percentage of angular fragments in the debitage population.

2b. Ratio of Split to Whole Flakes

Rationale

A higher ratio of split flakes to whole flakes could be an indication of the average impact velocity of hammerstones when flaking cores. As noted above, experiments have shown that for a given raw material, higher percussion forces tend to produce higher ratios of split to whole flakes (but also larger whole flakes).

Results

The bonobos had the lowest number of split to whole flakes (ratio of .095, 8.7% split flakes), with Gona hominins intermediate (ratio of .331, 24.9% split flakes) and the human sample the highest (ratio of .643, 39.2% split flakes). Each of these samples showed significant difference from the other two in their whole and split flake proportions at the .05 confidence level.

Whole and Split Flakes by Tool-maker

			Whole or split flake		
			split	whole	Total
Tool-maker	bonobo	Count	15	158	173
		% within Tool-maker	8.7%	91.3%	100.0%
	gona	Count	59	178	237
		% within Tool-maker	24.9%	75.1%	100.0%
	human	Count	157	244	401
		% within Tool-maker	39.2%	60.8%	100.0%
Total		Count	231	580	811
		% within Tool-maker	28.5%	71.5%	100.0%

Chi-Square Test: Whole and Split Flakes, All Tool-makers

	Value	df	Asymp. Sig. (2-sided)
Pearson Chi-Square	57.243[a]	2	.000
Likelihood Ratio	64.190	2	.000
N of Valid Cases	811		

a. 0 cells (.0%) have expected count less than 5. The minimum expected count is 49.28.

Chi-Square Test: Whole and Split Flake Proportions, Bonobo v. Gona

	Value	df	Asymp. Sig. (2-sided)	Exact Sig. (2-sided)	Exact Sig. (1-sided)
Pearson Chi-Square	17.796[b]	1	.000		
Continuity Correction[a]	16.716	1	.000		
Likelihood Ratio	19.135	1	.000		
Fisher's Exact Test				.000	.000
N of Valid Cases	410				

a. Computed only for a 2x2 table.

b. 0 cells (.0%) have expected count less than 5. The minimum expected count is 31.22.

Chi-Square Test: Whole and Split Flake Proportions, Bonobo v. Human

	Value	df	Asymp. Sig. (2-sided)	Exact Sig. (2-sided)	Exact Sig. (1-sided)
Pearson Chi-Square	53.509[b]	1	.000		
Continuity Correction[a]	52.066	1	.000		
Likelihood Ratio	62.038	1	.000		
Fisher's Exact Test				.000	.000
N of Valid Cases	574				

a. Computed only for a 2x2 table.

b. 0 cells (.0%) have expected count less than 5. The minimum expected count is 51.84.

Chi-Square Test: Whole and Split Flake Proportions, Gona v. Human

	Value	df	Asymp. Sig. (2-sided)	Exact Sig. (2-sided)	Exact Sig. (1-sided)
Pearson Chi-Square	13.522[b]	1	.000		
Continuity Correction[a]	12.893	1	.000		
Likelihood Ratio	13.861	1	.000		
Fisher's Exact Test				.000	.000
N of Valid Cases	638				

a. Computed only for a 2x2 table.

b. 0 cells (.0%) have expected count less than 5. The minimum expected count is 80.24.

Discussion

It would appear that the bonobos has the lowest hammerstone velocity, the Gona hominins intermediate, and the humans the highest velocities. As previously mentioned, experiments and analysis by Harlacker (2006) have demonstrated that the mean hammerstone velocities just prior to impact of the bonobos was 3.67 meters per second, and the hammerstone velocities of the experienced human sample was 7.12 meters per second. If the split to whole flake ratio is strongly correlated with hammerstone velocities, then the extrapolated mean hammerstone velocity of the Gona hominins could be estimated to have been a minimum of 5 meters per second (and perhaps more if, as we believe, the Gona hominins were selectively removing numbers of larger flakes, as discussed in section 26 below).

3. Assemblage: Raw Material Type

Rationale

Examining differential use of raw materials can give insights into possible selectivity of early hominin tool-makers. Raw materials in the Gona archaeological sample were divided three major types: lighter volcanics (especially trachyte), darker volcanics (especially rhyolite), and vitreous volcanics (very fine-grained). Subsequent geological classification by Jay Quade (reported in Stout et al. (2005), has subsequently separated Gona raw materials into a number of more discrete types. The distribution of raw material types was examined in the archaeological sample and the experimental sample (knapped by humans and bonobos) for possible differences among the tool-makers in the flaked cores (as a proxy for cobble selections).

The experimental sample of cobbles was selected in proximity to the EG sites from the 2.6 million year old conglomerate that lies stratigraphically just below the sites. At the time of occupation of these sites, this conglomerate would have been a readily accessible cobble source in the overall region as exposed gravel bars along stream courses, and the raw material types at the EG sites are found within this conglomerate. Selection was based upon sizes and shapes of cobbles deemed suitable for flaking (excluding very large or very small cobbles, spherical clasts, etc.), as well as the likelihood or expectation of good, fine-grained stone (normally indicated by a cobble with smooth cortex but without extensive pitting or incipient cracks). Selection was based on these criteria, and not upon observable indications of the rock type itself. No initial testing (flaking) was done prior to the experiments. In practice, only approximately one in fifty cobbles from these conglomerates was selected via this procedure. We were interested to see whether the Gona EG sites showed more selectivity than this sampling procedure.

Results

The experimental and archaeological samples can be placed in placed in one of three general raw material categories. The types observable among cores include, in order of prevalence, lighter volcanics (approximately two-thirds to three-quarters of each sample), darker volcanics (approximately one-quarter of each sample), and a small proportion (approximately 9%) of vitreous volcanic material in the Gona archaeological sample.

Raw Material Types for Cores: Experimental v. Archaeological

			Raw material			
			lighter volcanics	darker volcanics	vitreous volcanics	Total
Exp or arch	arch	Count	15	6	2	23
		% within Exp or arch	65.2%	26.1%	8.7%	100.0%
	exp	Count	47	17	0	64
		% within Exp or arch	73.4%	26.6%	.0%	100.0%
Total		Count	62	23	2	87
		% within Exp or arch	71.3%	26.4%	2.3%	100.0%

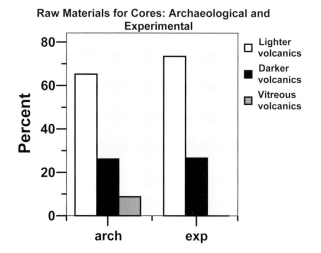

Raw Materials for Cores: Archaeological and Experimental

□ Lighter volcanics
■ Darker volcanics
▨ Vitreous volcanics

Chi-Square Test: Raw Material for Cores, Archaeological v. Experimental

	Value	df	Asymp. Sig. (2-sided)
Pearson Chi-Square	5.727[a]	2	.057
Likelihood Ratio	5.487	2	.064
N of Valid Cases	87		

a. 2 cells (33.3%) have expected count less than 5. The minimum expected count is .53.

Discussion

Although the two samples show some small differences in their raw material types, the assemblage differences are not significant at the .05 level. One difference is the presence at Gona of two cores in a vitreous volcanic material that was absent in the experimental sample. In other words, the experimental sample (already highly selected for "flakability") is largely similar to the Gona sample with regard to the major rock types flaked. As Stout *et al.* (2005) have shown, Gona hominins selected higher-quality cobbles compared to the proportions found in the stream conglomerates, especially the light trachytes. This pattern is also seen in the sample selected for experiments. More interesting, however, is the assessment of raw material quality from a knapper's perspective, which will be discussed in the next section.

4. Assemblage: Raw Material Quality

Rationale

The cores from the assemblages were assigned a quality-of-flaking value (excellent/good/fair) based on how fine-grained the rock type and how homogeneous the fracture surface. This was done from a stone-knapper's perspective to examine possible selectivity by early hominin tool-makers and to examine possible similarities or differences between the Gona archaeological sample and the experimental samples (bonobo and human) in terms of raw material quality. Again, the experimental sample was highly selected in the field based on size, shape, and cortex appearance but without any testing of the cobble (flaking) or examination of the interior rock quality.

Results

The archaeological and experimental samples do not significantly differ at the .05 confidence level. Nevertheless, there are some differences, for example, the Gona archaeological material has more "excellent" raw material.

Raw Material Quality for Cores: Archaeological and Experimental

			Raw Material Quality			
			excellent	good	fair	Total
Exp or arch	arch	Count	3	16	4	23
		% within Exp or arch	13.0%	69.6%	17.4%	100.0%
	exp	Count	2	47	15	64
		% within Exp or arch	3.1%	73.4%	23.4%	100.0%
Total		Count	5	63	19	87
		% within Exp or arch	5.7%	72.4%	21.8%	100.0%

Chi-Square Tests: Raw Material Quality for Cores, Archaeological and Experimental

	Value	df	Asymp. Sig. (2-sided)
Pearson Chi-Square	3.214[a]	2	.200
Likelihood Ratio	2.813	2	.245
N of Valid Cases	87		

a. 2 cells (33.3%) have expected count less than 5. The minimum expected count is 1.32.

Raw Material Quality for Cores: Experimental and Archaeological

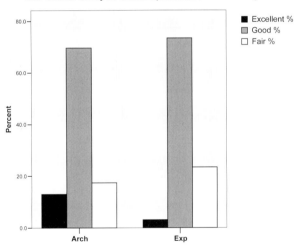

5. Cores: Original Form

Rationale

Cores were examined to identify what the original (blank) form was. In the early Oldowan, most cores were made on natural water-rounded river cobbles and this can readily be observed in the archaeological assemblage. However, sometimes heavily-reduced cores retain little or no evidence of their original form, and are assigned to an indeterminate category. Other, miscellaneous categories for original core form include flakes or flake fragments, tabular chunks, or broken cobbles.

Results

In all three samples, the dominant original forms for cores were cobbles. In addition at Gona, six cores were placed in the indeterminate category, with one more core made on a flake and another on a tabular chunk. The bonobos also primarily reduced whole cobbles but also made four simple cores from broken cobble fragments (broken in half during the knapping process).

Core Original Form by Tool-maker

			Original form					Total
			cobble	flake, flake fragment	tabular chunk	indet	broken cobble	
Tool-maker	bonobo	Count	29	0	0	0	4	33
		% within Tool-maker	87.9%	.0%	.0%	.0%	12.1%	100.0%
	gona	Count	15	1	1	6	0	23
		% within Tool-maker	65.2%	4.3%	4.3%	26.1%	.0%	100.0%
	human	Count	30	0	0	0	1	31
		% within Tool-maker	96.8%	.0%	.0%	.0%	3.2%	100.0%
Total		Count	74	1	1	6	5	87
		% within Tool-maker	85.1%	1.1%	1.1%	6.9%	5.7%	100.0%

Discussion

The archaeological and experimental populations are similar and not significantly different, although there is a higher proportion of "excellent" raw material observed in the archaeological sample from Gona than in the experimental samples. This is probably due to the Gona hominins testing the raw materials at the conglomerate sources before transporting cores to the floodplain sites for further reduction (which was deliberately not done for the experimental sample). It should be noted that selecting raw materials merely on their exterior characteristics (size, shape, cortex appearance), as was done for the experimental sample, yielded a sample with superior flaking characteristics overall (over 75% of the experimental sample assessed to have "good" to "excellent" flaking quality).

Core Original Form by Tool-maker

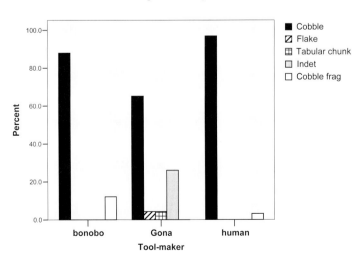

As the frequencies of original forms other than cobbles were small to absent, comparisons were also made between the tool-makers looking at cobbles v. "other" original forms (lumping flakes, tabular chunks, indeterminate, and cobble fragments). Chi-square tests indicate a difference between the Gona sample and the two experimental samples, bonobo and human, in the original forms of their cores (.042 and .002 levels of significance, respectively). The difference between the two experimental samples, bonobos and humans, was not significant.

Core Original Form by Tool-maker (Cobble v. Other)

			Original form lumped		Total
			Cobble	Other	
Tool-maker	bonobo	Count	29	4	33
		% within Tool-maker	87.9%	12.1%	100.0%
	gona	Count	15	8	23
		% within Tool-maker	65.2%	34.8%	100.0%
	human	Count	30	1	31
		% within Tool-maker	96.8%	3.2%	100.0%
Total		Count	74	13	87
		% within Tool-maker	85.1%	14.9%	100.0%

Chi-Square Tests: Core Original Form (Cobble v. Other), Bonobo v. Gona

	Value	df	Asymp. Sig. (2-sided)	Exact Sig. (2-sided)	Exact Sig. (1-sided)
Pearson Chi-Square	4.134[b]	1	.042		
Continuity Correction[a]	2.898	1	.089		
Likelihood Ratio	4.097	1	.043		
Fisher's Exact Test				.054	.045
N of Valid Cases	56				

a. Computed only for a 2x2 table.

b. 1 cells (25.0%) have expected count less than 5. The minimum expected count is 4.93.

Chi-Square Tests: Core Original Form (Cobble v. Other), Bonobo v. Human

	Value	df	Asymp. Sig. (2-sided)	Exact Sig. (2-sided)	Exact Sig. (1-sided)
Pearson Chi-Square	1.756[b]	1	.185		
Continuity Correction[a]	.738	1	.390		
Likelihood Ratio	1.882	1	.170		
Fisher's Exact Test				.356	.198
N of Valid Cases	64				

a. Computed only for a 2x2 table.

b. 2 cells (50.0%) have expected count less than 5. The minimum expected count is 2.42.

Chi-Square Tests: Core Original Form (Cobble v. Other), Gona v. Human

	Value	df	Asymp. Sig. (2-sided)	Exact Sig. (2-sided)	Exact Sig. (1-sided)
Pearson Chi-Square	9.467[b]	1	.002		
Continuity Correction[a]	7.331	1	.007		
Likelihood Ratio	10.105	1	.001		
Fisher's Exact Test				.003	.003
N of Valid Cases	54				

a. Computed only for a 2x2 table.

b. 1 cells (25.0%) have expected count less than 5. The minimum expected count is 3.83.

Discussion

With regard to the original form of cores, the presence at Gona of a number of cores (n=6) whose original blank form was indeterminate constitutes the major difference between the experimental samples and the archaeological sample. This proportion of Gona cores (26.1%) assigned to the "indeterminate" category with regard to original form strongly suggests that these cores were more heavily-reduced, obliterating most or all signs of the original blank type used. Most likely these were all cobbles, but in the absence of refitting this cannot be demonstrated conclusively. That Gona cores overall were more heavily reduced than the experimental samples will also become evident in further attributes examined below. Nevertheless, cobbles were by far the predominant original form in all samples in this study, with the Gona indeterminate core forms likely representing an interesting indication of higher intensity of core reduction.

6. Cores: Flaking Mode

Rationale

The dominant flaking mode was recorded to examine the major patterns of cobble reduction for the Gona and bonobo samples, thus focusing on the nonhuman subjects. The major flaking modes were:
a) unifacial (normal), or flaking on one face along a core edge (unidirectional flaking)
b) bifacial, or flaking on two faces along a core edge (bidirectional flaking)
c) unifacial plus bifacial, or flaking unifacially along one core edge and bifacially along another edge
d) unifacial alternate, or flaking unifacially on one face along one core edge and then unifacially on the opposite face from another edge

Results

Overall, both the bonobo and Gona samples showed a preponderance of unifacial flaking of cobbles (from 63.6 to 78.3% respectively), with relatively low incidence of the other flaking modes. These populations did not differ significantly in flaking mode at the .05

confidence level [nor was there a significant difference when all flaking modes (b) through (d) were combined as "other," resulting in a chi-square value of .253 with 0 cells having expected value of less than 5]. The table of results and chi-square test for all four flaking modes are presented below:

Cores: Mode of Flaking by Tool-maker

			Mode of flaking				
			unifacial (normal)	bifacial	uni -+ bifacial	uni-facial alter-nate	Total
Tool-maker	bonobo	Count	21	2	5	5	33
		% within Tool-maker	63.6%	6.1%	15.2%	15.2%	100.0%
	gona	Count	18	2	3	0	23
		% within Tool-maker	78.3%	8.7%	13.0%	.0%	100.0%
	human	Count	17	13	1	0	31
		% within Tool-maker	54.8%	41.9%	3.2%	.0%	100.0%
Total		Count	56	17	9	5	87
		% within Tool-maker	64.4%	19.5%	10.3%	5.7%	100.0%

Chi-Square Tests: Mode of Flaking, Bonobo v. Gona

	Value	df	Asymp. Sig. (2-sided)
Pearson Chi-Square	4.075a	3	.253
Likelihood Ratio	5.872	3	.118
Linear-by-Linear Association	2.789	1	.095
N of Valid Cases	56		

a. 6 cells (75.0%) have expected count less than 5. The minimum expected count is 1.64.

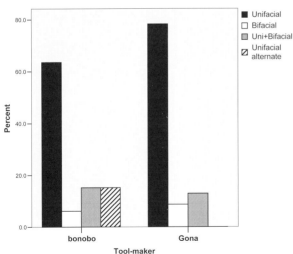

Cores, Mode of Flaking: Bonobo v. Gona

wards unifacial flaking could also have been a habitual norm among the hominins who produced these EG sites at Gona, or, at the very least, their tendency at these particular sites but perhaps variable at other sites and other times.

7. Cores: Invasiveness.

Rationale

This attribute is an assessment of how invasive flake scars are relative to core morphology (heavily, moderate, light). This assessment is a qualitative one, based on the distance flake scars have traveled across the core surfaces. Heavy invasiveness suggests that tool-makers are either driving larger flakes off of cores or heavily reducing cores, or both.

Results

The bonobo cores are characterized by higher percentages of light and moderate invasiveness, and low percentage of heavy invasiveness. The Gona cores have a preponderance of heavy invasiveness. Differences between the bonobo, Gona, and human samples were significant at the .05 level.

Invasiveness of Flaking by Tool-maker

			Degree of invasiveness			
			light, shallow	moderate	heavy, invasive	Total
Tool-maker	bonobo	Count	4	27	2	33
		% within Tool-maker	12.1%	81.8%	6.1%	100.0%
	gona	Count	0	7	16	23
		% within Tool-maker	.0%	30.4%	69.6%	100.0%
	human	Count	1	19	11	31
		% within Tool-maker	3.2%	61.3%	35.5%	100.0%
Total		Count	5	53	29	87
		% within Tool-maker	5.7%	60.9%	33.3%	100.0%

Discussion

Both the bonobo and Gona samples have a preponderance of unifacial flaking of cobbles. It can be argued that this is the simplest, easiest approach to reducing cobbles and producing sharp flakes. This tendency to-

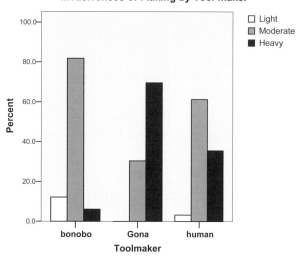

Invasiveness of Flaking by Tool-maker

Chi-Square Tests: Degree of Invasiveness, Bonobo v. Gona

	Value	df	Asymp. Sig. (2-sided)
Pearson Chi-Square	25.687[a]	2	.000
Likelihood Ratio	28.705	2	.000
N of Valid Cases	56		

a. 2 cells (33.3%) have expected count less than 5. The minimum expected count is 1.64.

Chi-Square Tests: Degree of Invasiveness, Bonobo v. Human

	Value	df	Asymp. Sig. (2-sided)
Pearson Chi-Square	9.369[a]	2	.009
Likelihood Ratio	10.123	2	.006
N of Valid Cases	64		

a. 2 cells (33.3%) have expected count less than 5. The minimum expected count is 2.42.

Chi-Square Tests: Degree of Invasiveness, Gona v. Human

	Value	df	Asymp. Sig. (2-sided)
Pearson Chi-Square	6.420[a]	2	.040
Likelihood Ratio	6.882	2	.032
N of Valid Cases	54		

a. 2 cells (33.3%) have expected count less than 5. The minimum expected count is .43.

Discussion

The bonobo cores have a low and moderate invasiveness because they are lightly reduced with small flakes being detached for the most part, whereas the Gona cores have a heavy invasiveness, indicating that they have been heavily and effectively reduced with larger flakes being detached overall. As mentioned previously, the experimental human sample intentionally tried to reduce most cores by about 50% of their original cobble mass in order to provide a controlled sample for comparisons with the bonobo and Gona samples. In this attribute, the human sample is intermediate between the more lightly flaked bonobo cores and the more heavily flaked Gona cores.

8. Cores: Percentage of Circumference Flaked

Rationale

The percentage of cobble circumference flaked is yet another criterion that can be used to determine how extensively an Oldowan core has been exploited. Although a very long, roller-shaped cobble can be extensively reduced and still retain a relatively low percentage of circumference flaked, most of the cobbles available to Gona hominins tend to fall into the disc and sphere categories of geological clast shape, so that their percentage of circumference flaked is reasonably indicative of the extent of flaking.

Results

The bonobo cores showed the lowest percentage of circumference flaked, with the human sample intermediate and the Gona sample having a very high percentage of circumference flaked. The Gona cores were flaked along nearly twice as much of the cobble circumference overall than the bonobo cores. The differences between the bonobos and both the Gona and human core samples were significant at the .05 level, but the difference between the Gona and human samples was not significant at this level.

Extent of Core Reduction by Tool-maker

Extent of core reduction

Tool-maker	Mean	N	Std. Deviation
bonobo	30.91	33	9.475
gona	58.67	23	23.799
human	45.81	31	9.924
Total	43.56	87	18.362

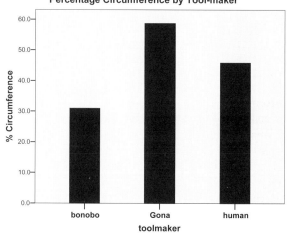

Percentage Circumference by Tool-maker

Test Statistics: Extent of Core Reduction, Bonobo v. Gona[a]

	Extent of core reduction
Mann-Whitney U	96.000
Wilcoxon W	657.000
Z	-4.855
Asymp. Sig. (2-tailed)	.000

a. Grouping Variable: Toolmkrcod

Test Statistics: Extent of Core Reduction, Bonobo v. Human[a]

	Extent of core reduction
Mann-Whitney U	152.000
Wilcoxon W	713.000
Z	-4.982
Asymp. Sig. (2-tailed)	.000

a. Grouping Variable: Toolmkrcod

Test Statistics: Extent of Core Reduction, Gona v. Human[a]

	Extent of core reduction
Mann-Whitney U	263.000
Wilcoxon W	759.000
Z	-1.684
Asymp. Sig. (2-tailed)	.092

a. Grouping Variable: Toolmkrcod

Discussion

The bonobos relatively low percentage of circumference flaked (~31%) is an indication of minimal reduction of cobbles, while the high percentage of circumference flaked in the archaeological sample (~59%) is an indication of much heavier reduction of cobble cores by the Gona hominin tool-makers. Again, the experimental human sample, which intentionally reduced most cores by about half their original mass, was intermediate (~46%).

9. Cores: Quality of Flaking

Rationale

From a flintknapper's perspective, cores were ranked by the quality of flaking. Higher quality of flaking denoted well-reduced cores, clean flake detachments, and crisp non-battered edges. Although a very subjective category, experience has shown us that this is nonetheless a useful category in assessing different levels of skill. Such cores are typically illustrated in Oldowan site reports. Quality categories were excellent, good, fair, and poor.

Results

The bonobo core sample was characterized by a modal value of fair-quality flaking, the Gona sample by good-quality flaking, and the human sample by excellent-quality flaking. Using all four quality categories, statistical testing showed differences between the human sample and the other two samples but not between the bonobo and Gona samples at the .05 level of significance, although 4 cells had expected counts less than 5. Testing after combining categories (good to excellent as "high" and poor to fair as "low"), on the other hand, points to a significant difference between the bonobo sample and the other two samples in this attribute, while the Gona and human samples show a much higher and relatively equivalent quality of flaking.

Quality of Flaking by Tool-maker

			Quality of flaking (skill)				
			excellent	good	fair	poor	Total
Tool-maker	bonobo	Count	1	15	16	1	33
		% within Tool-maker	3.0%	45.5%	48.5%	3.0%	100.0%
	gona	Count	1	18	4	0	23
		% within Tool-maker	4.3%	78.3%	17.4%	.0%	100.0%
	human	Count	14	12	5	0	31
		% within Tool-maker	45.2%	38.7%	16.1%	.0%	100.0%
Total		Count	16	45	25	1	87
		% within Tool-maker	18.4%	51.7%	28.7%	1.1%	100.0%

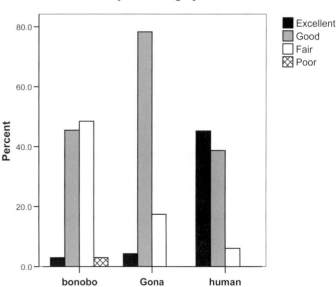

Quality of Flaking by Tool-maker

Chi-Square Tests: Quality of Flaking, Bonobo v. Gona, 4 categories

	Value	df	Asymp. Sig. (2-sided)
Pearson Chi-Square	6.907ᵃ	3	.075
Likelihood Ratio	7.574	3	.056
N of Valid Cases	56		

a. 4 cells (50.0%) have expected count less than 5. The minimum expected count is .41.

Chi-Square Tests: Quality of Flaking, Bonobo v. Human, 4 categories

	Value	df	Asymp. Sig. (2-sided)
Pearson Chi-Square	18.317ᵃ	3	.000
Likelihood Ratio	21.164	3	.000
N of Valid Cases	64		

a. 2 cells (25.0%) have expected count less than 5. The minimum expected count is .48.

Chi-Square Tests: Quality of Flaking, Gona v. Human, 4 categories

	Value	df	Asymp. Sig. (2-sided)
Pearson Chi-Square	11.648ᵃ	2	.003
Likelihood Ratio	13.576	2	.001
N of Valid Cases	54		

a. 1 cells (16.7%) have expected count less than 5. The minimum expected count is 3.83.

Quality of Flaking (2 categories) by Tool-maker

			Quality of flaking (2 categories)		
			excellent or good	fair or poor	Total
Tool-maker	bonobo	Count	16	17	33
		% within Tool-maker	48.5%	51.5%	100.0%
	gona	Count	19	4	23
		% within Tool-maker	82.6%	17.4%	100.0%
	human	Count	26	5	31
		% within Tool-maker	83.9%	16.1%	100.0%
Total		Count	61	26	87
		% within Tool-maker	70.1%	29.9%	100.0%

Chi-Square Tests: Quality of Flaking, Bonobo v. Gona, 2 categories

	Value	df	Asymp. Sig. (2-sided)	Exact Sig. (2-sided)	Exact Sig. (1-sided)
Pearson Chi-Square	6.734ᵇ	1	.009		
Continuity Correctionᵃ	5.357	1	.021		
Likelihood Ratio	7.124	1	.008		
Fisher's Exact Test				.012	.009
N of Valid Cases	56				

a. Computed only for a 2x2 table.

b. 0 cells (.0%) have expected count less than 5. The minimum expected count is 8.63.

Chi-Square Tests: Quality of Flaking, Bonobo v. Human, 2 categories

	Value	df	Asymp. Sig. (2-sided)	Exact Sig. (2-sided)	Exact Sig. (1-sided)
Pearson Chi-Square	8.873ᵇ	1	.003		
Continuity Correctionᵃ	7.373	1	.007		
Likelihood Ratio	9.258	1	.002		
Fisher's Exact Test				.004	.003
N of Valid Cases	64				

a. Computed only for a 2x2 table.

b. 0 cells (.0%) have expected count less than 5. The minimum expected count is 10.66.

Chi-Square Tests: Quality of Flaking, Gona v. Human, 2 categories

	Value	df	Asymp. Sig. (2-sided)	Exact Sig. (2-sided)	Exact Sig. (1-sided)
Pearson Chi-Square	.015ᵇ	1	.902		
Continuity Correctionᵃ	.000	1	1.000		
Likelihood Ratio	.015	1	.902		
Fisher's Exact Test				1.000	.592
N of Valid Cases	54				

a. Computed only for a 2x2 table.

b. 1 cells (25.0%) have expected count less than 5. The minimum expected count is 3.83.

Discussion

Examination of the distribution of flaking quality among the samples shows, in fact, a gradient among the three samples. The bonobo sample, characterized by cores that showing primarily fair (~49%) and good (~46%) flaking quality, showed the lowest overall level of skill of the three samples. The Gona sample showed a preponderance of good flaking (~78%), and the human sample was characterized by cores showing excellent (~45) and good (~39%) flaking quality. Thus, in terms of overall quality of flaking, the Gona sample was intermediate between the bonobo and human samples.

10. Cores: Edge Battering

Rationale

Core edges exhibiting heavy battering (hammerstone percussion marks, crushing, and small-scale step flaking) may be an indication of less skilled flaking and numerous unsuccessful attempts to remove flakes with a percussor. This criterion was developed after examination of the bonobo cores appeared to show much more edge battering than had been observed on archaeological cores or on experimental cores produced by humans. Edge battering was divided into four categories: none, low, moderate, and high.

Results

The bonobo sample is characterized by an appreciable percentage of cores (~33%) showing high levels of edge battering from hammerstone percussion. The modal value for Gona cores was a low degree of edge battering (~44%), but appreciable quantities of also showed moderate battering (~26%). The great majority of cores in the human sample (~84%) exhibited no edge battering. Interestingly, none of the cores from the Gona and human samples showed a high degree of edge battering. Each of the three tool-making populations was significantly different from the other two (at the .05 confidence level) in terms of core edge battering.

Core Edge Battering by Tool-maker

			Edge battering, microstepping, crushing				
			none	low	mod-erate	high	Total
Tool-maker	bonobo	Count	10	6	6	11	33
		% within Tool-maker	30.3%	18.2%	18.2%	33.3%	100.0%
	gona	Count	7	10	6	0	23
		% within Tool-maker	30.4%	43.5%	26.1%	.0%	100.0%
	human	Count	26	4	1	0	31
		% within Tool-maker	83.9%	12.9%	3.2%	.0%	100.0%
Total		Count	43	20	13	11	87
		% within Tool-maker	49.4%	23.0%	14.9%	12.6%	100.0%

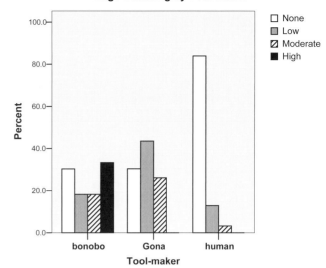

Edge Battering by Tool-maker

Chi-Square Tests: Core Edge Battering, Bonobo v. Gona

	Value	df	Asymp. Sig. (2-sided)
Pearson Chi-Square	11.098[a]	3	.011
Likelihood Ratio	14.997	3	.002
N of Valid Cases	56		

a. 2 cells (25.0%) have expected count less than 5. The minimum expected count is 4.52.

Chi-Square Tests: Core Edge Battering, Bonobo v. Human

	Value	df	Asymp. Sig. (2-sided)
Pearson Chi-Square	22.042[a]	3	.000
Likelihood Ratio	26.918	3	.000
N of Valid Cases	64		

a. 3 cells (37.5%) have expected count less than 5. The minimum expected count is 3.39.

Chi-Square Tests: Core Edge Battering, Gona v. Human

	Value	df	Asymp. Sig. (2-sided)
Pearson Chi-Square	16.254[a]	2	.000
Likelihood Ratio	17.071	2	.000
N of Valid Cases	54		

a. 2 cells (33.3%) have expected count less than 5. The minimum expected count is 2.98.

Discussion

Over half of the bonobo cores (~52%) showed either a moderate or high degree of edge battering. This indicates high numbers of unsuccessful hammerstone blows by the bonobos in attempts to remove usable flakes and a significantly lower level of skill than the Gona and human samples. Also, ~49% of bonobo cores showed little or no edge battering compared to ~74 % for Gona and almost all of the human cores (~97%). In terms of the criterion of edge battering, then, the Gona hominins are intermediate in skill level relative to the bonobos and humans.

11. Cores: Modified Leakey Typology

Rationale

Mary Leakey's (1971) typology of Oldowan cores is one of the most widely used systems, especially if one views these forms as core morphologies rather than functional classes. Her major categories of cores made on cobbles (her "heavy-duty tools") are choppers, discoids, heavy-duty scrapers, proto-bifaces, polyhedrons, discoids, heavy-duty scrapers, and spheroids. In this study, there were no polyhedrons, protobifaces, or spheroids in any of the samples. We added one new category, "casual core," for very minimally-flaked cobbles (probably closest to Mary Leakey's "modified and battered nodules and blocks" category). In addition, we decided to separate choppers into their two major categories, side choppers and end choppers, based upon whether the flaked edge was along the end or side of the cobble.

Results

The bonobo cores were characterized by an abundance of end choppers (~64%), in comparison with the Gona and human cores, which had predominantly side choppers (~65% and 71% respectively). The bonobo sample was significantly different from the other two samples in the numbers of end v. side choppers, while the Gona and human samples did not differ significantly at the .05 level. In view of the high proportion of choppers in each of the three tool-making samples (bonobos 82%, Gona ~74%, and humans 100%), they would all be assigned to the "Oldowan Industry" in Leakey's classification system (1971). The bonobo sample also has a number of cores assigned to the "casual core" category, and the Gona sample has small numbers of other core types (discoids, end choppers, heavy-duty scrapers, and one casual core).

Chi-Square Tests: Leakey Core Types, All Tool-makers

	Value	df	Asymp. Sig. (2-sided)
Pearson Chi-Square	43.131[a]	8	.000
Likelihood Ratio	46.596	8	.000
N of Valid Cases	87		

a. 9 cells (60.0%) have expected count less than 5. The minimum expected count is .53.

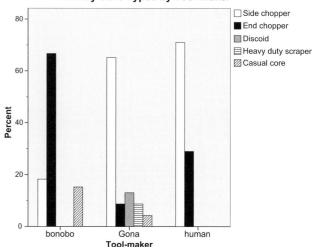

Leakey Core Types by Tool-maker

Chi-Square Tests: End and Side Choppers, Bonobo v. Gona

	Value	df	Asymp. Sig. (2-sided)	Exact Sig. (2-sided)	Exact Sig. (1-sided)
Pearson Chi-Square	18.968[b]	1	.000		
Continuity Correction[a]	16.379	1	.000		
Likelihood Ratio	20.771	1	.000		
Fisher's Exact Test				.000	.000
N of Valid Cases	45				

a. Computed only for a 2x2 table.

b. 0 cells (.0%) have expected count less than 5. The minimum expected count is 7.93.

Leakey Core Type by Tool-maker

			Mary Leakey core type					
			Side chopper	End chopper	Discoid	Heavy-duty scraper	Casual core	Total
Tool-maker	bonobo	Count	6	22	0	0	5	33
		% within Tool-maker	18.2%	66.7%	.0%	.0%	15.2%	100.0%
	gona	Count	15	2	3	2	1	23
		% within Tool-maker	65.2%	8.7%	13.0%	8.7%	4.3%	100.0%
	human	Count	22	9	0	0	0	31
		% within Tool-maker	71.0%	29.0%	.0%	.0%	.0%	100.0%
Total		Count	43	33	3	2	6	87
		% within Tool-maker	49.4%	37.9%	3.4%	2.3%	6.9%	100.0%

Chi-Square Tests: End and Side Choppers, Bonobo v. Human

	Value	df	Asymp. Sig. (2-sided)	Exact Sig. (2-sided)	Exact Sig. (1-sided)
Pearson Chi-Square	14.479[b]	1	.000		
Continuity Correction[a]	12.561	1	.000		
Likelihood Ratio	15.191	1	.000		
Fisher's Exact Test				.000	.000
N of Valid Cases	59				

a. Computed only for a 2x2 table.

b. 0 cells (.0%) have expected count less than 5. The minimum expected count is 13.29.

Chi-Square Tests: End and Side Choppers, Gona v. Human

	Value	df	Asymp. Sig. (2-sided)	Exact Sig. (2-sided)	Exact Sig. (1-sided)
Pearson Chi-Square	1.853[b]	1	.173		
Continuity Correction[a]	1.005	1	.316		
Likelihood Ratio	2.007	1	.157		
Fisher's Exact Test				.284	.158
N of Valid Cases	48				

a. Computed only for a 2x2 table.

b. 1 cells (25.0%) have expected count less than 5. The minimum expected count is 3.90.

Discussion

Although all three samples are dominated by core forms that would be designated as choppers in Mary Leakey's classification, the bonobo assemblage stands out in its high proportion of end rather than side choppers. This is a result of bonobos reducing a cobble significantly less, resulting in a preponderance of end choppers rather than side choppers. This is in contrast to the cores produced by the humans and Gona tool-makers, which tend to be more heavily reduced; many of these cores almost certainly started as end choppers, but with further, more efficient reduction, they were transformed into side choppers (in the human sample, in which reduction was held to approximately 50% of the original cobble, as well as in the Gona archaeological sample).

12. Cores: Weight

Rationale

Core weights in a given raw material can help show differences between assemblages that may be due to differences in the weights of the original cobble forms and/or differences in the amount of core reduction. For both of the experimental samples, an equivalent range of sizes and shapes of pre-selected cobbles (discussed above) was made available to the subjects.

Results

The bonobos produced the heaviest cores, averaging about 687 g, with a much larger standard deviation

than the other two samples as well. The Gona cores were the smallest (~226 g average), with the modern human sample intermediate (~385 g). Each tool-maker sample differed significantly from the other two in terms of final core weight at the .000 level.

Core Weight by Tool-maker

Weight in grams

Tool-maker	Mean	N	Std. Deviation
bonobo	686.82	33	328.682
gona	226.22	23	111.925
human	384.55	31	117.893
Total	457.34	87	290.953

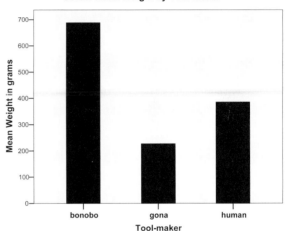

Cores: Mean Weight by Tool-maker

Test Statistics: Core Weight, Bonobo v. Gona[a]

	Weight in grams
Mann-Whitney U	35.000
Wilcoxon W	311.000
Z	-5.738
Asymp. Sig. (2-tailed)	.000

a. Grouping Variable: Toolmkrcod

Test Statistics: Core Weight, Bonobo v. Human[a]

	Weight in grams
Mann-Whitney U	179.500
Wilcoxon W	675.500
Z	-4.460
Asymp. Sig. (2-tailed)	.000

a. Grouping Variable: Toolmkrcod

Test Statistics: Core Weight, Gona v. Human[a]

	Weight in grams
Mann-Whitney U	105.000
Wilcoxon W	381.000
Z	-4.400
Asymp. Sig. (2-tailed)	.000

a. Grouping Variable: Toolmkrcod

Discussion

The bonobo cores are markedly heavier than the Gona or human samples. This is due to the fact that bonobos almost invariably chose larger cobbles for flaking, since their hands are large with long fingers and a very short thumb and are less dexterous than those of modern humans (and also, presumably, than those of Gona hominins as well), and also because the bonobo cobble cores were much less reduced than the Gona or human samples. The human cores are intermediate in weight between the bonobo and Gona cores, but, on average, are closer to the mean weight of Gona cores. A number of other attributes examined (above and below) suggest that the human cores are not as extensively reduced as the Gona cores; if they were, the mean weights would likely be very similar.

13. Cores: Length (Maximum Dimension)

Rationale

This measurement gives an indication of the size of cores based on the largest linear measurement. It also shows the minimum size of cobble blanks selected by early hominins. This measurement is also used to calculate core breadth/length ratios, as well as the ratio of the largest flake scar/core maximum dimension (discussed below).

Results

As with weight, the bonobo core maximum dimensions were the largest (mean ~116 mm), with humans intermediate (~93 mm), and the Gona cores the smallest (~81 mm); the bonobo cores also showed a much higher standard deviation. Each tool-making population produced cores whose mean maximum dimension differed significantly (at the .000 level) from the other two samples.

Cores: Maximum Dimension by Tool-maker

Length

Tool-maker	Mean	N	Std. Deviation
bonobo	115.85	33	18.171
gona	81.09	23	10.816
human	93.45	31	11.331
Total	98.68	87	20.083

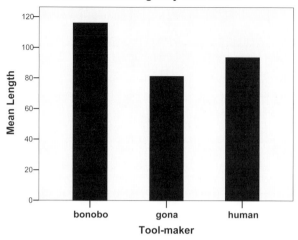

Cores: Mean Length by Tool-maker

Test Statistics: Core Maximum Dimension, Bonobo v. Gona[a]

	Length
Mann-Whitney U	39.000
Wilcoxon W	315.000
Z	-5.673
Asymp. Sig. (2-tailed)	.000

a. Grouping Variable: Toolmkrcod

Test Statistics: Core Maximum Dimension, Gona v. Human[a]

	Length
Mann-Whitney U	138.000
Wilcoxon W	634.000
Z	-5.019
Asymp. Sig. (2-tailed)	.000

a. Grouping Variable: Toolmkrcod

Test Statistics: Core Maximum Dimension, Gona v. Human[a]

	Length
Mann-Whitney U	147.000
Wilcoxon W	423.000
Z	-3.668
Asymp. Sig. (2-tailed)	.000

a. Grouping Variable: Toolmkrcod

Discussion

As with weight, the differences seen in core maximum dimension in the three tool-making samples can be explained. As the bonobos tended to choose larger cobbles (because of their hand morphology) and also reduced the cobble cores less extensively, their resultant cores tended to be larger. The human core sample was closer in size to the Gona cores. As previously discussed, the human cores are somewhat less reduced overall than the archaeological cores and thus somewhat larger.

14. Cores: Breadth

Rationale

Breadth was defined as the dimension of the core measured at a right angle to the length or maximum dimension. This measurement is also useful in determining breadth/length ratios and thickness/breadth ratios.

Results

The mean breadth of cores in the three samples differed greatly. As with core length, the bonobo core sample showed the greatest mean breadth (~90 mm), with the human sample intermediate (~75 mm) and the Gona sample the smallest (~61 mm). Each population differed from the other two at the .000 level of significance.

Cores: Breadth by Tool-maker

Breadth

Tool-maker	Mean	N	Std. Deviation
bonobo	90.06	33	16.948
gona	60.52	23	9.375
human	74.94	31	8.675
Total	76.86	87	17.182

Test Statistics: Core Breadth, Bonobo v. Gona[a]

	Length
Mann-Whitney U	37.500
Wilcoxon W	313.500
Z	-5.698
Asymp. Sig. (2-tailed)	.000

a. Grouping Variable: Toolmkrcod

Test Statistics: Core Breadth, Bonobo v. Human[a]

	Length
Mann-Whitney U	191.000
Wilcoxon W	687.000
Z	-4.309
Asymp. Sig. (2-tailed)	.000

a. Grouping Variable: Toolmkrcod

Test Statistics: Core Breadth, Gona v. Human[a]

	Length
Mann-Whitney U	89.500
Wilcoxon W	365.500
Z	-4.675
Asymp. Sig. (2-tailed)	.000

a. Grouping Variable: Toolmkrcod

Discussion

As with core length (maximum dimension), the larger core breadth seen in the bonobo sample seems to be primarily a function of their selecting larger cobbles for flaking. The greater breadth seen in the human cores relative to the Gona sample is probably due to the fact that the human cores are not as extensively reduced as the Gona cores; with further reduction of especially side choppers, breadth measurements would probably decrease accordingly.

15. Cores: Thickness

Rationale

Thickness was measured at a right angle to breadth, and represents the smallest of the three length/breadth/thickness measurements. Interestingly, this measurement on Oldowan cobble cores probably often represents a reasonable estimate of the thickness of the original cobble blank that was flaked.

Results

The Gona cores were the thinnest (~44 mm), while the bonobo and human samples were almost identical in terms of absolute thickness (~57 mm and ~56 mm respectively). Statistical tests indicated that the Gona sample differed from the bonobo and human samples (at the .001 and .000 confidence levels, respectively), while the bonobo and human samples were not significantly different (.824 significance level).

Cores: Thickness by Tool-maker

Thickness

Tool-maker	Mean	N	Std. Deviation
bonobo	56.76	33	14.908
gona	43.87	23	11.022
human	56.39	31	9.222
Total	53.22	87	13.238

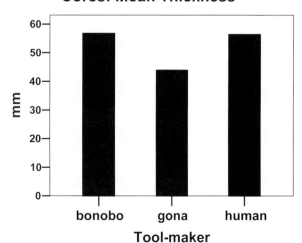

Cores: Mean Thickness

Test Statistics: Core Thickness, Bonobo v. Gona[a]

	Length
Mann-Whitney U	184.000
Wilcoxon W	460.000
Z	-3.257
Asymp. Sig. (2-tailed)	.001

a. Grouping Variable: Toolmkrcod

Test Statistics: Core Thickness, Bonobo v. Human[a]

	Length
Mann-Whitney U	495.000
Wilcoxon W	991.000
Z	-.222
Asymp. Sig. (2-tailed)	.824

a. Grouping Variable: Toolmkrcod

Test Statistics: Core Thickness, Gona v. Human[a]

	Length
Mann-Whitney U	113.000
Wilcoxon W	389.000
Z	-4.264
Asymp. Sig. (2-tailed)	.000

a. Grouping Variable: Toolmkrcod

Discussion

Core thicknesses on the Gona sample suggest that the early hominin tool-makers were selecting thinner cobbles than the experimental samples. This may have been done in part because thinner cobbles tend to be easier to flake, with thinner edges more amenable to flake detachment with a hammerstone. It should be noted that, although the bonobo cores had a mean thickness equivalent to the human cores, their absolute length was considerably greater. These point to an overall size and shape difference in the bonobo cores, a pattern that will be explored further below.

16. Cores: Ratio Breadth/Length

Rationale

This ratio is an indication of how relatively elongated a core is with regard to breadth. A low ratio indicates a more elongated core, whereas a higher ratio indicates a more equilateral core.

Results

Although all three populations have roughly similar breadth/length mean ratios (from about 75 to 81%), the difference between the means of the Gona sample (with the smallest ratio) and human sample (with the largest ratio) was significant at the .05 confidence level. There were no significant differences between the bonobo sample (with an intermediate breadth/length value) and the other two samples.

Cores: Breadth to Length Ratio

b_by_l

Tool-maker	Mean	N	Std. Deviation
bonobo	.7826	33	.11858
gona	.7493	23	.08815
human	.8077	31	.09408
Total	.7827	87	.10407

Test Statistics: Breadth to Length Ratio, Bonobo v. Gona[a]

	b_by_l
Mann-Whitney U	298.000
Wilcoxon W	574.000
Z	-1.357
Asymp. Sig. (2-tailed)	.175

a. Grouping Variable: Toolmkrcod

Test Statistics: Breadth to Length Ratio, Bonobo v. Human[a]

	b_by_l
Mann-Whitney U	448.000
Wilcoxon W	1009.000
Z	-.853
Asymp. Sig. (2-tailed)	.394

a. Grouping Variable: Toolmkrcod

Test Statistics: Breadth to Length Ratio, Gona v. Human[a]

	b_by_l
Mann-Whitney U	229.000
Wilcoxon W	505.000
Z	-2.230
Asymp. Sig. (2-tailed)	.026

a. Grouping Variable: Toolmkrcod

Discussion

This attribute is actually somewhat problematic as a point of comparison between different assemblages: this ratio can change dramatically as the axes of length and breadth change dimension in the process of core reduction, and can even reverse direction of change when continued flaking transforms the "length" into the "breadth." It should be noted that although the bonobo mean was not significantly different than the Gona and human samples, the bonobo sample was composed mainly of end choppers, with the worked edge roughly perpendicular to the long axis of the core, while the Gona and human samples were composed largely of side choppers, with the worked edge roughly parallel to the long axis of the core. As a hypothetical case, the reduction of an elongate cobble could give the following progression of breadth/length ratios and core typology as flaking proceeded: .74 (end chopper), .84 (end chopper), .95 (end chopper), .96 (side chopper) [after the worked "end" becomes the worked "side"], .79 (side chopper), .60 (side chopper), .51 (an extremely reduced side chopper). Thus, an end chopper can progress upward in breadth/length ratio until it is equidimensional, and then further working of the same edge as a side chopper will continue to reduce this ratio.

As the Gona and human samples were dominated by side choppers, continued flaking of their primary edges would tend to decrease the breadth/length ratio (as flake removals remove additional material from the breadth of the core). Thus, the fact that the Gona cores have a somewhat smaller mean breadth/length ratio than the human cores may reflect the more intensive reduction of the archaeological cores (with further reduction of the human side choppers tending to reduce their mean and moving it further toward the Gona value). Reduction of the bonobo end choppers, on the other hand, is acting to increase their breadth/length ratio to artificially place it within the human-Gona range; further flaking along the cobble end would tend to increase it further (until the core is equidimensional).

17. Cores: Modified Breadth/Length Ratio

Rationale

Perhaps a better indicator of core morphology would be to define length as the dimension perpendicular to the flaked edge and breadth as the dimension parallel to the flaked edge. This attribute would then discriminate fully between side choppers and end choppers. By doing this, the modified breadth:length ratio of end choppers would always be less the 1.0, while the modified breadth:length ratio of side choppers would be greater than 1.0. This new ratio was applied to the choppers from the three samples.

Results

The bonobo cores, with a preponderance of end choppers, had the lowest mean modified breadth/length ratio (~0.92), and the Gona cores, with a preponderance of side choppers, the highest (~1.27). The human cores were intermediate at ~1.16. Statistical tests showed a significant difference between the bonobo sample and both the Gona and human samples (at the .000 significance level), but not between the Gona and human core samples at the .05 level of significance.

Cores: Modified Breadth by Length

Mod breadth by length

Tool-maker	Mean	N	Std. Deviation
bonobo	.9211	27	.26246
gona	1.2745	17	.26293
human	1.1554	31	.24763
Total	1.0981	75	.28977

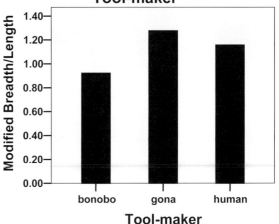

Cores: Modified Breadth/Length by Tool-maker

Test Statistics: Modified Breadth by Length, Bonobo v. Gona[a]

	Mod breadth by length
Mann-Whitney U	80.000
Wilcoxon W	458.000
Z	-3.603
Asymp. Sig. (2-tailed)	.000

a. Grouping Variable: Toolmkrcod

Test Statistics: Modified Breadth by Length, Bonobo v. Human[a]

	Mod breadth by length
Mann-Whitney U	194.500
Wilcoxon W	572.500
Z	-3.492
Asymp. Sig. (2-tailed)	.000

a. Grouping Variable: Toolmkrcod

Test Statistics: Modified Breadth by Length, Gona v. Human[a]

	Mod breadth by length
Mann-Whitney U	182.000
Wilcoxon W	678.000
Z	-1.757
Asymp. Sig. (2-tailed)	.079

a. Grouping Variable: Toolmkrcod

Discussion

These results show a likely continuum in the reduction of somewhat elongated cobbles from end choppers to side choppers. The bonobo sample, whose cores are characterized by less reduction, show a relatively low ratio indicating a dominance of end choppers, while the Gona sample and, to a lesser extent, the human sample, have higher ratios, in line with being more heavily reduced and having a dominance of side choppers. Thus, this modified breadth/length ratio more accurately reflects technological differences in the core assemblages in the samples than does the simple breadth/length ratio.

18. Cores: Ratio Thickness/Breadth

Rationale

This ratio was analyzed to help quantify the relative shape of core forms made on cobbles. As with the breadth/length ratio, this ratio treats a core as a geological clast irregardless of the orientation of the flaked edge.

Results

The bonobo core sample has the lowest thickness/breadth ratio (~.64), the human sample the highest (~.76), and the Gona sample an intermediate ratio (~.73). Statistical tests showed that only the bonobo and human samples differed at the .05 confidence level.

Cores: Thickness to Breadth Ratio by Tool-maker

th_by_b

Tool-maker	Mean	N	Std. Deviation
bonobo	.6439	33	.18172
gona	.7283	23	.15469
human	.7559	31	.10749
Total	.7061	87	.15796

Test Statistics: Thickness to Breadth Ratio, Bonobo v. Gona[a]

	th_by_b
Mann-Whitney U	280.500
Wilcoxon W	841.500
Z	-1.649
Asymp. Sig. (2-tailed)	.099

a. Grouping Variable: Toolmkrcod

Test Statistics: Thickness to Breadth Ratio, Bonobo v. Human[a]

	th_by_b
Mann-Whitney U	310.000
Wilcoxon W	871.000
Z	-2.707
Asymp. Sig. (2-tailed)	.007

a. Grouping Variable: Toolmkrcod

Test Statistics: Thickness to Breadth Ratio, Gona v. Human[a]

	th_by_b
Mann-Whitney U	316.500
Wilcoxon W	592.500
Z	-.700
Asymp. Sig. (2-tailed)	.484

a. Grouping Variable: Toolmkrcod

Discussion

The low ratio for bonobo cores relative to the human cores is probably a function of the bonobos using larger cobbles with larger absolute breadth and length than the human sample, although with similar thicknesses. The bonobos are also tending to make more end choppers and thus are tending to reduce the length, rather than the breadth, of these larger clasts as reduction proceeds.

19. Cores: Maximum Dimension Largest Scar

Rationale

The maximum dimension of the largest scar was measured; this was an absolute measurement, without consideration as to whether the scar was truncated by later flaking. This measurement can give an indication of the size of flakes being detached, although as reduction proceeds, it is likely that some large flake scars are truncated by later scars and thus their maximum dimension reduced.

Results

The bonobo flake scars had the highest mean (~62 mm), with human cores intermediate (~58 mm), and the Gona sample showing the lowest mean (~46 mm). Statistical tests showed that the means of the Gona flake scars differed significantly from both the other samples, while the human and bonobo samples showed very little difference.

Cores: Mean Length Largest Scar by Tool-maker

Length of largest scar

Tool-maker	Mean	N	Std. Deviation
bonobo	61.73	33	23.018
gona	45.91	23	10.352
human	57.84	31	13.902
Total	56.16	87	18.248

Test Statistics: Core Mean Length Largest Scar, Bonobo v. Gona[a]

	Length of largest scar
Mann-Whitney U	204.500
Wilcoxon W	480.500
Z	-2.918
Asymp. Sig. (2-tailed)	.004

a. Grouping Variable: Toolmkrcod

Test Statistics: Core Mean Length Largest Scar, Bonobo v. Human[a]

	Length of largest scar
Mann-Whitney U	505.500
Wilcoxon W	1066.500
Z	-.081
Asymp. Sig. (2-tailed)	.936

a. Grouping Variable: Toolmkrcod

Test Statistics: Core Mean Length Largest Scar, Gona v. Human[a]

	Length of largest scar
Mann-Whitney U	167.500
Wilcoxon W	443.500
Z	-3.309
Asymp. Sig. (2-tailed)	.001

a. Grouping Variable: Toolmkrcod

Discussion

The mean largest flake scar dimension for cores appears in part to be a function of the size of the cobble that was chosen as the blank and the size of the largest flakes removed, in combination with the amount of truncation of these scars by subsequent flaking (i.e., intensity of reduction). This latter factor, the truncation and even removal of earlier flake scars by later core reduction, appears to be a major factor reducing the mean of this attribute in the human core sample, as the largest flakes removed in the human sample for each core (77.5 mm) averaged approximately 15 mm larger than the largest bonobo flakes per core (62.2 mm). Thus, the dimensions of flake scars on the bonobo cores tend to be representative of the largest flakes removed (both approximately 62 mm), largely because their cores are less intensively reduced and thus flake scars tend to be more complete.

The human sample had largest mean scar value (57.8 mm), an underestimate for the largest flake from each core, which averaged approximately 77.5 mm. The relatively smaller size of flake scars on the Gona cores is likely due to the smaller core size and their more intensive reduction, which would have tended to truncate and reduce the size of flake scars on the cores. For more heavily reduced cores, then, the absolute size of the largest flake scar on a core has more limited utility in predicting dimensions of the flake populations removed.

20. Cores: Ratio of Largest Scar/Core Maximum Dimension

Rationale

This ratio is used to arrive at a quantifiable assessment of the invasiveness of the flaking. A ratio that is approaches 1.0 indicates that the largest flake scar is almost the same length as the maximum dimension of the core, in other words, denoting a very invasively flaked core.

Results

The bonobo cores were the least invasively flaked by this measure (largest scar-to-length ratio of approximately .53); the human sample was the most invasive (.62), with the Gona sample intermediate (.57). Statistical testing indicated that only the groups with the smallest and largest means, i.e. the bonobo and human samples, differed at the .05 level of significance.

Cores: Ratio of Largest Scar to Length by Tool-maker

lscar_l

Tool-maker	Mean	N	Std. Deviation
bonobo	.5278	33	.15330
gona	.5708	23	.12649
human	.6195	31	.13539
Total	.5718	87	.14420

Test Statistics: Core Ratio Largest Scar to Length, Bonobo v. Gona[a]

	lscar_l
Mann-Whitney U	286.500
Wilcoxon W	847.500
Z	-1.549
Asymp. Sig. (2-tailed)	.121

a. Grouping Variable: Toolmkrcod

Test Statistics: Core Ratio Largest Scar to Length, Bonobo v. Human[a]

	lscar_l
Mann-Whitney U	287.000
Wilcoxon W	848.000
Z	-3.016
Asymp. Sig. (2-tailed)	.003

a. Grouping Variable: Toolmkrcod

Test Statistics: Core Ratio Largest Scar to Length, Gona v. Human[a]

	lscar_l
Mann-Whitney U	262.000
Wilcoxon W	538.000
Z	-1.653
Asymp. Sig. (2-tailed)	.098

a. Grouping Variable: Toolmkrcod

Discussion

Although the size of the bonobo cores was the largest of the three samples, they were the least invasively flaked as assessed by this attribute. Even though the human cores were somewhat larger than Gona cores in terms of maximum dimension, their largest scar/length ratio was slightly higher. This may point to slightly more optimal flake production in the human sample (optimizing flake size per core), along with more intensive reduction of the Gona sample (truncating, and thus minimizing, the size of flake scars on the core).

21. Cores: Percentage of Original Cobble (Estimate)

Rationale

The technology exhibited in early Oldowan cores often makes it possible to arrive at reasonable estimates of the size of the original cobble clast and what percentage of that cobble mass remains in the final core form. This estimate is based on the presence and curvature of cortex on cores and extrapolating the continuation of cortical surface in areas that have been flaked, thereby reconstructing the original size and shape of the cobble. Obviously, cores that retain much of their cortex are better candidates for more accurate reconstruction of original cobble weight.

This subjective method is based on a great deal of experience in the analysis of Oldowan lithic technology. A blind test was conducted on an experimental sample of 25 cores (with known original cobble weights) to check our accuracy in inferring the original cobble weight of cores similar to those found at the Gona sites. The mean margin of error was plus or minus 5%, suggesting that this method is a useful guide to the intensity of core reduction.

As previously mentioned, the mass of the cobbles in the human experimental sample was intentionally reduced by approximately 50% to provide a baseline when analyzing the archaeological sample. This is reflected in the mean percentage of original cobble in the human sample.

Results

The bonobo cores were the least reduced (~70% of original cobble) and the Gona cores were the most re-

duced (~37%), with the human sample (~54%) intermediate. Statistical testing indicated that each of the three samples were significantly different from the other two at the .00 confidence level.

Cores: Percentage Original Cobble Remaining (Estimate) by Tool-maker

Percentage of original core left

Tool-maker	Mean	N	Std. Deviation
bonobo	70.30	33	16.490
gona	36.50	23	11.120
human	53.90	31	10.860
Total	55.50	87	18.850

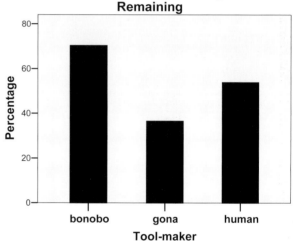

Cores: Estimate Percentage of Original Cobble Remaining

Test Statistics: Percentage Original Cobble, Bonobo v. Gona[a]

	Percentage of original core left
Mann-Whitney U	51.500
Wilcoxon W	327.500
Z	-5.520
Asymp. Sig. (2-tailed)	.000

a. Grouping Variable: Toolmkrcod

Test Statistics: Percentage Original Cobble, Bonobo v. Human[a]

	Percentage of original core left
Mann-Whitney U	199.500
Wilcoxon W	695.500
Z	-4.259
Asymp. Sig. (2-tailed)	.000

a. Grouping Variable: Toolmkrcod

Test Statistics: Percentage Original Cobble, Gona v. Human[a]

	Percentage of original core left
Mann-Whitney U	99.000
Wilcoxon W	375.000
Z	-4.612
Asymp. Sig. (2-tailed)	.000

a. Grouping Variable: Toolmkrcod

Discussion

The high mean percentage of original cobble remaining in the bonobo core sample is a clear indication of the low intensity of reduction (removing ~ 30% of the cobble mass), while the low mean percentage in the Gona sample is a clear indication of much higher intensity of reduction (removing 63% of the cobble mass). The ability to efficiently reduce a cobble and produce larger populations of usable flakes and fragments for a given range of raw materials is one measure of skill. Clearly the Gona hominins were able to reduce their cores more than twice as efficiently as the bonobos, suggesting better mastery of stone reduction.

22. Cores: Number of Flake Scars

Rationale

The number of flake scars (≥ 10 mm maximum dimension) on cores gives a minimum estimate of the number of flakes that have been removed from a core. Cores with low scar counts tend to be less reduced; cores with high scar counts often are heavily reduced. Of course, on more heavily reduced cores, later core reduction may remove earlier flake scars, so the total number of flakes removed can be considerably greater than the flake scar count estimate.

Results

The bonobo cores showed a much lower mean flake scar count (~5.5) than the Gona (~9.3) or human (~9.1) sample. Statistical testing showed that the bonobo core sample differed significantly from both the Gona sample and the human sample at the .00 confidence level. The Gona and human samples, however, were almost identical regarding mean flake scar number.

Cores: Number of Flake Scars by Tool-maker

Number of flake scars

Tool-maker	Mean	N	Std. Deviation
bonobo	5.48	33	3.537
gona	9.30	23	3.948
human	9.13	31	3.074
Total	7.79	87	3.903

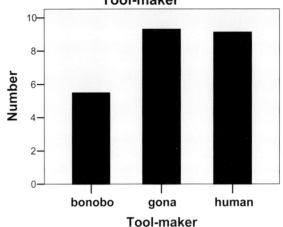

Cores: Mean Number of Flake Scars by Tool-maker

Test Statistics: Core Flake Scars, Bonobo v. Gona[a]

	Number of flake scars
Mann-Whitney U	145.000
Wilcoxon W	706.000
Z	-3.926
Asymp. Sig. (2-tailed)	.000

a. Grouping Variable: Toolmkrcod

Test Statistics: Core Flake Scars, Bonobo v. Human[a]

	Number of flake scars
Mann-Whitney U	178.000
Wilcoxon W	739.000
Z	-4.510
Asymp. Sig. (2-tailed)	.000

a. Grouping Variable: Toolmkrcod

Test Statistics: Core Flake Scars, Gona v. Human[a]

	Number of flake scars
Mann-Whitney U	354.000
Wilcoxon W	630.000
Z	-.044
Asymp. Sig. (2-tailed)	.965

a. Grouping Variable: Toolmkrcod

Discussion

The bonobo cores were much less heavily flaked, with between three and four fewer flake scars on average than the Gona and human samples. This is yet another indication of the low degree of reduction of bonobo cores relative to the Gona and human cores.

It is also useful to compare the actual number of

flakes removed from a core to the number of core flake scars, which is possible to do on the experimental samples. As an indication of the *total* number of flakes produced in a sample, the number of whole flakes, number of split flakes (judged from the number of left or right splits, whichever is greater), and the number of proximal snaps were added together to derive the minimum number of whole flakes produced. Via this procedure, the bonobo sample produced 177 reconstructed flakes, the Gona sample 216, and the human sample 368. The ratio of reconstructed flakes to cores was 5.36 for bonobos, 9.39 for Gona, and 11.87 for humans. It is likely that the Gona flake counts would be higher if all the debitage was represented, since the Gona cores are more heavily reduced than the human cores and because there is evidence (to be discussed below) that certain flakes in the reduction history of cores are not represented at the excavated sites in expected numbers.

23. Cores: Ratio of Steps & Hinges/Total Scars

Rationale

Flake scars terminating in steps and hinges are often viewed as representing misguided flaking, since the flakes did not feather off neatly from the core. (Steps are flake scars that terminate at a right angle break, usually producing a piece of debitage that would be classified as a proximal snap, while hinges are flake scars that terminate in a curved termination). The ratio of steps and hinges to total number of flake scars has been used by some archaeologists as an assessment of skill level. A low number and low ratio of steps and hinges to other scars has sometimes been interpreted as an indication of greater skill in flaking, whereas a higher number and higher ratio of steps and hinges has been interpreted as an indication of a lower level of skill.

Results

Surprisingly, the human 50% reduction sample had the highest ratio of steps and hinges to scars (~.31); the bonobos were intermediate (~.26) and the Gona sample the lowest (~.18). Statistical testing found that only the Gona and human samples showed significant differences at the .05 confidence level.

Cores: Step to scar ratio by tool-maker

step_scr

Tool-maker	Mean	N	Std. Deviation
bonobo	.2570	33	.21476
gona	.1839	23	.14908
human	.3083	31	.16444
Total	.2560	87	.18616

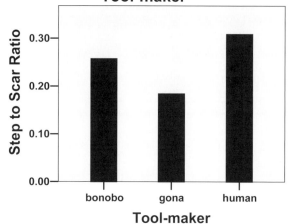

Cores: Mean Step to Scar Ratio by Tool-maker

Test Statistics: Core Step-to-Scar Ratio, Bonobo v. Gona[a]

	step_scr
Mann-Whitney U	304.000
Wilcoxon W	580.000
Z	-1.271
Asymp. Sig. (2-tailed)	.204

a. Grouping Variable: Toolmkrcod

Test Statistics: Core Step-to-Scar Ratio, Bonobo v. Human[a]

	step_scr
Mann-Whitney U	435.000
Wilcoxon W	996.000
Z	-1.033
Asymp. Sig. (2-tailed)	.302

a. Grouping Variable: Toolmkrcod

Test Statistics: Core Step-to-Scar Ratio, Gona v. Human[a]

	step_scr
Mann-Whitney U	202.000
Wilcoxon W	478.000
Z	-2.709
Asymp. Sig. (2-tailed)	.007

a. Grouping Variable: Toolmkrcod

Discussion

Initially, analysis of the results of this attribute was perhaps the most surprising of the entire study. If a high proportion of steps and hinges to flake scars is indeed an indication of lesser skill, then why would the human sample exhibit the highest proportion of step and hinge scars, when other attributes investigated here indicate greater flaking skill in the human sample than in the Gona or bonobo samples? There appear to be a number of factors that would help explain the observed pattern.

It is likely that the shape of the cobble chosen as a core blank could have a strong effect on the incidence of steps and hinges in an assemblage. Tool-makers selecting thinner cobbles or wedge-shaped cobbles with a thin edge that are much easier to flake might produce significantly lower proportions of steps and hinges. This could be the case with the Gona assemblage, in which relatively thinner, easier-to-flake cobbles may have been selected relative to those cobbles in the experimental samples. If there were enough type 1 flakes (flakes with a cortical striking platform and total cortex dorsal surface, normally the first flake to be removed from a cobble), this might be tested: the morphology of the cortex of that flake could be indicative of the ease of flaking of the cobble edge; unfortunately, there is only one type 1 flake from the Gona sites. Another test of clast shape would be refitting entire cobbles, but unfortunately, this is rarely possible in significant numbers at Oldowan sites.

In fact, investigation of the difference in core thickness may yield information pertinent to the patterns observed in experimental samples. The human cores tended to be thicker than the Gona cores, not only absolutely but also relative to core length (see below). This difference in relative thickness would tend to provide steeper edges for flaking and thus more opportunity to produce step or hinge flakes, increasingly so as a core is reduced and flaking forces are directed more steeply into the mass of the core. Although similar in absolute thickness to the human cores, the bonobo cores are thinner relative to their length and are not as heavily reduced, both of which factors may reduce the tendency for stepping to occur relative to the human sample.

As will be discussed below, in order to see how preferentially later stages of flaking might influence core and debitage characteristics (as this appears to be a major factor differentiating the human and Gona core samples), a subset of the human core sample (n=10, out of the total sample of 33 cores) was further reduced to a level more comparable to the Gona cores. Interestingly, these more heavily reduced cores had a step-to-scar ratio of .1565 (S.D. of .08), a ratio even lower than at Gona (.1839). The Mann-Whitney test did not find significant difference in step-to-scar ratios between the Gona sample and the human later stage sample.

It would thus appear that later stages of Oldowan cobble core reduction produce less step and hinge fractures, possibly because flakes are able to travel down established core ridges produced by previous flake removals. Earlier stages of cobble reduction, however, appear to produce relatively higher proportions of steps and hinges, possibly because flakes travel down curved cortical surfaces of cobble cores less successfully and terminate more frequently in steps or hinges.

In sum, the ratio of steps and hinges to total flake scars is probably not a reliable indicator of skill, but is likely influenced by a number of variables. Of special importance is the degree of reduction of the core, with more heavily reduced cores tending to exhibit fewer step and hinge scars. In addition, step and hinge scar frequencies are likely related to the overall ease of flaking the cobble, influenced in turn by variables such as presence of a thin edge for flaking, the absolute thickness of the core, and core thickness relative to length.

24. Cores: Edge Angle

Rationale

This measurement gives an indication of potential functional qualities of a core edge as well as an indication as to whether a given core could be easily flaked further. A flaked core edge will yield different angles depending on where you measure, so for this analysis the minimal edge angle was recorded. An edge angle of 70°, for example, suggests a possible cutting/chopping tool and a core that could still be reduced, assuming the core was still large enough for further reduction, whereas an angle of 95° suggests a non-functional edge for cutting or chopping and an exhausted core form.

Results

The bonobo cores showed the steepest angles (~83 degrees), the human sample showed the most acute angles (~69 degrees) with the Gona cores intermediate (~78 degrees). Statistical testing indicated that the human sampled differed from both the bonobo and Gona samples at the .000 confidence level, but that the bonobo and Gona sample were not different at the .05 level.

Test Statistics: Core Step-to-Scar Ratio, Gona v. Human Later Stages[b]

	step_scr
Mann-Whitney U	111.000
Wilcoxon W	166.000
Z	-.157
Asymp. Sig. (2-tailed)	.875
Exact Sig. [2*(1-tailed Sig.)]	.893[a]

a. Not corrected for ties.

b. Grouping Variable: Toolmkrcod

Cores: Core angle by tool-maker

Core angle, nearest 5 degrees

Tool-maker	Mean	N	Std. Deviation
bonobo	82.88	33	13.231
gona	78.26	23	6.676
human	68.87	31	9.722
Total	76.67	87	12.120

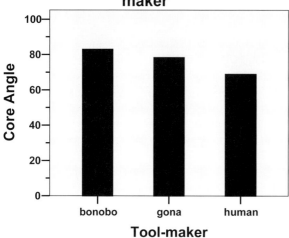

Cores: Mean Edge Angle by Tool-maker

Test Statistics: Core Edge Angles, Bonobo v. Gona[a]

	Core angle, nearest 5 degrees
Mann-Whitney U	305.500
Wilcoxon W	581.500
Z	-1.246
Asymp. Sig. (2-tailed)	.213

a. Grouping Variable: Toolmkrcod

Test Statistics: Core Edge Angles, Bonobo v. Human[a]

	Core angle, nearest 5 degrees
Mann-Whitney U	207.000
Wilcoxon W	703.000
Z	-4.126
Asymp. Sig. (2-tailed)	.000

a. Grouping Variable: Toolmkrcod

Test Statistics: Core Edge Angles, Gona v. Human[a]

	Core angle, nearest 5 degrees
Mann-Whitney U	148.500
Wilcoxon W	644.500
Z	-3.682
Asymp. Sig. (2-tailed)	.000

a. Grouping Variable: Toolmkrcod

Discussion

The ability of a stone knapper to maintain acute angles as flaking proceeds is one index of skill. By this criterion, the human sample showed the most skill, with the most acute angles, with the Gona cores intermediate (though not significantly different from the bonobo sample), and the bonobo cores showing the least skill. The exterior platform angle on whole flakes, to be discussed below, is perhaps an even better indicator of core edge angle, as every flake bears evidence of the core angle immediately prior to this flake detachment and can represent all stages of core reduction rather than just the cores final form (at which time it was abandoned or lost).

25. Cores: Percentage of Surface Cortex

Rationale

The amount of surface cortex on Oldowan cobble cores, relative to flaked core surface, can be a gross estimate of the amount of reduction of these cobbles. (A possible exception might be a roller-shaped cobble that is unifacially reduced along its long axis; even if the mass is reduced by over 50 percent, the surface cortex may still be ca. 80 percent). This attribute is estimated to the nearest 5% based on visual examination.

Results

The bonobo cores had the greatest mean cortex value at ~75 percent; the human cores were intermediate at ~62 percent, with the Gona core surfaces the most heavily flaked at ~53 percent. Statistical tests indicated that each of the three samples significantly differed from the other two samples at the .05 confidence level.

Cores: Percentage cortex remaining by tool-maker

Percentage of cortex

Tool-maker	Mean	N	Std. Deviation
bonobo	74.50	33	9.710
gona	53.48	23	16.130
human	61.61	31	11.280
Total	64.37	87	14.840

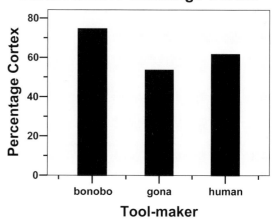

Cores: Mean Percentage Cortex

Test Statistics: Percentage Cortex, Bonobo v. Gona[a]

	Percentage of cortex
Mann-Whitney U	86.000
Wilcoxon W	362.000
Z	-5.009
Asymp. Sig. (2-tailed)	.000

a. Grouping Variable: Toolmkrcod

Test Statistics: Percentage Cortex, Bonobo v. Human[a]

	Percentage of cortex
Mann-Whitney U	202.000
Wilcoxon W	698.000
Z	-4.300
Asymp. Sig. (2-tailed)	.000

a. Grouping Variable: Toolmkrcod

Test Statistics: Percentage Cortex, Gona v. Human[a]

	Percentage of cortex
Mann-Whitney U	247.000
Wilcoxon W	523.000
Z	-2.009
Asymp. Sig. (2-tailed)	.045

a. Grouping Variable: Toolmkrcod

Discussion

Percentage of cortex on core surfaces is an index of how reduced a given cobble core is: 90 percent cortex would suggest minimal reduction, while 10 percent cortex would suggest very heavy reduction. The bonobo cores are the least reduced, and the Gona cores the most reduced. The extent of Gona core reduction, based on the low percentage of surface cortex, suggests a higher level of skill in the ability to remove substantial debitage as reduction proceeded.

26a. Flakes: Flake Types

Rationale

Classifying whole flakes into six distinct types (with a seventh indeterminate category) helps identify different stages of cobble reduction and preferential modes of flaking (Figure 23). These include:
Type 1: cortex platform; total cortex dorsal surface
Type 2: cortex platform; partial cortex dorsal surface
Type 3: cortex platform; non-cortex dorsal surface
Type 4: non-cortex platform; total cortex dorsal surface
Type 5: non-cortex platform; partial cortex dorsal surface

Type 6: non-cortex platform; non-cortex dorsal surface
Type 7: indeterminate; blown or punctiform platform or too weathered

As discussed by Toth (1982, 1985), examination of proportions of these flake types can also be employed to make predictions about what flake type populations would be expected in early stages (e.g. types 1, 2, 4, and 5 flakes) vs. later stages (e.g. types 3 and 6 flakes) of Oldowan cobble reduction, as well as about expected changes in flake type proportions in the event of hydrological winnowing during sedimentation and burial (e.g. preferentially winnowing away from a flaking locale the lighter types 3 and 6 flakes and leaving the heavier types 1, 2, 4, and 5). It is also possible to predict what flake types might be depleted in an assemblage if highly functional flakes were to be transported by hominins away from a flaking area (discussed below in section 27b). For the following analysis, the type 7 indeterminate flakes are omitted, since they are lacking technological information, usually missing identifiable striking platforms.

Results

The bonobo flake population is characterized by high proportions of type 2 flakes, with moderate proportions of flake types 3, 1, and 5 and low proportions of flake types 4 and 6. The Gona population has a predominance of flake types 3 and 2, with moderate proportions of type 6 flakes and low proportions of flake types 5, 4, and 1. The human sample, like the bonobo samples, has high proportions of flake type 2, but with more bifacial reduction also shows moderate proportions of types 5 and 3 flakes as well as type 1 flakes, and low proportions of flake types 6 and 4. Chi square tests indicated that all samples differed from one another at the .05 level of significance. The Kolmogorov-Smirnov test of cumulative frequency distributions indicate a significant difference between Gona and the other two samples, but found the human and bonobo sample were not significantly different at the .05 confidence level. The Chi-square test, which detected a significant difference between these two samples, was likely more sensitive to the specific features of the flake type distributions, particularly the higher frequencies of type 5 flakes and lower frequencies of type 1 flakes in the human sample relative to the bonobo sample.

Flake Type Frequencies by Tool-maker

			Flake Type						Total
			flake type 1	flake type 2	flake type 3	flake type 4	flake type 5	flake type 6	
Tool-maker	bonobo	Count	25	60	29	4	15	9	142
		% within Tool-maker	17.6%	42.3%	20.4%	2.8%	10.6%	6.3%	100.0%
	gona	Count	1	49	79	2	13	19	163
		% within Tool-maker	.6%	30.1%	48.5%	1.2%	8.0%	11.7%	100.0%
	human	Count	24	90	45	8	54	13	234
		% within Tool-maker	10.3%	38.5%	19.2%	3.4%	23.1%	5.6%	100.0%
Total		Count	50	199	153	14	82	41	539
		% within Tool-maker	9.3%	36.9%	28.4%	2.6%	15.2%	7.6%	100.0%

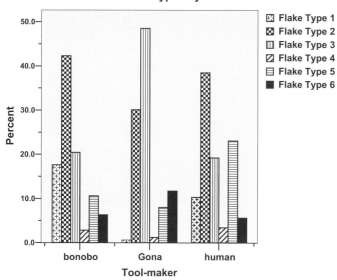

Flakes: Flake Types by Tool-maker

Legend:
- Flake Type 1
- Flake Type 2
- Flake Type 3
- Flake Type 4
- Flake Type 5
- Flake Type 6

Chi-Square Tests: Flake Types 1-6, Bonobo v. Gona

	Value	df	Asymp. Sig. (2-sided)
Pearson Chi-Square	49.582[a]	5	.000
Likelihood Ratio	55.761	5	.000
N of Valid Cases	305		

a. 2 cells (16.7%) have expected count less than 5. The minimum expected count is 2.79.

Kolmogorov-Smirnov Test: Flake Types 1-6, Bonobo v. Gona[a]

		Flake type number
Most Extreme Differences	Absolute	.292
	Positive	.000
	Negative	-.292
Kolmogorov-Smirnov Z		2.542
Asymp. Sig. (2-tailed)		.000

a. Grouping Variable: Toolmkrcod

Chi-Square Tests: Flake Types 1-6, Bonobo v. Human

	Value	df	Asymp. Sig. (2-sided)
Pearson Chi-Square	11.778[a]	5	.038
Likelihood Ratio	12.297	5	.031
N of Valid Cases	376		

a. 1 cells (8.3%) have expected count less than 5. The minimum expected count is 4.53.

Kolmogorov-Smirnov Test: Flake Types 1-6, Bonobo v. Human[a]

		Flake type number
Most Extreme Differences	Absolute	.123
	Positive	.008
	Negative	-.123
Kolmogorov-Smirnov Z		1.159
Asymp. Sig. (2-tailed)		.136

a. Grouping Variable: Toolmkrcod

Chi-Square Tests: Flake Types 1-6, Gona v. Human

	Value	df	Asymp. Sig. (2-sided)
Pearson Chi-Square	61.665[a]	5	.000
Likelihood Ratio	67.151	5	.000
N of Valid Cases	397		

a. 1 cells (8.3%) have expected count less than 5. The minimum expected count is 4.11.

Kolmogorov-Smirnov Test: Flake Types 1-6, Gona v. Human[a]

		Flake type number
Most Extreme Differences	Absolute	.180
	Positive	.180
	Negative	-.112
Kolmogorov-Smirnov Z		1.769
Asymp. Sig. (2-tailed)		.004

a. Grouping Variable: Toolmkrcod

Discussion

Inspection of the flake type distributions and the Kolmogorov-Smirnov test indicate strong similarities between the bonobo and human samples, with a preponderance of types 2, then 3, then 1 flakes. The human sample has more bifacial flaking, however, which can be seen especially in a higher proportion of type 5 flakes than in the bonobo (or Gona) samples.

The most striking difference between the Gona flake population and the bonobo and human population is the high percentage of Gona flakes with no dorsal cortex (types 3 and 6); for Oldowan technology this is usually an indication of later stages of flaking being preferentially represented. The flake population found at the EG sites at Gona suggests that the Gona cores were substantially flaked off-site and the partially reduced cores transported to the floodplain sites where they were further reduced. It is also likely that a subset of the debitage produced from flaking at the floodplain sites was subsequently transported off-site for potential use at another location. This pattern will be discussed in more detail in the following sections.

26b. Flakes: Simulated Selection of Most Useful Flake Types

We were interested in examining how flake populations could be altered by selecting only the most efficient flakes for cutting activities. Out of the total population of debitage produced by modern humans in these experiments (n=614), representing all stages of reduction, we selected the largest and sharpest pieces that, based on our experience in using stone tools, we would use for processing animal carcasses (skinning, dismembering, and defleshing). In this simulation, these usable flakes and fragments would then be carried off-site for later use, leaving behind the less useful debitage.

From the total sample of human debitage in these experiments, a sample of 166 debitage pieces (27.0% of the total) was thus selected for what was deemed to be their superior cutting utility. Of the 166 selected debitage pieces, the great majority consisted of whole flakes (n=126, or 75.9%). The flakes removed from the site in this simulation would constitute roughly half of the original flake population (51.6%). Of the 119 selected ("most usable") whole flakes that could be assigned to one of the six major types, the flake type breakdown was as follows in the table and graph below:

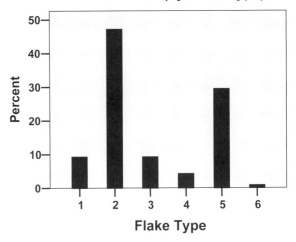

Most Usable Flakes (by Flake Type)

The graph below shows the percentage of each flake type that would have been removed in this simulation (larger, sharper flakes) for later use off-site. Note that in this simulation roughly half of the total flake population, and especially large percentages of types 1, 2, 4 and 5, would be transported away for later use. This would therefore leave behind elevated percentages of types 3 and 6.

MOST USABLE FLAKES

Flake Type	Frequency	Percent	Cumulative Percent
Type 1	11	9.2	9.2
Type 2	56	47.1	56.3
Type 3	11	9.2	65.5
Type 4	5	4.2	69.7
Type 5	35	29.4	99.1
Type 6	1	0.8	100.0
Total:	119	100.0	

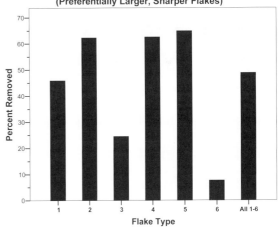

Simulation: Percent of Each Flake Type Transported Off-Site (Preferentially Larger, Sharper Flakes)

This simulation is based on selection of usable flakes from a *complete* assemblage of cobble reduction. At Gona, however, it appears that later stages of flaking are preferentially represented; in this case, the selection of usable flakes would be especially biased towards flake types 2 and 5 (common products of flaking heavily reduced cores, while types 1 and 4 are not produced), thereby inflating the percentages of flake types 3 and 6 even more in the residual assemblage.

It can be deduced from the tables and graphs above that selection and removal of more usable flakes from the general flake population would produce a residual flake population with depressed proportions of flake type 2 and flake type 5 (through flake transport away from the site by hominins) and elevated proportions of flake type 3 and, especially, flake type 6. The following three graphs show (1) the flake type distribution resulting from Oldowan flaking in the human sample, (2) a simulation of the flake types that would be left behind if more usable flakes were removed, and (3) the flake type distribution at the Gona EG sites, which is much closer to the residual population of flakes left behind in this simulation than to the entire human flake population. The Gona sites exhibit further depletion of Type 1 flakes, which may have resulted from Gona hominins 'testing' cobbles at the river cobbles and leaving behind these flakes when transporting cores to the floodplain sites.

Flake Types: Human Flake Population

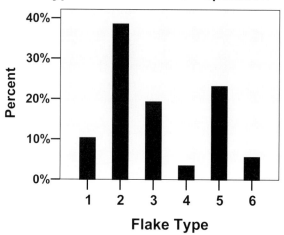

Simulation: Flake Types Left Behind (All Stages of Flaking)

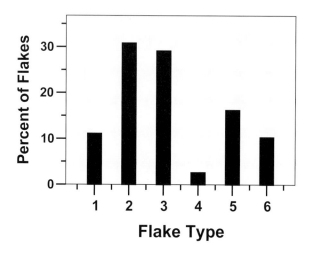

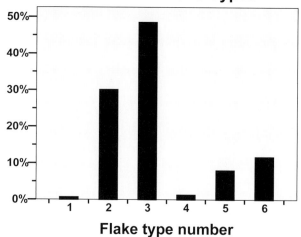

Gona EG Sites Flake Types

26c. Flakes: Simulation of Later Stages of Flaking

As analysis proceeded, it became clear that later stages of cobble reduction were typical of the Gona EG sites. In an attempt to replicate this pattern, ten of the experimental human cores (five unifacial chopper, five bifacial choppers) were further reduced to see what the later stage flake type population would look like. Interestingly, the more reduced cores produced had very similar mean attributes compared to the Gona cores (maximum dimensions, flake scar numbers, breadth/length ratios, percentage of circumference flaked, percentage of cobble left, percentage of cortical surface) and were not significantly different at the .05 confidence level). Visually, they also looked very much like the Gona cores.

The flake type frequencies produced by the human later stage of flaking was also compared to the Gona sample. Now, as with Gona, the human later stage sample showed higher percentages of type 3 to type 2 flakes (~28% v. 11%); however, unlike the Gona sample, type 5 flakes still outnumbered type 6 flakes (~31% v. 28%).

Flake Type Frequencies, Gona v. Human Later Stages

			Flake Type						Total
			1	2	3	4	5	6	
Tool-maker	gona	Count	1	49	79	2	13	19	163
		% within Tool-maker	.6%	30.1%	48.5%	1.2%	8.0%	11.7%	100.0%
	human-lat	Count	1	11	28	1	32	29	102
		% within Tool-maker	1.0%	10.8%	27.5%	1.0%	31.4%	28.4%	100.0%
Total		Count	2	60	107	3	45	48	265
		% within Tool-maker	.8%	22.6%	40.4%	1.1%	17.0%	18.1%	100.0%

Flake Types, Gona v. Human Later Stages

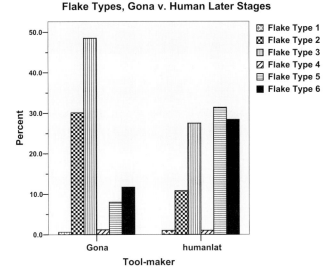

Test Statistics: Breadth/Length, Gona v. Human Later Stages[a]

	Breadth by Length
Mann-Whitney U	8436.000
Wilcoxon W	23661.00
Z	-.944
Asymp. Sig. (2-tailed)	.345

a. Grouping Variable: Toolmkrcod

Flake Thickness/Breadth, Gona v. Human Later Stages

th_by_b

Tool-maker	Mean	N	Std. Deviation
gona	.3688	174	.11522
humanlat	.3099	104	.11295
Total	.3467	278	.11769

Test Statistics: Flake Thickness/Breadth, Gona v. Human Later Stages[a]

	th_by_b
Mann-Whitney U	6451.000
Wilcoxon W	11911.00
Z	-4.004
Asymp. Sig. (2-tailed)	.000

a. Grouping Variable: Toolmkrcod

The later stage flake population was then compared with the Gona sample with regard to maximum dimension, breadth/length ratio, and thickness/breath ratio. The only significant difference at the .05 confidence level was the thickness/breadth ratio, indicating that Gona flakes were somewhat thicker relative to flake breadth, on average, than the human later stage flakes.

Flake Maximum Dimension, Gona v. Human Later Stages

Maximum dimension

Tool-maker	Mean	N	Std. Deviation
gona	41.37	174	15.037
humanlat	40.96	104	13.073
Total	41.22	278	14.311

Test Statistics: Flake Maximum Dimension, Gona v. Human Later Stages[a]

	Maximum dimension
Mann-Whitney U	8961.500
Wilcoxon W	24186.500
Z	-.133
Asymp. Sig. (2-tailed)	.894

a. Grouping Variable: Toolmkrcod

Flake Breadth/Length, Gona v. Human Later Stages

Breadth by Length

Tool-maker	Mean	N	Std. Deviation
gona	1.0047	174	.27968
humanlat	1.0711	104	.40945
Total	1.0295	278	.33500

26d. Flakes: Simulation of Later Stages of Flaking and Subsequent Removal of Larger, Sharper Flakes

We then proceeded to see how the later stage flake population would change (and compare/contrast to the Gona sample) if we were to remove the largest, sharpest flakes. The graph below shows the proportion of each flake type removed in this simulation, with a high proportion of flake type 5 removed, as well as quantities of flake types 6, 4 and 3. This simulates early hominins selecting and transporting the most useful flakes off-site for future use. Interestingly, the residual flake populations produced in this experimental simulation (the second graph below) more closely matched the Gona flake population (the third graph below): after this simulated transport and removal of more useful flakes from the flake population, type 3 and type 6 flakes outnumbered type 2 and type 5 flakes among the flakes left behind, producing a flake type distribution much closer to that in the Gona sample.

In sum, these simulations strongly suggest that the Gona archaeological sites have a somewhat complex history: first cobbles were significantly reduced before transport to the floodplain sites, then later stages of cobble reduction took place at the sites, then many of the largest and sharpest flakes were transported away from the sites.

Simulation: Flake Types Transported Away from Human Later Stages Assemblage

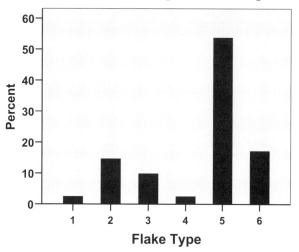

Simulation: Flake Types Left Behind in Human Later Stage Assemblage (after selection and transport)

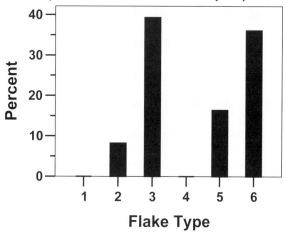

Flake Types: Gona EG Sites

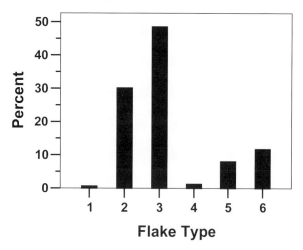

The working hypothesis presented here describes a behavioral pattern of tool-making hominins that could explain the nature of the EG Gona sites:

1) Cobble blanks were selected by Gona hominins in the gravel bars of the paleo-stream system. The cobbles are probably tested at the gravel bars, and the less suitable raw materials immediately discarded. For the better raw materials, it would appear that an appreciable amount of the mass of each cobble, perhaps on average one-half of the cobble mass, had been flaked off-site. So by the time the Gona cores were transported to the EG sites, they may have retained, on average, only about half the original cobble mass. The flakes produced away from the EG floodplain sites would have been dominated by flake type 2 if unifacial flaking is carried out, with type 5 flakes as well if bifacial flaking is carried out. This suggests that there may have been considerable tool use by the hominin group(s) prior to their occupation of these floodplain sites, as the extent of previous core reduction would appear to be far beyond mere testing of raw materials at the river gravel sources.

2) The partially flaked cores are transported to the EG floodplain sites (probably with several percussors) and are reduced further, the resultant cores now averaging about one-quarter to one-third of the mass of the original cobble blanks.

3) Some of the cores (perhaps more than half) are subsequently transported off-site, along with the percussors and many of the larger, sharper flakes (preferentially types 2 and 5). The removal of many of these key flakes would make archaeological refitting much more difficult. The remaining flake population would be dominated by type 3 flakes, and to a lesser extent, type 2 flakes (produced by unifacial flaking) as well as type 6 flakes, and to a lesser extent, type 5 flakes (produced by bifacial flaking) (e.g., similar to the EG site flake type distribution).

This scenario, then, would go a long way toward explaining major EG Gona site patterns, including the flake types represented at the EG sites (with their elevated levels of types 3 and 6) as well as the low frequency of refitting pieces, as due to hominins flaking cores prior to bringing them to the EG sites and then transporting some flakes away after on-site flaking. What is clear from this assessment is that Gona hominins 2.6 million years ago showed a complex behavioral pattern of core reduction and transport of flaked artifactual materials on the landscape, and only portions of the entire history of stone flaking and technological behavior of Gona hominins occurred in the excavated site areas.

27. Flakes: Location of Cortex.

Rationale

The location of cortex on dorsal surfaces of flakes (top, bottom, left, right, center, all, etc.) gives important information about core morphology prior to flake removal, as well as about the order and procedure of flake removal. Location of cortex can give important information about core morphology and stage of flaking. Location of cortex was divided into all (totally cortical); left; right; bottom; top; center; other; left and right; circumference; and no cortex.

Results

The bonobo sample showed high percentages of flakes with no cortex (~29%), all cortex (~18%), right cortex (~17%), and left cortex (~17%), and bottom cortex (~12%). The Gona flake sample had very high percentages of flakes with no cortex (~57%), left cortex (~16%), and right cortex (~13%). The human sample was dominated by roughly equal proportions of flakes with right cortex (~24%), no cortex (~23%), and left cortex (~21%) but also had numbers with bottom cortex (~14%) and all cortex (~13%).

Flakes: Location of Cortex by Tool-maker

			all cortex	left	right	bottom	top	center	other	left and right	circumference	no cortex	Total
Tool-maker	bonobo	Count	29	26	27	19	2	3	0	3	4	45	158
		% within Tool-maker	18.4%	16.5%	17.1%	12.0%	1.3%	1.9%	.0%	1.9%	2.5%	28.5%	100.0%
	gona	Count	5	26	22	7	1	2	7	0	0	94	164
		% within Tool-maker	3.0%	15.9%	13.4%	4.3%	.6%	1.2%	4.3%	.0%	.0%	57.3%	100.0%
	human	Count	32	52	58	35	3	3	0	2	4	55	244
		% within Tool-maker	13.1%	21.3%	23.8%	14.3%	1.2%	1.2%	.0%	.8%	1.6%	22.5%	100.0%
Total		Count	66	104	107	61	6	8	7	5	8	194	566
		% within Tool-maker	11.7%	18.4%	18.9%	10.8%	1.1%	1.4%	1.2%	.9%	1.4%	34.3%	100.0%

Header spanning "Location of cortex" covers columns: all cortex, left, right, bottom, top, center, other, left and right, circumference, no cortex.

Discussion

The high percentage of Gona flakes with no dorsal cortex (over half of the flakes) is a clear indication of later stages of flaking being preferentially represented at the sites (and also Gona cores being more heavily reduced than the experimental samples). The Gona core also has lower percentages of flakes with all cortex and bottom cortex, again because early stages of flaking are generally missing from these archaeological sites.

28. Flakes: Scar Patterning

Rationale

The scar patterning on the dorsal surface of a flake yields information about core morphology and the history of previous flake removals on cores before the flake being analyzed was struck off. Categories include plain (all cortex), same platform (unidirectional), non-cortex (featureless, no scar), opposed scars (bidirectional), transverse ("crest" flake), blade-like, convergent, subradial, and other.

Results

Compared to the other two samples, the bonobos had elevated percentages of plain (all cortex) flakes. The Gona sample had elevated percentages of flakes with opposed scars (bidirectional), while the human sample had higher percentages of same platform (unidirectional) flakes.

Flakes: Scar Pattern by Tool-maker

			Flake scar pattern									Total
			plain (cortex)	same platform, simple	non-cortex, feature-less	opposed scars	trans-verse, crest	blade-like, same	conver-gent	subradial	other	
Tool-maker	bonobo	Count	30	100	6	13	5	0	1	1	2	158
		% within Tool-maker	19.0%	63.3%	3.8%	8.2%	3.2%	.0%	.6%	.6%	1.3%	100.0%
	gona	Count	4	120	6	21	1	1	1	3	15	172
		% within Tool-maker	2.3%	69.8%	3.5%	12.2%	.6%	.6%	.6%	1.7%	8.7%	100.0%
	human	Count	32	179	14	4	7	2	0	4	2	244
		% within Tool-maker	13.1%	73.4%	5.7%	1.6%	2.9%	.8%	.0%	1.6%	.8%	100.0%
Total		Count	66	399	26	38	13	3	2	8	19	574
		% within Tool-maker	11.5%	69.5%	4.5%	6.6%	2.3%	.5%	.3%	1.4%	3.3%	100.0%

Discussion

The higher percentages of plain (all dorsal cortex) flakes observed in the bonobo sample is a reflection of early stages of flaking (flake types 1 and 4), and less reduction of cores. For the opposite reason (later stages of flaking preferentially represented), plain flakes are rare in the Gona sample. The higher percentage of flakes with opposed scars at Gona may be a reflection of debitage from the few discoidal cores represented at the site.

29. Flakes: Shape

Rationale

Flake shape gives some indication of core morphology, flake scar patterning, and functional feasibility. For example, for cutting activities, parallel-sided and convergent flakes are often ideal flake shapes, while divergent flakes are often less suitable. Flakes were divided into the following shapes: convergent; divergent; irregular; déjeté (skewed) left (oriented with platforms at top and dorsal surface facing up); déjeté (skewed) right; oval; and parallel-sided.

Results

The bonobo flake sample, relative to the Gona and human samples, had elevated percentages of oval and divergent flake shapes, and lower percentages of parallel flakes. The Gona sample had higher percentages of parallel and irregular shaped flakes, and lower percentages of oval flakes. The human sample showed higher percentages of parallel and convergent flakes, and lower percentages of divergent and irregular flakes. Each toolmaker group significantly differed from the other two groups at the .05 confidence level.

Flakes: Shape by Tool-maker

			Flake shape							Total
			conver-gent	divergent	irregular	dejete left	oval	parallel	dejete right	
Tool-maker	bonobo	Count	22	30	17	14	47	17	11	158
		% within Tool-maker	13.9%	19.0%	10.8%	8.9%	29.7%	10.8%	7.0%	100.0%
	gona	Count	27	20	31	17	22	40	17	174
		% within Tool-maker	15.5%	11.5%	17.8%	9.8%	12.6%	23.0%	9.8%	100.0%
	human	Count	43	21	19	36	44	58	23	244
		% within Tool-maker	17.6%	8.6%	7.8%	14.8%	18.0%	23.8%	9.4%	100.0%
Total		Count	92	71	67	67	113	115	51	576
		% within Tool-maker	16.0%	12.3%	11.6%	11.6%	19.6%	20.0%	8.9%	100.0%

Chi-Square Tests: Flake Shape, Bonobo v. Gona

	Value	df	Asymp. Sig. (2-sided)
Pearson Chi-Square	25.797[a]	6	.000
Likelihood Ratio	26.301	6	.000
N of Valid Cases	332		

a. 0 cells (.0%) have expected count less than 5. The minimum expected count is 13.33.

Chi-Square Tests: Flake Shape, Bonobo v. Human

	Value	df	Asymp. Sig. (2-sided)
Pearson Chi-Square	27.785[a]	6	.000
Likelihood Ratio	28.210	6	.000
N of Valid Cases	402		

a. 0 cells (.0%) have expected count less than 5. The minimum expected count is 13.33.

Chi-Square Tests: Flake Shape, Gona v. Human

	Value	df	Asymp. Sig. (2-sided)
Pearson Chi-Square	13.570[a]	6	.035
Likelihood Ratio	13.514	6	.036
N of Valid Cases	418		

a. 0 cells (.0%) have expected count less than 5. The minimum expected count is 16.65.

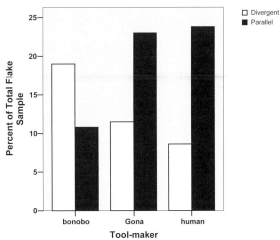

Flake Shape: Divergent v. Parallel by Tool-maker

Chi-Square Tests: Divergent or Parallel Flake Shape, Bonobo v. Gona

	Value	df	Asymp. Sig. (2-sided)	Exact Sig. (2-sided)	Exact Sig. (1-sided)
Pearson Chi-Square	9.847[b]	1	.002		
Continuity Correction[a]	8.660	1	.003		
Likelihood Ratio	9.981	1	.002		
Fisher's Exact Test				.002	.002
N of Valid Cases	107				

a. Computed only for a 2x2 table.

b. 0 cells (.0%) have expected count less than 5. The minimum expected count is 21.96.

Discussion

The most interesting pattern that emerged from the analysis of this attribute was that the bonobos had a high percentage of divergent flakes, while the Gona and human samples showed higher percentages of parallel flakes, within their respective flake assemblages.

Bonobos:
    a. Divergent: 19.0%
    b. Parallel: 10.8%

Gona:
    a. Divergent: 11.5%
    b. Parallel: 23.0%

Human:
    a. Divergent: 8.6%
    b. Parallel: 23.8%

Chi-Square Tests: Divergent or Parallel Flake Shape, Bonobo v. Human

	Value	df	Asymp. Sig. (2-sided)	Exact Sig. (2-sided)	Exact Sig. (1-sided)
Pearson Chi-Square	16.969[b]	1	.000		
Continuity Correction[a]	15.458	1	.000		
Likelihood Ratio	17.069	1	.000		
Fisher's Exact Test				.000	.000
N of Valid Cases	126				

a. Computed only for a 2x2 table.

b. 0 cells (.0%) have expected count less than 5. The minimum expected count is 19.02.

Chi-Square Tests: Divergent or Parallel Flake Shape, Gona v. Human

	Value	df	Asymp. Sig. (2-sided)	Exact Sig. (2-sided)	Exact Sig. (1-sided)
Pearson Chi-Square	.747[b]	1	.387		
Continuity Correction[a]	.458	1	.499		
Likelihood Ratio	.744	1	.388		
Fisher's Exact Test				.454	.249
N of Valid Cases	139				

a. Computed only for a 2x2 table.

b. 0 cells (.0%) have expected count less than 5. The minimum expected count is 17.70.

Parallel-sided flakes tend to be excellent cutting tools, while divergent (usually side-struck) flakes are, on average, less functionally useful. The fact that the Gona and human flakes contain over two times the proportion of parallel-sided flakes is probably a function of better flaking skills, producing longer flakes relative to breadth (discussed below), and showing the ability to follow ridges between flake scars on cores to produce more parallel flakes.

The low percentages of oval-shaped flakes (often cortical) in the Gona sample is almost certainly a function of the early stages of flaking (i.e. flake types 1 and 4, and early stages of flake types 2 and 5) not being well-represented at the archaeological sites. Oval-shaped flakes are much better represented in the bonobo and human samples, where all stages of reduction are present. The prevalence of oval and divergent flakes in the bonobo sample would appear to reflect earlier stages of flaking, with flakes removed from more highly cortical surfaces and less following of ridges in the flaking process.

The predominance of asymmetrical left-skewed flakes relative to right-skewed in the human sample (14.8% and 9.4%, respectively) could be an indication of preferential right handedness in the knappers; this pattern deserves more consideration in the future (through experimentation and analysis of archaeological assemblages).

30. Flakes: Platform Battering/Stepping

Rationale

This attribute recorded the relative degree of battering (as well as microstepping and crushing) along the striking platform and proximal dorsal surface of flakes (none, light, moderate, heavy). Such damage is the equivalent to the attribute of battering on core edges (attribute no. 10), since the striking platform of a flake was part of the core edge prior to flake detachment. And, as with core edge battering, this attribute was developed specifically after seeing some bonobo flakes with unusually heavy battering.

Results

The bonobo sample showed the highest degree of battering, with moderate/high levels at 8.8% (eleven times as much as the human sample). The Gona sample was intermediate, with moderate/high levels at 5.9%, and the human sample very low, at 0.8%. In addition, the human sample had higher percentages of flakes with no battering (89.3%) relative to Gona (78.1%) and bonobos (75.9%). The human sample was significantly different than both the bonobo and Gona sample at the .05 confidence level, although the bonobo and Gona samples were not significantly different.

Flakes: Platform Battering & Stepping by Tool-maker

			Platform battering				Total
			none	low	moderate	high	
Tool-maker	bonobo	Count	120	24	10	4	158
		% within Tool-maker	75.9%	15.2%	6.3%	2.5%	100.0%
	gona	Count	132	27	7	3	169
		% within Tool-maker	78.1%	16.0%	4.1%	1.8%	100.0%
	human	Count	218	24	1	1	244
		% within Tool-maker	89.3%	9.8%	.4%	.4%	100.0%
Total		Count	470	75	18	8	571
		% within Tool-maker	82.3%	13.1%	3.2%	1.4%	100.0%

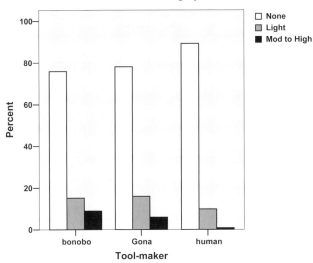

Flake Platform Battering by Tool-maker

Chi-Square Tests: Flake Platform Battering, Bonobo v. Gona

	Value	df	Asymp. Sig. (2-sided)
Pearson Chi-Square	1.051[a]	3	.789
Likelihood Ratio	1.054	3	.788
N of Valid Cases	327		

a. 2 cells (25.0%) have expected count less than 5. The minimum expected count is 3.38.

Chi-Square Tests: Flake Platform Battering, Bonobo v. Human

	Value	df	Asymp. Sig. (2-sided)
Pearson Chi-Square	20.100[a]	3	.000
Likelihood Ratio	20.760	3	.000
N of Valid Cases	402		

a. 3 cells (37.5%) have expected count less than 5. The minimum expected count is 1.97.

Chi-Square Tests: Flake Platform Battering, Gona v. Human

	Value	df	Asymp. Sig. (2-sided)
Pearson Chi-Square	13.638[a]	3	.003
Likelihood Ratio	13.939	3	.003
N of Valid Cases	413		

a. 4 cells (50.0%) have expected count less than 5. The minimum expected count is 1.64.

Discussion

If low frequencies of platform battering/microstepping/crushing is taken as an indication of highly skilled flaking, then the human sample is the most skilled, the bonobo sample the least skilled, and the Gona sample intermediate. The lower incidence of platform battering in the human sample indicates that the human subjects were more successful at flake detachment, leading to "cleaner" core edges (and thus subsequent flake platforms) with less battering.

31. Flakes: Weight

Rationale

Flake weight is a rough indication of the size of the flake (length x breadth x thickness) and often the stage of flake reduction (early stages of flake reduction tend to have heavier flakes) (Sahnouni *et al.*, 1997). In addition, for a given Oldowan core form, mean flake weight may be an indication of knapping skill level: larger flakes with longer cutting edges can be detached with a higher level of skill.

Results

The Gona flakes had the lowest mean weight (~18 g) with the bonobo sample intermediate (~21g), and the human sample a significantly higher mean weight (~31g). The human sample was significantly different from both the bonobo and Gona samples at the .00 confidence level, but the bonobo and Gona samples did not significantly differ from each other at the .05 level.

Flakes: Mean Weight by Tool-maker

Weight in grams

Tool-maker	Mean	N	Std. Deviation
bonobo	20.84	158	50.845
gona	17.94	174	23.983
human	30.74	244	37.616
Total	24.16	576	38.859

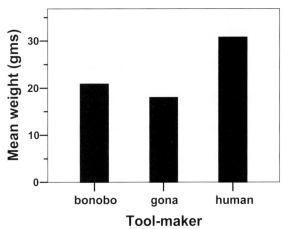

Flake Weight by Tool-maker

Test Statistics: Flake Weight, Bonobo v. Gona[a]

	Weight in grams
Mann-Whitney U	12819.000
Wilcoxon W	25380.000
Z	-1.063
Asymp. Sig. (2-tailed)	.288

a. Grouping Variable: Toolmkrcod

Test Statistics: Flake Weight, Bonobo v. Human[a]

	Weight in grams
Mann-Whitney U	13434.500
Wilcoxon W	25995.500
Z	-5.138
Asymp. Sig. (2-tailed)	.000

a. Grouping Variable: Toolmkrcod

Test Statistics: Flake Weight, Gona v. Human[a]

	Weight in grams
Mann-Whitney U	15846.000
Wilcoxon W	31071.000
Z	-4.423
Asymp. Sig. (2-tailed)	.000

a. Grouping Variable: Toolmkrcod

Discussion

The Gona sample showed the lowest mean flake weight; this is probably because later stages of flaking were preferentially represented at the EG site, with core size reduced, and also because it is possible that cobbles selected for cores at Gona may have been, on average, smaller than the bonobo and human sample. Also, as has been discussed, it is likely that larger, sharper flakes produced on the floodplain sites were subsequently taken away, which could also significantly lower the mean flake weight.

The bonobo mean is interesting, since these subjects were preferentially selecting larger cobbles and reducing them less than the Gona and human samples. One would think that larger flakes would therefore be produced, but, interestingly, the mean bonobo weights are similar to the Gona sample and significantly less than the human sample. This is almost certainly an indication of lower skill levels in the bonobo subjects in detaching flakes, and the lower impact velocities of their hammerstones.

The human sample, not surprisingly, suggests greater skill in detaching large flakes. The mean flake weight is 1.5 times as great as the Gona sample, and 1.7 times as great as the bonobo sample. This is probably due to high impact velocities and better exploitation of acute core edge angles and core morphologies.

32. Flakes: Maximum Dimension

Rationale

Flake maximum dimension (as opposed to oriented length) is the longest linear measurement that can be taken on a flake. It is partially a reflection of the size of the core from which a flake is detached (the maximum dimension of the flake cannot be larger than the maximum dimension of the core prior to flake detachment), and possibly the level of skill of the knapper. From a given core size and morphology, a skilled knapper can often remove larger flakes than a less skilled knapper, producing more cutting edge and a larger, more comfortable flake to use.

Results

As with the data of flake weight, the Gona sample had the smallest flakes (~41 mm), the bonobo sample was intermediate (~45 mm), and the human sample the largest (~53 mm). The human sample showed significant differences with both the bonobo and Gona samples at the .00 confidence level, although the bonobo and Gona samples were not significantly different at the .05 level.

Report

Maximum dimension

Tool-maker	Mean	N	Std. Deviation
bonobo	45.29	158	19.879
gona	41.37	174	15.037
human	52.56	244	17.915
Total	47.19	576	18.303

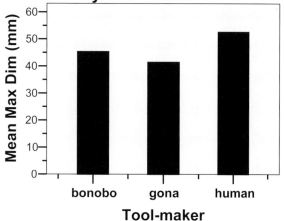

Mean Flake Maximum Dimension by Tool-maker

Test Statistics: Flake Maximum Dimension, Bonobo v. Gona[a]

	Weight in grams
Mann-Whitney U	12357.000
Wilcoxon W	27582.000
Z	-1.591
Asymp. Sig. (2-tailed)	.112

a. Grouping Variable: Toolmkrcod

Test Statistics: Flake Maximum Dimension, Bonobo v. Human[a]

	Weight in grams
Mann-Whitney U	14054.500
Wilcoxon W	26615.500
Z	-4.590
Asymp. Sig. (2-tailed)	.000

a. Grouping Variable: Toolmkrcod

Test Statistics: Flake Maximum Dimension, Gona v. Human[a]

	Weight in grams
Mann-Whitney U	13471.500
Wilcoxon W	28696.500
Z	-6.372
Asymp. Sig. (2-tailed)	.000

a. Grouping Variable: Toolmkrcod

Discussion

The results of the analysis of this attribute are similar to those for flake weight in the previous section. Bonobos showed a lower degree of skill in removing large flakes (even though their cobble blanks were larger). The human flakes showed a higher skill level, producing flakes that were on average 2.4 centimeters larger than the bonobo flakes and 1.2 cm larger than Gona flakes. The fact that Gona flakes appear to represent later stages of flaking (and therefore smaller cores), and the fact that Gona hominins may have been removing significant numbers of larger, sharp flakes from the excavated EG sites, suggests that the Gona mean flake size might be more similar to the human sample if all flakes from all stages of core reduction were represented at the archaeological sites. If so, then the skill level of Gona hominins in consistently producing larger flakes may have been similar to the human subjects, with well-struck hammerstone blows.

33. Flakes: Thickness

Rationale

The maximum thickness of flakes (often at the striking platform or bulb of percussion) helps describe the morphology of a flake. Clearly, as overall size of flakes goes up, the thickness tends to increase as well. Some archaeologists would argue that thinner flakes in an assemblage can be an indication of greater skill (in theory producing more cutting edge per weight of debitage).

Results

Surprisingly, the bonobo flakes were the thinnest, averaging 10.4 mm. The Gona flakes were intermediate (12.6 mm) and the human flakes the thickest (13.8 mm). The bonobo sample was different from the Gona and human sample at the .00 confidence level, although the Gona and human sample were not significantly different at the .05 level.

Flakes: Thickness by Tool-maker

Thickness

Tool-maker	Mean	N	Std. Deviation
bonobo	10.44	158	7.369
gona	12.64	174	6.126
human	13.84	244	7.343
Total	12.54	576	7.130

Mean Flake Thickness by Tool-maker

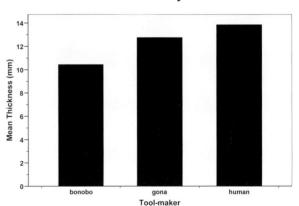

Test Statistics: Flake Thickness, Bonobo v. Gona[a]

	Thickness
Mann-Whitney U	9737.500
Wilcoxon W	22298.500
Z	-4.599
Asymp. Sig. (2-tailed)	.000

a. Grouping Variable: Toolmkrcod

Test Statistics: Flake Thickness, Bonobo v. Human[a]

	Thickness
Mann-Whitney U	12939.500
Wilcoxon W	25500.500
Z	-5.577
Asymp. Sig. (2-tailed)	.000

a. Grouping Variable: Toolmkrcod

Test Statistics: Flake Thickness, Gona v. Human[a]

	Thickness
Mann-Whitney U	19296.500
Wilcoxon W	34521.500
Z	-1.589
Asymp. Sig. (2-tailed)	.112

a. Grouping Variable: Toolmkrcod

Discussion

The bonobo flakes were the thinnest. Although some lithic analysts might argue that this suggests higher skill, observation of bonobo flaking showed that they would often "chew" on an edge by hammerstone percussion until a flake came off. This repeated percussion of an edge would normally produce observable attrition to the core edge (battering, microstepping, crushing), but when a flake was finally detached it tended to be relatively small, side-struck, and thin.

The Gona flakes tended to be absolutely thicker (2.2 mm thicker on average) than the bonobo flakes, even though the bonobo flakes were absolutely larger in maximum dimension (4 mm larger on average). The human flake sample was also thicker (3.4 mm on average) than the bonobo flakes. The thicker flakes produced by Gona hominins and the human knappers are probably a function of higher impact velocities (being able to strike further away from a core edge and still have fracture, producing a thicker striking platform), and the ability of the Gona and human tool-makers to follow ridges more systematically on cores, creating flakes with a thicker, often triangular cross-section.

In sum, the Gona and human samples suggest better flaking skills than the bonobo sample. This is a case where assumptions about skill (thinner flakes meaning better flaking skill) are not confirmed by experimentation and archaeological analysis.

34. Flakes: Ratio of Breadth/Length

Rationale

Breadth/length ratios on flakes gives an indication of how relatively elongated ("end struck") or short ("side struck") a flake population is. Flakes were oriented by bisecting the bulb of percussion and delineating an imaginary rectangle around the flake. Oriented length was the longest measurement parallel to the bulb axis (roughly perpendicular to the striking platform), while breadth was the largest measurement perpendicular to the bulb axis. A ratio of less than 1.0 denotes an end-struck flake; a ratio of greater than 1.0 denotes a side-struck flake.

Results

The bonobo mean flake breadth/length ratio tended to be side-struck (1.22), the Gona flakes intermediate and averaging equidimensional (1.00), and the human sample slightly end-struck (~0.98).

The bonobo sample showed significant differences from the Gona and human sample at the .00 confidence level, although the Gona and human sample were not significantly different at the .05 confidence level.

Flakes: Breadth/Length Ratio by Tool-Maker

b_by_l

Tool-maker	Mean	N	Std. Deviation
bonobo	1.2203	158	.40523
gona	1.0047	174	.27968
human	.9799	244	.38247
Total	1.0533	576	.37531

Test Statistics: Flake Breadth/Length, Bonobo v. Gona[a]

	b_by_l
Mann-Whitney U	9190.000
Wilcoxon W	24415.000
Z	-5.216
Asymp. Sig. (2-tailed)	.000

a. Grouping Variable: Toolmkrcod

Test Statistics: Flake Breadth/Length, Bonobo v. Human[a]

	b_by_l
Mann-Whitney U	12073.500
Wilcoxon W	41963.500
Z	-6.330
Asymp. Sig. (2-tailed)	.000

a. Grouping Variable: Toolmkrcod

Test Statistics: Flake Breadth/Length, Gona v. Human[a]

	b_by_l
Mann-Whitney U	19089.000
Wilcoxon W	48979.000
Z	-1.757
Asymp. Sig. (2-tailed)	.079

a. Grouping Variable: Toolmkrcod

Discussion

The bonobo flakes were significantly different from the Gona and human samples, being appreciably more side-struck (and, as previously mentioned, significantly thinner). It would appear that the bonobos, compared to the Gona and human tool-makers, were not as skilled at producing longer flakes relative to breadth. Again, this is probably due to lower hammerstone impact velocities generated by the bonobos, as well as to their tendency to remove flakes right behind a previous flake scar (rather than directing the blow along a ridge between scars, which generally allows the flaking blow to travel further and detach at a greater distance from the point of percussion).

35. Flakes: Ratio of Thickness/Breadth

Rationale

This ratio is used to determine how relatively thick or thin flakes are in cross-section. In bifacial and blade industries a low thickness/breadth ratio is often interpreted as a function of high skill levels. In the Oldowan, however, a higher thickness/breadth ratio may indicate that early hominin tool-makers were following ridges on cores to produce longer flakes; these flakes tend to be more triangular in cross-section and may have a higher thickness/breadth ratio.

Results

The bonobo flake sample had the lowest mean thickness/breadth ratio at ~.26. The human sample was intermediate at ~.33, while the Gona archaeological sample was the highest at ~.37. Statistical tests showed that each tool-maker group differed significantly from the other two groups at the .000 confidence level.

Flake Thickness/Breadth by Tool-maker

th_by_b

Tool-maker	Mean	N	Std. Deviation
bonobo	.2600	158	.10920
gona	.3688	174	.11522
human	.3325	244	.12623
Total	.3236	576	.12548

Test Statistics: Flake Thickness/Breadth, Bonobo v. Gona[a]

	th_by_b
Mann-Whitney U	6347.500
Wilcoxon W	18908.50
Z	-8.471
Asymp. Sig. (2-tailed)	.000

a. Grouping Variable: Toolmkrcod

Test Statistics: Flake Thickness/Breadth, Bonobo v. Human[a]

	th_by_b
Mann-Whitney U	12276.50
Wilcoxon W	24837.50
Z	-6.152
Asymp. Sig. (2-tailed)	.000

a. Grouping Variable: Toolmkrcod

Test Statistics: Flake Thickness/Breadth, Gona v. Human[a]

	th_by_b
Mann-Whitney U	16870.50
Wilcoxon W	46760.50
Z	-3.579
Asymp. Sig. (2-tailed)	.000

a. Grouping Variable: Toolmkrcod

Discussion

The bonobo sample had the thinnest flakes relative to breadth, followed by the human and Gona samples. Thus, bonobo flakes were not only absolutely thinner but also relative to the overall mass of the flake. The fact that the human and Gona samples had significantly larger thickness/breadth ratios is likely a result of a greater tendency to follow ridges in reducing the core. It is interesting that the Gona flakes are even thicker relative to their breadth than the human sample, despite being absolutely somewhat thinner (though not significantly so). In any case, in Oldowan flaking, a higher ratio of flake thickness to breadth appears to be an indication of greater skill and precision in reducing cores.

36. Flakes: Ratio of Platform Thickness/ Platform Breadth

Rationale

The ratio of platform thickness/platform breadth gives an indication of the morphology of the striking platform. A higher ratio denotes a thicker platform (often associated with hard-hammer percussion), whereas a lower ratio denotes a thinner platform (associated, for example, with soft-hammer percussion in later technologies. The flake populations include flake types 1 through 6 for each sample, omitting flake type 7 whose platform tends to be missing or punctiform.

Results

The platforms of the bonobo flakes were proportionately the thinnest relative to breadth (~.34). The human sample was intermediate at (~.39), and the Gona sample was the thickest (~.44). Statistical testing showed that the all three samples differed significantly from the other two.

Flakes: Platform Thickness/Platform Breadth by Tool-maker

pth_by_b

Tool-maker	Mean	N	Std. Deviation
bonobo	.3385	104	.16978
gona	.4374	142	.15316
human	.3828	202	.14054
Total	.3898	448	.15581

Test Statistics: Flake Platform Thickness/ Platform Breadth, Bonobo v. Gona[a]

	pth_by_b
Mann-Whitney U	4424.000
Wilcoxon W	9884.000
Z	-5.370
Asymp. Sig. (2-tailed)	.000

a. Grouping Variable: Toolmkrcod

Test Statistics: Flake Platform Thickness/ Platform Breadth, Bonobo v. Human[a]

	pth_by_b
Mann-Whitney U	8227.000
Wilcoxon W	13687.000
Z	-3.106
Asymp. Sig. (2-tailed)	.002

a. Grouping Variable: Toolmkrcod

Test Statistics: Flake Platform Thickness/ Platform Breadth, Gona v. Human[a]

	pth_by_b
Mann-Whitney U	11309.00
Wilcoxon W	31812.00
Z	-3.341
Asymp. Sig. (2-tailed)	.001

a. Grouping Variable: Toolmkrcod

Discussion

It would appear that the bonobos were striking off the thinnest flakes and the Gona hominins the thickest flakes, with the human sample intermediate. As will be discussed below (regarding flake thickness/breadth ratios) it is possible that Gona hominins were able to identify prominent ridges on cores and systematically detach flakes along these ridges. The cross-section of many of the flakes produced would be triangular.

37. Flakes: Number of Platform Scars

Rationale

Number of platform scars is an attribute often used in archaeology to examine patterns of platform preparation and faceting (e.g. biface thinning flakes and Levallois flakes and points). (In this study, the cut-off for

platform scar count was 1 mm or larger in the scar's maximum dimension). In the Oldowan, a larger number of mean platform scars are usually associated with core forms such as bifacial choppers, bifacial discoids, and polyhedrons. It is sometimes useful to examine the means of all flakes (types 1-6) as well as the means of flakes only with non-cortical platforms (types 4-6). A preponderance of unifacial flaking of cobbles (platforms scars= 0, or cortical) will clearly bring down the total mean values for platform scars, so that by also considering non-cortical platforms (types 4-6), an appreciation of the platform scar patterning produced by the non-cortical platform flakes could also be assessed.

Results

When considering all flake types, the bonobo and Gona flakes had the same average number of platform scars (.32), while the human sample had a larger mean number (.47). When considering only flake types 4-6 (i.e., with non-cortical platforms), however, the human sample had the lowest value, with a mean platform scar count of ~1.3. The bonobo and Gona samples were very similar, with a mean count of ~1.6. Statistical testing indicated that only the bonobo and human sample were significantly different at the .05 confidence level.

Flakes: Number of Platform Scars, Types 1-6, by Tool-maker

No. platform scars

Tool-maker	Mean	N	Std. Deviation
bonobo	.32	142	.698
gona	.32	162	.736
human	.47	234	.998
Total	.38	538	.854

Test Statistics: Platform Scars, Types 1-6, Bonobo v. Gona[a]

	No. platform scars
Mann-Whitney U	11473.000
Wilcoxon W	21626.000
Z	-.054
Asymp. Sig. (2-tailed)	.957

a. Grouping Variable: Toolmkrcod

Test Statistics: Platform Scars, Types 1-6, Bonobo v. Human[a]

	No. platform scars
Mann-Whitney U	14776.000
Wilcoxon W	24929.000
Z	-2.282
Asymp. Sig. (2-tailed)	.022

a. Grouping Variable: Toolmkrcod

Test Statistics: Platform Scars, Types 1-6, Gona v. Human[a]

	No. platform scars
Mann-Whitney U	16865.000
Wilcoxon W	30068.000
Z	-2.372
Asymp. Sig. (2-tailed)	.018

a. Grouping Variable: Toolmkrcod

Flakes: Number of Platform Scars (Types 4-6) by Tool-maker

No. platform scars

Tool-maker	Mean	N	Std. Deviation
bonobo	1.57	28	.690
gona	1.55	33	.869
human	1.31	75	.677
Total	1.42	136	.736

Test Statistics: Platform Scars, Types 4-6, Bonobo v. Gona[a]

	No. platform scars
Mann-Whitney U	426.500
Wilcoxon W	987.500
Z	-.586
Asymp. Sig. (2-tailed)	.558

a. Grouping Variable: Toolmkrcod

Test Statistics: Platform Scars, Types 4-6, Bonobo v. Human[a]

	No. platform scars
Mann-Whitney U	798.000
Wilcoxon W	3648.000
Z	-2.371
Asymp. Sig. (2-tailed)	.018

a. Grouping Variable: Toolmkrcod

Test Statistics: Platform Scars, Types 4-6, Gona v. Human[a]

	No. platform scars
Mann-Whitney U	1048.500
Wilcoxon W	3898.500
Z	-1.644
Asymp. Sig. (2-tailed)	.100

a. Grouping Variable: Toolmkrcod

Discussion

For all flake types, the human sample had a higher platform scar count since more of the flaking was on bifacial cores; the preponderance of unifacial flaking in the bonobo and Gona samples was due to the preponderance of unifacial flaking of cobbles, producing more flakes with cortical butts. These flakes would give a flake scar count of 0, thereby lowering the mean platform scar count.

For flake types 4-6, the human flake sample had the lowest platform scar count. It is possible that the human knappers were more proficient at isolating core flake scars as potential striking platforms, thus the lower platform scar count. This difference is especially seen between the human and bonobo sample. In sum, a higher platform scar count, considering Oldowan technology, does not seem to be indicative of better stone-working skills.

38. Flakes: Number of Dorsal Scars

Rationale

The number of scars (10 mm or larger) on the dorsal surface of flakes is an indication of the morphology and extent of flaking of a core at the time the flake was detached. For a Mode 1 industry, higher numbers of dorsal scars usually implies more heavily reduced cores, with less cortex and more flake scars. Size is also a consideration: at a given stage of core reduction, a larger flake will, on average, have more dorsal scars than a smaller flake.

Results

When all flake types are examined, the bonobo sample shows the lowest mean number of scars (2.14), the human sample intermediate (2.22) and the Gona sample the highest (3.07). Statistical tests showed that the Gona sample differs from the experimental groups at the .000 level of significance, while the bonobo and human samples were not significantly different.

All Flakes: Number of Dorsal Scars by Tool-maker

No. dorsal scars

Tool-maker	Mean	N	Std. Deviation
bonobo	2.14	158	1.732
gona	3.07	169	1.580
human	2.22	244	1.576
Total	2.45	571	1.669

Test Statistics: All Flakes, Number of Dorsal Scars, Bonobo v. Gona[a]

	No. dorsal scars
Mann-Whitney U	8703.000
Wilcoxon W	21264.000
Z	-5.563
Asymp. Sig. (2-tailed)	.000

a. Grouping Variable: Toolmkrcod

Test Statistics: All Flakes, Number of Dorsal Scars, Bonobo v. Human[a]

	No. dorsal scars
Mann-Whitney U	18331.500
Wilcoxon W	30892.500
Z	-.847
Asymp. Sig. (2-tailed)	.397

a. Grouping Variable: Toolmkrcod

Test Statistics: All Flakes, Number of Dorsal Scars, Gona v. Human[a]

	No. dorsal scars
Mann-Whitney U	14278.000
Wilcoxon W	44168.000
Z	-5.434
Asymp. Sig. (2-tailed)	.000

a. Grouping Variable: Toolmkrcod

Discussion

The Gona sample had a significantly higher flake dorsal scar value compared to the bonobo and the human sample. This is almost certainly due to the fact that the Gona flakes appear to represent later stages of the reduction of cobble cores, with the subsequent flakes having much less cortex and higher dorsal scar counts. The bonobo and human samples are not statistically different, as both samples represent all stages of cobble reduction and thus have a greater percentage of dorsal cortex and subsequently relatively fewer dorsal scars than the Gona sample.

39. Flakes: Ratio of Dorsal Hinges & Steps/Total Dorsal Scars

Rationale

As discussed with cores (attribute no. 23), the ratio of hinge and steps to total scars on dorsal surfaces of flakes has been used as an indicator of knapping skill. Traditionally, a lower ratio has been interpreted as higher skill, a higher ratio interpreted as lower skill.

Results

The Gona flake sample had the lowest mean step/hinge to scar ratio, at ~.05. The human sample was intermediate (~.10) with the bonobo sample showing the highest ratio (~14). Statistical testing showed that the Gona sample differed significantly from both the bonobo and human samples at the .05 confidence level; the bonobo and human sample, however, did not show significant difference at the .05 level.

Flakes: Step+Hinge to Scar Ratio by Tool-maker

Flake step to scar ratio

Tool-maker	Mean	N	Std. Deviation
bonobo	.1350	129	.23160
gona	.0450	171	.11857
human	.1049	211	.19514
Total	.0924	511	.18736

Test Statistics: Flake Step to Scar Ratio, Bonobo v. Gona[a]

	Flake step to scar ratio
Mann-Whitney U	9090.000
Wilcoxon W	23796.000
Z	-3.641
Asymp. Sig. (2-tailed)	.000

a. Grouping Variable: Toolmkrcod

Test Statistics: Flake Step to Scar Ratio, Bonobo v. Human[a]

	Flake step to scar ratio
Mann-Whitney U	12943.500
Wilcoxon W	35309.500
Z	-.954
Asymp. Sig. (2-tailed)	.340

a. Grouping Variable: Toolmkrcod

Test Statistics: Flake Step to Scar Ratio, Gona v. Human[a]

	Flake step to scar ratio
Mann-Whitney U	15609.500
Wilcoxon W	30315.500
Z	-3.156
Asymp. Sig. (2-tailed)	.002

a. Grouping Variable: Toolmkrcod

Discussion

As with the step/hinge to scar ratio on cores, the flake ratio indicated that the Gona sample was the outgroup, with less than half of the stepping and hinging seen of the dorsal scars of the bonobo and human samples. Again, this was surprising, since conventional wisdom would suggest that the Gona hominins, with a lower step-to-scar ratio, were the most skilled of the three groups.

In order to test whether the later stages of flaking preferentially represented at the Gona sites were influencing this attribute, we then examined step-to-scar ratios on the later stages of reduction of the human sample. As with core step-to-scar ratios, the mean step-to-scar ratio on flakes produced in the later stages of reduction in the human sample was very similar to the Gona sample mean (.059 and .046 respectively), with no significant difference between the samples (Mann-Whitney 2-tailed Asymp. Sig.=.447).As mentioned previously, it is also possible that the lower ratio in the Gona sample could be partially due to the hominins intentionally selecting thinner, morphologically easier cobbles to flake, making it easier to produce flakes without stepping or hinging throughout the reduction sequence.

40. Flakes: Exterior Platform Angle ("Core Angle")

Rationale

This angle, measured with a goniometer, is formed between the striking platform and the dorsal surface, essentially showing the morphology and angle of the core edge before flake detachment. A skilled stone knapper knows, however intuitively, that acute angles (less than 90 degrees) are required to flake stone efficiently, so that more acute exterior platform angles in Oldowan assemblages may be an indication of more skilled flaking and intentionally making cognitive decisions in selecting such angles. Such acute angles can sometimes also be maintained by the continuous unifacial flaking of a flat, thin cobble, where flakes keep detaching on the underside of the cobble (in an "outrepassé-like" manner); an acute edge angle can often be maintained in this manner. (In some later technologies, such as bifacial thinning of handaxes or projectile points or blade manufacture, knappers often intentionally steepened edge angles to

strengthen edges for soft hammer, punch, or pressure flaking, but this is not a consideration in this study).

Results

The bonobo flake sample showed the highest mean external platform ("core") angles at ~84 degrees. The Gona sample was intermediate (~82 degrees) with the human sample the lowest (~76 degrees). Statistical testing showed that the human sample differed significantly from both the bonobo and Gona sample at the .00 confidence level; differences between the bonobo and Gona sample were not, however, significant at the .05 level.

Flakes: Exterior Platform ("Core") Angle by Tool-maker

Core angle, nearest 5 degrees

Tool-maker	Mean	N	Std. Deviation
bonobo	83.81	97	15.939
gona	81.77	150	11.986
human	76.36	195	14.699
Total	79.83	442	14.456

Test Statistics: Exterior Platform Angle, Bonobo v. Gona[a]

	Core angle, nearest 5 degrees
Mann-Whitney U	6546.500
Wilcoxon W	17871.500
Z	-1.340
Asymp. Sig. (2-tailed)	.180

a. Grouping Variable: Toolmkrcod

Test Statistics: Exterior Platform Angle, Bonobo v. Human[a]

	Core angle, nearest 5 degrees
Mann-Whitney U	6679.500
Wilcoxon W	25789.500
Z	-4.110
Asymp. Sig. (2-tailed)	.000

a. Grouping Variable: Toolmkrcod

Test Statistics: Exterior Platform Angle, Gona v. Human[a]

	Core angle, nearest 5 degrees
Mann-Whitney U	10529.000
Wilcoxon W	29639.000
Z	-4.498
Asymp. Sig. (2-tailed)	.000

a. Grouping Variable: Toolmkrcod

Discussion

The human sample showed significantly more acute external platform ("core") angles than the bonobo and Gona samples. This suggests that the human subjects had superior cognitive abilities to recognize and exploit acute core edges in the production of flakes relative to the bonobos and Gona hominins. Interestingly, although the mean angles for the bonobo and Gona samples were similar, it is clear that the Gona hominins were more skilled stone knappers, producing flakes similar in over-all morphology to the human sample and reducing cores much more efficiently and heavily than the bonobo tool-makers.

41. Flakes: Interior Platform Angle ("Bulb Angle")

Rationale

This angle, measured with a goniometer, is formed between the striking platform (at the point of percussion) and the ventral surface on the bulb of percussion. For most raw materials (including those in this study), most interior platform angles are greater than 90°: a bulb of percussion with a higher angle (e.g. 130°) would suggest a prominent bulb of percussion, while a lower angle (e.g. 95°, would suggest a more diffuse bulb of percussion. Certain raw materials, however, such as some limestones (not part of this study), probably a factor of their softness and the physics of fracture, tend to produce acute interior platform angles and obtuse exterior platform angles. (Sahnouni *et al.*, 1997).

Results

The bonobo sample and the least obtuse interior platform angles (~98 degrees), with the human sample intermediate (~106 degrees) and the Gona sample the most obtuse (~108 degrees). Statistical testing indicated that the bonobo sample differed significantly from both the Gona and human samples at the .00 confidence level, but that the Gona and human sample did not differ significantly at the .05 confidence level.

Flakes: Interior Platform (Bulb) Angle by Tool-maker

Bulb angle, nearest 5 degrees

Tool-maker	Mean	N	Std. Deviation
bonobo	97.56	133	17.961
gona	108.45	153	12.540
human	106.33	226	16.241
Total	104.68	512	16.276

Test Statistics: Flake Bulb Angle, Bonobo v. Gona[a]

	Bulb angle, nearest 5 degrees
Mann-Whitney U	6620.500
Wilcoxon W	15531.500
Z	-5.121
Asymp. Sig. (2-tailed)	.000

a. Grouping Variable: Toolmkrcod

Test Statistics: Flake Bulb Angle, Bonobo v. Human[a]

	Bulb angle, nearest 5 degrees
Mann-Whitney U	10795.500
Wilcoxon W	19706.500
Z	-4.477
Asymp. Sig. (2-tailed)	.000

a. Grouping Variable: Toolmkrcod

Test Statistics: Flake Bulb Angle, Gona v. Human[a]

	Bulb angle, nearest 5 degrees
Mann-Whitney U	16442.500
Wilcoxon W	42093.500
Z	-.814
Asymp. Sig. (2-tailed)	.416

a. Grouping Variable: Toolmkrcod

Discussion

Exactly why the bonobo internal (bulb) platform angles are different from the Gona and human sample is not clear. Perhaps lower hammerstone impact velocities applied closer to core edges, produce less pronounced bulbs of percussion and therefore, less obtuse platform angles than flakes produced by higher impact velocities further from the core edges. Controlled experiments with these raw materials should resolve this question.

42. Flakes: Percentage Dorsal Cortex ("Cortex Index")

Rationale

The amount of dorsal cortex on flakes (estimated as a percentage) can, like flake types, give indications of what stages of cobble reduction are represented in the debitage and how heavily reduced cobbles were. A mean "cortex index" for each assemblage can then be arrived at and assessed.

Results

The mean percentage of dorsal cortex on flakes in the bonobo and human sample were almost identical (~40%). The cortex index for Gona was much less (~12%), over three times less cortex on average than the experimental samples. Statistical testing showed that the Gona sample differed significantly from the bonobo and human sample at the .00 confidence level, but the bonobo and human sample were not significantly different at the .05 confidence level.

Flakes: Percentage of Dorsal Cortex by Tool-maker

Percentage of cortex

Tool-maker	Mean	N	Std. Deviation
bonobo	39.70	158	39.300
gona	12.20	166	20.600
human	40.30	244	35.500
Total	31.90	568	35.200

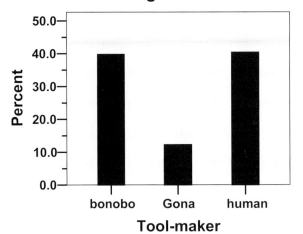

Flakes: Percentage Dorsal Cortex

Test Statistics: Percentage of Dorsal Cortex on Flakes, Bonobo v. Gona[a]

	Percentage of cortex
Mann-Whitney U	7544.00
Wilcoxon W	21405.000
Z	-6.963
Asymp. Sig. (2-tailed)	.000

a. Grouping Variable: Toolmkrcod

Test Statistics: Percentage of Dorsal Cortex on Flakes, Bonobo v. Human[a]

	Percentage of cortex
Mann-Whitney U	18720.000
Wilcoxon W	31281.000
Z	-.495
Asymp. Sig. (2-tailed)	.621

a. Grouping Variable: Toolmkrcod

Test Statistics: Percentage of Dorsal Cortex on Flakes, Gona v. Human[a]

	Percentage of cortex
Mann-Whitney U	10300.000
Wilcoxon W	24161.000
Z	-8.743
Asymp. Sig. (2-tailed)	.000

a. Grouping Variable: Toolmkrcod

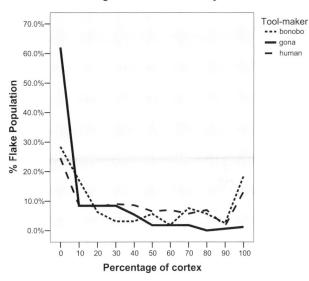

Percentage of Dorsal Cortex by Tool-maker

Discussion

This "cortex index" of mean percentage of dorsal cortex is an important way of assessing whether complete or partial reduction of Oldowan cobble cores took place at an archaeological site. Interestingly, the bonobo and human samples (representing complete reduction of cobble cores) had essentially the identical cortex index (40%), even though the human cores were reduced about 16 % more than the bonobo cores and produced 1.6 times as many whole flakes and 2.2 times as much overall debitage compared to the bonobo cores. It would appear that reducing cobbles by about 30% (bonobo cores) or by about 46% (human cores) can both yield the same cortex index.

The Gona flakes, on the other hand, tell a very different story. The astonishingly low cortex index of about 12% is a clear sign that significant numbers of flakes retaining significant cortex are missing from the archaeological sites. Even though the Gona cores (cobbles reduced about 64%) are somewhat more reduced than the human cores (about 17% more) and markedly more reduced than the bonobo cores (about 34% more), the fact that the Gona cores are more reduced and later stages of flaking are preferentially represented, this cannot explain the cortex index discrepancy: there should still be more cortex represented on the Gona flakes.

It would appear that, on average, perhaps half of the reduction of Gona cores took place off-site, and that subsequent flaking took place at the archaeological sites. But even these later stages of flaking of these Oldowan cobble cores should produce higher cortex indices than that seen at the Gona sites: the Gona cores were discarded with, on average, about 54% of their surfaces still retaining cortex (compared to 75% for bonobos and 62% for the human sample). As discussed previously, the most parsimonious explanation is that out of the debitage produced by later stages of reduction hominins selected larger, sharper flakes (and fragments) and subsequently transported them off-site. These larger flakes would tend to have higher percentages of cortex on their dorsal surfaces, so removing appreciable amounts of these flakes would markedly reduce the cortex index, as is seen at the Gona EG sites.

The fact that many large cortical flakes are missing from the sites, and that the majority of remaining flakes have little or no cortex, makes refitting much more difficult. Cortex in refitting studies is analogous to the straight-edged border pieces in a jigsaw puzzle; refitting Gona artifactual materials back together is comparable to doing a jigsaw puzzle lacking almost all border pieces as well as many other internal pieces; there will by definition be many gaps in the puzzle no matter how much one perseveres.

The human later stage flakes had a very similar mean percentages of dorsal cortex compared to the Gona hominin flake sample (~16% and ~12%, respectively). Statistical testing showed no significant difference between these two populations at the .05 confidence level.

Flakes: Percentage of Dorsal Cortex, Gona v. Human Later Stages

Percentage of cortex

Tool-maker	Mean	N	Std. Deviation
gona	12.23	166	20.638
humanlat	16.25	104	35.367
Total	13.78	270	27.277

Test Statistics: Flakes, Percentage of Dorsal Cortex, Gona v. Human Later Stages[a]

	Percentage of cortex
Mann-Whitney U	8188.000
Wilcoxon W	22049.000
Z	-.804
Asymp. Sig. (2-tailed)	.421

a. Grouping Variable: Toolmkrcod

If we remove the larger, sharper flakes from the later stage population of human flakes, the percentage of cortex goes down even further, with a mean virtually identical to that of the Gona sites (12.2%). This data is presented in the tables below.

Flakes: Percentage of Dorsal Cortex with Larger, Sharper Flakes Removed

Percentage of cortex

Tool-maker	Mean	N	Std. Deviation
gona	12.23	166	20.638
Human later stages with removal	12.22	63	40.259
Total	12.23	229	27.367

Test Statistics: Dorsal Cortex Percent, Gona v. Human Later Stages with Removal[a]

	Percentage of cortex
Mann-Whitney U	4604.500
Wilcoxon W	6620.500
Z	-1.647
Asymp. Sig. (2-tailed)	.100

a. Grouping Variable: Toolmkrcod

The percentage of dorsal cortex can be a powerful indicator of stages of cobble reduction represented at Oldowan sites. Cortex indices of less than 20% strongly suggest that only later stages of cobble reduction are represented (unless very large cobbles are being very heavily reduced). The very low cortical index at the Gona EG sites appears to represent later stages of flaking (also indicated by the flake types at these sites), very likely lowered even more by off-site transport of useful, also more highly cortical, flakes. Once again, this pattern strongly suggests a more complex tool-making and tool-using behavioral pattern than many palaeoanthropologists have probably suspected. An appreciation of this complexity is only possible when the archaeological assemblages are compared to experimentally-generated ones.

SUMMARY

This study has had two major foci: 1) an examination of the stone tool-making skills of our nearest living relatives of the genus *Pan*, and 2) an examination of the stone tool-making skills and tool-related behaviors of the early hominins (possibly *A. garhi*) that produced the Gona EG sites some 2.6 Ma. This was accomplished through a 3-way comparison of the stone tool-making products of bonobos, of early hominin tool-makers at Gona (Ethiopia), and, as a point of reference, of modern humans.

This comparison entailed detailed analysis of 42 different attributes observable in the artifacts produced by these three species—looking at characteristics of their overall artifact assemblages, their cores, and their flakes. The process of this comparison has also provided a valuable, critical examination of these attributes of analysis themselves—many of which are commonly used by archaeologists, others less commonly so—in order to see what kinds of information and insights they can yield regarding tool-making behaviors and, overall, which ones are most useful in assessing tool-making skill. The experimental approach presented here thus shows a methodology that can be used to assess tool-making skills in Oldowan assemblages. With good control over raw material type, this approach can be applied across time, using objective criteria of flaked stone artifact attributes.

The major results and findings of this study can be summarized in three major categories:

Ape Stone Tool-making

1. We have established, in an experimental setting, a tradition of stone tool-making and tool-using in African apes, *Pan paniscus*.
2. This tradition, begun in 1990, has recently become transgenerational, with offspring of one of our two original subjects now beginning to flake stone.
3. The transmission of these skills now has a strong cultural component within the bonobo social group.
4. This experimental program is on-going and is providing valuable information regarding the initial acquisition and long-term development of stone tool-making skills among apes, and may provide insight into development of stone tool-making proficiency among early stone tool-making ancestors.
5. The two adult bonobos have become adept at making stone tools and producing conchoidally-flaked artifacts with many similarities to Oldowan cores and flakes, although their overall stone assemblages remain demonstrably different in many attributes from Oldowan assemblages.
6. The bonobo stone tool-making products can be used as a valuable frame of reference for evaluating artifacts at Early Stone Age sites.

Artifact Attribute Patterns

1. Out of eleven attributes deemed to be a function of knapping skill, the patterns at Gona were very similar to those in the human sample in six attributes, intermediate and significantly different from the human and bonobo patterns in two attributes, and intermediate but not significantly different from the bonobos in three other attributes.
   a. The Gona sample tended to cluster with the human sample in these skill-related attributes:
      i. Assemblage composition
      ii. Core types
      iii. Number of flake scars on cores
      iv. Flake shape (parallel v. divergent)
      v. Flake breadth/length (endstruck v. sidestruck)
      vi. Overall quality of flaking
   b. The Gona sample was intermediate between and significantly different from the human and the bonobo samples in two other skill-related attributes:
      i. Core edge battering
      ii. Ratio of core largest scar/core maximum dimension
   c. The Gona sample was intermediate between the human and bonobo samples, but not significantly different from the bonobo sample, in three other skill-related attributes:
      i. Core edge angle (the Gona mean was less than, but not significantly different from, the bonobo mean)
      ii. Flake edge angle (exterior platform angle) (the Gona mean was intermediate but not significantly different from the bonobo mean)
      iii. Flake platform battering (again, the Gona mean was intermediate but not significantly different from the bonobo mean)
2. For two other attributes (flake weight and maximum dimension), the Gona and the bonobo samples clustered, but each apparently had low values for these attributes for different reasons:
   a. The bonobo sample showed lower weights and maximum dimension as a function of lower skill
   b. The Gona sample showed lower flake weight and smaller maximum dimension of flakes as a function of later stages of flaking and highly reduced cores.
3. In ten other attributes deemed to be related to stage of core reduction, the Gona sample differed significantly from the two experimental samples (the bonobo and the human 50% reduction samples) largely as a function of later stage of flaking being preferentially represented at the Gona sites. The human 50% reduction sample tended to be intermediate in most of these attributes, with values between the Gona and bonobo samples, as the human cores in this sample were intermediate in their degree of reduction. Comparison of the human 50% reduction sample with the Gona assemblage highlights the relatively high degree of reduction at these early, 2.6 million year old Gona sites. (In the human later stages sample, these attributes tended to become much closer to the Gona values). The attributes related to degree of reduction which distinguished the Gona sample included:
   a. Invasiveness of flaking
   b. Percentage of circumference flaked
   c. Core weight
   d. Core maximum dimension
   e. Core breadth
   f. Modified breadth/length ratios
   g. Percentage of original cobble
   h. Percentage of surface cortex on core
   i. Number of flake dorsal scars
   j. Percentage of flake dorsal cortex

Early Archaeological Sites at Gona

1. The Gona archaeological sites at 2.6 Ma show a remarkable level of skill in flaking stone (with regard to efficiently reducing cobbles and producing usable flakes), in many ways more similar to the pattern produced by modern humans than by the experienced bonobo tool-makers;

2. The dominant flaking mode of both the Gona hominins and the bonobos entailed the unifacial reduction of lava cobbles. This suggests that unifacial reduction was a simple yet efficient way of detaching flakes from a cobble core. The Gona hominins appear to have preferentially selected absolutely thinner cobbles than the experimental samples; thinner, discoidal cobbles would have been easier to reduce, producing more debitage per core

3. Gona hominins reduced their cores much more heavily than the bonobo sample, with more invasive flake scars and less remaining cortex on cobble cores, suggesting that the prehistoric hominins were more skilled at removing numerous flakes from cores.

4. The Gona artifact assemblages suggests that hominins transported partially flaked cores to the archaeological sites, heavily reduced the cores at the sites, and carried away from the sites many of the more useful, larger and sharper flakes. The removal of these useful flakes is a strong indication that such flakes were an important part of the Gona hominins' technological repertory. This also suggests that much of the tool-using behavior of these hominins occurred "off-site," perhaps in part associated with animal carcass processing on the landscape, as indicated by cut-marked bones found in the same stratigraphic level.

5. Regarding the attributes related to skill, the fact that the Gona sample tended either to cluster strongly with the modern human sample or to be intermediate between the human and bonobo samples, suggests that:
   a. Gona hominins were remarkably skilled knappers;
   b. The early stone tool-makers at Gona had a sound cognitive understanding of the principles of stone fracture;
   c. The Gona tool-makers had the biomechanical abilities to impart well-directed, forceful hammerstone blows when detaching flakes.

6. Based upon the ratio of split to whole flakes, Gona hominins appear to be intermediate between bonobos and modern humans in the hammerstone forces generated in detaching flakes.

7. Bonobos produced more end choppers, while the Gona hominins produced more side choppers, almost certainly due to the Gona hominins' ability to reduce cobbles more extensively than the bonobos (i.e., with continued flaking, an end chopper will usually grade into a side chopper).

8. The Gona hominins were selective in their acquisition of raw material, obtaining higher proportions of stone cobbles with excellent-to-good flaking qualities (fine-grained and isotropic) than are found on average in the nearby river gravels. This selection was probably facilitated by testing the cobbles at the gravels by removing flakes to examine the internal characteristics of the potential core. Such selectivity for better-quality raw materials has also been independently reported from Gona sites (Stout *et al.*, 2005). In choosing raw materials for the experiments reported here, approximately 1 in 50 (or 2%) of the cobbles within the Gona conglomerates were deemed suitable for flaking based upon their superficial qualities. Testing of cobbles by Gona hominins at the gravel sources would have tended to enhance this selectivity even more.

Criteria for Assessing Skill

Based upon the study presented here, criteria for assessing higher levels of skill from Oldowan assemblages would include the following attributes for cores and debitage:

Cores
1. More side choppers versus end choppers (in general, more heavily reduced cores)
2. More flake scars per core
3. More reduced cores/less cortex on cores
4. More acute edge angles on cores
5. More invasive flaking on cores
6. Less edge battering on cores
7. Higher "quality of flaking"

Debitage
1. More debitage per core
2. Longer (and thicker) flakes relative to breadth
3. Larger, heavier flakes
4. Less platform battering/stepping
5. More parallel-sided flakes, less divergent flakes
6. More acute exterior platform angle
7. Lower percentage of cortex on flakes
8. Higher ratio of split to whole flakes
9. Fewer fragments, more whole flakes and broken flakes

In applying some of these criteria, particularly those that are also related to the degree of core reduction (e.g., amount of core cortex, number of flake scars, amount of debitage), it would be best to proceed cautiously with regard to an assignment of lower level of skill to the assemblage, as an individual site may not always indicate maximum ability of the hominin tool-makers but may also vary according to special circumstances (for instance, there may be lesser degree of core reduction at some sites very near raw material sources). A "lower skill" assessment of an assemblage should preferably involve diverse, multiple criteria within any one assemblage and, ideally, should also refer to region-wide artifact patterns to take into account maximum skill levels in a particular area in a general time period and factor out possible site-specific idiosyncrasies, such as raw material differences that can impact flaking qualities. Indication of high level of skill according to multiple attributes, however, would support greater stone tool-making proficiency and skill among the prehistoric tool-makers.

CONCLUSION

In sum, the Gona EG sites show that by 2.6 Ma the earliest known stone tool-making hominins, possibly *Australopithecus garhi*, had already evolved the cognitive and biomechanical capabilities to efficiently flake stone and produce sharp-edged flakes and fragments with stone percussors. This new behavior pattern, which appears to be roughly contemporaneous with cut-marked bones from the Middle Awash, may mark a major adaptive shift towards more carnivorous behavior through scavenging and/or hunting larger mammals.

In addition to a high degree of skill in stone tool-making, relatively complex tool-related behaviors among the Gona hominins are also indicated by this study, involving testing cobbles at gravel sources, substantially reducing cores prior to bringing them to these sites, and subsequently transporting numbers of flakes away from the sites, most probably for use at some other location(s). Thus, the Gona patterns do not suggest purely expedient tool-making and tool-using behaviors: these artifact assemblages do not appear to represent immediate responses to immediate, local opportunities but instead appear to involve some degree of planning, perhaps anticipation of future needs, and spatial displacement of different phases of tool-making and tool-using in different locations in the landscape.

The modern bonobo subjects were able to flake stone in a manner that would clearly be recognized as artifactual in a prehistoric context. However, the lithic assemblage produced thus far by the bonobos, despite showing some similarities to early Oldowan artifacts (e.g. chopper-dominated), exhibits important differences from Oldowan archaeological occurrences in Africa. While some individual flakes or cores may resemble Oldowan assemblages, the overall bonobo artifact assemblage shows marked differences in a number of features. If such a bonobo assemblage were discovered in a prehistoric context, with so many distinct differences from early Oldowan artifact assemblage in so many attributes, particularly ones associated with skill, it might be assigned to a "Pre-Oldowan" stage of technology.

Thus, this bonobo-generated sample may give clues regarding possible earlier, yet-to-be-discovered archaeological sites produced by early hominins. On the other hand, early hominins may have developed their cognitive and biomechanical capabilities prior to the need to flake stone, and the early Gona sites may in fact represent the earliest stages of hominin flaked stone technology. Future fieldwork and analysis should shed light on this question.

ACKNOWLEDGMENTS

We would like to thank the following organizations for their support of this project: The Department of Antiquities and the National Museum of Ethiopia, both under Ethiopia's ARCCH (Authority for Research and Conservation of Cultural Heritage), the Language Research Center of Georgia State University in Atlanta, the Great Ape Trust of Iowa, and the Afar people of the Gona study area. Funding for this project came from the National Science Foundation, the L.S.B. Leakey Foundation, the Wenner-Gren Foundation for Anthropological Research, the National Geographic Society, the Stone Age Institute, the Center for Research into the Anthropological Foundation of Technology (CRAFT) at Indiana University, and Friends of CRAFT, Inc. We thank Richard Klein for statistical advice. Finally, we would like to thank our bonobo subjects, Kanzi and Panbanisha, for their cooperation and enthusiasm in this study.

Appendix 1a. Assemblage attributes: results of statistical tests.

Attrib. No.		Test	All (3-way)	Arch v. Exp	B v. G	B v. H	G v. H	Out-lier?	Group(s)?	All signif. differ at .05 level?	Gradient?
ASSEMBLAGE											
1	Assemblage Composition (Lithic Class)	Chi-Square			0.000	0.000	0.788	B	G+H	n	n
2a	Debitage Breakdown	Chi-Square			0.000	0.000	0.000	n	n	y	n
2b	Whole v. Split Flake Proportions	Chi-Square	0.000		0.000	0.000	0.000	n	n	y	B→G→H
3	Rock Type (exp/arch, cores)	Chi-Square		0.057	NA	NA	NA	NA	~(G+Exp)	NA	NA
4	Raw Material Quality (exp/arch, cores)	Chi-Square		0.200	NA	NA	NA	NA	G+Exp	NA	NA

B=bonobos, G=Gona, H=humans
Exp=Experimental (B+H), Arch=Gona
NA=not applicable

Appendix 1b. Assemblage attributes: comments on overall patterns and results.

Attrib. No.	ASSEMBLAGE	COMMENT
1	Assemblage Composition (Lithic Class)	Bonobos outlier with much higher core %, lower flake and fragment percents
2a	Debitage Breakdown	Bonobos high flake %
2b	Whole v. Split Flake Proportions	Split flakes: Humans highest, Gona intermediate, bonobos lowest
3	Rock Type (exp/arch, cores)	Experimental and archaeological rock types similar
4	Raw Material Quality (exp/arch, cores)	Experimental and archaeological rock quality similar

Exp=Experimental (B+H), Arch=Gona

Appendix 2a. Core attributes: results of statistical tests.

Attrib. No.	CORES	Test	All (3-way)	Arch v. Exp	B v. G	B v. H	G v. H	Out-lier?	Group(s)?	All signif. differ at .05 level?	Gradient?
5	Cores: Original Form	None			0.045	0.198	0.003	G	B+H	n	
6	Cores: Flaking Mode	Chi-Square			0.253		0.003		B+G	y	B→H→G
7	Cores: Invasiveness	Chi-Square			0.000	0.009	0.04	n	n	y	B→H→G
8	Cores: % Circumference	Mann-Whitney			0.000	0.000	0.092	B	~(G+H)	n	B→G→H
9a	Cores: Quality of Flaking (4 categories)	Chi-Square	0.000		0.075	0.000	0.003	H	n	n	B→G→H
9b	Cores: Quality of Flaking (2 categories)	Chi-Square	0.003		0.009	0.003	0.902	B	G+H	n	B→G→H
10	Cores: Edge Battering/Stepping	Chi-Square			0.011	0.000	0.000	n	n	y	B→G→H
11	Cores: Modified Leakey Type	Chi-Square			0.000	0.000	0.029	n	n	y	
11(b)	Cores: Modified Leakey Type (End v. Side Ch.)	Chi-Square	0.000		0.000	0.000	0.173	n	G+H	n	B→H→G
12	Cores: Weight	Mann-Whitney			0.000	0.000	0.000	n	n	y	G→H→B
13	Cores: Maximum Dimension (L)	Mann-Whitney			0.000	0.000	0.000	n	n	y	G→H→B
14	Cores: Breadth	Mann-Whitney			0.000	0.000	0.000	n	n	y	G→H→B
15	Cores: Thickness	Mann-Whitney			0.001	0.824	0.000	G	B+H	n	G→B→H
16	Cores: B/L Ratio	Mann-Whitney			0.175	0.394	0.026	n	G+B, B+H	n	G→B→H
17	Cores: Modified B/L Ratio	Mann-Whitney			0.000	0.000	0.079	B	~(H+G)	n	B→H→G
18	Cores: Th/B Ratio	Mann-Whitney			0.099	0.007	0.484	(~B)	~(G+H)	n	B→G→H
19	Cores: Max D of Largest Scar	Mann-Whitney			0.004	0.936	0.001	G	H+B	n	G→H→B
20	Cores: Largest Scar/Maximum D	Mann-Whitney			0.121	0.003	0.098	n	B+G, G+H	n	B→G→H
21	Cores: % Original Clast	Mann-Whitney			0.000	0.000	0.000	n	n	y	G→H→B
22	Cores: # Flake Scars	Mann-Whitney			0.000	0.000	0.965	B	G+H	n	B→H→G
23	Cores: Step/Scar Ratio	Mann-Whitney			0.204	0.302	0.007	n	G+B, B+H	n	G→B→H
23(b)†	Cores: Step/Scar Ratio, Gona v. Human Later†	Mann-Whitney					0.893†	n	G+H later†		
24	Cores: Edge Angle	Mann-Whitney			0.213	0.000	0.000	H	G+B	n	H→G→B
25	Cores: % Cortex	Mann-Whitney			0.000	0.000	0.045	n	n	y	G→H→B

B=bonobos, G=Gona, H=humans
†=Human later stages of reduction

Appendix 2b. Core attributes: comments on overall patterns and results.

Attrib. No.	CORES	COMMENT
5	Cores: Original Form	Mostly cobbles for all; Gona more indeterminate forms, as more reduced
6	Cores: Flaking Mode	Bonobos v. Gona comparison: Bonobos + Gona, no significant difference
7	Cores: Invasiveness	Gradient B→H→G, all differ, humans 50% reduction sample intermediate
8	Cores: % Circumference	Gradient B→H→G, humans 50% reduction sample intermediate, bonobos ~outlier
9a	Cores: Quality of Flaking (4 categories)	Gradient, Gona intermediate (but many cells < 5)
9b	Cores: Quality of Flaking (2 categories)	Bonobos outlier, G + H higher quality, Gona intermediate
10	Cores: Edge Battering/Stepping	Bonobos high, Gona intermediate, humans low
11	Cores: Modified Leakey Type	Bonobos very high end % choppers, Gona, humans very high % side
11(b)	Cores: Modified Leakey Type (End v. Side Ch.)	Bonobos very high end % choppers, Gona, humans very high % side
12	Cores: Weight	All differ: Gona low, humans intermediate, bonobos high
13	Cores: Maximum Dimension (L)	All differ: Gona low, Humans intermediate, bonobos high
14	Cores: Breadth	All differ: Gona low, humans intermediate, bonobos high
15	Cores: Thickness	Gona outlier (thinner), B + H group (thicker)
16	Cores: B/L Ratio	Gona ratio < human; bonobos intermediate
17	Cores: Modified B/L Ratio	Bonobos lowest (end choppers); Humans + Gona higher, more reduced (side choppers)
18	Cores: Th/B Ratio	Bonobos lowest, Gona intermediate, humans highest
19	Cores: Max D of Largest Scar	Gona outlier, smallest ratio (more reduced cores); H 50% control sample + B group
20	Cores: Largest Scar/Maximum D	Gona intermediate, B (lowest) + H (highest) extremes differ
21	Cores: % Original Clast	All differ: Gona lowest, human 50% reduction sample intermediate, bonobos highest
22	Cores: # Flake Scars	Bonobo lowest, outlier, humans + Gona group, higher
23	Cores: Step/Scar Ratio	Bonobo intermediate; Gona + human 50% reduction control sample extremes differ
23(b)†	Cores: Step/Scar Ratio, Gona v. Human Later†	No significant difference between Gona and human later stages sample
24	Cores: Edge Angle	Human lowest, outlier, Gona intermediate, groups with bonobos
25	Cores: % Cortex	All differ; Gona lowest (more reduced), bonobo highest (least reduced)

B=bonobos, G=Gona, H=humans
†=Human later stages of reduction

Appendix 3a. Flake attributes: results of statistical tests.

Attrib. No.	FLAKES	Test	All (3-way)	Arch v. Exp	B v. G	B v. H	G v. H	Out-lier?	Group(s)?	All signif. differ at .05 level?	Gradient?
26	Flakes: Flake Types 1-6	Chi-Square			0.000	0.031	0.000	n	n	y	y
	Flakes: Flake Types 1-6	Kolmog.-Smirnov			0.000	0.136	0.004	G	B+H	n	y
27	Flakes: Location of Cortex										
28	Flakes: Scar Pattern										
29	Flakes: Flake Shape	Chi-Square	.000		0.000	0.000	0.035	n	n	y	n
29(b)	Flakes: Flake Shape, parallel v. divergent	Chi-Square	.000		.002	.000	.387	y	G+H	n	
30	Flakes: Platform Battering/Stepping	Chi-Square			0.789	0.000	0.003	H	B+G	n	H→G→B
31	Flakes: Weight	Mann-Whitney			0.288	0.000	0.000	H	B+G	n	G→B→H
32	Flakes: Maximum Dimension	Mann-Whitney			0.112	0.000	0.000	H	B+G	n	G→B→H
33	Flakes: Thickness	Mann-Whitney			0.000	0.000	0.112	n	n	n	B→G→H
34	Flakes: B/L Ratio	Mann-Whitney			0.000	0.000	0.079	B	G+H	n	H→G→B
35	Flakes: Th/B Ratio	Mann-Whitney			0.000	0.000	0.000	n	n	y	B→H→G
36	Flakes: Platform Th/Platform B Ratio	Mann-Whitney			0.000	0.002	0.001	n	n	y	B→H→G
37	Flakes: No. Platform Scars (Types 1-6)	Mann-Whitney			0.957	0.022	0.018	H	B+G	n	(B+G)→H
37(b)	Flakes: No. Platform Scars (Types 4-6)	Mann-Whitney			0.558	0.018	0.100	(~H)	~(B+G)	n	H→G→B
38	Flakes: No. Dorsal Scars, All Flakes	Mann-Whitney			0.000	0.397	0.000	G	B+H	n	B→H→G
39	Flakes: Step & Hinge/Scar Ratio	Mann-Whitney			0.000	0.340	0.002	G	B+H	n	G→H→B
39(b)†	Flakes: Step & Hinge/Scar Ratio, Gona v. Human Later Stages†	Mann-Whitney					0.447†		G+H†		H→G→B
40	Flakes: Core Angle (Ext. Plat. Ang.)	Mann-Whitney			0.180	0.000	0.000	H	B+G	n	B→H→G
41	Flakes: Bulb Angle (Int. Plat. Ang.)	Mann-Whitney			0.000	0.000	0.416	B	H+G	n	G→(B+H)
42	Flakes: % Dorsal Cortex ("cortex index")	Mann-Whitney			0.000	0.621	0.000	G	B+H	n	
42(b)†	Flakes: % Dorsal Cortex ("cortex index"), Gona v. Human Later Stages†	Mann-Whitney					0.421†		G+H†		
42(c)†	Flakes: % Dorsal Cortex ("cortex index"), Gona v. Human Later Stages† w/removal	Mann-Whitney					0.100†		G+H†		

B=bonobos, G=Gona, H=humans
†=Human later stages of reduction

Appendix 3b. Flake attributes: comments on overall patterns and results.

Attrib. No.	FLAKES	COMMENT
26	Flakes: Flake Types 1-6	All samples differ
26	Flakes: Flake Types 1-6	B + H group (high types 2, 3 and 1, H also 5); Gona differs in higher % types 3 and 6
27	Flakes: Location of Cortex	Gona: low % with all cortex, high % with no cortex (later stages, heavily reduced)
28	Flakes: Scar Pattern	All: mainly simple pattern; bonobo relatively high in plain, Gona in opposed
29	Flakes: Flake Shape	All differ, bonobos high oval + divergent; Gona and humans high parallel + convergent
29(b)	Flakes: Flake Shape, parallel v. divergent	Gona and humans group, more parallel flakes; bonobos more divergent flakes
30	Flakes: Platform Battering/Stepping	Humans lowest, outlier, Gona intermediate; bonobos highest
31	Flakes: Weight	Human flakes heaviest; Gona lightest (small, later stages), bonobo intermediate
32	Flakes: Maximum Dimension	Human flakes largest; Gona smallest (later stages), bonobo intermediate
33	Flakes: Thickness	Bonobo flakes thinnest, Gona intermediate, humans thickest
34	Flakes: B/L Ratio	Bonobo flakes sidestruck; gona and human more endstruck
35	Flakes: Th/B Ratio	Bonobo flakes thinnest, Gona flakes thickest relative to breadth
36	Flakes: Platform Th/Platform B Ratio	Bonobo flake platforms thinnest, Gona flakes thickest, relative to platform breadth
37	Flakes: No. Platform Scars (Types 1-6)	Bonobo and Gona group, low number, humans highest
37(b)	Flakes: No. Platform Scars (Types 4-6)	Humans lowest, Gonas and bonobos high
38	Flakes: No. Dorsal Scars, All Flakes	Bonobos lowest, group with humans; Gona highest
39	Flakes: Step & Hinge/Scar Ratio	Gona lowest, humans intermediate, bonobos highest
39(b)†	Flakes: Step & Hinge/Scar Ratio, Gona v. Human Later Stages†	Human later stage flakes and Gona nearly identical in step & hinge/scar ratio
40	Flakes: Core Angle (Ext. Plat. Ang.)	Human sample smallest angle; Gona intermediate; bonobos highest
41	Flakes: Bulb Angle (Int. Plat. Ang.)	Bonobo sample smallest angle; humans and Gona larger, group
42	Flakes: % Dorsal Cortex ("cortex index")	Gona very low % cortex (later stages); humans 50% reduction + bonobos much higher
42(b)†	Flakes: % Dorsal Cortex ("cortex index"), Gona v. Human Later Stages†	Human later stage flakes and Gona very similar in cortex index
42(c)†	Flakes: % Dorsal Cortex ("cortex index"), Gona v. Human Later Stages† w/removal	Human later stage flakes with removal and Gona nearly identical in cortex index

B=bonobos, =Gona, H=humans
†=Human later stages of reduction

REFERENCES CITED

Asfaw, B., White, T., Lovejoy, O., Latimer, B. & Simpson, S. (1999). *Australopithecus garhi*: a new species of early hominid from Ethiopia. *Science* 284 (5414): 629-634.

Dominguez-Rodrigo, M., Pickering, T., Semaw, S., & Rogers, M. (2005). Cutmarked bones from Pliocene archaeological sites at Gona, Afar, Ethiopia: implications for the function of the world's oldest stone tools. *Journal of Human Evolution* 49:109-121.

Harlacker, L. (2006). *The Biomechanics of Stone Tool-Making: Kinematic and Kinetic Perspectives on Oldowan Lithic Technology.* Ph.D. Dissertation, Anthropology Department. Bloomington, Indiana: Indiana University.

Holloway, R.L. (2000). Brain. In: *Encyclopedia of Human Evolution and Prehistory, Second Edition* (E. Delson, I. Tattersall, J.S. Van Couvering, & A.S. Brooks, Eds.), pp. 141-149. New York: Garland Publising, Inc.

Hovers, E. (2003). Treading carefully: site formation process and Pliocene lithic technology. In: *Oldowan: Rather More than Smashing Stones: First Hominid Technology Workshop, Treballs d'Arqueologia, 9* (J. Martinez Moreno, R. Mora Torcal, & I. de la Torre Sainz, Eds.), pp. 145-158. Bellaterra, Spain: Universitat Autonoma de Barcelona.

Leakey, M. (1971). *Olduvai Gorge Volume 3: Excavation in Beds I and II, 1960-1963.* New York: Cambridge University Press.

Sahnouni, M., Schick, K. & Toth, N. (1997). An experimental investigation into the nature of faceted limestone "spheroids" in the Early Paleolithic. *Journal of Archaeological Science* 24:701-13.

Savage-Rumbaugh, S. & Lewin, R. (1994). *Kanzi: The Ape at the Brink of the Human Mind.* New York: John Wiley & Sons.

Savage-Rumbaugh, S., Toth, N. & Schick, K. (2006). Kanzi learns to knap stone tools. In: *Primate Perspectives on Behavior and Cognition* (D. Washburn, Ed.), pp. 279-291. Washington: American Psychological Association.

Schick, K. (1986). *Stone Age Sites in the Making: Experiments in the Formation and Transformation of Archaeological Occurrences.* Oxford: British Archaeological Reports.

Schick, K., Toth, N., Garufi, G.S., Savage-Rumbaugh, E.S., Rumbaugh, D, & Sevcik, R. (1999). Continuing investigations into the stone tool-making and tool-using capabilities of a bonobo (*Pan paniscus*). *Journal of Archaeological Science* 26:821-832.

Semaw, S. (1997). *Late Pliocene Archaeology of the Gona River Deposits, Afar, Ethiopia.* Ph.D. Dissertation, Anthropology Department. New Brunswick: Rutgers University.

Semaw, S. (2000). The world's oldest stone artefacts from Gona, Ethiopia: their implications for understanding stone technology and patterns of human evolution. *Journal of Archaeological Science* 27:1197-1214.

Semaw, S., Renne, P., Harris, J.W.K., Feibel, C.S., Bernor, R.L., Fesseha, N. & Mowbray, K. (1997). 2.5-million-year-old stone tools from Gona, Ethiopia. *Nature* 385:333-336.

Semaw, S., Rogers, M.J., Quade, J., Renne, P.R., Butler, R.F., Dominguez-Rodrigo, M., Stout, D., Hart, W.S., Pickering, T. & Simpson, S.W. (2003). 2.6-million-year-old stone tools and associated bones from OGS-6 and OGS-7, Gona, Afar, Ethiopia. *Journal of Human Evolution* 45:169-177.

Stout, D., Quade, J., Semaw, S, Rogers, M.J. & Levin, N.E. (2005). Raw material selectivity of the earliest stone toolmakers at Gona, Afar, Ethiopia. *Journal of Human Evolution* 48)4):365-380.

Suwa, G., White, T.D., & Howell, F.C. (1996). Mandibular postcanine dentition from the Shungura Formation, Ethiopia: crown morphology, taxonomic allocations, and Plio-Pleistocene hominid evolution. *American Journal of Physical Anthropology* 101:247-282.

Toth, N. (1982). *The Stone Technologies of Early Hominids at Koobi Fora: An Experimental Approach.* Ph.D. Dissertation, Anthropology Department. Berkeley: University of California.

Toth, N. (1985). The Oldowan reassessed: a close look at early stone artifacts. *Journal of Archaeological Science* 12:101-120.

White, T.D., Asfaw, B., & Suwa, G. (2005). Pliocene hominid fossils from Gamedah, Middle Awash, Ethiopia. *Transactions of the Royal Society of South Africa* 60(2):79-83.

CHAPTER 7

RULES AND TOOLS: BEYOND ANTHROPOMORPHISM

SUE SAVAGE-RUMBAUGH AND WILLIAM MINTZ FIELDS

ABSTRACT

This chapter presents the perspectives of a cognitive psychologist (SSR) and a cultural anthropologist (WMF) in assessing and interpreting the acquisition of skill in the stone toolmaking behavior of modern bonobos (Pan paniscus) in an experimental setting. These perspectives are presented as personal narratives based upon memory, notes, and video documentation.

A COGNITIVE PSYCHOLOGIST'S PERSPECTIVE (SSR)

This is a first person narrative account of the initiation and development of knapping in bonobos. It is drawn from memory, notes and video documentation. It surely leaves out much of what actually happened from the viewpoint of the bonobo knappers themselves. Yet it includes, from the perspective of intimately knowledgeable, Homo sapiens observers, the salient behavioral transitions in skill development. Like all narrative accounts, this one relies upon the insight, intuition and analysis of the observers, who are, in this case, participant observers in the classical anthropological tradition. Narrative accounts, by definition, describe events. They do not predict events, nor do they focus upon quantitative data. Good narrative accounts serve as valuable explanatory tools, permitting hypotheses to be formulated and tested, when and if events similar to those of the narration occur again. It is through movement between the processes of analytic and categorical description, coupled with hypothesis formation, prediction, data collection and finally theory

formation, that the scientific understanding of behavior progresses. Because this work focuses upon the long-term rearing effects of a small number of nonhuman primate individuals, ethnographic narrative account is the appropriate research tool.

Strict empiricists (MacPhail, 1987; Heyes, 1998) dismiss narrative accounts of nonhuman primate behavior, treating them as "anecdotes." They assert that such accounts are based on mere "interpretation" rather than "actual data." Moreover, these empiricists argue that any explanatory account of nonhuman primate behavior is inevitably infused with anthropomorphism because we ourselves are primates. This fact alone is believed to make objective accounts impossible. All descriptive statements regarding the motivational states of nonhuman primates are held to be inappropriate, as they attribute some form of consciousness and/or intentionality to nonhuman beings.

By contrast, we assert that to claim that a monkey who is engaging in a certain posture and facial expression is indeed threatening another monkey (or even a human observer) is a perfectly legitimate scientific statement when made under the appropriate conditions. The interpretation of "threat" can be validated by the ensuing behavior of the other monkey, or by the human observer should the threat be directed toward them.

Going one step beyond the description above, one might also say that the monkey intended to threaten the human observer and we might offer as "proof" the fact that the threat was followed by attack when the threat was ignored. However, the empiricist would disagree with the term "intended," maintaining that the behaviors of threat and attack could better be explained in terms of stimulus and response. "Intentionality," according to many radical behaviorists, must be exclusively reserved for human beings.

The difficulty with this view is that it is a "speciesist" argument. It limits the role of conscious intent to one species, Homo sapiens. One can just as readily explain human threat and attack as stimulus-response behavior, leaving out any discussion of intentionality. But were we to do so, human behavior would become meaningless, for most human beings have lent meaning through either expressed or inferred intent. As human beings, we tend to think, attack and threaten for very specific reasons, which we expound upon verbally. Because the empiricist does not know how to ask a monkey its intent, he or she concludes that the safest theoretical position is that the competency for intentional behavior, and explanation thereof, be limited, by caveat, to Homo sapiens.

This anthropocentric perspective overlooks two things. First, much human behavior, while explained, is not rational. The reasons given for threat and attack behaviors are often illogical and frequently, in humans, words and actions fail to coincide. Thus revealing that explained "intentions" do not necessarily explain behavior. If they did, we would barely need a psychology of the human mind. We not only permit human beings to explain their intent, we insist upon it to such a degree that individuals will offer socially acceptable explanations that have little relevance to the actual behaviors observed. In such instances, we could correctly say that "anthropomorphism" is present and clouds our view of the real explanation of behavior. In this case, anthropomorphism means the interpretation of another person's behavior based upon ones own thoughts and feelings. Such interpretations of the behavior of other human beings are as equally problematic and/or valid as are interpretations of the behavior of nonhuman primates.

Therefore, it is important to recognize that the inherent fallacy of anthropomorphism is not the species to which it is applied, but the way in which a loose subjective account fails to authentically describe the facts of the observed behavior at hand. Loose description can occur for any species, human or nonhuman. Likewise, and much more important to recognize for this chapter, is the fact that legitimate, accurate and historically informed description is equally valid for human and nonhuman beings. To say that a monkey intends to threaten another, is a statement easily validated by observing the situation. To say 'why' the monkey or the person intended to threaten another is one step removed from the immediate context, but no less valid if this statement is based upon historical analytic observations of the broader social context, where the social context is well studied and understood across time. Descriptions of intentions are no more or less valid for a given species.

Many scientists all too easily accept that we shall never be able to perceive the world of any non-human being in an adequate manner. While it is certainly true, for example, that human beings lack echo-location and thus are unable to perceive the world as does a bat, it is also the case that some human beings are deaf and blind and therefore unable to perceive the world as do others of their own species. Nonetheless blind persons frequently speak of having "seen a friend" and deaf persons will relate accounts they "heard" from others. These are not mere "manners of speech," they are statements which reflect the perceived feelings of the speaker. That is, the blind person feels as if they have seen something and even though their sensory input is different - and this feelings leads to a sensation they term "seeing."

The empiricist argument fails on a second count. It overlooks the fact that data cannot be gathered on behavior that is emerging spontaneously. During these conditions, one cannot know what form or course emergent processes will take (Savage-Rumbaugh et al., in press). One can film the behavior -- if one knows when it is going to occur or when salient events in the emergent process will take place. However, when the behavior is spontaneous and not produced by a designated environmental stimulus, filming is difficult to accomplish without cameras following the organism wherever it goes. Once a behavior has emerged and been closely observed, should this emergent process repeat itself in a predictable manner, plans can be made for data collection.

Such was not the case for tool production in bonobos. The first knapper, Kanzi did not follow the experimental trajectory that was prescribed for him. His younger sister, Panbanisha provided no opportunity for replication, as her trajectory did not repeat Kanzi's. This could have been the result of individual differences or sex differences. Such is impossible to determine without a 25-year rearing study of two additional bonobos. However, Panbanisha's alternative emergent trajectory might not be attributable to sex or individual differences, but rather to the fact that Kanzi's experience affected Panbanisha. Kanzi was the first knapper in the bonobo group, and there can be only one first knapper. Panbanisha had the opportunity to learn from Kanzi's successes and mistakes and to observe a bonobo knapping style. Kanzi had no opportunity to learn from the failures of another knapper, nor any opportunity to observe another bonobo.

To complicate the issue even further, personalities, competitiveness, and pride began to enter the experimental area. These were variables that moved the question far beyond the initial one of "can they make stone tools and how do they learn to do so." Thus in order to provide the uninitiated reader with a modicum of insight into how it is that a nonhuman species began to knap, a narrative account is essential. Without such a background, empirical statements generated by data collected on bonobo tool production exist in a meaningless vacuum. When readers cannot understand the environment that generated cultural transmission of tool manufacture from Homo sapiens to Pan paniscus they must inevitably assume that the nonhuman artifacts produced by Kanzi and Panbanisha are the result of tedious instrumental shaping, which results in a competency sans awareness. They have no basis on which to draw the proper conclusion. It is the goal of this chapter to correct that problem and to enable the reader to access for themselves the cognitive competencies and awarenesses that guide Kanzi's and Panbanisha's knapping behavior.

Whatever flaws, bumps and singular views this approach may entail, there are unique and valuable insights offered through a synthetic historical narrative. It provides the reader with a framework through which he or she can begin to understand the complexity of affairs that currently surround tool production in bonobos. By inference, we can legitimately infer that a far greater set of complex affairs surely surrounded the tool production of australopithecine and/or early Homo groups who flaked, not intermittently, like Kanzi and Panbanisha, but as a central component of cultural group survival.

This chapter is the product of a joint authorship, one author working from the methodological demands of experimental psychology and the other author applying the techniques of ethnography to a society of nonhuman primates. These perspectives, perhaps not surprisingly, result in distinctly different narrative accounts. We have only partially succeeded in being able to merge these different frameworks of thought and question into a single unified approach. Thus we offer below a historical narration of our experimental attempts to introduce stone tool manufacture to a nonhuman species, followed by descriptive narrative of the anthropological cross-cultural experience of suddenly encountering and living among a group of bonobos who produce stone tools. It is our hope that these multiple perspectives, coupled with the accompanying artifact-oriented chapter by Toth, Schick, and Semaw will provide the reader with a far more comprehensive picture of this project than has heretofore been made available. In addition, we hope it will represent the beginning a truly interdisciplinary approach to the study of human origins, as contrasted with multidisciplinary approaches that characterize the current condition of the science.

How the Project Began

The idea that a living ape might become a stone knapper was first suggested to this author by Nicholas Toth at a Wenner-Gren Conference in Portugal on the relationship of tool and language. My initial reaction was that such would not be possible. We recognized however that in the realm of behavior -- where nearly anything is possible and where the expectancy effect operates with an unseen hand -- such a view was not a sufficient reason to reject a study. So we welcomed Nicholas Toth and Kathy Schick to the Language Research Center and asked how they might wish to go about teaching a bonobo to knap stone.

It is worth noting that, as we proceeded, it was my unexpressed belief that:

a) the bonobos would not want to knap

b) that they were possibly not sufficiently manually supple

c) that toolmaking must have evolved long after a simple language emerged and that given what we then saw as the simplicity of their language, toolmaking was much too advanced for them

d) finally, that knapping was so difficult for us at the Language Research Center, that we could not possibly expect the bonobos to be able to accomplish it.

However, science is replete with the incorrect ideas and theories of those who have attempted to practice it, and we knew that we were no exception. When Toth and Schick arrived, having never worked with living apes and especially bonobos, we set aside our doubts and began working with the ideas they presented to us.

They wished to begin with an adolescent male bonobo named Kanzi. Toth and Schick had studied living Homo sapiens stone toolmakers in New Guinea, and in that group only adult males flaked stone and began to do so around adolescence. Thus Kanzi was the right age and the right sex to serve as prime stand-in for nonliving hominid knappers. Interestingly, for the human knappers in New Guinea, flaking stone was inevitably a group social activity. All learning and all practice took place in a social setting, in which the sounds of the stone against stone, the comparisons of work, discussions of the product, etc. inevitably took place. In addition, the skill of each knapper was known by the others and the tools, when finished, became essential implements to group survival.

We mention these critical aspects of human flaking solely to point out that they were absent, by necessity, from Kanzi's experience. Had they been present Kanzi's skills would surely have emerged more rapidly and with greater proficiency than they did. Thus in evaluating and comparing Kanzi to ourselves or to our earliest ancestors, we must recognize that Kanzi is at cultural disadvantage. This is not because he is a bonobo, but because he does not belong to a cultural group for whom knapping is an essential social and survival skill. Knapping is therefore not likely to rivet his attention and motivate him to excel, as it would if it were a necessary survival skill. This is not to suggest that a knapping tradition could not occur among bonobos. Quite the contrary, we believe that while such as not been observed in the field, it is certainly possible. However, we must temper our comparisons between Kanzi and early human knappers, by reminding ourselves that Kanzi probably would behave very differently if he were reared in a knapping culture.

How, we asked would Kanzi learn to knap? Toth explained that we could build a box that would require Kanzi to make a stone tool to open it by cutting a string holding a door flap closed. We could put something inside the box and then, from outside Kanzi's enclosure, Toth would make a stone flake and demonstrate its use while Kanzi observed. From the start, Kanzi watched as Toth picked up rocks, knocked off a flake and used it to cut the string on a box that held some grapes. Toth then handed the grapes to Kanzi. Toth repeated this activity a few times and then the "string box" was brought into Kanzi's enclosure.

Kanzi did not pick up rocks and start to knap. So one of us (Sue) put a rock in each of Kanzi's hands and urged him to do so. He attempted to knap as he had observed Toth do, but made the typical novice mistake of bringing the rocks together with a horizontal motion (more or less in a 'clapping' action). He also used very little force. When Toth knaps he appears to use very little force as well. In fact, to the naïve observer, the way in which flakes fall off in the hands of an accomplished knapper appears magical. When knapping stone, one cannot succeed by simply hitting rocks together. It is a difficult skill to master, just as it is learn to play the violin. One cannot simply pick up a violin, draw a bow back and forth across it and produce music. Neither can one begin hitting rocks together and produce a tool. One must learn to use one stone as a hammer and the other as a base or "core." The hammer stone must strike the core at just the right angle and just the right speed to produce even a single flake. The first flake production determines how and where the hammer must strike to produce the next flake. The hammer blows must be precisely aimed and timed. They must be produced with a more or less controlled throwing motion rather than a hitting motion, as appears to the untrained eye.

Kanzi's rocks did not fall apart and he quickly concluded that he was unable to make a flake. We continued to demonstrate, and tempted him with extra special incentives in the box, but Kanzi refused to try after his initial failure. He appeared sensitive to failure and to resent being repeatedly encouraged to do something that he was too difficult for him. Perhaps this is too anthropomorphic an interpretation of his reaction but it has been reported that "Every researcher with apes has learned that they will balk and simply stop working if problems encountered are in any manner beyond them" (Savage-Rumbaugh et al., in press). This "balking" might best be understood as a refusal to engage in behavior whose outcome is either unpredictable or whose goal cannot be achieved.

As a result, it was decided to have Toth make flakes for Kanzi and hand them to him, to enable Kanzi to open the box. Kanzi appreciated this and readily took the tool and retrieved the incentive. Seeing how proficiently Kanzi employed the stone flake that was given to him, Toth and Schick wondered whether Kanzi actually understood something about the properties of stone tools and whether he could determine a good flake from a bad one. That is, would Kanzi be able to judge the sharpness and strength of a variety of flakes and choose the best tool or would he use any flake to open the tool site. Toth had not demonstrated flake selection and Kanzi had no previous experience with stone tools. If he were to know good flakes from poor ones already, this would suggest a preliminary understanding of stone geometry.

The tool site was baited and Kanzi was given an array of flakes to choose from. From the start he ignored the duller flakes and selected the sharper ones, generally testing them in his mouth before beginning to use them on the string. The flakes were generally small, about 2 inches in length, and it was difficult to hold them in his large hands as he cut the string. He could not cut it swiftly and easily in one motion as he did not hold the flake at a downward angle and pull it towards him as did Toth. Instead he tried to push the flake through the string, and he so doing began to wear the string away fiber by fiber. However, he quickly hit upon the idea of doing this with two hands instead of one. (Toth had employed only one.) Kanzi began to put the index finger of his left hand under the string and pull it taunt and then to "saw" the string with the rock chip. His sawing motions were limited to one direction (towards himself) rather than back and forth, but he focused on one part of the string until it separated. He did this very patiently, employing the chip even when the string was hanging only by a thread.

Development of Kanzi's Knapping Skill

Although Toth and Schick had hoped that Kanzi would immediately begin to flake, he was not dissuaded. As they left, they asked us to continue trying to encourage Kanzi to knap. After all, they reminded us, early hominids surely had not begun to do this in just one day. Everyday, for that first week after Toth and Schick left, we baited the tool site several times and demonstrated flaking for Kanzi. However, there was one difference between demonstrations by those of us who were not expert knappers and Toth. It was not easy for us. The stone flakes did not fall as if by magic. We had to use quite a bit of force as we did not know how to focus the blows or to aim in the precise way that Toth did. Hence we failed a lot and it often took many blows to produce a flake. In addition, our flakes were not as sharp as Toth's. We continued to encourage Kanzi to knap repeatedly offering him rocks. Sometimes he took the rocks and made a few apparently half-hearted attempts at hitting them together in the midplane, as if he were clapping, and then gave up. When we asked him to continue he would either hand the rocks to us or walk away and ignore us.

Kanzi is so strong, that even if he were not employing the correct technique, he had sufficient strength to break the rocks if could just employ it. So we verbally encouraged him to hit the rocks together "harder" and tried to demonstrate this. In response to our verbal suggestion he would produce one or blows that sound loud and strong, but then give up again. No matter how tenaciously we worked to in interest him or to assist him, there was no change in Kanzi's behavior for approximately 2 weeks.

Then just as we were beginning to conclude that Kanzi was not going make progress in this endeavor, we began to hear a very loud repetitive banging noise coming from the group room. We peeked in to see Kanzi, a very determined look on his face, hitting the stones together with as much force as he could muster, over and over and over again, until his arms were too tired to permit him to continue. While chips were not flying off the rocks because of the horizontal clapping action of both hands, stone powder was being created with each blow. In fact, Kanzi the first bonobo knapper, sat amongst a veritable cloud of stone powder, produced by his own efforts. Kanzi apparently had decided to try and "open" the rocks by his sheer strength alone. We use the term "decided" because nothing had changed in the situation but Kanzi himself. It took about 15 minutes to achieve the production of his first chip, a tiny piece less than 1/4 in width and just a sliver of thickness. But he used it nonetheless. However it was so small that it crumbled in his large hands before he could cut the string. Kanzi sighed, but made another flake, with similar effort, but more quickly.

That day was something of a watershed for Kanzi in that he realized it was possible for him to make a flake. We infer this from the observation that ever after, each time the tool site was baited Kanzi would pick up the stones and begin to knap. He no longer needed encouragement or demonstration. When he did not succeed, we would remind him that it was possible for him to do so. Kanzi would appear to reflect upon this and then return to the task. He also began to look the rocks over carefully and to select those with the better flaking potential according to material and angles.

All of his initial flakes were very small, and many of them broke as he tried to cut the string. He was not dissuaded and would make as many small flakes as he needed in order to achieve the task. However he did not change the plane of his blow or try to make large flakes, even when strongly encouraged to do so. Observations such as these, made across time, are neither anecdotal, nor anthropomorphic. They go beyond anthropomorphism. To say that Kanzi "realized it was possible for him to make a flake" is not a simple statement about how we would interpret his behavior at one point in time - were he a human being. Rather it is a statement, constructed across a long span of time from a multiplicity of observations. It is supported, in time, by observations of one sort before the event and observations of a different sort after the event. The nature of these observations allow them to be contrasted in ways that can be said to empirically justify a statement about Kanzi's cognitive capacity to realize changes in his own competencies in this task.

Kanzi remained content with the smallest flake and would use it until it wore away. However, the time and effort required to produce such a small flake was often very great. Often Kanzi would knap away at the stone for 10 to 15 minutes before producing the tiniest flake, and then this flake would wear away before he was finished cutting the string. He might have to exert this effort two or three more times before the string was cut. Kanzi did not appear to enjoy the effort that flaking required. Flaking was obviously tiring and difficult for him. He had to hold the stones and strike them without hitting his fingers, and he was able to successfully avoid getting chips into his eyes as well.

After a few months Kanzi began to employ his left and right hands differently. He would rest his left hand, with a stone in it, against his abdomen and then effect the blow on this abdomen anchored stone as "substrate" -- with the other stone in his right hand. This enabled him to position the stones more securely and was the beginning of a bimanual differentiation of blows, with one hand acting as the stabilizer and the other as the "actor." By resting his left hand against his stomach, he could also grip the stone without wrapping his entire hand around it, thereby lessening the odds of striking his fingers. As bonobo fingers are approximately twice as long as our own, the holding of the stone presents, for their anatomy, a unique problem.

The innovation from "clapping" blows of the mid-plane, to those which required one hand to act as a stabilizer and the other as a hammer, made it possible for Kanzi to knap longer and to produce stronger blows. Unfortunately however, because his left hand rested against his abdomen for support, the force of his blows were now partially absorbed by this own body mass. Consequently, it took even more forceful blows to produce a chip in this manner. Kanzi continued to flake, and he seemed somewhat more contended with his new-found position, but he still searched for ways to make the process simpler. The problems that Kanzi was forced to solve were not those that could be solved for him by observing Toth, because his anatomy was not human. Humans have shorter fingers and longer thumbs, which make it relatively easy for us to hold smallish stones and knap then without hitting our fingers. We also lack bonobo strength and cannot easily produce any chips by hitting the rocks together in the mid-plane. How are we to interpret Kanzi's newly-found solution -- as imitation, as trial and error, or as a reasoned attempt to solve a recurring problem of physical anatomy of knapping as joined to the bonobo form? While empiricists would suggest trial and error, we find that explanation incomplete. Kanzi was not randomly attempting to position the stone differently about his body. He was, instead, responding to the dynamic physics of the task at hand. When he used mid-plane blows, chips were rarely produced. If, instead of hitting the rocks together, he hit one rock against another, he was more likely to produce a chip. However it was difficult for him to do this holding the rock in front of his body. He would sometimes hit his fingers, or knock the core out of his hand. By placing the core against his body, he eliminated these problems. Kanzi did not randomly try a whole set of workable and nonworkable set of solutions, but only a few different stone, hand and body positions, indicating that he understood the physical constraints inherent in the situation.

Kanzi's Innovations in Toolmaking

However, given the difficulties that Kanzi continued to encounter as a result of his anatomy, it was, in retrospect, not surprising that he arrived at an altogether different solution. One day, after the box was baited, Kanzi just sat and looked at, resting his elbow on one knee and his head in his hand. We were surprised, since he had always begun to flake in response to an incentive being placed in the box. However Kanzi did not appear disinterested on this occasion. Quite the contrary -- he assumed the classic position of the "thinker" and remained frozen for some time, his eyes fixated upon the rocks in front of him. Such a pose was most extraordinary for Kanzi. Finally Kanzi picked up only one stone instead of two. He held it in his right hand and rose into a full bipedal stance instead of sitting to knap as he normally did. We could not imagine what Kanzi was intending to do. Then he raised his arm and threw the rock onto the hard tile floor with great force causing it to shatter into more flakes than he had produced from all of his bimanual knapping combined. Kanzi at once selected a large sharp flake and cut the string on the tool site within seconds.

Kanzi had come up with a technique none of us had demonstrated for him. And it was far more efficient and produced better flakes than the one we had encouraged him to use. It is not anthropomorphic to infer that Kanzi "thought his way" to this solution. This sudden action, situated as it was, within the context of Kanzi's past knapping failures and successes, his style of placing the rocks on his abdomen, his great efforts to make tiny flakes, his intense fixation on the rocks, followed by a sudden change in technique -- all provide data supporting this inference. Moreover, were such actions taken by a Panda bear or an otter (both of whom can hold stones and sit upright), the conclusion would be the same. It is not Kanzi's physical resemblance to us that drives these conclusions, but rather the integration and nature of his behavior displayed across historical time.

We were delighted with Kanzi's innovation. It clearly demonstrated that for Kanzi, the end result of producing a flake was well understood. Even more importantly, it revealed that he could invent a method of flake production on his own. Critically, his innovation

did not result from trial and error or even from play. It was a direct response to difficulties he encountered using a human technique with a bonobo anatomy and it was far more efficient, for his purposes, than what he had been taught to do. This meant that living apes have the potential to begin production of simple stone tools without human intervention. While it is true that we designed a task for Kanzi which required a tool, that fact is simply a byproduct of his captive environment. One can easily imagine bonobos being placed in a setting where the hunting of meat became essential, for either cultural or environmental reasons. Individuals with broken, worn or small canines could begin to utilize sharp flakes and then to throw stone to make flakes as needed. As Toth et al. (1993) suggest, the earliest stone tools in the archaeological record may be more difficult to identify than previously assumed, since flakes made by throwing are not as readily distinguished from natural rock fractures as are knapped flakes.

Nonetheless, in spite of Kanzi's informative triumph, Toth emphasized the need for Kanzi to utilize bimanual percussion, however difficult, so that the work of a living ape might be adequately compared with hominid flakes that were certainly not produced by throwing. When one throws a stone and it breaks open into many pieces all at once, it is not possible to reconstruct what was in the mind of the toolmaker at the time. However when the flakes are removed, one at time through knapping, the entire process can be directly reconstructed by putting the stone back together. The pieces of the stone become a three-dimensional puzzle, and as the puzzle is worked (backward as the stone is reconstructed), it is possible to see just what the original knapper saw, the striking platforms as they appeared to him, and the selections he made at each point. Accomplished knappers such can readily detect the expertise and thought process present in the mind of the original toolmaker, through these methods.

Some archaeologists (Davidson and Noble, 1993) have argued that simple flakes and even early Acheulean handaxes reflect only a dim awareness that striking rocks together can result in flakes that have sharp edges. However, as we have learned from Kanzi, throwing the stones on a hard surface produces a similar result and is much more efficient. Therefore, the question of why our hominid ancestors elected to knap, rather than throw, arose. There is only one value of knapping over throwing if one desires a flake. Knapping is a far more precise process and depending upon the skill of the knapper, it can be employed to produce flakes of desired size and shape. However, such intentional production of sizes and shapes would appear to require planning and skills not attributed to the earliest hominids by most anthropologists.

The difference in our enthusiasm over Kanzi's novel solutions and Toth and Schick's interest in the products of Kanzi's actions may highlight some epistemological differences between psychology and archaeology as a subdiscipline of anthropology. Psychologists begin with questions about the mind, its contents and the nature of learning. Archaeological anthropologists begin with evolutionary questions of the origins of human mind. Some psychologists look for continuity of learning processes across species, others postulate a sharp divide between human and nonhuman learning. Anthropologists attempt to situate the human/nonhuman divide in evolutionary time and to look for some archaeological evidence, such a fire-making to support their position. When fossil and artifactual evidence change coincidentally, as in the case of the appearance of stone tools, speculations regarding the emergence of new cognitive capacities appear warranted. Psychologists rely heavily upon detailed experimental methods of "task presentation" to investigate learning and cognition. Anthropologists interested in fossils and human cognition must reply upon inference from fossil and artifact discovered in context. Anthropologists working and living in hunting and gathering societies are able to learn and speculate about the lifestyle that early hominids might have utilized, relying on participant observation and ethnographic techniques. These methodological differences, driven as they are by the actual physical and social material available for study in the different disciplines, result in fundamentally different orientations toward mind. That of the psychologist is oriented around questions of individual capacity and development. That of the anthropologist is oriented around questions of cultural change across time, and archaeologists necessarily must also rely upon physical prehistoric evidence for their inferences.

Thus there was delight in the processes of mind which Kanzi displayed, but no direct means of linking such processes to the archaeological record unless Kanzi produced stone flakes through bimanual percussion. The process of one mind, while it may reflect potential and creativity, cannot be equated with culture as process of group adaptation.

Thus, in spite of Kanzi's innovative solution, we joined together in an empirical decision. As stone tools in the early archaeological record show evidence of manufacture by bimanual knapping rather than by throwing, in this phase of our "experiment," Kanzi would be required to knap and would not be permitted to throw. While our logic for this decision was impeccable, it was not clear how we would achieve this goal. Our first approach was simply to ask Kanzi to knap rather than to throw. Most students of animal behavior would look askance at such a decision, and concentrate instead on effective shaping procedures. However, psychologists working with human subjects typically request participation of a specific sort during an experiment. Kanzi was not human, but he could understand verbal requests. Such requests served, as they do with human beings, a much more simple way to achieve a particular behavior than shaping techniques.

He understood our request and accommodated our expressed desire by hitting the rocks together a few times without achieving a flake. Then, as if to emphasize the effectiveness of his own technique, he slammed the rock on the floors, producing an array of flakes, and then looked directly at us and gestured toward his accomplishment. We tried a few times to insist verbally that Kanzi knap rather than throw. However, having once demonstrated the efficiency of his technique, Kanzi proceeded to ignore us. When we baited the site, he did not bother to knap or to listen to our imploring. He threw the rock, made his chip, obtained his incentive, and then walked away, all in a matter of minutes.

Kanzi had obviously had made up his mind regarding the relative efficiencies of the technique we taught and preferred (for its comparability to toolmaking responsible for the early archaeological record) versus the one he devised and preferred, on the rational basis of his own toolmaking efficiency. Perhaps an empiricist would wish to argue that we were being too anthropomorphic in our interpretation of Kanzi's behavior. However, Kanzi did not have to make his choice so clear. He had several months of practice and reward for knapping and only one experience with throwing. Yet that experience, preceded as it was by thoughtful steady gazing upon the rocks, changed his behavior unequivocally from that time forward. If reward-based experience were driving his behavior, he should have fallen back on knapping, not really understanding what he had done.

Realizing that we were at a major impasse in the work, we attempted to devise an experimental method to force Kanzi to abandon his own solution. (In retrospect, and with our current knowledge of Kanzi's cognitive awareness, we might have done best to simply explain to Kanzi that we were performing an experiment in which we needed him to knap rather than throw. This method seems to be one that Kanzi's prefers us to employ to engage his cooperation at the current time.)

Thus we determined to alter the environment so as to make Kanzi's technique less efficient than our own. To this end we carpeted the entire group room floor with blankets, so that any stone would bounce against the softer surface, rather than flake into pieces. Kanzi entered the room and observed the soft blanket covering on the floor. We presented Kanzi with a baited box and with stones, assuming that he would throw a stone on the floor only to find that it bounced off the blankets and not shatter as he had intended. He proceeded to make a few half-hearted attempts to throw stones onto the blanketed floor, but he did so with noticeably less force and enthusiasm than he had thrown stones onto the hard tile floor, as if he anticipated less success in the current circumstance. The stones bounced off the blanketed floor without shattering or producing any useful flakes. We then encouraged him to knap with both hands, and he then made a few attempts at bimanual knapping. After a short time attempting to knap the stone, Kanzi then got up and attempted again to throw the stones on the blanketed floor. After a few futile attempts to shatter the stones in this way, he walked to the edge of the room and carefully pulled a few of the carpet blankets loose from their tape to form a hole in the blanket covering, thus revealing the hard tile floor. He then threw a stone into this hole and succeeded in producing a number of flakes.

From his initial reaction to the blanket-covered floor, it was clear that Kanzi had surmised, just as we did, that a stone thrown against a soft surface would not shatter. He had no previous experience throwing stones against soft surfaces, nor had he observed anyone do so. Yet he seemed to have a cognizance of the properties of the materials he had been working with and he behaved accordingly. Some empiricists might object to this interpretation, but it is not one based on simple anthropomorphic tendencies. The anthropomorphic assumption of the experimenters planning the procedure was that Kanzi would throw the stone on to the carpet without realizing that it would bounce. Kanzi, in this case, went "beyond anthropomorphism" in his first reaction to the blanketed floor and in his eventual solution to the problem presented to him.

Why did we think that he would not realize the new properties of the changed situation, and why did we assume that he would simply throw the stone, find it would not break, and then simply revert back to bimanual percussion? We did not grasp Kanzi's comprehension of the physics of about the task in which he was engaged. The story of Kanzi's life is one of "experimenters" and "care-takers" repeatedly underestimating his cognizance of the situation at hand and his overall intelligence (Savage-Rumbaugh and Lewin, 1994). Even though we have made similar mistakes in the past, we continue to make them in new situations. We did not generalize what we had learned about Kanzi's linguistic abilities to his potential toolmaking ability. In the terms of learning theorists, we "failed to generalize to a new situation." Such failure to generalize is often characterized as the hallmark of animal thought, as contrasted with human thought (Rumbaugh, 2003).

While we deliberated what our next experimental step should be, Kanzi's facility with throwing enabled him to make large flakes quickly and easily. His quick fashioning of large, efficient cutting flakes permitted us, in turn, to increase the thickness of the string Kanzi was required to cut. Initially, the string was ¼" thick. With the larger sharper flakes produced by throwing, Kanzi was easily able to cut it. Slowly the diameter of the string was increased until it was over an inch in thickness. For such a thick string or rope Kanzi needed very large, strong, sharp flakes. Kanzi responded by selecting his material even more carefully and throwing it much harder. He began to ignore completely the small chips he hasd earlier worked so hard to produce. Anything under ¾" came to be treated as debris by Kanzi.

During the "throwing phase," we noticed that initially Kanzi displayed no arm bias or preference. However, within a few days he settled upon the right hand and never again utilized his left for this task. A review of the video tape of this short transitional period revealed the origin of the right arm bias. When Kanzi stood bipedally and employed his left hand to throw, his right arm rather automatically moved upward and forward and across his body, mirroring in a slightly delayed manner, the motion of his left arm. However, the same throwing motion with his right arm did not evoke a mirroring motion of the left hand, which rested in the normal position beside his body while Kanzi threw. This difference in the motion of the opposing hand during throwing was observed only when Kanzi stood fully erect while simultaneously executing a throwing motion of considerable force. The need for the right hand and arm to follow the motion of the left hand and arm, but not the inverse indicated the existence of a neurological basis for the development of Kanzi's right-handed preference. It is not known, of course, whether a similar constraint existed among early hominids for they were more proficient bipeds than Kanzi. The follow-through movement of the right hand-arm occurred as though it were part of the locomotor pattern of motion. That is, if Kanzi were brachiating, and he moved the left hand forward to catch a branch, the right hand would need follow in patterned precision. Kanzi was apparently able to inhibit this primitive motor pattern when leading with the right but not when leading with the left hand. This fact might support Calvin's view that it was precision throwing which placed pressure on the nervous system for extreme hemispheric specialization and for the development of the rapid sequencing capacities that underlie music, grammatical construction, dance and many other activities thought to be exclusively human.

When the weather grew warmer, it permitted us to once again attempt to exert empirical control over Kanzi's knapping methodology. The outdoor play area had a yard covered with bark, and we planned to place stone tools there while also blocking Kanzi's access to the indoor stone floor. We believed that this procedure would force Kanzi to abandon his throwing technique and revert to bimanual percussion. Bark, unlike carpet cannot be removed, for under it one finds only more bark. We quickly found that we were mistaken and that we had yet again underestimated Kanzi's ingenuity and creativity.

The first time he encountered the need to make a stone tool in this new location he visually surveyed the entire area, looking for a hard surface against which to throw the stone. He noted the large pole which held up the chain link cover and threw the rock at the base of this round pole. The rock glanced off. Kanzi then looked around the area again. It seemed that there was nothing left for him to do but to try bimanual percussion. There were some steel tables in the enclosure and a small pool, but no horizontal flat hard surfaces like the tile floor indoors. This fact did not trouble for Kanzi for more than a few minutes. Again he paused in thought, then calmly walked over to the rocks in the enclosure, selected a large stone from the group positioned it directly in front of him. Then he picked up a second stone and threw it against the first! It did not shatter immediately, but within 3 more throws, Kanzi had produced a nice large flake, without knapping! Kanzi had expanded his throwing technique to include two stones, one as substrate and one as hammer.

This innovation required a much more precise aim that simply throwing onto a hard floor. Initially, Kanzi missed the target stone quite frequently. However, after a few days of practice he became as proficient at this technique as he had been at throwing a single rock onto the floor.

Interestingly, Kanzi did not stand bipedally when throwing one stone against the other as he had done when he threw rocks against the floor. Instead he assumed a quadrupedal stance about two to three feet away from the target stone cobble. By this time, he had settled firmly into a right-handed technique. His left hand was used to support his body in a tripedal stance as he threw. Kanzi's remarkable ability to visually see the pieces of stone fly apart as the blow shattered the rock became even more apparent with a bark floor. Small chips flew rapidly away from the stone in all directions and into the bark, seemingly becoming invisible. On many trials all we observers could discern was one stone smashing against the other and the remains of the impact on the stones themselves. The flakes seemed to vanish. Kanzi however, must have been able to see flakes as they flew -- several feet in different directions in the bark. For, immediately upon breaking the stone, he would head toward a precise location and quite often, without any visual searching behavior whatsoever, he would pick up an excellent flake that was partially hidden the bark. Daily and with great ease, Kanzi located chips that flew into bark and completely out of site to us. His visual capacity to perceive rapidly flying small objects was clearly considerably more developed than ours. What evolutionary advantage, we wondered, does this skill provide bonobos in their forest habitat and why have we lost such a capacity?

Kanzi's solution of throwing one stone against another to overcome the carpet of bark yet again demonstrated his ability to come up with a functional innovation that had not been taught or even demonstrated for him. In many ways this solution was more impressive than the innovation of throwing, or moving the carpet aside. Throwing required Kanzi to consider the hammer stone, its trajectory and the target stone. This technique required the consideration of two stones, their relative positions and trajectory. This was also a technique used on occasion by our hominid ancestors to break stones too large to knap bimanually. Clearly, if

Kanzi were in the forest and he needed stone tools to survive, he would be able to produce them. Toth's initial demonstrations served to reveal to Kanzi that rocks could be broken and that such breakage resulted in sharp edges. In the forest, knowledge of the properties of stone could arise by other means. One would only need to observe a rock break when it fell and to note the sharp edges produced. Kanzi's behavior to date would suggest quite clearly that he possessed the capacity to reason his way to stone flake production.

We do not pretend to suggest that Kanzi's behavior has answered the question of how our ancestors began to make stone tools. We do believe however, that Kanzi's behavior casts doubt upon the commonly held assumption that hominids before 2.5 million years ago did not have stone tool cultures because they are not cognitively competent to do so.

Kanzi's second success in foiling our experimental attempts to force bimanual percussion left us puzzled. If we were going to provide Kanzi with a least two stones, as knapping required, how could we keep him from throwing one against the other when his method produced sharp flakes more efficiently than knapping? Of course, we could reward him only for knapping flakes produced by knapping rather than throwing, but denying him access to the tool site until he had knapped. From the perspective of simply artifact comparison, it did not matter what motivated Kanzi to flake bimanually, only that he did so. However, it was also the case that to the extent we were retracing -- even in some minimal and artificial sense -- the emergence of stool tool production in ancestral hominids, it would be certainly inappropriate to force a less efficient technique upon Kanzi by arbitrary means. Equally important was the fact that Kanzi's techniques had been self-generated. He understood them and preferred them. If he were in the wild he would have continued to employ them as long as they were effective for the desired ends. We needed to design a situation that called for knapped flakes rather than simply sharp flakes.

We attempted to visualize a situation that might have induced a similar need in our ancestors, had they, as had Kanzi, hit upon the idea of throwing one stone against another and found that throwing produced perfectly acceptable sharp flakes. At first, it seemed that the only possible reason would be the need for a more precise tool or one with a predetermined shape. Such specific shapes could not be achieved consistently except by intentional design and systematic flaking with the geometry of the desired flake clearly in mind. Kanzi's geometrical needs were simple, a large sharp edge. More rapid forceful well-aimed blows achieved that goal quite well. Of course if stones were rare and had to be carried long distances, it would become essential to get the maximum number of flakes from each stone and to get a single flake when needed. One could not afford to waste material by simply smashing it apart.

Many archaeologists have assumed that the earliest knapped flakes did not reflect any intent to produce a flake of a specific size or form. However, Kanzi's efficient throwing technique cast strong doubt upon that assumption and upon that of Davidson, as well, who suggests that even Acheulean handaxes were not a product produced by intentional effort on the part of the knapper. Archaeologists believe that the main purpose of flakes and handaxes was for skinning and butchering meat. Most stone tools are found near old lake beds, where groups congregated, perhaps attracted by large hoofed stock that came there to drink.

There is another property of water that might affect stone tool production. If prey were killed by drowning and then brought to shore, it could be too heavy to move very far. Butchering around water would inevitably result in tools falling into water. Moreover, throwing one rock against another would not work.

We began to wonder how being around water might have affected tool manufacture and the need for handaxes and decided to put Kanzi's rocks into his small wading pool which was approximate two-and-a-half feet deep and eight feet wide by ten feet long. If he tried to throw one stone at another while the rocks were in the water, it would not work. He would either have to get and take the rocks out to throw them, or employ bimanual percussion. We thought he would take the rocks out, but it nonetheless this situation would pose an interesting problem for him. This experiment was set up on a very hot summer day, a time when Kanzi enjoyed being in his pool anyway. Kanzi entered the tool site area and quickly noted that the rocks were not in their usual location. He looked around and quickly spied them in his pool. He stepped and stood bipedally in the water looking down at the rocks. He picked up one stone and looked at another on the bottom of the pool and raised his hand as if to throw. Had he done so, the water would have prevented success. However he did not do, he paused, leaned down and picked up a second stone and began to percuss bimanually while standing. Even when he was bimanually percussing before he began to throw, he had not stood bipedally while doing so. When he finished his chip it fell into the water. Kanzi saw it, leaned down into the water to retrieve it, and then stepped out of the pool to open the tool site. Clearly Kanzi could have taken his rocks out of the pool and thrown them. He had been throwing now for nearly eight months without any bimanual percussion. His shift from throwing to bimanual percussion was as precipitous and dramatic as was his shift away from percussion to throwing. It was also hardly possible to conclude that Kanzi could not reason sufficiently to move the rocks out of the pool. Kanzi easily carried rocks long distances. When we went into the forest, he would often place rocks in his backpack and carry them to the tool site located in the forest. When playing alone in his enclosure he would frequently gather and move rocks. Moreover, we noted later that if we left rocks both inside and outside the pool, Kanzi would occasionally stand inside the pool and percuss or throw a rock obtained from the water toward one that was already on the ground.

But the most critical factor that emerged from this "experiment" was that Kanzi's percussion techniques were now quite different than his earlier efforts. No longer did he "clap" the stone together in the midplane. His left hand stabilized the anvil and the right hand produced a glancing blow with the hammer stone against the anvil, much as might a human knapper. Immediately and easily Kanzi produced a sizeable flake. It appeared that the "throwing period" had provided Kanzi with some very new abilities that were, all at once, utilized when he began to percuss while standing in water. He could now throw with force and while aiming with precision. In the water, his left hand provided the stationary platform for the anvil stone, and his right hand delivered blow of the hammerstone with considerably greater force and precision. Kanzi did not actually let go of the hammerstone but the motion was more of a controlled throw, as opposed to the holding and hitting motion he previously employed. Equally important, Kanzi seemed to have learned something about the geometry of knapping as he now tended to strike toward the edges of the core rather than the center. He seemed to understand the need to knock chips off the edge rather than simply "split the rock." Apparently he had acquired these skills while throwing, even though throwing itself did not require the confluence of the abilities that he now brought to the task of knapping.

Not only was Kanzi now a much more efficient knapper, these innovations made the task more enjoyable for him. For the first time since he had innovatively developed the throwing technique, he began to elect to knap rather than throw, even when the stones were no longer in the water. Sometimes he alternated between these techniques, but he slowly began to prefer knapping. The preference for knapping appeared to be a function of his increased understanding that flakes were produced by hitting the core in a precise way. Once he understood the basic principles of where to hit in order to produce a flake, he also realized that it was necessary to hold the core and rotate it to achieve the best striking platform for each blow. Clearly, core rotation and orientation for the next blow was not possible unless one held the core. Kanzi also began to make multiple flakes from the same core and would now make them readily upon request, until the core was reduced to rubble, even when no incentive was in the tool site.

Having finally achieved, all of the basic skills that knapping required, Kanzi settled down into a pattern of right-handed bimanual percussion and continued to improve his technique. He began to pay increasing attention to the angle of his blows and to the striking platform for each blow. He also began to rotate the striking platform after each blow to achieve the best striking surface, and he paid close attention to the surface he selected. As his ability to flake increased, he came to ignore small and medium-sized flakes, attending only to the larger ones, as he now wanted not just any tool, but an effective and efficient tool. His increased competency provided him the luxury of desiring a really effective flake.

Panbanisha, Kanzi's younger sister was not initially a subject in the experiment devised by Toth and Schick. Kanzi started as the initial subject in this study due to his greater age (nine years old at the experiment's inception) and greater strength. However, because Panbanisha was with Kanzi much of the time, she often showed an interest in his activities. When we encouraged her to knap, though, we encountered the halfhearted attempts that characterized Kanzi's earliest behavior, before he was able to obtain, on this own, his very first flake. Moreover, when these pale attempts produced no flakes after only a few attempts, Panbanisha would put the rocks down and refuse any further attempts. Consequently we ended up making tools for her and handing them to her for nearly a year, with no improvement in either interest or competency on her part. There were no moments of epiphany, no sudden solutions, no throwing, nothing. Panbanisha would simply make a few meager inept attempts to knap, and then hand the rocks to the experimenter. She did not seem to like the hardness of the rocks, the sound or anything at all about knapping. No amount of encouraging her could change this behavior.

We continued occasionally to place incentives in the tool site for her and to make flakes for her that she could utilize, however this only seemed to increase her dependence upon us rather than to motivate her to achieve knapping on her own. Then during a visit of Toth and Schick, we happened to notice that Schick knapped a bit of stone as she was explaining a point about knapping. Until this time, all knapping in front of the bonobos had been done by Toth. However we noticed that Panbanisha was sitting quietly yet taking great notice of Schick (a female knapper) and so we asked her to demonstrate knapping for Panbanisha. Panbanisha continued to watch Schick with great attention, though her glances were from the side and intermittent, almost as if she were shy. When asked to knap herself, she politely refused but continued to observe. Panbanisha lacks the over ebullient enthusiasm that Kanzi brings to anything upon which he focuses his attention, and she prefers display only skills which can execute competently, especially in front of others.

Later, after Schick left, she began to practice her knapping with the intent of making a flake. That is, instead of tapping the stones together in a gestural manner to illustrate that she was complying with our request, she began concentrating on the stones them-

selves and the goal of producing a flake. From this point forward, she always selected one stone as the hammer and the other as the core. She often held the core with one foot and the hammer with the opposing hand. Unlike Kanzi as this stage, she began to rotate the core, looking for the best striking platform, illustrating that she had some understanding of the properties of the core's platfrom that had not arisen from direct knapping experience, as had been the case with Kanzi.

Using these methods, she frequently produced fairly large flakes. She did not move gradually from small to large flake as had Kanzi, but produced a variety of sizes from the start because she focused upon the edges of the core rather than simply hitting the rocks together in the midplane with considerable force as Kanzi had done. Since she did not have Kanzi's strength, this technique, had she employed it, would not have been effective for her in any case. From the beginning, Panbanisha also employed glancing downward blows, using the hammerstone as a true hammer, rather than as another hard surface as Kanzi had done.

Schick was not the first female knapper that Panbanisha had observed. All of her caretakers were female and all of them had repeatedly demonstrated knapping for her. However none of them were expert knappers, nor did they have the status of Schick, an important outside female visitor who was especially interested in the bonobos, who filmed them and who, much to Panbanisha's surprise was an expert knapper. Somehow this experience seemed to legitimize the activity of knapping for Panbanisha in a way that the knapping of Toth had not done. This sensitivity to role-related tasks was not something that we had previously recognized in Panbanisha. While we were certainly aware that she was quieter, far less rough in play than Kanzi, and preferred very different toys, we had not internalized the significance of the degree to which she took females as her role models. We also had no awareness of the fact that the expertise and status of the female model was itself could be an important component of the desire to emulate.

These extremely simple observations alone illustrate starkly many flaws in the classical experimental approach to the study of novel emergent behaviors in apes. These findings most probably hold true for all complex and highly intelligent organisms. Had we assumed that shaping was needed to induce knapping in Kanzi, we would have gained no understanding whatsoever of the ingenuity and comprehension of physics that Kanzi would bring to the task. By shaping him, and reporting on the success of our procedures, we would have simply verified our own anthropomorphic bias of "man the toolmaker," and revealed that an ape needed considerable training that our early ancestors could not possibly received.

If we had compounded this error by increasing our N, to improve our reliability, and put Panbanisha through the same training regimen that we employed for Kanzi, we would have leaned even less. We would have come closer to authenticating our predetermined views of the grandeur of human mind as contrasted with the paucity of that of the apes. (See Povinelli, 1996 for a classic description of the current anthropological thought regarding the difference between human and ape minds and for experimental methods that serve to verify this conclusion).

The emergence of stone knapping in Panbanisha also produced another unexpected result, jealousy on Kanzi's part. Prior to the time, he was the undisputed stone tool knapper of the group. He received profuse praise and attention for this activity. He was repeatedly filmed and photographed, and much time was spent with him to encourage him in these endeavors. While he did not mind Panbanisha knapping now and then, if she did so with vigor or for an extended period of time, he would often interrupt her by displaying towards her. He also began to make tools and leave them for her so that she did not need to make her own tool to open the baited box.

Thus we encountered not only the need for a female of high status and ability to motivate Panbanisha, we also encountered responses on Kanzi's part which, had they occurred in a natural cultural setting in which flintknapping was linked to survival and way of life, would clearly have lead to role dichotomization of stone knapping. Significantly, such role dichotomization would not be based on physical on mental sex-linked differences. Panbanisha began knapping at a higher skill level than Kanzi and produced larger, sharper flakes. However her ability was not the deciding factor regarding whether or not she would become a skilled knapper. Kanzi intervened on two fronts, one by directly interfering with her attempts from time to time and two, by providing her with tools. These actions, though relatively infrequent on Kanzi's part, were sufficient to cause Panbanisha to diminish tool production efforts. Certainly, whenever visitors were present and Kanzi and Panbanisha were housed together, only Kanzi produced flakes as presents for visitors. Thus many people have a piece of stone flaked by a bonobo as a reminder of their visit to the Language Research Center, but all such mementos have been flaked by a male bonobo.

Toward a New Methodology in Ape Research

We have learned more than we ever suspected possible from the simple question of whether or not bonobos could learn to flake stone. The answer to that question is yes, but if that were all we had learned, it would mean relatively little. The most important findings have emerged without experimental design or prediction. Had observation not been a part of our methodology, or had observation been limited to predefined classes of behavior, derived from the limited perspectives with which we began this undertaking,

we would have learned almost nothing of importance. These simple facts argue convincingly for a new methodology to be applied to the study of complex behaviors in nonhuman animals. We must move away from the delimited paradigms employed by empiricists. These paradigms begin with the faulty assumption that we cannot know the minds of other species, and therefore we will be fooled if we attempt to attribute thought or intentionality to other than ourselves. This is anthropomorphism in the extreme and if we do not move beyond it, we will only continue to glory in false self-fulfilling distinctions between ourselves and other complex beings that are more like us than we have dared to admit.

This new methodology for the investigation of complex cultural, linguistic and tool behavior in other species should:

a) include long-term observations that span important developmental processes

b) take place in group settings because no complex behavior evolves in a social vacuum

c) entail flexible observational schema that do not limit what is seen, understood and recorded

d) incorporate the perspective offered only by historical narrative

e) have a specific behavioral goal that is clear to participants and researcher, but permit flexibility in the achievement of that goal.

f) utilize participant-observations approaches when possible

g) strive to be free of anthropomorphic biases which characterize our species and which had traditionally prevented us from understanding other species

h) recognize that the majority of learning is not always manifest quickly or under precise experimental conditions comfortable to the experimenter who operates under fixed time constraints in which a behavior must occur or not occur in a given set of trials or presentations

i) maintain a flexible give-and-take between observed and observer and between what is searched for and what is found

ETHNOGRAPHIC FACTS: AN INSIDE POINT OF VIEW (WMF)

The above account of the emergence of Kanzi's knapping skills reflects a psychological bias, focusing as it does, upon skill emergence and the interface between "experiment" and emergent behavior. The "we" is employed, on the one hand, to reflect the fact that a number of different people observed and participated in Kanzi's journey to proficient knapping across the 8-year period it took him to achieve the status of accomplished bonobo knapper. It is employed on the other to represent a sort of "communal agreement on the events observed" and to lend an objectivity and depersonalization to the account. Objectivity in description is considered essential to basic psychological method. It is a critical orientation and precept of the discipline. By design therefore, the psychological perspective inevitably leaves aside is the subjective experience of both Kanzi and his observers. It focuses tightly upon skill emergence, leaving all else aside.

By contrast, the anthropological tool of ethnography utilizes as foreground all that psychology pushes to the background. The subjective experience of the participant observer, presented in the first person, is the central vehicle through which all else is expressed. Ethnography, as a tool of anthropology, acknowledges and celebrates the change that occurs in observer and observed. This is because of a deep understanding and recognition of the role of culture in the interpretation of all behavior. While many psychologists, especially evolutionary psychologists, speak of culture and admit to the role it can play in shaping the lens of observation, there is still a general failure in the field to understand culture in the deeper sense as a basic force driving group interaction at an often-unconscious level. Only in moving between cultures, can this unseen hand begin to be recognized, and even then the recognition, as it begins to occur, so changes the observer that the more assimilated and knowledgeable he or she becomes, the more difficult it is not to have one's vision bent by the lens of the culture. Thus, the "transition period," as one moves into a new culture is frequently the most difficult time and yet the most important, if one in retrospect can sincerely grapple with the changes which occurred during this time.

These changes are necessarily subjective and personal, for the lens of culture must operate at this level. To explain it rather lightly -- Culture is not how you wear your blue jeans or how you drink your tea. Culture is why you wear your blue jeans the way you do and why you drink your tea the way you do and how you and others feel about it as you do it. That is, culture is not so much about what you do or how you do it, but rather why you do it the way you do and why you feel as you do when you do it and what others feel about you as you do it. This kind of information is clearly absent from the account of Kanzi's knapping emergence presented above. The participants are sufficiently emic to seemingly be unaware of anything other than the need to account for Kanzi's actions in a formal manner and to justify their use of terms. While worthwhile, such an account does not draw the reader into the world of knapping bonobos. It may leave the reader with a sense of what Kanzi can do and how he came to do it, but little else. What is Kanzi really like, what is it like to be with him and to knap tools with him? How does he feel about the process, and what role does it play in his life? To answer questions such as these, a different perspective is needed. In order to fill this gap, the second author, who arrived after Kanzi had become an accomplished knapper, offers the reader an ethnographic account, written in the first person, and filled with the subjective "stuff" of what it is to be a part of Kanzi's world.

Impressions of a Bonobo Knapping Culture

When I met Kanzi several years ago, I felt as though I already knew him. I had seen the NHK (Japanese Broadcasting Corporation) documentary illustrating all of his abilities, and like a fan meeting a celebrity, we projected a romantic illusion of who we thought Kanzi was. Kanzi, on the other hand, knew I was green and ignorant and had so much to learn. In reflection, I believe he also saw a potential in me and sought to guide me. A potential, I might add, that took human others much longer to detect, including myself. He knew that I did not understand him, but he believed I could. I suspect he picked me to become one of his many spokespersons and ambassadors, a position I am honored to assume.

Those early misconceptions, beliefs, and thoughts of mine were contoured through four-field anthropology. I came to Kanzi believing that his linguistic competencies were no more than an antecedent of human language and tool expression; and therefore, associatively, he must have in his cognitive possession a type of proto-culture. And so I arrived at the Language Research Center (LRC) with the hopes of conducting the first pseudo-ethnographic interview with a non-human primate, a kind of cuteness which today I find so offensive in others. Quickly, it became very clear to me that Kanzi and his family, while they are not human, are in fact persons, and the entire notion of antecedents to human language, culture, and tools is quite faulty. Moreover, the notion of proto anything emerged as ridiculous and absurd. In time, it became quite clear that the description set which might be applied to Kanzi and his family could not be exclusively interpreted in terms of biological change over time. Eventually and fortunately, in those first days with the bonobos, my subjectivity took over. In my common experience and perception there were violations and exceptions to everything I had been taught. Epistemologically, I fell apart. Kanzi and his family, socially and culturally, violated my deeply held beliefs about the world.

Today I have reassembled myself, and let me state from the outset, I am biased. I have a new cultural bias, for I am a part of a Pan/Homo cultural world in which I now have a non-human child and Kanzi is my son's uncle. I have an emic perspective. There is extreme subjectivity in my perception of Kanzi and his family that only a postmodernist can appreciate. After Varela, known for his studies of cognition, consciousness and mind, I take subjective experience quite seriously. However, in this moment, I will try and step back in time to an empirical past that I believe is informative and speaks to issues that may lay a foundation upon which I may persuade you to consider that Kanzi and his family are toolmakers in the very sense that humans make and use tools.

Nicholas Toth and Kathy Schick came to Kanzi's world in May of 1990. Many describe Toth as teaching Kanzi how to make stone tools by showing him how to do it. Often I am shocked at conferences by questions about Kanzi's stone toolmaking, trapped in the rhetoric of trials, respondents, operants, and even worse, "monkey see, monkey do." Ultimately, this trajectory of questioning focuses on how long it took Savage-Rumbaugh to shape Kanzi's behavior so he could flake tools and on whether I think he is conscious of what he is doing. As if I could prove my audience is conscious by the questions they ask. I am not being facetious here. The questions I am asked are almost identical and appear to have the quality of a response to a stimulus. These questions have less creative variability than Neanderthal tools. It would appear that these inquiries are generated from a perspective learned long ago before science informed the questioner's opinions - a kind of first cause scrutiny that they do not apply to their own disciplinary assumptions. Therefore, I tend always to be astonished by the distance between the human world and the local world of the bonobos I know. It would seem that the entire world is a 1960's psychologist primed with terms like training, reinforcers, and cueing, and armed with an offensive anti-intellectualism and amateurish speak, dated and stale. In addition, what seems to be commonly missing on the part of many humans in their attribution of Kanzi is the idea of his personhood. If those in my audience could understand Kanzi has a point of view, that he has "beliefs and feelings about beliefs and feelings" (Dennett, 1998), then the form and genre of my explanations would not be so tedious and protracted. What justification do humans have in assuming that Kanzi "does not have control over his thoughts" (Donald, 2001) or, worse, does not have a mind of consciousness? The only fact in evidence is that Kanzi's brain is smaller than mine. For if you cannot accept a component of desire, ego, pride, and the "impulse of sheer delight," then you cannot understand Kanzi nor his ability to flake stone tools and use them. Such Cartesianism will ultimately relegate the reader to interpretations of Kanzi's abilities through radical behaviorism. This is a non-informing dead end.

Toth did show Kanzi how to flake stone tools and Kanzi admires Toth, if not loves him. This is notable, for Toth has spent very little time with Kanzi and I am sure has never slept in the colony room with him, cared for him, or nursed him to health in sickness. Yet the mention of Toth coming to the laboratory can get Kanzi so excited. I remember the first time I was present with both of them, I could see in Kanzi's eyes the happiness to see Toth. I did not understand it. A friend explained, "Oh, Kanzi is just excited because he knows when Toth comes he is going to get prestige foods" (anonymous to protect the foolish). However, over the years, I have come to know that this is not true. For the foods Kanzi gets when Toth comes are no different from the ones he can have everyday (Figure 1).

The truth is Kanzi likes Toth. Kanzi respects him. Kanzi's status seems quite elevated in the presence of Toth. But why? Well, Toth is a likable fellow; however, there is something about percussing two stones together which is extremely visual, powerful, and dominant. Toth has prestige, competence, and a silent demeanor which I read as confident and high-ranking - a person who can be in charge and get things done. Bonobos like that, especially in males. Moreover, it is clear that flaking stone to make a sharp edge is transformative. Kanzi knows what a knife is. He knows that a smooth rock cannot cut rope or hide. We have tried it. And to me, there is just a little magic in taking two non-cutting objects and transforming a piece of one into a powerful cutting edge. You know, I am impressed with

Figure 1

Kanzi, the male tool-making bonobo.

Toth myself. I have been flaking stone for about five years, and I am amazed at what Toth can do because my efforts have led to more blood and bruises than works of tools. My feeling is that Kanzi enjoys the feeling of status as much as anyone else and can be a little chauvinistic about it. It seems clear to me that Kanzi has a high level of desire to flake tools because he has an investment in the prestige of the outcome. While it is fun to make and use tools, it also makes you feel good to be able to do something for which everybody admires you, and especially something your sister Panbanisha doesn't seem to do as well as you. Or does she?

Panbanisha, Kanzi's younger sister and the biological mother of my (cultural) bonobo "son," is one of my close friends. We have spent a good deal of time together, and she is the first individual, human or non-human, with whom I ever made rock tools. Our circumstances were somewhat different from the formal expositions of the talent that has made Kanzi so famous. Our efforts to make tools were of a necessity. We were out in the woods one hot summer day, cooling off at Oranges, just past Flatrock (locations within wooded grounds of the LRC). At the Oranges location, metal drums filled with juice and snacks are covered with hide, and the only way to get in the drums is to cut the hide with a knife. The food had been left in the drums from a previous filming episode with Kanzi. The source rocks for making tools, all the good ones, had been removed back to the lab.

We were very hungry and thirsty. We didn't have a knife and there were very few rocks there from which to make tools. Panbanisha and I started looking for cores and cobbles, but all we could find were quartz-like rocks, which I tried to percuss bimanually with those glancing blows. We didn't care if we had thick striking platforms or prominent bulbs of percussion. We were so hungry and thirsty. Sue was no help, she fell asleep snoozing away. I kept trying to produce a flake with a sharp edge, but the substrate was not lending itself to my technique. My hands were bleeding and sore. Panbanisha put her hand on mine and I gave her the rock. She lifted her arm and she slammed the quartz into a large flat rock on the ground, which shattered the rock into lots of sharp pieces. We each grabbed a stone flake, tested it for sharpness, and ran up the hill to cut open the four or five canisters filled with refreshment. They worked just great. We ripped the hide open, got our food, and we didn't cut ourselves. It was some of the best food we ever had. After stuffing ourselves, we were so exhausted from toolmaking we fell asleep. When we awoke, I asked Panbanisha if she liked to flake stone, and she peeped "Yes." I was surprised, because I thought she had no interest in toolmaking or stones. She had appeared to prefer weaving and stringing beads, painting, and grooming. When we got back to lab, I was ready to give her some good rocks the first chance we had.

The following Saturday, Sue was out of town, and Kanzi had gone to P-Suke building. P-Suke, for whom the building is named, was with us at the Main building. He is the biological father of Nyota - a nice, polite, entertaining guy, but wild-caught and raised in captivity in Japan. It is hard to have a conversation with him. He knows how to scream but has no interests in stone toolmaking. If you provide him with rocks, he might hand you one for food, but he does not seem to have any desire to manipulate the rocks. Therefore, he went out to the play yard to eat lettuce while I gave Panbanisha her rocks. These were chert, different from those bad rocks we found at Oranges. Panbanisha seemed to select the rocks she thought were best and sat down on a blanket and began to use bimanual percussion to produce flakes. I was shocked. Where was the throwing technique? In about ten minutes, Panbanisha handed me a large sharp knife and then went to keyboard and uttered, "PINEAPPLE." I got a pineapple and started to cut it with a store bought knife and Panbanisha uttered at the keyboard, "KNIFE." I told her I was using a knife and then it occurred to me to ask, "Do you want me to use the rock knife you made?" She immediately responded and enthusiastically. So I cut the pineapple with Panbanisha's rock knife, and it was a much better knife than those cheap things we have in the kitchen.

The next day, I wanted to capture Panbanisha's two-handed technique on film. Unfortunately, those nice chert rocks were missing and the only ones I could find were those old bad quartz rocks. So I told her I was sorry and gave them to her anyway. Panbanisha looked at her selection for a long time. Then she picked two rocks that I didn't think looked very promising. In the next moment, she stood up and slammed the rocks down unto the hard tile floor shattering the stones everywhere. She looked around at the scattered debris, selected a tool and handed it to me. It was a little knife but Panbanisha thought we could cut "PEACHES" with it. We did, but it didn't last long. My confusion was over her reversion back to the throwing technique, as just the day before she had used bimanual percussion. Well, let me emphasize with embarrassment, my assumption was pejorative with respect to her technique such that I viewed it as qualitatively retrogressive, as you will see.

The next week, Panbanisha, Nyota, Sue and I were out in the woods at Oranges again. This time Sue had the metal drums baited just for us. Apparently, in her sleep the previous visit, she had taken notice of our real time adaptation. Panbanisha and I were looking for rocks from which to make knives. Panbanisha seemed to be digging in the ground. As I looked over at the hole she had dug, she had located a tool flaked on another day which was as useful as the day it was made. I looked to see if there was one for me, and I found one. As we two hungry people ran off with our knives to open the drums, Sue hollered out, "You both cheated!" Panbanisha and I were amused and really did not care, as we were hot, tired, buggy, and hungry. However, as I laid there resting, I began to think about the various techniques Panbanisha was applying in solving these issues of getting food. What would we do in the wild in real situations? Well, this was real. We were hungry. Our hands hurt. Bugs, spiders, and scorpions had all taken bites out of us. Panbanisha's leg hurt from a fall. My feet had blisters all over them. We were covered with mud and the only food we had was in this forest. It is true our snacks and juice were commercially prepared, but their richness of sugar, fat, and carbohydrate was not available to us without a knife. We had to manufacture a cutting edge or find one. More importantly, we were not even sure what we would find in the metal containers, but we were hungry and in a hurry.

As weeks passed, I began to remember what I already knew: rocks are different and therefore it is reasonable that techniques are different. Efficiency of results is the goal we were seeking. Aesthetics and style? Well, we just did not care. It is delightful to think that Kanzi makes Oldowan tools, but I will tell you, when you are hungry, dirty, bleeding and tired, it does not matter. However, what was clear in reflection is that Panbanisha was evaluating her resources and more authentically utilizing them than I had previously realized. My presumption of a superior knapping technique obscured my ability to see what was really happening. As my false assumptions were washed away, I realized Panbanisha could make tools or scavenge for them. She selected her substrate carefully and applied the techniques that she thought would give the greatest result with the least effort. This meant she could bimanually flake, use the throwing technique invented by her brother, or use an available tool.

In 1999, Roger Lewin, published his 4th addition of Human Evolution: An illustrated Introduction. I always enjoy reading introductory anthropology texts and I was really looking forward to Lewin's new perspectives. He had co-written *Kanzi: The Ape at the Brink of the Human Mind* (1994) with Sue, and Toth was making significant contributions to Lewin's new undergraduate textbook. I felt that the Kanzi research had been perpetually misunderstood and mischaracterized. I was excited that Lewin who knew Kanzi and Toth would be able to adequately inform the truth of the matter; however I was astonished when I read Lewin's quote of William McGrew's theoretical question, "When in human evolution did our ancestors cease behaving like apes?" Then, commenting "In other words, given the opportunity and motivations, could an ape make Oldowan tools?" Lewin reports and interprets Toth's work in this way:

"Toth had an opportunity to test this experimentally, when he collaborated with Sue Savage-Rumbaugh, of Georgia State University. Savage-Rumbaugh, had spent 10 years working with a male bonobo, Kanzi, who had learned to use a large vocabulary of words displayed on a computerized keyboard and who understood complex spoken English sentences. Toth encouraged Kanzi to make sharp stones flakes in order to gain access to a box that was secured with string. Kanzi was an enthusiastic participating the experiment over a period of several years. Despite being shown the percussion knapping technique, however, he never used it. Sometimes Kanzi produces flakes by knocking cobbles together, but without the precision inherent in the Oldowan technique; often he would simply smash the cobble by throwing it at another hard object, including the floor. Kanzi knew what he needed (sharp flakes) and figured out ways to obtain them (banging or throwing rocks), but he was not an Oldowan tool maker" (Lewin, 1999, pp. 133-134).

My astonishment arises from the fact that I think the tone of this exposition is misleading. First, Kanzi had to cut through a thick cord. It was hardly a string, and therefore, the tool had to be sharp. Second, the research is ongoing and today, Kanzi is 25 years old and Toth and Schick continue to work with Kanzi. I find it curious that Lewin treats the Kanzi research as if the experiment has been completed, and he may interpret the verdict. It seems quite striking that Lewin appears to have completely overlooked the empirical context informing Toth and Schick's investigations, as revealed in the following quote from the first report on Kanzi's toolmaking:

"Our strategy has been to motivate Kanzi to want a sharp-edged cutting tool (to cut through a cord or membrane to get into a box containing the desired reward), to show him the basic principle of producing sharp stone flakes, and then allow him to work out his own ways of producing his tools from an assortment of rocks provided" (Toth et al., 1993).

Since Kanzi and Panbanisha have both done this, that is, worked out a way of producing sharp-edged stone tools because they want to, exactly what the investigators sought, why jump to judgment? While I do not for a moment support Lewin's hypothesis that Oldowan technologies are a litmus test for 'when our ancestors quit behaving like apes,' I take issue with the inference that undergraduates most likely will assume from this textbook, namely that, based upon the Toth evidence, one may conclude that Kanzi and his kin are biologically incapable of the technical dimension ascribed to Oldowan tools. My objection is the arbitrary and radical line of demarcation between apes and humans in terms of their capabilities.

More recently, Toth and Schick have provided Kanzi and Panbanisha with Gona rocks from Ethiopia. These cobbles are hard and require bimanual percussion to flake. Sue and I have nearly killed ourselves trying to make rock knives with them; however, Kanzi and Panbanisha both make very useful sharp knives with this material. Moreover, they seem to prefer these rocks, which appear to be more suitable for toolmaking.

They are beautiful rocks, the color and size of baked potatoes. They are dense and many of their shapes are easy to hold and are obviously just the hammer you wanted, the ends pointed enough for those glancing blows. Moreover, when these rocks fracture, they seem homogenous and tend to break in more predictable ways. The trade-off, however, is this: it takes a lot of strength and skill to break these rocks efficiently.

Toolmaking with the Gona rocks has been recorded on videotape. Kanzi, from my perspective is a thoughtful and excellent knapper. We are awaiting the interpretation by Toth and Schick of these collections of cobble reductions. The effort has been exciting as Panbanisha has been participating equally with her brother. Well, almost!

One day while filming, Kanzi was busy making his rock tools. He had made some good useful knives and he showed them to me with pride. When it was time for Panbanisha to make some tools, Kanzi went to the middle test room to eat some grapes. We brought Panbanisha out. As usual, she examined her rocks carefully and then began using two hands to make her tool. At first, the effort was anemic compared to what I have observed when she was alone. Then things started to change with a little encouragement from Sue. As Panbanisha progressed with enthusiasm, she started making the sounds that a good knapper makes when flaking success is imminent. At that moment, Kanzi rushed into the Group Room and stopped Panbanisha from making her tool. It became very clear to us that sound was informing how and where the percussion ought to be delivered. Most importantly, however, the sound encouraged us onward, predicting the moment of success! Kanzi seemed very jealous of Panbanisha and Kanzi simply was not going to let her knap anymore. The rest of that afternoon Panbanisha just sort of slapped the rocks together and acted like it was just too hard and she could not do it. I knew better, but it made her older brother happy.

A week later both Schick and Toth visited the lab. Schick, like Toth, is an artist and craftsperson when it comes to stone toolmaking. Schick began knapping. Panbanisha's eyes were as big as saucers. She had only seen Toth make tools. But here was a woman making them too. This really gave Panbanisha the desire to ignore her brother's intimidation, and, from that day forward, she has enthusiastically asserted her right to stone tool manufacture. Interestingly, Kanzi has deferred to her. This is particularly important to Panbanisha's youngest baby Nathan. His attention is most often directed towards what Panbanisha is doing. The experience for Nathan certainly primes the future for his competence in stone toolmaking and use.

The relationship between tools and language seems clear to me. And this is readily observed among the LRC's non-English competent apes' abilities compared with Kanzi, Panbanisha, Nyota, and Nathan. Even with Nyota, who is almost four years old, I have far more success in getting him to make stone tools than I can with P-Suke who is approximately 24 years old. From casual observation it would appear that aside from just the aspect of English as a common medium between Nyota and me, making stone tools just makes more sense to Nyota than it does P-Suke. The bonobos who were not raised by a human language speaker simply organize their communication, culture, and tools in another way, and we are pressed for a common basis of understanding to penetrate the cognitive walls of meaning which different biases erect. However, the most prominent feature of toolmaking is the desire to make them, and while the reward is sometimes food, it is just as often prestige, status, or delight.

Certainly, I have always thought of myself as an Oldowan toolmaker, and I have been confident my tools would be classified as mode I; however, this might be a foolish assumption, and I might have to rethink the characteristics of my tool manufacture. I consider this because, if Kanzi's and Panbanisha's tool sets are not Oldowan, as Lewin claims, then neither are mine. For I cannot tell the difference between a knife I have made and the ones they have made. I must confess my motivation to make stone tools has been about adapting to the challenges that we face; namely, we need a cutting edge that will cut hide or rope. So perhaps our flaking is undirected, and we slap rocks together until we get something that can cut. However, this is counter-intuitive to what happens. Remember, both Kanzi and Panbanisha use different knapping styles at different times with different kinds of rocks. Moreover, there seems to be a melody to the sounds of percussion that guides the flaker to success. It is almost if the stones speak to us as to how and where to strike. That dull low sound means you have to hit harder or find a new spot. As the pitch increases to the sweeter sound, there is a pitch of success which serves as a guiding light to these Pan/Homo stone knappers. Often, the primary attending goal is to produce a cutting edge quickly that we are able to hold easily. While we admire the beauty of our rock knives, we have never sought to contour the style of the knife, but rather its beauty has been in its utility.

Our rules and conventions for tools are quite simple: make a cutting edge that you can hold and that is sharp enough to cut rope or hide. Kanzi, Panbanisha, and I all use our right hand to hold the hammer and it is clear that we all have preferences for certain kinds of hammer stones. This is particularly obvious in Kanzi. While Panbanisha and I will often switch between a hammer stone and cobble, Kanzi demonstrates a marked preference for certain rocks as hammers. When he finds a hammer, he seems to continue to use it as a hammer. In terms of quality and style, our only criteria and motivation is utility. As a rule, we now all use bimanual percussion, Kanzi's throwing technique, and scavenging. It just depends on the substrate and circumstances. Since we have lots of Gona rocks, we are all very proud of our stone tool products and we enjoy the activity despite the fact we know our hands will bleed and hurt later. We seem to have a singular mindset about what rock tools can do for us and this is based upon the empirical scripts that have been superimposed upon us. We are not a natural population and therefore the pressures and stresses are quiet different. I would argue that as of today, as a rule, creativity and invention do not often play a big part in our straightforward utilitarian toolmaking and use because we have perfected our technique for producing a product appropriate to demands of our Pan/Homo world. For our stone tool technology is a mature one.

However, on one particular day after a lot of toolmaking and filming, Kanzi took one of his rock tools, dare I call it an awl, and he used it to scribe a lexigram upon a metal sheet, just as he uses chalk upon the floor to write and draw. Then he picked up a keyboard and indicated he had written "MILK." This is so typical of Kanzi's inventiveness and is characteristic of his personality. My feeling is that if we were less a laboratory population with more choice and options, faced with real pressures of survival, we would see Kanzi's playful inventiveness directed towards technological expression meeting the challenges of frank survival. However, I might add, if we simply stressed a certain cultural technology, we might observe " . . . mastering the concepts of searching for acute angles on cores . . . and producing acute edged bifacial and poly-facial cores typical of many Oldowan assemblages" (Schick et al., 1999).

But our toolmaking is Pan/Homo, for by design Kanzi was left to his own cultural ways to produce a sharp-edged cutting tool, and his styles have influenced all of us. Today, after years of experience with the bonobos, I am quite certain that we could imitate technology of Oldowan cultures; however, from the Oldowan social perspectives, we are the other. Our adaptation is different because our environment is different. From a postmodernist's view, there is nothing universal about the cultures that produced Oldowan technologies, and therefore, we would not expect them to merely emerge. The absence of Oldowan features in our stone tools is meaningless, unless one assumes that God is broadcasting Oldowan algorithms and you simply have to have the right kind of humanlike brain to access this universal.

From my perspective, both Kanzi and Panbanisha are better rock knappers than I am. They seem to more accurately self-monitor for success and endure longer periods of rehearsal than me. Their expertise is also evidenced in the fact they have fewer wounds from knapping than I do, for we have all not only met the goal of wanting to produce sharp-edged cutting tools, but we can quickly produce them when we need them. I have found Panbanisha, when she is all alone in the play yard, practicing bimanual percussion when she happened to find rocks with which to work. No one was there. She was merely rehearsing and practicing for herself. When she heard me coming, she put the rocks away and attended to me. When I left, she returned to rehearsal with the rocks. This has often suggested a certain level of self consciousness and I believe Panbanisha often practices and performs rehearsals of her activities before she actually tries them in front of cameras and audiences. Whatever emotional interpretation may be applied to her private knapping episodes, I believe it is clear she has a desire to make stone tools and engages in private practice sessions to this end.

Toolmaking and tool use are merely one aspect of the spectrum of competencies that are natural expressions of Kanzi and his family's world. These bonobos are cultural beings who live in a cultural-English-linguistic world, adapting to the stresses and challenges of their world, expressing themselves through the complex opportunities that we offer them and those they create for themselves. Comparative work using Oldowan standards is a useful and interesting exercise when examining the tool expressions of Kanzi and Panbanisha, though their stone tool-related activities constitute just one aspect of their cultural world. It is significant to note that stone tool-making is a craft that Kanzi and Panbanisha have learned as adults; stone tool technologies were not aspects of their ontogeny. They have grown up with modern things such as televisions, VCRs, blenders, mixers, cars, books, computers, and a rich tool-and-gadget set of the information age. Their interests in making stone tools, when they could just as easily have used a store-bought knife, is a choice based in delight, in their involvement and interest in the task at hand. Kanzi's and Panbanisha's only research requirement has been that they have a desire to produce a sharp-edged cutting piece of rock to use as a tool. This is what they have done, and they have done it with finesse.

Kanzi and Panbanisha have acquired their cultural agency and expressions of competency have in much the same manner that human children acquire their language and culture. Paradigms of training, shaping and reinforcement, and arbitrary standards fail to inform the essential truths of these matters. The behaviors of bonobos who employ language and tools and who employ them together deserve an audience of interdisciplinary thinkers who can authentically embrace the discovery they have offered us.

REFERENCES CITED

Davidson, Iain & Noble, William (1993). Tools and language in human evolution. In (K.R. Gibson and T. Ingold, Eds) *Tools, Language and Cognition in Human Evolution.* Cambridge: Cambridge University Press, pp. 363-388.

Dennett, Daniel (1998). Brainchildren: essays on designing minds. Cambridge: MIT Press.

Donald, Merlin (2001). *Mind So Rare: The Evolution of Human Consciousness.* New York: Norton.

Heyes, Celia, M. (1998). Theory of mind in nonhuman primates. *Behavioral and Brain Science,* **21**: 100-134.

Lewin, Roger (1999). *Human Evolution: An Illustrated Introduction, 4th ed.* Oxford: Blackwell Science.

Macphail, Euan, M. (1987). The comparative psychology of intelligence. *Behavioral and Brain Sciences* **10**:645-696.

Povinelli, D.J. (1996). Growing up ape. *Monographs of the Society for Research in Child Development* (Vol. 61, No. 2, Serial No. 247):174-189.

Rumbaugh, Duane, M. (2003). *Intelligence in apes and other rational beings.* Hartford: Yale University Press.

Savage-Rumbaugh and Lewin, Roger (1994). *Kanzi: The ape at the brink of the human mind.* New York: Wiley & Sons.

Savage-Rumbaugh, E. S., Toth, N., & Schick, K. (2006). Kanzi learns to knap stone tools. In: *Primate Perspective on Behavior and Cognition*, D.A. Washburn (Ed.), pp. 279-291.

Schick, Kathy D., Toth, Nicholas, & Gary Garufi (1999) "Continuing Investigations in the Stone Tool-making and Tool-using Capabilities of a Bonobo *(Pan paniscus)." Journal of Archaeological Science* **26**:821-832.

Toth, Nicholas, Schick Kathy D., Savage-Rumbaugh, Sue, Sevcik, Rose A., & Duane Rumbaugh (1993) "Pan the Tool-Maker: Investigation in the Stone Tool-Making and Tool-Using Capacities of a Bonobo *(Pan paniscus). Journal of Archaeological Science* **20**: 81-91.

242 ◄ *The Oldowan: Case Studies Into the Earliest Stone Age*

CHAPTER 8

SEX DIFFERENCES IN CHIMPANZEE FORAGING BEHAVIOR AND TOOL USE: IMPLICATIONS FOR THE OLDOWAN

BY KEVIN D. HUNT

ABSTRACT

Chimpanzee positional behavior, diet, activity budget and canopy use differ between males and females. Contrary to expectations based on body size and the demands of pregnancy and lactation, females have a lower quality diet than males. Males ate more fruits, especially those harvested from large trees, ate at larger patches, ate terrestrial items more often, ate more piths, and ate more meat. Females ate more invertebrates, more small-patch fruit, more seeds, and more leaves. Items eaten by females were high in protein and high in calcium compared to males. Female-selected foods required greater handling times. Items selected by males contained high proportions of sugars or digestible hemicellulose, were found in large patches, or could be harvested from the ground. Most differences appear to be due to higher male social rank, since they parallel differences between high and low ranking males. Female selection of leaves does not follow rank-effect predictions, but is attributed the nutritional demands of pregnancy and lactation. This pattern suggests that low-ranking individuals — including females compared to males, and juveniles compared to adults — are under greater pressure to reduce handling times than are high-ranking individuals, since individuals with high-handling time diets would realize the highest return from "short cuts." Such sex differences might have been found in early hominins, given their greater body-size dimorphism. If so, early hominin males may have concentrated on terrestrially gathered food items, nutrient-dense foods, large food items, and other easy-to-process resources. It follows that females were more arboreal, ate foods lower in nutrient density, ate smaller foods, and selected foods that required greater processing times. Paralleling chimpanzee sex differences, female hominins likely used tools more often. Even among earliest toolkits we should expect to find female tools specialized for processing low-return food resources that require substantial handling times, and tools that can be used arboreally. Early hominin males likely utilized tools designed to harvest terrestrial items. A wooden digging stick/spear/club useful for harvesting underground storage items, utilizable in spearing prey, or in group defense against conspecifics, and heavy enough to serve as a club, seems a likely early hominin tool.

KEY WORDS:

Sex differences, Division of Labor, Early hominin diet, Chimpanzee, Arboreality, Food Processing

INTRODUCTION

Living humans are so profoundly dependent on technology, even in comparison to the most adept non-human tool users, that analogies between *Homo sapiens* and proto-tool-users may seem pointless. Fire is a complicating factor. Even the most technologically simple human toolkits, toolkits hardly different than those of chimpanzees (McGrew, 1992), are used in the context of fire (Wrangham *et al.*, 1999). Apes, by contrast, offer a technologically simple anchor point from which we might extrapolate toward modern humans to model selective pressures that acted on the first hominin stone toolmakers. Among the biological pressures that may have influenced the form of the earliest human toolkit is the timing of the origin of sex differences in foraging strategies.

ARE HUMANS UNIQUELY UNIQUE?

Our concept of ourselves as unique underwent a shift in perspective after World War II. The New Physical Anthropology increasingly drew on theory and data from biology (Cartmill *et al.*, 1986). Accordingly primatology, perhaps a more natural fit in biology, established itself within bioanthropology. As information about primate ecology and behavior mounted, humans seemed less unique. However unique living humans are, it began to seem likely that early hominins were not so special, and that while our early ancestors may have been quite unusual apes, they were not *uniquely* unique, but just *Another Unique Species* (Foley, 1987). As human capacities for tool use, language, self-concept, and complex social interactions blurred into ape capacities, human paleontologists shortened their lists of human traits considered unique to perhaps a score: relative brain size (including correlated traits such increased cognitive capacity, and lengthened life-history variables such as age at maturation), bipedality, high heat tolerance (*sensu* Wheeler, 1991, including nakedness and sweating), high diet quality (and consequent small guts), social network size, use of composite tools, sexual division of labor (including mutualistic exchange between the sexes), and language capacity. Human uniqueness has led some to question the utility of *referential models* for the origin of tool use, i.e., models that draw on a single referent, and others may even argue that humans are unusual enough that even conceptual models — models that draw on rules linking selective pressures to adaptations (*sensu* Tooby and DeVore, 1987) — are questionable. I consider that parallels between humans and chimpanzee culture and ecology are profound (McGrew, 1992; Whiten *et al.*, 1999), but nevertheless I will approach my analysis with these cautions in mind.

Sex Differences in Diet and Foraging Strategy as Division of Labor in Human Societies

Students of human foraging noted that the pattern of *sexual division of labor* was rather consistent across cultures. That is, tasks that females took on in one culture tended to be female tasks in others as well. In a survey of 185 societies (Murdock and Provost, 1973), 23 activities were found to be performed mostly by males, and nine activities were seen to be predominantly female tasks. Many of these activities, such as net making or ore-smelting, can be dismissed as unimportant for early tool-users. They will be excluded from discussion here. Among tasks that early hominins might have performed, Murdock and Provost found that males more often engaged in hunting large fauna, woodworking, fowling, stoneworking, bone/horn/shell working, mining and quarrying, bonesetting and other surgery, butchering, and honey-collecting. Females were more likely to engage in gathering small aquatic fauna, gathering vegetal foods, and preparing vegetal foods.

The root-cause of these differences was debated. Perhaps, it was argued, tasks had sex-specific costs and benefits, and each sex allocated time and energy to tasks according to the net benefit to that sex. Emerging from the welter of factors that were proposed as influencing time- and energy-allocation strategies was the likelihood that nursing, and to a lesser extent other child-care duties (Brown, 1970; Sanday, 1973; Gough, 1975; Parker and Parker, 1979) shift female foraging strategies toward tasks that are compatible with infant care. Heavy physical labor was eventually dismissed as incompatible with nursing. As a further burden, proximity to a safe infant cache was seen as important (Brown, 1970). Damping sex differences is the capacity for humans, like other primates, to perform at least low-risk subsistence activities while carrying infants, and the fact that older offspring may be quite independent. While Mead (1949) was rightly dismissive of many of Malinowski's (1913) explanations of sex differences, she recognized that there is a reproductive basis for the capacity of males to better afford "sudden spurts of energy" compared to females (Mead, 1949: 164). Two selective pressures are profound: infant survival depends on mother having access to calories for lactation that are both *consistent* and *adequate*. To nourish infants mothers cannot suspend nutrient acquisition for long. Paternal physical condition is freer to vary without directly affecting the survival of their offspring. Males may be quite active for short bursts, and then inactive during recovery. Cross-sectional geometry of long bones suggest that a pattern of higher male activity existed at least as early as the Middle Paleolithic (Ruff, 1987).

In her review of research on sexual division of labor, Brown (1970) found that "repetitive, interruptible, non-dangerous tasks that do not require extensive excursions" are most compatible with child care. The net nutritional value of the resources acquired via these tasks, she contended, is greater than that contributed by males. Whereas some recent research questions the axiom that gathering is necessarily a more reliable strategy than hunting (Hurtado *et al.*, 1985; Hurtado and Hill, 1990), the vital status of gathering has been recognized consistently (Tanner and Zihlman, 1976; Zihlman, 1978, 1981), even if it has not been empirically demonstrated. Twenty-five years after Brown (1970), Hurtado *et al.* summarized their work on division of labor in the Hiwi and Ache as follows: "Women seem to have solved the problem of obtaining energy and allocating time to raising offspring by adopting strategies which increase male productivity, by relying on male provisioning and by spending time and effort in activity types that are readily compatible with childcare and expose the young to minimum risks."

Are Sex Differences in Diet and Foraging Strategy a Uniquely Human Feature?

Surprisingly quickly, data on chimpanzee foraging stripped away much of what was considered unique about human sex differences in food-getting strategies. Although the null hypothesis that male and female apes might have *no* sex differences is a null hypothesis that is, in Eckhardt's (1981) words, "nuller than most," the extent of ape sex differences was unexpected. At first implicitly (Goodall,1968) and then explicitly (McGrew, 1979) sex differences in chimpanzee diet, habitat use, tool use, and other foraging behavior was articulated. In particular, McGrew (1979) recognized a long list of sex differences that have been confirmed in subsequent work. There is a strong bias for females to harvest invertebrates (McGrew, 1979), and ants in particular are gathered arboreally almost exclusively by females (Nishida, 1973; Nishida and Uehara, 1980; Uehara, 1986, 1987). Males are hunters: in 48 of 49 cases at Gombe where sex was identified, it was a male that killed mammalian prey (McGrew, 1979). McGrew noted that more mobile and wide-ranging male chimpanzees hunted, and females engaged in activities that he characterized as gathering, a pattern he found similar to that of modern humans. A similar trend was observed at Taï, Ivory Coast where 281 of 331 identified hunters were male, and 31 of 38 successful hunters were male (Boesch and Boesch, 1989). In a review of chimpanzee sex differences in morphology, life-history variables and patterns of affiliation, Hiraiwa-Hasegawa (1987) pointed out the need for a rigorous, evolutionary approach.

McGrew (1979, 1981) considered the meaning of these differences for human tool-use origins. Noting that chimpanzee females use tools to capture prey, whereas males do so only rarely, he suggested that human tool use originated in "solitary female-foraging activities, [not] hunting" (McGrew, 1979: 461). Throwing among chimpanzees is rare, he noted; it is engaged in not for hunting, but for defense or aggression. Throwing is a male behavior (N=44; Goodall, 1968, reviewed in McGrew, 1981). In contrast to most scenarios for the evolution of hunting, McGrew (1981) argued that early hominins were poorly adapted for competing with large carnivores for carcasses, and therefore that scavenging was an insignificant part of their food-getting behavior. He concluded that females were more likely to have invented *facilities*, such as lures and traps, which are principally gathering devices, not hunting devices, and that carrying devices were likely invented by and used by females first.

The timing of the habitual use of carrying devices may be the most important unanswered question in the origin of hominin stone tool use. Efficient transport of stone tools and/or raw materials is a significant barrier to stone tool use. The issue of carriage is a more troublesome than it might appear at first glance. Without carrying devices, keeping track of stone tools while engaged in two-handed gathering activities, which I will argue below was a critical early hominin strategy, is difficult. Tools would have to be left on the ground during arboreal foraging, necessitating returning to the cache site. This retrieval cost can be large, if in the course of unbroken gathering the forager is several trees and many meters away. If tools are carried in the hands during travel, they curtail the normal primate feed-as-you-go strategy, which involves frequent use of both hands for gathering, even when terrestrial. As each shrub, tree, herb, or clump of grass is encountered, tools must be deposited on the uneven, leaf-littered ground, and must be rediscovered and retrieved later — or the resource must be ignored. Tool carriage, in other words, must push the early hominin food-collection strategy toward larger, more compact food patches. One solution to this problem, assuming early hominins had no carrying devices, might be long-term stone tools caches. This strategy implies a proto-home base, and such a central place foraging is widely rejected for early hominins (Sept, 1992).

The lack of containers or carrying devices among wild chimpanzees suggests that carrying devices are not readily conceived by a chimpanzee-like mind, in which category I would include early hominins, though like many composite devices, they are utilized readily enough when provided by humans (e.g., the bonobo Kanzi makes frequent use of a backpack). Materials from which to construct a carrier are not as available as one might expect. Woven devices for early hominins can be dismissed immediately. Animal products look promising, but absent tanning or other quite complicated processing they decay quickly. In drier habitats where decomposition might be slowed, when hides and other animal tissues are not immediately eaten they dry to the consistency of plywood. The conception of carrying devices seems outside early hominin intellectual capacities. Such speculation, however, is incomplete and unsatisfying. A container would dramatically decrease the costs for tool use, increase the practical size of the toolkit, allow for transporting raw materials, and allow delayed consumption of some foods. This may be a crucial missing datum as we speculate on the origin of tool use. Here I will assume that carrying devices were not part of the toolkit of the first stone tool users, but may have been invented as early as the first-appearance of *Homo erectus* (*sensu lato*).

Living Apes Best Inform Our Conceptual Models

While living humans may be uniquely unique, and therefore somewhat uninformative for reconstructing the origin of tool use, the gap between apes and fossil hominins is manageable. If we accept that our closest relatives, the apes, are more adept tool-users than once recognized (McGrew, 1992; Schick *et al.* 1999, Toth *et al.* 1994), and if we further accept that early hominins — rather than *Homo* (de Heinzelin *et al.*, 1999) — were the first stone tool users, apes and hominins are arguably similar enough to justify even referential models. That is, as Wynn and McGrew (1989) suggested, it seems likely that the first stone tool-makers had cognitive and manipulative abilities that were quite similar to those of living apes, with only bipedalism as a notable difference (Wynn and McGrew, 1989). Nor is bipedalism particularly confounding. Chimpanzees use tools most often when sitting. Among hominins, bipedalism is temporally disassociated from stone tool use, having appeared three and a half million year before the first stone tools. Indeed, in a comprehensive review of bipedalism origins, Rose (1991) found that among active scholars, a link between tool use and bipedalism is promoted merely as a preadaptation for tool carriage, not as a cause for bipedalism. In short, habitual bipedalism is unlikely to have much altered the dynamics of tool use and tool manufacture from the condition seen in apes.

I will look to the apes for the components of a conceptual model that will consider the effects that food-getting strategies, relative arboreality, social dynamics, and sex differences on patterns of early hominin tool use.

What is the Source of Sex Differences?

Sex differences in early hominins, if such were present, likely stemmed from the same source that dictates sex differences in living nonhuman primates. Differences between the sexes may arise for four reasons (Hunt *et al.* in review): (1) The 'sexes have different **reproductive demands**' hypothesis posits that females must gestate, lactate and (among anthropoids) carry offspring, while males may guard territories, guard females, or both. These different demands mean females and males have different dietary needs and different locomotor costs, which in turn will affect foraging behavior and anatomy. (2) The **'social rank'** hypothesis holds that sex differences may arise when one sex is consistently socially dominant, and thus free to monopolize highly desired food items. (3) The **'body size'** hypothesis stresses that when the sexes differ in body size, their nutritional needs, their mobility in the canopy, and their ability to open food items yields further sex differencess. (4) The **'paternal investment'** hypothesis indicates that, when males have high confidence in paternity or significant inclusive fitness benefits, they may defer to mates and/or offspring at feeding sites as a parental investment. Such competition avoidance may either reducing reduce or increasing increase sex differences. Each of these selective pressures may have affected early hominins.

Tool Use Can Only Be Understood in the Context of Other Ecological Variables

Conceptual models rely on identifying linked variables. For example, among frugivorous primates, incisor breadth is correlated with the diameter of the species' average food item (Lucas *et al.*, 1986). I will make use of many such links as I discuss early hominin tool use. Identifying differences between humans, apes and early hominins is our first order of business.

Whereas humans use tools in a wide variety of contexts, even among chimpanzee populations that use tools, and not all do, stone tools are used in only a few contexts. Only one is very common, nutcracking, and stone tools are used to process only a handful of species (Boesch and Boesch-Achermann, 2000). Likewise, among early hominins, stone tool use and manufacture must have begun as a relatively minor food processing technique in an already complicated and well-integrated foraging regime. Tool use could only have arisen if it was compatible with foraging strategies already in place. That is, stone tools must have increased net caloric return when it was introduced into an already long list of food-getting strategies. It cannot have reduced the efficiency of in-place strategies beyond some critical point, and costs must have been low relative to benefits. Among possible costs are transport effort, caloric expenditure during tool use, costs of searching for raw material, costs of manufacture, risks of injury, risk of predation, and risks of intraspecific agonism. Benefits depend on the encounter rate of items requiring stone tools, and include the increase in calories harvested per unit time using tools, and decreases in risks during harvesting. Costs and benefits will differ according to the diet, habitat use and foraging strategies. For example, a central-place forager that included meat in its diet might cache tools at its home base. If carcasses were carried home to share, tool use would require no additional transport costs, no additional exposure to predators, and little risk of tool loss (and therefore raw material and labor loss). For a nomadic, arboreal, frugivorous, forest-living ape, risks and costs are quite different. As we sift through possible the selective pressures acting on early hominins, we must consider that tool use likely arose in hominins that were principally frugivores.

ANALYSIS

What Was the Early Hominin Diet? Trace Element Evidence

Sponheimer and Lee-Thorp (1999) found that $\delta^{13}C$ values for three *Australopithecus africanus* specimens were most similar to a fossil hyena, suggesting that they were strongly dependent on C_4 plants, or animals that ate them. Note that among the animals that eat such plants are termites. One hominin had a C_3 signal suggesting it had consumed fruit, herbs or leaves. Australopithecines, these data suggest, were generalists compared to sympatric species, exhibiting $\delta^{13}C$ range more variable than 18 of 19 comparison taxa, among them baboons, vervets and *Notochoerus capensis*. Australopithecine $\delta^{13}C$ levels, however, were not unique: monkeys that range into open habitats, vervets and baboons, had similar levels. The authors concluded that australopithecines ate between 25-50% C$_4$ foods. Australopithecine δ^{13}C levels then are consistent with Backwell and d'Errico's (2001) interpretation of bone artifacts as termiting tools, even if the link is highly inferential.

Early Hominin Diet Inferred From Dental Size, Tooth Shape and Dental Microwear

Kay (1985) and Teaford and Unger (2000) found that in extant primates large molar areas correlate with high proportions of seeds, nuts, or other hard items in the diet. Molar areas also correlate with percentage of fruit in the diet (Lucas *et al.*, 1985). Compared to living primates, early hominins are 'megadont' — their cheek teeth are large (McHenry, 1984; Kay, 1985). In a regression of cheek-tooth area against body weight among living apes, *A. afarensis* fell 22% and *A. africanus* 32% above the regression line (Kay, 1985). Other australopithecines have similarly large molar surface areas, including *A. anamensis* and *Ardipithecus ramidus*, compared to chimpanzees (Teaford and Ungar, 2000). Thus, tooth size suggests a diet high in fruit, in seeds, or both.

Early hominin tooth shape accords well with dental dimensions. Early hominins lack long shearing crests that are correlated with leaf eating among extant hominoids (Teaford and Ungar, 2000). Teaford and Ungar interpret molar morphology in early hominins as evidence against carnivory. Instead, tooth shape suggests a diet of fruit or hard-coated seeds.

Thick enamel (Teaford and Ungar, 2000) is found among living frugivores such as capuchins and orangutans (Kay, 1985), whereas terrestrial primates such as gorillas and baboons have thinner enamel. In comparisons of closely related dyads such as gorilla/chimpanzee or siamang/gibbon, frugivores had thicker enamel. Across the primates, folivores have the thinnest enamel, whereas primates that consume hard, brittle foods have the thickest enamel, and frugivores are intermediate. Enamel microstructure (decussation) also suggests hard-object feeding (Kay, 1985; Teaford and Ungar, 2000). Thick enamel in early hominins suggests a diet of hard-husked fruit and/or hard-coated seeds.

Incisor size is correlated with food item diameter (Lucas *et al.*, 1985) and other physical characters. Small incisors among *A. anamensis*, *A. afarensis* and *A. africanus*, similar in relative size to gorillas, suggests they fed on smaller fruits than do chimpanzees and orangutans (Hylander, 1975; Teaford and Ungar, 2000).

Molar microwear has not been analyzed in early hominins, but *A. africanus* microwear (Walker, 1981; Teaford and Walker, 1984; Grine and Kay, 1988; Kay and Grine, 1988; Teaford, 1994) suggests they were frugivores. Kay and Grine (1988) found that *A. africanus* microwear feature width fell between howlers and capuchin monkeys. Pit:scratch frequency comparisons place them between orangutans and chimpanzees (Kay and Grine, 1988). Table 1 presents feeding records for these four taxa. Using these diet data, Hunt (1998) offered a best-guesstimate early hominin diet by averaging the diets of the species with microwear signatures most similar to early hominins(updated in Table 1). The results suggest that fruit made up nearly half the early hominin diet, that leaves were a critical food item, and that piths, insects, flowers seeds and bark were also included in the diet.

Early hominins, even the less robust species, had considerably thicker mandibular corpora than living hominoids (Chamberlain and Wood, 1985; recent finds reviewed in Teaford and Ungar, 2000). Among living primates, the Pitheciini (*Chiropotes*, *Cacajao* and *Pithecia*) have the most robust mandibles (Kinzey, 1992). Their diet is high in hard-husked fruits and seeds (Anapol and Lee, 1994; Kinzey, 1992; Boubli, 1999).

A. afarensis incisor microwear data seem indicate a lowland gorilla-like wear pattern that included gritty plant parts, perhaps grass stems, roots and rhizomes, in addition to fruits (Ryan and Johanson, 1989). The undulating pattern of wear on *A. afarensis* incisors indicates a stripping function, consistent with leaf stripping (Puech and Albertini, 1984; Puech, 1992). I argue that these data may also suggest that fruits eaten by early hominin were so small that they were ingested without incisal processing with the incisors. This would mean that leaves, as the second most common item in the diet,

Table 1

Species	Insects	Leaf	Meat	Fruit	Piths/ Herbs	Flowers	Bark	Seed
Pan[1]	5.6	10.3	1.0	57.0	22.6	0.7	0.0	0.0
Alouatta palliata[2]	0.0	64.0	0.0	12.0	0.0	18.0	0.0	0.0
Pongo[3]	1.0	26.0	0.0	58.0	–	—	13.0	2.0
Cebus apella[4]	15.0	0.0	0.0	51.8	3.5	1.3	0.0	16.1
Early hominins?	**5.4**	**25.1**	**0.3**	**44.7**	**6.5**	**5.0**	**3.3**	**4.5**

Table 1 — Early hominin diet as suggested by molar microwear

[1]Hunt (1989); feeding time, based on 3,891 feeding records of *Pan troglodytes schweinfurthii* at Mahale.
[2]Glander (1978); feeding time
[3]Rodman (1984); feeding time, Kutai, Kalimantan 40,022 min. observation
[4]Janson, 1985; proportion of total kJ intake. Nectar made up 12.3% of kJ intake.

would be the only microwear signature in the incisors.

The trace element analysis reviewed above suggests directions for fine-tuning. High C_4 levels in *A. africanus* (Sponheimer and Lee-Thorp, 1999) pushes the early hominin diet in the direction of that of at least some chimpanzees. Mahale chimpanzees included a very high proportions of the grass *Pennisetum purpureum* (Hunt, 1989) in their diet in 1986-87, very near the 25% level the early hominin C_4 data suggest. Chimpanzees also consume termites, presumably a C_4 food.

Although all data bearing on early hominin diet are not in complete accord, taken together they produce a rather consistent picture. Weighing each of these lines of data, we may synthesize them to conclude that a) approximately half of the early hominin diet was fruit, principally small-diameter, hard-husked or seedy fruit (e.g., *Grewia*, *Harungana*), that b) leaves made up as much as a quarter of the diet (note incisor microwear data above) and were an important fall-back food, c) that seeds and hard husked fruits (given dental morphology and microwear) were the second most important dietary item (though not the second highest in proportion), d) that grasses were perhaps equally important (also consistent with microwear data, I argue), e) gritty food items, perhaps underground storage organs, made up at least part of the diet, and f) that early hominin diets were quite varied and included insects, meat, herbs, blossoms and bark, in addition to the staples suggested above.

Evidence Suggests Early Hominins Were Woodland-Living and Dependent on Arboreal Foods

The broad, shallow and cone-shaped (Schmid, 1983) torso of *A. afarensis* resembles that of chimpanzees, and is unlike the barrel shape characteristic of *Homo*. Hunt (1992) interpreted this shape as a stress-reducing adaptation that decreases compression on the ribcage during unimanual suspension (i.e., arm-hanging). Hunt (1991a, b, 1992) argued that the raised-arm set of the chimpanzee shoulder joint is an adaptation to arm-hanging and/or brachiation (arm-swinging locomotion), since no other behavior requires the complete abduction of the humerus. He argued that other positional behaviors hypothetically linked to the arm-raised set of the ape shoulder joint either a) do not require full abduction (e.g., vertical climbing), b) are found in monkeys that are incapable of full abduction, yet are nevertheless capable vertical climbers, or c) on closer examination, actually do involve some form of arm-hanging or brachiation, either unimanual suspension, actual brachiation or some other arm-raised suspensory positional behavior (e.g., quadrumanous, or four-handed, climbing).

The arm-raised set to the joint is manifested most clearly in chimpanzees in the observation that when resting they often spontaneously raise one or both arms

above the head. When grooming, they often fully abduct the humerus to engage in hand-clasp grooming. The scapular glenoid fossa of australopithecines is also cranially oriented (Robinson, 1972; Stern and Susman, 1983), giving the shoulder joint an upward tilt intermediate between *Pan* and humans. Inouye and Shea (1997) showed that smaller humans have more uptilted joints, and went on to argue that this is evidence the feature has no function. I argue instead that the allometry itself is an adaptation. It suggests that humans have an evolutionary history of contrasting behaviors between smaller and larger individuals; specifically, it suggests smaller individuals were better adapted to unimanual armhanging than larger individuals.

Early hominins show evidence of vertical climbing adaptation. Origin and/or attachment areas of biceps, latissimus dorsi, extensor carpi radialis and brachioradialis muscles, used to perform a pull-up action during vertical climbing, were large. Although there is evidence of only very limited great toe gripping adaptations, the hip shows evidence of a long moment arm for the hamstrings (Stern and Susman, 1983), which would increase the power of thigh extension, presumably during vertical climbing. Inferred large deltoid muscles, if analogous to those in chimpanzees, were used to raise when reaching out to pluck fruits while arm-hanging and during vertical climbing.

The convex joint surface of the *A. afarensis* and *A. africanus* medial cuneiform indicates a rudimentary gripping capacity for the big toe (Stern and Susman, 1983; Deloison, 1991; pers. obs.; Clarke and Tobias, 1995). The diameter of a support which the reconstructed grip the foot could accommodate was considerably less than that of extant apes. The early hominin gripping capacity would be useful for gripping moderate-sized (•5 cm) supports during vertical climbing. Evidence for a large, ape-like peroneus longus muscle suggests powerful great-toe flexion. Together these features suggest early hominins ascended using narrower supports than do living apes, or vertical-climbed less often. I argue they utilized shorter trees that less often required vertical climbing, and that the climbing bouts were shorter in duration.

A large calcaneus in *A. afarensis* (Latimer and Lovejoy, 1989) suggests that terrestrial locomotion was nearly exclusively bipedal. A large calcaneus, however, is no barrier at all to arboreality (Hunt, 1998).

A plantar set, or at least greater mobility (Latimer and Lovejoy, 1990) of the ankle joint, compared to humans, would have allowed greater plantar flexion (pointing the toe) in early hominins. Gombe and Mahale chimpanzees plantar-flexed their feet when they used their toes grip a branch to support body weight with the hind limb in tension. If females were more arboreal, and more suspensory, one would expect greater plantar flexion among females, and indeed this has been observed (Stern and Susman, 1983). Consistent with the suspensory anatomy of the ankle, early hominins have long, curved pedal phalanges (Tuttle, 1981). Gripping with the lateral four toes but not involving the great toe was seen in Tanzanian chimpanzees in concert with unimanual suspension (arm-hanging). Such pedal gripping increased stability among slender terminal branches (Hunt, 1994b). Australopithecine fingers are curved, robust, and exhibit flexor sheath ridges, evidence that they had a powerful, chimpanzee-like grip, a capacity used by apes during arboreal arm-hanging and vertical climbing.

A more convex articular surface of the proximal tibia and a anterior-posteriorly compressed distal articular surface of the femur in *A. anamensis* (Leakey *et al.*, 1995), *A. afarensis* (Stern and Susman, 1983) and *A. africanus* (Berger and Tobias, 1996) suggests an emphasis on flexibility rather than stability in the hindlimb. Flexibility is useful during arboreal locomotion.

Although the os coxae of australopithecines are human-like in appearance, the pelvis is considerably wider than necessary for giving birth to an ape-sized neonatal head, or indeed even a human-sized neonatal head. Hunt (1994b, 1998) suggested the wide pelvis lowers the center of gravity, thereby improving balance on unstable substrates such as tree branches. Short hindlimbs have been speculated to serve the same function (Kummer, 1991). Such a broad pelvis decreases locomotor efficiency and increases skeletal and ligamentous stress. Short hindlimbs further decrease locomotor efficiency (Jungers, 1982, 1994).

These features were present for more than four million years among early hominins, thereby rendering it extremely unlikely that they persisted without function. Short hindlimbs, a wide pelvis, and flexibility of the knee are maladaptive in an exclusively terrestrial biped. The null hypothesis that early hominin behavior anticipated that of their descendents for millions of years, rather than that it reflected their own anatomy is, to quote Eckardt (1981) again, nuller than most.

In summary, recent finds only re-emphasize conclusions made a decade ago: "although bipedalism may have been virtually the only terrestrial locomotor mode in [early hominins], poor bipedal mechanics and compromises that improve arboreal competence suggest a role for locomotor bipedalism that is relatively reduced compared to modern humans early hominids may not have been reluctant, half-evolved bipeds, but rather they had a fully evolved, unique adaptation for both terrestrial and arboreal bipedal gathering that was unlike that of any extant species, including humans. The persistence of arm-hanging features in later hominids…suggests that this adaptation may have remained relatively unchanged, even in *Homo habilis*…until the evolution of a more locomotion-oriented, near-modern postcranial morphology in *Homo erectus*." (Hunt, 1994b).

Evidence Suggests the Early Hominin Habitat was an Open Forest/Woodland/Savanna Mosaic

At all early hominin sites published to date, associated fauna (Table 2) include both dry, open habitat taxa (elephants, giraffe, hyena) and taxa that are largely confined to more densely wooded, waterside or well-watered habitats (hippos, mustelids, colobus monkeys). Even colobus monkeys are range into open canopy woodlands. In short, there is no evidence early hominins were forest living, and much evidence that, compared to chimpanzees, they were a dry-habitat, woodland taxon. The tool-using *A. garhi* habitat may have been drier and more open than earlier early hominins. Although a finer-grained analysis that closely considers species frequencies may ultimately improve the resolution of our habitat description, at present the best we can say is that early hominins lived near relatively open forests of as yet undetermined density and canopy height, and that they also lived near more open woodland and savanna habitats. The faunal list is very similar to that of the dry-habitat chimpanzees at Semliki, Uganda (Hunt and McGrew, 2002).

Chimpanzees are Ascent-Minimizers

Hunt (1994a) found evidence that when budgeting energy expenditures, larger chimpanzees are more tightly constrained to minimize vertical climbing than are smaller individuals (Table 3). He used multiple regression to factor out social rank effects and showed that large chimpanzee males fed lower in the forest canopy, were found on the ground more often, utilized shorter-stature species of trees, and ascended significantly less frequently than did small males. This observation conforms to theoretical expectations that vertical ascents are energetically more costly for larger than for smaller animals. Since early hominins — like chimpanzees — were large compared to other primates, we expect that large male early hominins were particularly constrained to minimize ascents. Arboreal resources were more costly to acquire for males compared to females, due to energetic constraints, and were more dangerous to acquire, since the risk of falls was greater. Very large species are under some pressure to reduce climbing to the absolute minimum; that is, to become terrestrial full-time. If early hominins budgeted their energy expenditure as do chimpanzees, they climbed as little as they could, but as much as they had to.

Why Climb Into Trees, Then? A Lesson From The Chimpanzee

Mahale chimpanzees spent 61% of their active period on the ground (N=11,896). While it may be tempting to characterize chimpanzees as terrestrial primates, it would obscure the fact that they are utterly dependent on arboreal food resources. Despite the precariousness of movement and support among the terminal branches

Table 2

Closed Habitat Fauna						
	hippo	crocodile	turtle	mustelid	colobus monkey	wild boar[8]
Sahelanthropus[1]	X	X	X	otters	X	X
Orrorin[2]	X	X	X	otters	X	X
Ardipithecus[3]	X	X	X	otters	X	X
A. anamensis[4,t]	X	X	X	X	X	X
A. afarensis[6]	X	X	X	X	X	X
A. garhi[7]	X	X	X	X	—	—
Semliki	**X**	**X**	**X**	**X**	**X**	**(other suids)**

Dry Habitat Fauna						
	equid	elephant	giraffe	hyena	kob	baboon	rhino
Sahelanthropus[1]	X	X	X	X	X	—	-
Orrorin[2]	X	X	X	X	X	—	X
Ardipithecus[3]	-	X	X	X	—	X	-
A. anamensis[4]	X	X	X	X	X	X	X
A. afarensis[5]	X	X	X	X	X	X	X
A. garhi[6]	X	X	X	—	X	X	—
Semliki	**(historic)**	**X**	**(Holocene)**	**X**	**X**	**X**	**—**

Table 2 — Fauna associated with early hominins, compared to fauna sympatric with chimpanzees in a dry habitat (X = presence; — absence)

[1]Vignaud *et al.*, 2002
[2]Pickford and Senut, 2001
[3]WoldeGabriel *et al.*, 1994
[4]Leakey, M. *et al.*, 1998
[5]Macho *et al.*, 2003
[6]Johanson *et al.*, 1982
[7]de Heinzelin *et al.*, 1999
[8]*Nyanzachoerus*

Table 3

Large Males	Small Males
Ground	Above ground
Lower in canopy	High in canopy
Small trees	Large trees
Vertical climb less	Vertical climb more

Table 3 — Body size and chimpanzee habitat use (after Hunt 1994a)

Table 4

	Feature	Behavior Inferred	Inferred Stratum Use
1.	Dietary generalist trace element signal	Omnivory	Arboreal or terrestrial possible
2.	Molar, premolar area	Fruit, seed eating	Some arboreality required
3.	Molar morphology	Fruit, seed eating	Some arboreality required
4.	Enamel thickness	Fruit, seed eating	Some arboreality required
5.	Enamel microstructure	Hard foods, likely seeds	Arboreal or terrestrial possible
6.	Molar microwear	Fruit diet	Arboreal feeding
7.	Mandibular robusticity	Hard-husked fruit or seed diet	Indeterminant
8.	Incisor microwear	Folivorous diet	Arboreal or terrestrial possible
9.	Incisor size	Small fruit, seed or leaf diet	Arboreal feeding
10.	Large biceps, brachioradialis etc.	Elbow, arm extension	Climbing
11.	Large deltoid	Abduction	Suspensory feeding, climbing
12.	Torso shape	Arm-hanging posture	Arboreal armhanging
13.	Scapula shape, raised-arm set	Arm-hanging posture	Arboreal armhanging
14.	Curved fingers	Arm-hanging posture	Arboreal armhanging
15.	Short (vs. apes) fingers	Gripping small supports only	Indeterminant
16.	Robust fingers, powerful grip	Arm-hanging, vertical climbing	Arboreal armhanging
17.	Plantar set of ankle	Hindlimb suspension	Arboreal armhanging
18.	Long toes	Arboreal support gripping	Arboreal climbing, armhanging
19.	Curved toes	Arboreal support gripping	Arboreal armhanging
20.	Large calcaneus	Terrestrial bipedalism	Bipedal locomotion
21.	Long forelimbs, brachial index	Arm-hanging, fruit harvesting	Arboreal climbing, armhanging
22.	Large biceps, brachioradialis	Vertical climbing locomotion	Arboreal climbing
23.	Hamstring mechanical advantage	Vertical climbing locomotion	Arboreal climbing
24.	Flexible knee joint	Arboreal gripping	Arboreal armhanging, climbing
25.	Slightly divergent great toe	Climbing, gripping branches	Arboreal climbing
26.	Powerful great toe grip	Vertical climbing locomotion	Arboreal climbing
27.	Wide pelvis	Improves arboreal balance, decreases locomotor endurance	Arboreal feeding
28.	Short hindlimbs	Improves arboreal balance, decreases locomotor endurance	Arboreal feeding
29.	Reduced pelvic ligaments	Terrestrial locomotor endurance reduced in re *Homo*	Walking range small
30.	Habitat reconstruction	Savanna and woodland	Indeterminant

Table 4 — Early hominin morphological features and inferred behaviors

of trees, Mahale chimpanzees spent 7.1% of their active period there (N=840). They did so for one reason only: their food is there-Mahale chimpanzees fed 87.8% of time they were among terminal branches. Thirty-two percent of the Mahale chimpanzee active day was spent in the tree core (i.e., any part except terminal branches; N=3835), and 50% of that time was spent feeding. Feeding made up only 28% of their terrestrial activity. All told, 70% of chimpanzee activity in trees was feeding. Chimpanzees enter trees not because they prefer arboreality, but because their more preferred food, fruit, is found there (Hunt, 1998).

It is assumed that in more open habitats chimpanzees might spend more time on the ground. We have a convenient test, since it is more open at Gombe than Mahale (Collins and McGrew, 1985). Gombe spent *less* of their time on the ground than Mahale chimpanzees, 47.2% (N=3,056) versus 60.7% at Mahale. Nutritional demands do not decrease when the habitat is more open.[1]

Early Hominins Are Sexually Dimorphic

Sexual dimorphism of approximately 50% (f/m body mass) is suggested for *A. anamensis* (Ward *et al.*, 2001), a level somewhat greater than that posited by

McHenry (1992) for australopithecines as a whole. Recently Reno *et al.* (2003) are flimsy (Ramos and Hunt, *in prep.*) suggested sexual dimorphism in early hominids was not significantly different from that of modern humans. Their body mass dimorphism reconstruction was based on femoral head diameters estimated from other skeletal elements. Reference to the elements themselves yields a dimorphism estimate around 70%, or in line with previous estimates.

Early Hominin Ecomorphology

Reviewing early hominin morphology and inferred ecology, evidence that they were a semi-arboreal, frugivorous species is pervasive (Table 4). Associated fauna suggest their habitat included forest. As a primatologist, it is difficult to imagine such apes foregoing resources in the forest, whether they spent the majority of their time in more open areas or not. Baboons, for example, use both forest and savanna when the two are contiguous. Of 29 notable early hominin features (Table 4), ten are adaptations to arboreal suspensory behavior, eight are associated with vertical climbing, and eight more suggest at least some arboreal behavior. Two suggest both greater arboreal balance and reduced locomotor endurance. It is significant that among living apes, more arboreal species (*Hylobates* spp., orangutans) are more bipedal, albeit it in the trees, than more terrestrially adapted species (*Pan* spp., Gorilla), which characteristically knucklewalk on the ground (Hunt, 1991a). Terrestriality *per se* seems not to encourage bipedality.

Of 30 lines of evidence pertaining to early hominin diet and habitat use, 22 are consistent only with an arboreal lifeway. None suggests exclusive terrestriality, since only the very most open habitats, where woody plants grow to only two or three meters in height, could allow hominins to gather fruit without climbing trees. There is much to suggest early hominins were dependent on arboreal food resources, in particular ripe fruit, and that they spent considerable time in the trees, though almost certainly less time than extant chimpanzees or Asian apes.

In summary, early hominins were an ape unlike any living ape (Table 5), but they were not uniquely unique. Other than bipedalism, they share their most significant attributes with one ape or another. Early hominins were Great Ape sized (similar to *Pan* or *Pongo*), had a Great Ape level of sexually dimorphism (like *Gorilla* and *Pongo*), occupied a habitat at the very limit of ape dryness and openness (perhaps slightly outside the *Pan* range), spent more time on the ground than chimpanzees but less than mountain gorillas, ate fruit (like *Pan* or *Pongo*), piths (like *Pan* or *Gorilla*), hard objects such as nuts and seeds (like *Pongo*) and supplemented their diet with leaves, like all apes. Their claim to uniqueness would be that they utilized underground storage organs, as no ape does, and walked on the ground bipedally.

Sexual Dimorphism Implies A Hamadryas–Or Gorilla-Like Organization

Great sexual dimorphism among early hominins suggests high levels of male-male competition. If hominins were ripe fruit specialists, as are chimpanzee (Wrangham *et al.*, 1998) and orangutans (Leighton, 1993), as both referential modeling and dentofacial morphology suggest, they lacked strong female bonds (Wrangham, 1980, 1986). In the absence of female bonding, sexually dimorphic primates such as hamadryas baboons and gorillas have instead strong bonds between males and females. Orangutan- or gorilla-like sexual dimorphism implies that males defended breeding units relying not on strength in numbers as is the case in chimpanzees, with their male-bonded community social organization, but in groups small groups of two or three, and in the case of gorillas sometimes as a single individuals, where individual fighting skill rather than group coordination is important. The implied social system is similar to that of hamadryas baboons, where one-male breeding units coalesce into larger groups via bonds between several males. Females are either coerced into maintaining strict proximity to males (hamadryas), or are forced into proximity to protect infants from infanticidal extragroup males (gorillas). As the number of bonded males in a social group increases, successful defense depends on the strength of male bonds and an effective multi-male defensive strategy, rather than body mass, and sexual dimorphism is lower. A hamadryas or gorilla-like social system fits best with early hominins ecomorphology (Wrangham, 1986). In both hamadryas and gorillas breeding units travel as a rather tightly coordinated group, rather than in dispersed and fluid feeding parties, as are seen in chimpanzees. Single- or all-female groups seem precluded. All-male groups, however, are found in such societies, and are significant threats to infants and to the integrity of breeding units.

If early hominins had a hamadryas or gorilla-like social system, it would profoundly affect their foraging strategies. Rather than individual or small-group foraging parties, the entire social group must forage as a unit. In the gorilla female-choice society, females tend to limit copulation to a single male. In consequence, males have low sperm competition, small testes (Harcourt *et al.*, 1981) and high confidence in paternity. Males with high confidence in paternity can receive inclusive fitness benefits if they defer to females at feeding sites in order to increase female reproductive success. Such deference means that sex differences will be driven more by body size effects and reproductive needs effects than to by rank effects. Peripheral to arguments presented here, but possibly of interest, visible estrus (sexual swellings) is unlikely in a female-choice social organization, and copulation rates are expected to have been low.

Table 5

Species	Female[1] Mass (kg)	Sexual Dimorph.	Neocortical ratio[2]	Home Range (ha)	Habitat[1]	Aboreality	Diet[1]
Hylobates	6.6	96.6[1]	2.1	49[3]	For	100%[1]	Frt, Leaf
Pongo	35.7	45.7[1]	-	70[3]	Moist For	90%[1]	Frt, Leaf
G. g. gorilla	71.5	47.2[1]	2.7	2170[4]	For	30%[5]	Pith, Frt
Pan	41.4	75.4[6]	3.2	2150[3]	For, Wood	50-60%[7]	Frt, Leaf
Hominins	25-55[8]	50-75[8]	>3.2	?	Wood?	10-30%?	Frt? Seed?

Table 5 — Ecology and physical attributes of apes and early hominins
[1]Fleagle, 1999; Frt=Fruit,; For=Forest, Wood=Woodland; Dimorphism = female mass/male mass
[2]Dunbar, 1993
[3]Dunbar, 1992
[4]Tutin, 1996
[5]Estimated in Hunt, 2004
[6]Wrangham and Smuts, 1980
[7]Gombe and Mahale range; Hunt, 1989, Table 5.31
[8]McHenry, 1992; Ward *et al.*, 2001

Sex Differences in Chimpanzee Diet

In a review of chimpanzee diet data and presentation of new data Hunt and colleagues (Hunt, 1993; Hunt *et al.* in review) noted that female chimpanzees eat more invertebrates (mostly termites, some ants, a small proportion of unidentified invertebrates), more seeds, and more *Garcinia huillensis* fruit than males. These differences were largely due to rank-effects, since female selection of invertebrates and seeds paralleled rank differences among males. Females ate a wider variety of leaves, and ate them slightly more often than males (Goodall, 1986). Female selection of leaves was argued to be a result of reproductive demands, rather than rank effects, since high ranking males were shown to have eaten more leaves than low ranking males. Furthermore, Hunt *et al.* reviewed evidence that even in species where females were dominant to males, females tended to eat more leaves. Among three guenon species, *C. nictitans*, *C. pogonias* and *C. cephus,* females ate a higher proportion of protein-rich leaves when pregnant than otherwise (Gautier-Hion, 1980). Cook and Hunt (1998) suggested that this protein/calcium preference among females extends to humans.

Males ate more piths and stems (predominantly stems of *Pennisetum* grass), more meat, more *Dioscorea* spp. fruits, more *Harungana madagascarensis* fruit (a small, palatable fruit found in large patches), more *Psychotria peduncularis* fruit (a large-patch,1 one m tall shrub), and more miscellaneous fruit (*Cordia* spp. mostly). Hunt and colleagues showed that high ranking males ate more fruit and less pith than low ranking males, suggesting to them that females ate lesser amounts of fruit due to male-female competition. In support, they showed that when females were in parties with males the proportion of fruits in their diet dropped dramatically. Hunt's short-term observations were consistent with long-term records at Gombe that showed that males engaged in hunting dramatically more often than females. Males took the prey in 288 of 336 records (Goodall, 1986; McGrew, 1992). Among notable sex differences were that males were overwhelmingly more likely to take dangerous colobus monkeys, whereas females were biased toward juvenile and infant bush pigs and the least dangerous prey, bushbuck.

Sex Differences in Chimpanzee Positional Behavior

Hunt *et al.* found that female chimpanzees used a flex-legged sitting posture, engaged in unimanual suspension (arm-hanging), squatted and transferred (slow suspensory movement among terminal branches) more often than males. Hunt (1992) found that high ranking males monopolized larger, more stable perches, including large-based, hammock-like interwoven tangles of branches. Males knucklewalked and sat with legs extended significantly more often. Females had a more diverse positional repertoire than males, eleven positional modes constituting 90% of male positional behavior, versus fifteen modes for females.

Armhanging was found to be a fruit gathering and, to a lesser extent, leaf gathering posture; 88% of all armhanging was observed during fruit gathering (Hunt, 1989). Counter-intuitively, males ate more fruit than females, but armhung less. The cause for this incongruity is that males found a way to eat fruit when sitting, whereas females ate fruit while armhanging significantly more often than males. Males were significantly more likely to utilize armhanging when feeding on fruits in large trees with multiple feeding sites, whereas females tended to use armhanging when feeding in small fruit trees.

Sex Differences in Chimpanzee Canopy Use

Female chimpanzees are more arboreal than males (Doran, 1993; Hunt, 1993), and females fed and moved among terminal branches significantly more often, 21.5% of the time, versus 15.6% of the time for males (Hunt *et al.* in review). Whereas feeding males spent nearly half of their time on the ground (48.5%), feeding females spent only 35.9%, a significant difference.

Sex Differences in Chimpanzee Activity Budgets

Females spent more time feeding than males, more time resting, and less time grooming. Hunt *et al.* attributed this difference to male monopolization of low-handling time food items. Female travel was significantly more often in the context of moving between feeding sites, whereas male knucklewalking episodes ended more often in rejoining a social group.

Sex Differences in the Order Primates

In their review, Hunt *et al.* found that among species where females are higher ranking than males, females devote less of their activity budget to feeding than males. Among species where males are dominant, females tended to spend more time feeding. Four taxa did not conform to this pattern. In vervet, gelada, gorilla and orangutan, males allocated more of their time-budget to feeding than did females. In primate species in which sexual dimorphism was similar to that in early hominins, either males exhibited a selectivity for fruit, females showed a selectivity for leaf, or both.

Of 24 taxa where sex differences were reported, females showed a preference for flowers, leaves and invertebrates in 17 cases, classes of items that tend to contain high levels of protein and calcium. Males selected these items in only two species (vervet, capuchin). Among four primate species that eat meat (bonobo, baboon, capuchin and chimpanzee) males ate more meat; in no cases did females eat more meat.

Explaining Chimpanzee Sex Differences

There is a pattern to sex differences in chimpanzee diet and behavior (Hunt *et al.*, in review; Table 6). Fruits eaten in abundance by females were those found in small, isolated trees. Travel costs likely decrease energy return from such small patches. Many of these same fruits are small-diameter, thus requiring more whole-body movements and more picking motions per unit weight harvested. Invertebrates are small, are often concealed, and often require tools to harvest. They are a low-risk food item. Whereas they require little energy to harvest, compared to meat they have a low nutrient return per unit time. Blossoms and shoots must be picked individually and are therefore likewise a high handling-time item. Seeds must be gathered individually and opened individually. Often they require the use of tools. In short, female diet lists are rich in high-handling time food items.

In contrast, items eaten more often by males tended to be large fruits, items that could be harvested without climbing because the items are found on the ground or in low trees or bushes, and items found in large patches. Fruits, compared to most other dietary items, are calorie-rich. Some patches were large because the trees themselves were large, but others constituted a large patch because numerous smaller trees were found in dense stands. Meat is harvested in large packages, compared to invertebrates. A supporting phenomenon was observed by Goodall (1986), who reported that Gombe females ate more insects than males at most times of the year, but during the brief periods of greatest abundance, when the resource is large and concentrated, males ate them more often (Goodall, 1986: 258). Males, it seems, specialize on dietary items that have low handling times, either because the foods are in large packages, or because they are nutrient-dense, or because they are themselves large. Hunt *et al.* suggest that males spent less time feeding because they specialized on items that could be harvested quickly.

Despite their lower quality diet, females spent more time feeding and did it with more arboreal and acrobatic positional behaviors. They harvested foods arboreally, which requires greater energy expenditure because this demands ascents, greater effort to stabilize postures due to irregularly placed and unstable supports, and the need for challenging arboreal locomotion. Positional modes used significantly more often by females, armhanging and squatting, are more acrobatic than the predominant male positional mode, sitting. Armhanging is of particular interest, since it is a distinctively ape positional mode. The only obvious reason for females to work harder to get worse food is that they may be forced into such a regime by competition from males.

Leaves are also a high handling time item, since they must be picked individually and since chimpanzees appear to be very selective about which individual leaves are acceptable. Leaf eating therefore is slow going. Despite this, evidence suggests that female preference for leaves is due to reproductive demands. Females select leaves for their high density of protein and calcium.

Hunt *et al.* concluded that across the primates, females include more fruit in their diet when their social rank allows it, but that female preference for leaves is independent of social rank. Female diets resemble those of males more in species where males have a high confidence in paternity, and therefore might be deferring to females as a form of paternal investment. While male deference might reduce sex differences, males must maintain good condition to defend against interloper, infanticidal extragroup males.

Chimpanzees must also be under selection for paternal investment, since they defend a territory on which females depend, and their reproductive success in

Table 6

Male	Female
Large patches	Small, isolated patches
Terrestrial	Arboreal
Large items	Small items
Plant products with sugars, digestible hemicellulose	Plant products high in protein, calcium[1]
Few secondary compounds	Secondary compounds[2]

Table 6 — Sex differences in chimpanzee food item characteristics.

[1]Females ate more leaves, which have relatively high levels of protein (Gautier-Hion, 1980) and calcium (Leighton, 1993; Rogers *et al.*, 1990)

[2]Blossoms, seeds and leaf can contain high levels of secondary compounds (McKey *et al.*, 1981; Marks *et al.*, 1988)

Table 7

Item	Greater in:	Cause(s) in *Pan*	Inferred hominin sex difference[2]
Invertebrates	Female	Rank, Repro	Females eat more invertebrates
Seeds	Female	Rank, Repro	Females eat more seeds
Leaf	Female	Repro, Rank	Females eat more leaf
Small-patch fruit	Female	Rank	S.U., P.I.
Small diameter fruit	Female	Rank	S.U., P.I. (or Females > due to Size)
Piths	Male	Repro, Size	Males eat more pith due to Size
Large diameter fruit	Male	Rank, Repro	Males > (Size, Rank) (small P.I., S.U. effect)
Terrestrial fruit	Male	Rank, Repro, Size	Males > terrestrial fruits due to Size, Rank
Large-patch fruit	Male	Rank, Repro, Size	S.U.
All fruit	Male	Rank, Repro	Males eat more fruit due to Repro, Rank
Meat	Male	Rank, Repro, Size	Males due to Size, Repro, Rank
Sit (extended)	Male	Rank	P.I.
Walking speed	Male	Repro	S.U.
Knucklewalk	Male	Repro	S.U.
Terrestrial	Male	Rank, Repro, Size	Males more terrestrial due to Size, Rank
Stand	Female	Rank	Males stand more due to > terrestriality
Arm-hang	Female	Rank	Females armhang more due to > arboreality
Transferring	Female	Rank	Females transfer more due to > arboreality
Squat	Female	Rank	Females squat more due to > arboreality
Terminal branches	Female	Rank	Females use t.b. more due to > arboreality
Arboreality	Female	Rank	Females more arboreal due to Size
>Time spent feeding	Female	Rank, Repro	Possibly greater in females due to Repro
Rest	Female	Male Repro	P.I, S.U.
Groom	Male	Repro	Females greater; protection from infanticide
Travel between feeding patches	Female	Repro	S.U.

Table 7 — Statistically significant chimpanzee sex differences, inferred causes, and inferences for early hominins[1]

[1]After Hunt *et al.*, in review. **Rank** = rank effects, **Repro** = reproductive demands, **Size** = body size effects,

[2]**P.I.** = no differences or small differences inferred due to high paternal investment, **S.U.** = no differences inferred because groups are presumed to travel as a single, unfissionable unit.

almost entirely dependent on the fecundity of females within their community range. Chimpanzees differ from gorillas because they must compete with intragroup males as well as extragroup males. They are expected to behave somewhat more selfishly than gorilla males, gibbon males, or perhaps even orangutan males, all of which have societies where males have higher confidence in paternity.

A Conceptual Model for Early Hominin Foraging Patterns

Table 7 summarizes the conceptual components of an early hominin foraging model, including which of rank-effects, body size-effects, reproductive demands effects and paternal investment might drive early hominin sex differences. Early hominin social units, at least as reconstructed here, were cohesive groups quite unlike the fluid, fission-fusion grouping seen among chimpanzees. This means that chimpanzee sex differences that are allowed by or result from males traveling in all-male groups, from males and females traveling separately, or from females foraging alone would not be found in early hominins.

There are no grounds to suggest that early hominin reproductive demands differed from those of chimpanzees and other living primates. Accordingly, early hominins might be expected to conform to the trend in chimpanzees for female food lists to have more high-handling time food items. If so, early hominin females might be expected to eat more invertebrates, more seeds, and more arboreal food items, compared to males, due to rank effects. Females are expected to eat more leaves due to reproductive demands. Males, free from the demands of pregnancy and lactation, and possibly required to engage in vigorous group defense, might be expected to monopolize low-handling time, high nutrient density items like fruit. Higher social rank but also larger jaw gapes and greater strength would reinforce the tendency for males to eat larger fruits and other larger items.

The greater costs and risks entailed in vertical climbing are argued to lower the net value of arboreal food items for males, and to increase the value of terrestrial fruits and other terrestrially harvested foods, such as piths. This effect is theoretically multiplied by rank effects, which are also expected to press females to be more arboreal and males more terrestrial. That is, males have a doubled reason to select fruits that can be harvested terrestrially: terrestrial fruits would be more valuable to them due to greater ascent costs and climbing risks for arboreal fruits, and males might monopolize terrestrial feeding sites due to their higher rank, even if the items were more valuable to females. Although Hunt *et al.* found no significant sex difference in vertical climbing between male and female chimpanzees, it would have been dramatically greater among female early hominins. Body size differences, rank differences and defense imperatives together reinforce one

another to suggest profound differences in arboreality between the sexes. This, I argue, is the explanation for allometry in shoulder morphology among early hominins (Inouye and Shea, 1997); it is an adaptation to sharp differences in arboreality between the sexes.

This in turn predicts greater frequency among females of squatting, arm-hanging, transferring, and other suspensory behaviors, those particularly among terminal branches. Arboreality would also provide females some protection from predators and, perhaps even more critically, some measure of protection from a sudden rush from an extragroup infanticidal male. Female arboreality would force larger and therefore less arboreally maneuverable males into a slower stalking strategy, leaving intragroup males time to come to the defense of mother and infant. Terrestriality would place intragroup males in an advantageous position to defend offspring and females from males or predators that must approach on the ground, assuming open habitats. Males might be expected to approach terrestrially even in closed habitats, as we know is the case with chimpanzees (Goodall, 1986).

Unencumbered by nursing and more capable of overcoming prey defense due to large body size, males might be expected to capture more meat than females. Meat-sharing as a paternal investment might be expected to reduce differences in consumption rates somewhat compared to chimpanzees, but even so males likely ate more meat.

These observations receive some support from the fossils. Where we can compare male and female skeletal elements, the female fossils have many more features associated with arboreality (Stern and Susman, 1983), suggesting that the trend for male terrestriality seen in chimpanzees was even more exaggerated in early hominids.

Sex Differences In Chimpanzee Tool Use

At the three sites (Gombe, Mahale, and Taï) where chimpanzee tool-using has been observed most frequently, female chimpanzees used tools more often than males. Boesch and Boesch (1981, 1984, 1990) observed that females cracked more nuts per minute, needed fewer blows to crack nuts, cracked more total nuts, and were more competent with heavier hammers (Boesch and Boesch, 1981, 1984, 1990). Males and females crack *Coula* nuts in equal proportions when the nuts are dry, and therefore easy to open, but when they are fresh and difficult to open, females opened nearly twice as many as did males (Boesch and Boesch, 1984). Cracking nuts in trees is a skill acquired relatively late, and which requires complicated coordination of two, three and four limbs. Females used tools in this challenging context over ten times more often than males (Table 8). Differences between males and females were greatest for *Panda* nuts, the most difficult nuts to crack. Panda nuts require both an adequate anvil and a stone hammer to open. Stones hammers are rare, which means nut-crackers must invest time in carrying hammers to anvils, and they must remember hammerstone locations. Females were 2.4 times as likely to engage in Panda cracking as males (Table 8). As a general expression of lesser male competence, males were also more likely to choose a tool that was inefficient for the task at hand. In short, the more difficult the technique, the more likely it was that females were the ones that did it.

McGrew (1979) found that females fished for termites three times as many hours as males (166.3 v. 50.8), had more than twice as many individual termiting bouts (372 v. 123), and spent 3 times as great a proportion of their active period termite fishing (4.3% v. 1.4%). Females were more likely to dip ants using an ant wand, a collecting regime that requires delicate two-handed coordination. Seventy-five percent of females were seen to ant-dip, but only 45% of males did so. Fecal samples confirmed a sex difference (56% of female samples contained insect parts, 27% of male samples). Similar observations were made at Taï (Boesch and Tomasello, 1998) and Mahale (Nishida and Uehara, 1980; Uehara, 1984). In addition to using tools to open nuts arboreally, females also harvested ants arboreally much more frequently than males(Nishida, 1973; Nishida and Hiraiwa, 1982).

Drinking tools are thought to be used more often by females than males (Sugiyama, 1995). Two cases of unusual tool use were engaged in by females, a tool used to prey on a squirrel (Huffman and Kalunde, 1993), and stepping sticks (Alp, 1997).

Males seem to use wooden probes to extract marrow (Boesch and Boesch, 1989), but since females eat meat less often than males, it is not clear that this is a meaningful sex difference.

Stones (66%) and other objects thrown at perceived threats are the only tools used more often by males than females, but it is a profoundly male behavior: all 44 throws mentioned by Goodall were by males (Goodall, 1964; McGrew 1981).

Of the tools commonly used by chimpanzees such as missiles, termite fishing tools, and nut-cracking tools, tools that could be called collecting tools were used by females between 1.3 and 11.3 times more often than males (Table 8). Missiles, in contrast, were used by males alone. Females use tools more often, in more different ways, in more difficult circumstances, and more innovatively than do males.

The Female Diet List Makes Tool Use Particularly Valuable to Females

Whether it is termiting tools, or hammers and anvils, females use tools with greater frequency, greater competence, and obtain more calories from their use. Hunt (1993) suggested that the reason for this difference is that sex differences in diets and foraging behavior serve to make tool-use relatively more valuable to females.

Termites, small fruits, shoots, blossoms, seeds and Panda nuts are quite different food items, but each requires a considerable time- and/or energy- investment before payoff, compared to items preferred by males. The male diet list, in comparison, contains items that have low handling times. Large fruits have a high volume per surface area, which means that gathering and opening individual fruits yields greater mass per item. Larger items require fewer harvesting motions per unit weight, which decreases both gathering motions and between-feeding-site locomotion. As a source of protein, meat is a calorically dense (compared to leaves) and large(compared to insects). Fruits are calorically dense, and therefore require less harvesting investment than lower-quality items. Foods found in large patches (e.g., *Harungana*, grass stems) require lowered travel investment and allow increased harvesting rates. Terrestrial fruits require lesser investment in ascents, arboreal movement and balance during harvesting.

The more strenuous foraging regime and lower quality diet in females means they must work harder but end up with what is, by most measures, a worse diet. The male diet list means that males are, compared to females, more effective time minimizers. Females, by virtue of their lower feeding rates, are in a position to benefit tremendously from tool use, whereas male food handling times are already relatively low. When females reduce feeding time, they reduce energy expenditures and free up time to reallocate on care of offspring and productive foraging.

Females benefit from tool use more than males because it allows them to compensate for male monopolization of food items such as meat by using tools to

Table 8

| | Coula[1] | | Panda[1] | Termites[2] | Missiles[2] |
	ground	tree			
Female	336	68	92	372	0
Male	255	6	37	123	44
F/M	1.3	11.3	2.4	3.0	—

Table 8 — Sex differences in chimpanzee tool use, pooled data

[1]Boesch and Boesch, 1984
[2]McGrew, 1979

add nutritionally similar items such as nuts and termites to their diet. Since meat is preferable to termites, males have little motivation to use tools to harvest invertebrates or crack nuts.

In addition to these rank effects, body size effects are expected. Greater body mass and therefore greater strength means that some food items that females cannot open or process with the teeth or hands can be processed by males without tools.

Females appear to have compensated for their lesser strength and for being excluded from preferred foods and preferred feeding sites by increasing the time they allocate to foraging, and by using positional modes (e.g. arm-hanging and transferring) that allow them access to less desirable feeding sites where there was little competition from males. Tool-use partly ameliorates this disadvantage for females.

Reproductive demands also have a role in shifting the balance toward female food items. McGrew (1992) identified reproductive demands as determining lower female hunting rates. Females are less free than males to make large, short-term energy or time investments which would put them in poor condition, whereas males are freer to engage in bouts of intense activity that result in short-term energy deficits, followed by dramatic, short-term increases in feeding and resting budgets. Since hunting is often unsuccessful and usually dangerous, it is a food-getting strategy that has two risks: the risk of investing energy in the hunt without a return, and the risk of injury. Regarding the former, chimpanzees hunt less often when their core food, fruit, is in low supply, and hunt more when fruit availability is high, when they can afford to fail (Mitani and Watts, 2001). Nutrient demands of pregnancy, lactation and infant carriage are more constant than male reproductive demands. Foods that can be located more consistently,

even if they are poor quality, are more valuable to females.

In short, female chimpanzees use tools more than males because they receive a disproportionate benefit when they lower food handling-times, gain access to items unavailable without tools, or avoid dangerous food-getting practices such as hunting.

Connecting the Dots: Inferred Sex Differences in Early Hominin Tool Use and Diet

The thrust of much of McGrew's (1992) innovative reasoning concerning chimpanzee tool use is that in order to use chimpanzees as an effective model we must be thoughtful to separate what chimpanzees do from what they are capable of doing. They use termiting tools, missiles, pry bars, seat sticks and tooth probes. We know from the lab that bonobo chimpanzees *can* produce a sharp edge from a cobble by flaking (Toth *et al.*, 1993; Schick *et al.*, 1999), and they can make effective use of carrying devices. Most presume that they do not manufacture stone tools in the wild because they are incapable of doing so. McGrew and his colleagues (e.g., Wynn and McGrew, 1989) make an implicit argument that chimpanzees do not make cutting tools because they would not increase net nutrient intakes. In support, in Gabon where stones are common chimpanzees do not use them to crack nuts (McGrew *et al.*, 1997). In other words, there may be jobs stone tools can do for chimpanzees, but the costs of using them make them impractical. In order to use tools full-time, chimpanzees would have to cache and retrieve them, carry them throughout their daily foraging regime, or discover (or remember) them just as they are needed. Because chimpanzees feed arboreally, carrying would necessitate constant short-term caching and retrieval. Arboreal walking and leap-

Table 9

Task	Sex of human forager	Sex of *Pan* forager
Hunting large fauna	Male	Male
Woodworking	Male	Females termite tools are sometimes wood
Fowling	Male	Either (Goodall, 1986: 262-3, 293)
Stoneworking	Male	Females use stones to open nuts
Bone/horn/shell working	Male	—
Mining and quarrying	Male	—
Bone-setting and other surgery	Male	Females use medicinal plants more
Butchering	Male	Male?
Honey-collecting	Male	~Equal (Goodall, 1986: 255)
Gathering small aquatic fauna	Male	—
Gathering vegetal foods	Female	Female
Preparing vegetal foods	Female	Female

Table 9 — Modern human and chimpanzee sex differences

ing would interfere with stone tool carriage. Early hominins faced these and other influences and constraints.

Table 9 lists tasks for which there are sex differences in modern humans, and identifies where similar sex differences have been observed in chimpanzees. We might have great confidence that when tasks have the same pattern of sex differences in humans and chimpanzees, early hominins had similar differences. Conceptual modeling can fine tune these extrapolations.

Among both humans and chimps (Table 9), males hunt the larger fauna. The mass of the prey, however, differs over an order of magnitude. Chimpanzee hunt prey up to perhaps 15-20 kg, which they can dismember without stone tools. Stone tools for butchery would have no utility unless prey size were exceed the size for which brute strength could dismember the carcass. Chimpanzees are sympatric with prey that is presumably attractive to them. For instance, they eat bushpig piglets, but they ignore adult bushpigs (Goodall, 1986). This suggests that the reason chimpanzees do not take larger prey is not because they cannot butcher them, but because they cannot kill them. As has long been speculated (Brace, 1968), the likeliest first hominin tool was a stout, hand held stick which could be used to dig up underground storage organs, fend off predators, deter aggressive conspecifics, and dispatch largish prey as a club or spear. I argue that chimpanzees do not carry such a tool not because they cannot manufacture the tool, but for the same reason they abandon nutcracking hammers, even though they are rare and valuable. Carrying a stick would hamper arboreal activity, and if left on the ground there is a retrieval cost. Chimpanzees conceive of detaching smaller plant material to make termite and ant fishing tools. A spear/digging stick/club is conceptually similar, only larger. Chimpanzees are enormously powerful and easily break one and two centimeter branches when making nests. Breaking off a five centimeter sapling is possible for male chimpanzees, and presumably early hominins. Such a stick, spear or club might last for years, so a significant investment to shape it with the hands and teeth might pay off.

Males, I argue, would find stick carriage less costly than would females, since they were hypothetically more terrestrial and therefore would bear fewer caching and retrieval costs. Males would also find spear use more beneficial, since they were more likely to encounter prey and predators, and in a better position to dispatch them since they were not burdened by carried infants.

Among chimpanzees stoneworking (Table 9) is unknown. Perhaps stone tool use is a less sophisticated precursor for stoneworking: stone tools is mostly a female affair. Females chimpanzees are more deft, better able to choose a correct tool, and better able to master tool use that requires sequential tasks (Boesch and Boesch, 1981, 1984; work reviewed in McGrew, 1992). This suggests that stone tools for shaping and sharpening spears were likely a male tool. In the absence of carrying devices, presumably but not certainly out of the early hominin cognitive realm, carrying raw materials and flakes would be costly and possibly impractical. Males might have engaged in opportunistic woodworking, discarding or abandoning flakes after spear-sharpening. Because caching and retrieving tools is time consuming and energy expensive, tool carriage is still expensive, even for males. When stone tools began to be carried rather than used opportunistically, tool carriage of a single, all-purpose tool that could serve both woodworking and butchering duty seems more likely than a large, diverse toolkit. Carrying devices, of course, would change that balance.

Whereas chimpanzee males range independent of females and offspring, and might therefore increase range even further to increase encounter rates with prey, male early hominins hypothetically traveled in unfissionable gorilla-like units. Butchering is a male activity among humans, but this may be due in part to the absence of females in the early stages of processing, in addition to rank-effects. For proto-tool-users, prey were most likely to be encountered by the group, and likely encountered on the ground. We know that early prey were large enough (de Heinzelin *et al.*, 1999) that butchering tools would have had some utility. Among early hominins, males were likely the hunters, and so had first access to meat. If similar to chimpanzees, males monopolized the larger, easier to process body parts. Males were large enough to manage some dismembering tasks manually, and in any case would have been in a position to command the larger, time-minimizing portions. Females were more likely to increase access and decrease handling times by using tools to process body parts males deferred to them. Butchery tools, then, were more likely to have been female tools. Scrapers used for extracting the last bits of nutritional residue from items seem to fit clearly within a female tool-use strategy. Marrow is an embedded food much like nuts, more time-intensive to harvest, and therefore likely to have been a female food. Given large body (and jaw) size, males could have processed some bones without tools. For tasks for which bone, horn and shell would be more available or better suited, the same trends should apply.

Among chimpanzees both males and females take nestlings, whereas among humans males are more likely hunters. Use of nets, traps and snares placed at dispersed sites and therefore requiring long distance travel to monitor may account for human sex differences. With no sex difference in fowling among chimpanzees, and the expectation that male and female early hominins did not differ in day-range, we are left with only arboreality as a factor. Birds harvested terrestrially (e.g. guinea fowl) are more likely to be encountered and killed by males, birds that are encountered arboreally would be likely preyed upon by females. Since chimpanzees kill and dismember birds without tools, and nets and traps were unlikely, this resource likely affected tool use little.

Mining and quarrying (Table 9) requiring strength or large body size would seem to be a male early hominin task. Since most early hominin tools were made of cobbles, rather than mined raw materials, this task was probably insignificant.

Among chimpanzees, females used medicinal plants more often than did males (Wrangham and Goodall, 1987). Whereas bone setting and other surgery is quite different from collecting medicines, if there is any such thing as early hominin medicine, it seems unlikely that females would be more common practitioners.

Among humans, honey collecting is likely engaged in by males because it is risky. Since early hominins were more competent arborealists, sex differences were probably less pronounced.

"Gathering" of small aquatic fauna fits most closely with female gathering habits (McGrew, 1981), but its male bias among living humans leaves the issue undecided. Gathering and preparing vegetal foods was likely a female pursuit among early hominins, as it is among both humans and chimpanzees, and as fits with a conceptual model of females specializing on more highly processed and less desirable food items.

If early hominins harvested termites, as seems likely (Sponheimer and Lee-Thorpe, 1999; Backwell and d'Errico, 2001), a chimpanzee analogy suggests that termite fishing would be a female activity.

Female chimpanzees and female humans each process and gather more plant material than males. Tools used for such activities are much more likely to be female tools. Again, it seems unlikely that female proto-tool users would have persistent tool carriers, given the demands of arboreal harvesting, but could have used tools opportunistically, or used them in confined areas where plant resources are persistent and tools could be cached without great travel cost.

Tools used for any arboreal activity whatsoever seem clearly female tools, both from analogy with chimpanzees, and reconstructed canopy use. Tools used to strip bark or score trees to gather exudates would be female tools.

Experimental evidence (Toth, 1985) suggests that toolkits might differ as in Table 10. In cases where raw materials are carried, males are more likely carriers, since they are larger, more terrestrial, and in less danger when encumbered. Some tools are ambiguous. Hammerstones might be used as missiles. If so, the pronounced tendency for male chimpanzees to throw missiles in defense or in intragroup aggression (McGrew pers. comm., 1992) suggest this as a possibility for male use, but inferences above suggest this tool would be in the hands of females most of the time. Cleavers may have been used (Toth, 1985) for hide slitting, an inferred female activity, but also for heavy duty butchery, possibly a male activity. Woodworking, if the inference that males are more likely to carry spears and to shape and sharpen them, would seem to be a male activity.

Table 10

Tool	Inferred use (Toth, 1985)	Sex
Chopper	Flake production	Either
Polyhedron	Flake production	Either
Bifacial discoid	Flake production	Either
Core scraper	Flake production	Either
Cleaver	Hide slitting, heavy duty butchery, woodworking	Either?
Acute chopper	Woodworking	Male
Large acute flake scraper	Heavy duty butchery, light woodworking	Male
Handaxe	Heavy duty butchery	Male
Pick	Heavy duty butchery, defense	Male
Unmodified stone	Missile	Male?
Small acute flake scraper	Hide scraping	Female
Steep flake scraper	Light woodwork, hide scraping	Female
Flake	Hide slitting, heavy and light butchery	Female
Hammerstone	Bone breaking, nut cracking	Female
Anvil	Bone breaking, nut cracking	Female

Table 10 — Early hominin tool use and manufacture sex differences

CONCLUSION

Early hominins are different enough from living apes that early prospects that they might be slotted into one or another ape socioecological niche have been disappointed. However, with fine-grained conceptual modeling, we can reconstruct more than might appear at first glance. Diet, social system, foraging strategies, and male-female differences are susceptible to reconstruction. We might have predicted that relatively small-brained early hominins would have exhibited few sex differences in foraging strategies. Among the inferences we can draw from analogy with living primates are:

- Dental microwear, tooth morphology, relative dental dimensions, facial morphology and trace-element analysis suggest that half of the early hominin diet was fruit, principally small-diameter, hard-husked or seedy fruit, and that leaves made up less of the diet than in living apes (but still perhaps as much as a quarter of the diet) and were an important fall-back food. It is also likely that seeds were the second most important dietary item (though not the second highest in proportion), that animal protein including both invertebrates and meat were an important part of the diet (though not clearly more important than among chimpanzees), and that grass stems and pithy items including underground storage organs were important.
- Early hominins were partly arboreal, collected small fruits using an arm-hanging bipedalism, slept in trees at night, and retreated to trees when threatened by predators.
- Faunal lists suggest early hominins were adapted to and lived in a rather open woodland habitat, not a closed canopy or forest habitat.
- Chimpanzees seek to minimize ascents, and early hominins likely did too, more so for large individuals — adult males — yielding a clear sex difference in arboreality.
- Chimpanzees climb trees and arm-hang when feeding on their most important food, fruit. Early hominins were likely the same.
- Early hominin females were 70% the body weight of males, suggesting that early hominins had a gorilla-like social system with no bonds between females, strong bonds between breeding males and females in their breeding group, and secondary bonds between males that allowed group-male defense against attacker males attempting to displace breeding males or kill infants.
- Female chimpanzees engage in more acrobatic and varied locomotor and postural modes and do so among smaller branches, compared to males. Early hominins were likely similar.
- Female chimpanzee diets differ from male diets principally due to social rank differences, but females preferentially pursue protein independent of rank. Early hominins were likely the same.

- Female chimpanzee tool use patterns suggests early hominins females used tools in a greater variety of circumstances, obtained more calories than males from their use, and used them with greater frequency.
- Early hominins likely used stone tools for tasks that had a low return, for processing smaller items, for tasks that required finer work, and for tasks that required smaller tools.
- Early hominin males were more likely to be the makers of large, heavy duty tools, and tools used to process high-return items.
- Early hominin males were likely to have used tools appropriate for first-access to carcasses, and for heavy-duty butchery tasks.
- Early hominin females were more likely to have used tools for fine butchery at the end of carcass processing, for hide slitting (bite sized pieces of hide might yield some last nutrients, if chewed long enough), for marrow harvesting, for scraping, for processing plant foods (e.g., pith harvesting, underground storage organ peeling), for nut-cracking, and for any arboreal food-getting activities.
- Early hominin males are more likely to have used tools for dispatching prey, and as missiles or clubs to deter predators or aggressive conspecifics.
- Even the extremely simple toolkit most expect among early hominins should be expected to have differed between the sexes.

Rather than sex differences having developed after stone tool use began, it seems most likely that sex differences evolved in the common ancestor of humans and chimpanzees, if not even earlier. Chimpanzee-like sex differences likely continued and were elaborated upon as the foraging regime in early hominins became more sophisticated and dependent on tools.

ACKNOWLEDGMENTS

I am grateful to the chimpanzees of Mahale and Gombe for their forbearance. In particular, I thank Chausiku, Ntologi, Pulin, Lubulungu, Aji, Fifi, Goblin, Spindle and Gremlin. Richard W. Wrangham, William C. McGrew, and C. Loring Brace engaged me in energetic discussions of many of the concepts elaborated here. I am particularly appreciative that WCM and RWW gave the manuscript a very detailed vetting. They corrected any number of embarrassing errors. My intransigence in some areas, despite their best efforts, means errors and misjudgments likely remain. I am profoundly indebted to Jane Goodall and Toshisada Nishida for their extraordinary generosity in allowing me to parasitize their lives' works in 1986-1987, and to RWW for badgering them until they did. I am grateful to the Mahale Mountains Wildlife Research Centre, the staff of the MMWRC, the Gombe Stream Research Centre, the village of Mugambo, and to the generous people of the United Republic of Tanzania for their hospitality during my Tanzanian field work. My editors,

Kathy Schick and Nicholas Toth made numerous helpful suggestions. Preliminary versions of this work were presented in talks to the Chicago Academy of Science, December 11, 1991, to the American Association of Physical Anthropology, April 15, 1993, to the Exploring African Prehistory Workshop, Indiana University, May 1, 1993, to the Society of Africanist Archaeologists (SAfA), 12th Biennial conference, Indiana University, April 29, 1994, and to the Recent Advances in Primate Locomotion Wenner-Gren symposium, University of California, Davis, March 27, 1995.

REFERENCES CITED

Alp, R. (1997). "Stepping-sticks" and "seat-sticks": new types of tools used by wild chimpanzees (*Pan troglodytes*) in Sierra Leone. *American Journal of Primatology* **41**, 45-52.

Anapol, F. and Lee, S. (1994). Morphological adaptation to diet in Platyrrhine primates. *American Journal of Physical Anthropology* **94**, 239-262.

Backwell, L. R., and d'Errico, F. (2001). Evidence of termite foraging by Swartkrans early hominids. *Proceedings of the National Academy of Sciences* **98**, 1358-1363.

Berger, L. R. and P. V. Tobias (1996). A chimpanzee-like tibia from Sterkfontein, South Africa and its implications for the interpretation of bipedalism in Australopithecus. *Journal of Human Evolution* **30**(4): 343-348.

Boesch C. and Tomasello, M. (1998). Chimpanzee and human cultures. *Current Anthropology* **39**, 591-614.

Boesch, C. and Boesch, H. (1989). Hunting behavior of wild chimpanzees in the Taï National Park. *American Journal of Physical Anthropology* **78**, 547-573.

Boesch, C., and Boesch, H. (1981). Sex differences in the use of natural hammers by wild chimpanzees: a preliminary report. *Journal of Human Evolution* **10**, 585-93.

Boesch, C., and Boesch, H. (1984). Possible causes of sex differences in the use of natural hammers by wild chimpanzees. *Journal of Human Evolution* **13**, 415-440.

Boesch, C. and Boesch, H. (1989). Hunting behavior of wild chimpanzees in the Taï National Park. *American Journal of Physical Anthropology* **78**: 547-573.

Boesch, C., and Boesch, H. (1990). Tool use and tool making in wild chimpanzees. *Folia Primatologica* **54**: 86-99.

Boesch, C. and Boesch-Achermann, H. (2000). *The Chimpanzees of the Taï Forest: Behavioural Ecology and Evolution.* Oxford: Oxford University Press.

Boubli, J.P. (1999). Feeding ecology of black-headed uacaris (*Cacajao melanocephalus melanocephalus*) in Pico da Neblina National Park, Brazil. *International Journal of Primatology* **20**, 719-749.

Brown, J.K. (1970). A note on division of labor by sex. *American Anthropologist* **72**, 1073-1078.

Cartmill, M., Pilbean, D. and Isaac, G. (1986). One hundred years of Paleoanthropology. *American Scientist* **74**, 410-420.

Chamberlain, A.T. and Wood, B.A. (1985). A reappraisal of variation in hominid mandibular corpus dimensions. *American Journal of Physical Anthropology* **66**, 399-405.

Clarke, R.J., and Tobias, P.V. (1995). Sterkfontein Member 2 foot bones of the oldest South African hominid. *Science* **269**, 521-524.

Collins, D.A. and McGrew, W.C. (1988). Habitats of three groups of chimpanzees (*Pan troglodytes*) in western Tanzania compared. *Journal of Human Evolution* **17**, 553-574.

Conklin-Brittain, N.L., Wrangham, R.W., Hunt, K.D. (1998.) Dietary response of chimpanzees and cercopithecines to seasonal variation in fruit abundance. II. Macronutrients. *International Journal of Primatology* **19**: 971-998.

Cook, D.C. and Hunt, K.D. (1998). Sex differences in trace elements: status or self-selection? In (A. Grauer and P.L. Stuart, Eds.) *Gender in Palaeopathological Perspective.* Cambridge: University Press, pp. 64-78.

Deloison, Y. (1991). Les Australopithèques marchaient-ils comme nous? In (Y. Coppens and B. Senut, Eds.) *Origine(s) de la Bipédie chez les Hominidés.* Paris: Cahiers de Paléoanthropologie, pp. 177-186

Doran, D.M. (1993). Sex differences in adult chimpanzee positional behavior: the influence of body size on locomotion and posture. *American Journal of Physical Anthropology* **91**, 99-116.

Dunbar, R.I.M. (1993). Coevolution of neocortical size, group size and language in humans. *Behavioral Brain Sci.* **16**: 681-735.

Eckhardt, R. B. (1981). Comment on variation in the subsistence activities of female and male pongids: new perspectives on the origins of hominid labor division. *Current Anthropology* **22**, 248.

Fleagle, J.G. (1999). *Primate Adaptation and Evolution,* 2nd Ed. San Diego: Academic Press

Foley, R.A. (1987). *Another Unique Species: Patterns in Human Evolutionary Ecology.* New York: Wiley.

Gautier-Hion, A. (1980). Seasonal variations of diet related to species and sex in a community of *Cercopithecus* monkeys. *Journal of Animal Ecololgy* **49**, 237-69.

Goodall, J. (1964). Tool-using and aimed throwing in a community of free-living chimpanzeees. *Nature* **201**, 1264-1266.

Goodall, J. (1986). *The Chimpanzees of Gombe: Patterns of Behavior.* Cambridge: Harvard University Press, .

Goodall, J. van Lawick. (1968). The behavior of free-living chimpanzees in the Gombe Stream Reserve. *Animal Behaviour Monographs* **1**(3):165-311.

Gough, K. 1975. The origin of the family. In (D.H. Spain, Ed.) *The Human Experience.* Homewood, Illinois: Dorsey Press.

Grine, F.E. and Kay, R.F. (1988). Early hominid diets from quantitative image analysis of dental microwear. *Nature* **333**: 65-68.

Harcourt, A.H., Harvey, P.H., Larson, S.G., and Short, R.V. (1981). Testis weight, body weight and breeding system in primates. *Nature* **293**, 55-57.

de Heinzelin, J., Clark, J.D., White, T., Hart, W., Renne, P., WoldeGabriel, G. , Beyene, Y. , and Vrba. E. (1999). Environment and behavior of 2.5-million-year-old Bouri hominids. *Science* **284**, 625-629.

Hiraiwa-Hasegawa, M. (1997). Development of sex differences in nonhuman primates. In (M.E. Morbeck, A. Galloway and A.L. Zihlman, Eds.) *The Evolving Female: A Life-History Perspective.* Princeton: Princeton University Press, pp. 69-75

Huffman, M.A. and Kalunde, M. S. (1993). Tool-assisted predation on a squirrel by a female chimpanzee in the Mahale Mountains, Tanzania. *Primates* **34**, 93-98.

Hunt, K.D. (1989). *Positional Behavior in* Pan troglodytes *at the Mahale Mountains and Gombe Stream National Parks, Tanzania.* Ann Arbor: University Microfilms.

Hunt, K.D. (1991a). Positional behavior in the Hominoidea. *International Journal of Primatology*, **12**(2): 95-118.

Hunt, K.D. (1991b). Mechanical implications of chimpanzee positional behavior. *American Journal of Physical Anthropology* **86**, 521-536.

Hunt, K.D. (1992a). Positional behavior of *Pan troglodytes* in the Mahale Mountains and Gombe Stream National Parks, Tanzania. *American Journal of Physical Anthropology* **87**, 83-107.

Hunt, K.D. (1992b). Social rank and body weight as determinants of positional behavior in *Pan troglodytes*. *Primates* **33**, 347-357.

Hunt, K.D. (1993). Sex differences in chimpanzee foraging strategies: implications for the australopithecine toolkit. *American Journal of Physical Anthropology*, Supplement **16**, 112.

Hunt, K.D. (1994a). Body size effects on vertical climbing among chimpanzees. *International Journal of Primatology* **15**: 855-865.

Hunt, K.D. (1994b). The evolution of human bipedality: ecology and functional morphology. *Journal of Human Evolution* **26**, 183-202.

Hunt, K.D. (1998). Ecological morphology of *Australopithecus afarensis*: traveling terrestrially, eating arboreally. In (E. Strasser, J.G. Fleagle, H.M. McHenry and A. Rosenberger, Eds.) *Primate Locomotion: Recent Advances*. New York: Plenum, pp. 397-418.

Hunt, K.D. (2004). The special demands of Great Ape locomotion and posture. In: *Evolutionary Origin of Great Ape Intelligence*, A.E. Russon and D. Begun (eds.). Cambridge University Press, pp. 629-699.

Hunt, K.D. and W. C. McGrew, W.C. (2002). Chimpanzees in dry habitats at Mount Assirik, Senegal and at the Semliki-Toro Wildlife Reserve, Uganda. In (C. Boesch, G. Hohmann and L.F. Marchant, Eds.) *Behavioural Diversity in Chimpanzees and Bonobos*. Cambridge: Cambridge University Press, pp. 35-51.

Hunt, K.D., Nishida, T. and Wrangham, R.W. (in review). Sex differences in the chimpanzee positional behavior, activity budget and diet: relative contributions of rank, reproductive demands, and body size. American Journal of Physical Anthropology.

Hunt, K.D. and Ramos, G.L. (in preparation). Sexual Dimorphism in Australopithecus afarensis: indistinguishable from that of modern humans?

Hurtado, A.M. and Hill, K.R. (1990). Seasonality in a foraging society: variation in diet, work effort, fertility, and sexual division of labor among the Hiwi of Venezuela. *Journal of Anthropological Research* **46**, 293-347.

Hurtado, A.M., Hawkes, K., Hill, K.R. and Kaplan, H. (1985). Female subsistence strategies among Ache Hunter-gatherers of Eastern Paraguay. *Human Ecology* **13**, 1-28.

Hylander, W.L. (1975). Incisor size and diet in anthropoids with special reference to Cercopithecidae. *Science* **189**, 1095-1098.

Inouye S.E. and Shea B.T. (1997). What's your angle? Size correction and bar-glenoid orientation in ''Lucy'' (AL 288-1). *International Journal of Primatology* **18**, 629-650.

Jungers, W.L. (1982). Lucy's limbs: skeletal allometry and locomotion in Australopithecus afarensis. *Nature* **297**, 676-678.

Jungers, W.L. (1994). Ape and hominid limb length. *Nature* **369**, 194.

Kay, R.F. (1985). Dental evidence for the diet of Australopithecus. *Annual Review of Anthropology* **14**, 315-341.

Kay, R.F., and Grine, F.E. (1988). Tooth morphology, wear and diet in Australopithecus and Paranthropus from Southern Africa. In (F. Grine, Ed.) *Evolutionary History of the Robust Australopithecines*. Chicago: Aldine, pp. 427-447.

Kinzey, W.G. (1992). Dietary and dental adaptation in the pitheciinae. *American Journal of Physical Anthropology* **88**, 499-514.

Kummer, B. (1991). Biomechanical foundations of the development of human bipedalism. In (Y. Coppens and B. Senut, Eds.) *Origine(s) de la Bipédie chez les Hominidés*, Cahiers de Paléoanthropologie. Paris: CNRS, pp. 1-8.

Latimer, B.M. and Lovejoy, C.O. (1989). The calcaneus of *Australopithecus afarensis* and its implications for the evolution of bipedality. *American Journal of Physical Anthropology* **78**, 369-386.

Latimer, B.M. and Lovejoy, C.O. (1990). Hallucal tarsometatarsal joint in *Australopithecus afarensis*. *American Journal of Physical Anthropology* **82**, 125-133.

Leakey, M.G., C.S. Feibel, I. McDougall, and A. Walker. 1995. New four-million-year-old hominid species from Kanapoi and Allia Bay, Kenya. *Nature*, **376**: 565-571.

Leighton, M. (1993). Modeling dietary selectivity by Bornean orangutans: evidence for integration of multiple criteria in fruit selection. *International Journal of Primatology* **14**: 257-313.

Lucas, P.W., Corlett, R.T. and Luke, D.A. (1986). Patterns of post-canine tooth size in anthropoids. *Zeitschrift für Morphologisches Anthropologie*, **76**, 253-276.

Macho, G.A, Leakey, M. G. Williamson, D. K. and Y. Jiang, Y. (2003). Palaeoenvironmental reconstruction: evidence for seasonality at Allia Bay, Kenya, at 3.9 million years. *Palaeogeography, Palaeoclimatology, Palaeoecology*, **199**: 17-30.

Malinowski, B. (1913). *The Family among the Australian Aborigines: a Sociological Study*. London: University of London Press.

Marks, D.L., Swain, T., Goldstein, S., Richard, A., and Leighton, M. 1988. Chemical correlates of rhesus monkey food choice: the influence of hydrolyzable tannins. *Journal of Chemical Ecology* **14**: 213-235.

McGrew, W.C. (1979). Evolutionary implications of sex differences in chimpanzee predation and tool use. In (D.A. Hamburg and E.R. McCown, Eds.) *The Great Apes*. Menlo Park: Benjamin/Cummings, pp. 441-463

McGrew, W.C. (1981). The female chimpanzee as a human evolutionary prototype. In (F. Dahlberg, Ed.) *Woman the Gatherer*. New Haven: Yale University Press, pp. 35-73.

McGrew, W.C. (1992). *Chimpanzee Material Culture: Implications for Human Evolution*. Cambridge Univ. Press: Cambridge.

McHenry, H.M. (1984). Relative cheek tooth size in Australopithecus. *American Journal of Physical Anthropology.* **64**: 297-306.

McGrew, W.C., Ham, R M, White, L J T, Tutin, C E G, Fernandez, M. (1997). Why don't chimpanzees in Gabon crack nuts? *International Journal of Primatology* **18**, 353-374,

McHenry, H.M. (1985). Implications of postcanine megadontia for the origin of Homo. In: E. Delson, ed. *Ancestors: the Hard Evidence*, pp 178-183. New York: Alan R. Liss.

McHenry, H.M. (1992). Body size and proportions of early hominids. American *Journal of Physical Anthropology* **87**, 407-431.

McKey, D.B., Gartlan J.S., Waterman P.G., and Choo G.M. (1981). Food selection by black colobus monkeys (*Colobus satanas*) in relation to plant chemistry. *Biol. J. Linn. Soc.* **16**:115-146.

Mead, M. (1949). *Male and Female: A Study of the Sexes in the Changing World.* New York: William Morrow.

Mitani J.C. and Watts D.P. (2001). Why do chimpanzees hunt and share meat? *Animal Behaviour.* **61**: 915-924.

Murdock, G.P. and Provost, C. (1973). Factors in the division of labor by sex: A cross-cultural analysis. *Ethnology* **12**, 203-225.

Nishida, T. (1973). The ant-gathering behaviour by the use of tool among wild chimpanzees of the Mahale Mountains. *Journal of Human Evolution* **2**, 357-370.

Nishida, T., and Hiraiwa, M. (1982). Natural history of a tool-using behaviour by wild chimpanzees in feeding upon wood-boring ants. *J. Hum. Evol.* **11**:73-99.

Nishida, T. and S. Uehara. (1980). Chimpanzees, tools and termites: another example from Tanzania. *Currrent Anthropology* **21**, 671-672.

Parker, S., and Parker, H. (1979). The myth of male superiority: rise and demise. *American Anthropologist* **79**, 289-309.

Puech, P.-F. (1992). Microwear studies of early African hominid teeth. *Scanning Microscroscopy* **6**, 1083-1088.

Puech, P.-F., H. Albertini. (1984). Dental microwear and mechanisms in early hominids from Laetoli and Hadar. *American Journal of Physical Anthropology* **65**, 87-92.

Ramos, G. and Hunt, K.D. in prep. Sexual dimorphism in *Australopithecus afarensis* was similar to that of gorillas.

Reno P.L, Meindl R.S, Melanie A. McCollum M.A., and Lovejoy C.O. 2003. Sexual dimorphism in *Australopithecus afarensis* was similar to that of modern humans. *Proceedings of the National Academy of Science.* 100: 9404-9409

Robinson, J.T. (1972). *Early Hominid Posture and Locomotion.* Chicago: University of Chicago Press.

Rose, M.D. (1991). The process of bipedalization in hominids. In (Y. Coppens and B. Senut, Eds.) *Origine(s) de la Bipédie chez les Hominidés.* Paris: Éditions du Centre National de la Recherche Scientifique.

Ruff, C.B. (1987). Sexual dimorphism in human lower limb bone structure: relationship to subsistence strategy and sexual division of labor. *Journal of Human Evolution* **16**, 391-416.

Ryan, A.S. and Johanson, D.C. (1989). Anterior dental microwear in *Australopithecus afarensis*: comparisons with human and nonhuman primates. *Journal of Human Evolution* **18**: 235-268.

Sanday, P.R. (1973). Toward a theory of the status of women. *American Anthropologist* **75**, 1682-1700.

Schick K.D., Toth N., Garufi G., Savage-Rumbaugh E., Rumbaugh D. and Sevcik R. (1999). Continuing investigations into the stone tool-making and tool-using capabilities of a Bonobo (Pan paniscus). *Journal of Archaeological Sciences* **26**, 821-832.

Schmid, P (1983) Eine Reconstrucktion des skelettes von A.L. 288-1 (Hadar) und deren konsequenzen. *Folia Primatologica* **40**, 283-306.

Sept, J.M. (1992). Was there no place like home? A new perspective on early hominid archaeological sites from the mapping of chimpanzee nests. *Current Anthropolology* **33**: 187-207.

Sponheimer, M. and Lee-Thorp, J.A. (1999). Isotopic evidence for the diet of an early hominid, Australopithecus africanus. *Science* **283**, 368-370.

Stern, J T, Jr. and Susman, R.L. (1983). The locomotor anatomy of *Australopithecus afarensis*. American *Journal of Physical Anthropology* **60**, 279-317.

Sugiyama, Y. (1995). Drinking tools of wild chimpanzees at Boussou. *American Journal of Primatology* **37**, 263-269.

Tanner, N.M. and Zihlman, A.L. (1976). Women in evolution. Part 1: Innovation and selection in human origins. *Signs* **1**, 585-607.

Teaford M.F. and Ungar P.S. (2000). Diet and the evolution of the earliest human ancestors. *Proceedings of the National Academy of Science* 97,13506-13511.

Teaford, M.F. (1994). Dental microwear and dental function. *Evolutionary Anthropology* **3**, 17-30.

Teaford, M.F. and Walker, A. (1983). Dental microwear in adult and still-born guinea pigs (*Cavia porcellus*). *Archives of Oral Biology* **28**, 1077-1081.

Tooby, J. and DeVore, I. (1987). The reconstruction of hominid behavioral evolution through strategic modeling. In (W.G. Kinzey, Ed.) *The Evolution of Human Behavior: Primate Models.* Albany: State University of New York Press, pp. 183-237.

Toth N., Schick K.D., Savage-Rumbaugh E.S., Rose A. Sevcik, R.A. and Rumbaugh, D.M. (1993). *Pan* the tool-maker: investigations into the stone tool-making capabilities of a bonobo (*Pan paniscus*). *Journal of Archaeological Science* **20**, 81-91.

Toth, N. (1985). The Oldowan reassessed: a close look at early stone artifacts. *Journal of Archaeological Science* **12**, 101-120.

Tutin, C.E.G. (1996). Ranging and social structure of lowland gorillas in the Lopé Reserve, Gabon. In (W.C. McGrew, L.F. Marchant, and T. Nishida, Eds.), *Great Ape Societies.* Cambridge: Cambridge University Press, pp. 58-70.

Tuttle, R.H. (1981). Evolution of hominid bipedalism and prehensile capabilities. *Philosophical Transactions of the Royal Society* B **292**, 89-94.

Uehara, S. (1984). Sex differences in feeding on *Camponotus* ants among wild chimpanzees in the Mahale Mountains, Tanzania. *International Journal of Primatology* **5**, 389.

Walker, A.C. (1981). Diet and teeth — dietary hypotheses and human evolution. *Philosophical Transactions of the Royal Society* B **292**, 57-64

Ward, C.V., Leakey, M.D. and Walker, A.C. (2001). Morphology of *Australopithecus anamensis* from Kanapoi and Allia Bay, Kenya. J*ournal of Human Evolution* **41**, 255-368.

Wrangham, R. W. (1986). Evolution of social structure. In (B.B. Smuts, D.L. Cheney., R.M. Seyfarth, R.W. Wrangham, and T.T. Struhsaker, Eds.) *Primate Societies.* Chicago: University of Chicago Press, pp 282-296.

Wrangham, R.W. (1980). An ecological model of female-bonded primate groups. *Behaviour* **75**, 262-299.

Wrangham, R.W., and Goodall, J. (1987). Chimpanzee use of medicinal leaves. In (P.G. Heltne, and L.A. Marquardt, Eds.), *Understanding Chimpanzees.* Cambridge: Harvard University Press, pp. 22-37.

Wrangham R.W., Jones J.H., Laden G. Pilbeam, D., and Conklin-Brittain, N.L. (1999). The raw and the stolen - Cooking and the ecology of human origins. *Current Anthropology* **40**, 567-594.

Wynn, T. and McGrew, W.C. (1989). An ape's view of the Oldowan. *Man* **24**: 383-398.

Zihlman, A.L. (1978). Women and evolution. Part II: subsistence and social organization among early hominids. *Signs* **4** 4-19.

Zihlman, A.L. (1981). Women as shapers of human adaptation. In (F. Dahlberg, Ed.) *Woman the Gatherer.* New Haven: Yale University Press, pp. 75-120.

CHAPTER 9

OLDOWAN TOOLMAKING AND HOMININ BRAIN EVOLUTION:
THEORY AND RESEARCH USING POSITRON EMISSION TOMOGRAPHY (PET)

BY DIETRICH STOUT

ABSTRACT

Attempts to understand the paleopsychological and neuro-evolutionary significance of early stone tools have long suffered from a scarcity of hard evidence regarding the actual neural substrates of stone toolmaking skill. The Positron Emission Tomography (PET) pilot study of Stout *et al.* (2000), together with preliminary results from ongoing follow-up research, are beginning to readdress this problem by providing a new avenue of experimental inquiry for human origins researchers. In these studies, PET was used to identify the regions of the brain that display increased activity during simple Oldowan-style (Mode I) flake production. Although results are preliminary pending further analysis, robust evidence of activation in the primary sensorimotor cortices surrounding the central sulcus, in the visual cortices of the occipital lobe, and in the cerebellum has already been observed. These activations reveal the relatively intense visuomotor demands of stone knapping and highlight those regions of the brain that would have been the most likely targets of selection on knapping skill. Somewhat less definitive evidence of superior parietal activation further suggests that higher-level visual association and spatial cognition may also be involved. Available evidence does not indicate the recruitment of prefrontal planning and problem solving regions, nor show any clear overlap between toolmaking and language processing networks. Results from the PET research, although preliminary, are already relevant to numerous hypotheses concerning the cognitive and evolutionary implications of early stone tools.

INTRODUCTION

What role might early stone tools have played in the evolution of the human mind? This is an old question, but one of enduring interest. In the past, researchers have generally approached the issue by attempting to define the cognitive demands of stone tool manufacture. This has been done in the relatively casual or "common-sense" language of archaeologists (e.g. Belfer-Cohen & Goren-Inbar, 1994; Chase, 1991; Gowlett, 1984; Isaac, 1986; Karlin & Julien, 1994; Tobias, 1979) as well as through more explicit reference to psychological theory (Mithen, 1996; Parker & Gibson, 1979; Robson Brown, 1993; Wynn, 1989). Some workers have even attempted to identify the neuroanatomical foundations for tool-behavior, usually in order to demonstrate some direct co-evolutionary connection with language abilities (Calvin, 1993; Greenfield, 1991; Wilkins & Wakefield, 1995).

Despite the quality and quantity of consideration devoted to the issue, the link between tools and cognition in human evolution remains tentative and controversial. Part of the reason is a lack of direct evidence regarding the relationship between tool-behavior and brain function. What is needed is concrete evidence regarding the actual neurophysiological underpinnings of stone toolmaking skill. The technology of Positron Emission Tomography (PET), initially suggested as a tool for the study of stone tools and cognition by Toth & Schick (1993), and applied for the first time in research (Stout *et al.*, 2000) discussed below, provides the opportunity to collect just this kind of evidence.

HOMININ "PALEOPSYCHOLOGY"

Information about the workings of the brain and mind can come from two main sources: the study of neuroanatomy/neurophysiology and the observation of behavior. This is as true in human evolutionary studies as it is in neuroscience and psychology, although the available data and degree of experimental control in the former are obviously much more limited. Direct but greatly limited evidence of protohuman neuroanatomy is provided by endocasts of hominin cranial fossils (Falk, 1980; Holloway, 1995; Tobias, 1991) while more detailed but indirect evidence comes from comparative studies of modern primate brains (Gannon *et al.*, 1998; Preuss *et al.*, 1999; Semendeferi & Damasio, 2000). Observation of modern non-human primate behavior also provides an important comparative perspective (McGrew, 1992; Savage-Rumbaugh & Lewin, 1994; Rumbaugh *et al.*, 1996; Tomasello & Call, 1997). With respect to the "observation" of pre-modern behavior, it is the reconstructive work of Paleolithic archaeologists that provides the best source of data. The use of these behavioral data to explore pre-modern mental characteristics and capacities might, for lack of a better term, be called *hominin paleopsychology*. Stone artifacts are one major source of information in this challenging undertaking.

Holloway (1981a) refers to prehistoric stone tools as "fossilized behavior". Although far from ideal, durable stone artifacts do represent one of our best indicators of prehistoric behavior and cognition. This is due in part to practical issues of preservation and recovery, but also to the nature of chipped stone technology itself. At a theoretical level, tool-behavior rivals language as a hallmark of human cognition (Preston 1998). In fact, Schlanger (1994: 143) argues that "even if… we could actually observe a Palaeolithic band *in vivo*, it would be highly informative and rewarding to study their ubiquitous material actions and products". Added to such theoretical considerations is what Pigeot (1990) refers to as "the privileged nature of lithic technology" - the fact that each percussive act produces a distinct physical trace. Although great care is needed to avoid over-interpretation of archaeological sites or individual artifacts, stone tools do present a unique and valuable opportunity to investigate prehistoric cognition.

The Paleopsychology of Stone Tools

Archaeologists interested in early cognition commonly hold the view, expressed by Gowlett (1992: 341), that "striking a flake from a cobble - is relatively simple. To strike a sequence of flakes, in such a way that each one helps in the removal of others, demands more ability…as in control by the brain." This is consistent with a traditional emphasis on internal mental representation and explicit cognition as the defining characteristics of advanced, distinctly human intelligence. Correspondingly less emphasis is placed on "lower-level" processes such as perception and action.

Among human origins researchers, discussion tends to center on such concepts as the "imposition of arbitrary form" (Holloway, 1969) and the use of mental (Clark, 1996) or procedural (Gowlett, 1984) "templates" in tool production. These criteria are used to informally compare and evaluate the cognitive complexity of industrial complexes, as in the "opportunistic" Oldowan (Isaac, 1981) or the "more complicated and patterned" Acheulean (Schick & Toth, 1993). Although both informative and useful, the use of such "intuitive criteria" (Robson Brown, 1993) to evaluate lithic technologies obviously leaves many more specific questions unanswered. In order to achieve a more complete appreciation of the cognitive implications of stone tools, researchers have tended to borrow from one or another branch of psychological theory. Developmental psychology has generally been the most popular, including the constructivist developmental stages of Piaget and Inhelder (e.g. 1969) and elaborations of the nativist modularity first proposed by Fodor (1983).

Piaget

An early and influential application of Piagetian psychology to Paleolithic archaeology is that of Parker and Gibson (1979). These authors argue, not only that Piaget's developmental stages may be used to evaluate the cognitive sophistication of early toolmakers, but that "certain projective and Euclidean preconcepts…arose as adaptations for stone-tool manufacture" (p. 375). This conclusion is presented as part of a broader, recapitulationist, model of cognitive evolution that sees sensorimotor, symbolic, intuitive and linguistic capacities as primary and secondary adaptations to intelligent tool use. The scope of the model is not such that much attention is paid to specific tool types or industries; rather the attempt is made to understand the overall pattern of human cognitive evolution.

This general model has been reworked and elaborated over the years (e.g. Gibson, 1983; Parker & Milbrath, 1993), with greater emphasis being placed on social interaction, learning, and planning. Most recently it has been presented as a three stage explanation of cognitive evolution in ancestral hominoids, *Homo erectus*, and *Homo sapiens*, through selection on extractive foraging "apprenticeship," joint attention and declarative planning respectively (Parker & Mckinney, 1999).

In contrast to the sweeping theoretical work of Gibson, Parker and colleagues, Wynn (1989) applies Piagetian theory to a more detailed analysis of specific lithic evidence. In so doing, he expresses the view that "Selection for intelligence does not appear to have been closely tied with stone tools" (p. 98) and cautions that such tools can provide evidence only of minimum capacities. Nevertheless, he finds Piagetian theory useful in evaluating hominin intelligence, concluding that Oldowan technology displays evidence of preoperational intelligence like that of modern apes, whereas late

Acheulean hominins (c. 300,000 b.p.) had achieved fully modern operational intelligence. Early Acheulean intelligence is characterized as transitional between these.

Modularity

Of course, Piaget's developmental theory has itself been controversial. Fodor (1983) proposed that the mind is not entirely "constructed" during development, but also contains innate, domain-specific, input "modules". Although Fodor's modules were envisioned simply as input/output processors, with cognition as a kind of "black box," analogous concepts have been more generally applied by other researchers. Gardner (1983), for example, enumerated seven distinct "intelligences" in the human mind, while evolutionary psychologists (e.g. Cosmides & Tooby, 1994) see a multitude of specific cognitive adaptations. Chomsky (1972) and Pinker (1994) have taken a similarly "modular" view of language. Unlike the "domain-general" intelligence of the constructivists, it is a prediction of strict modularity that different modules should have discrete neural foundations.

Robson Brown (1993) applied this strict concept of modularity to the analysis of stone tools in much the same way as Wynn (1989) utilized Piagetian theory. Working with two Mode I assemblages from Zhoukoudian, Robson Brown inferred and evaluated aspects of a "spatial intelligence", including mental rotation, element recognition, discrimination of oblique lines, and visual attention. From her analysis, she concluded that the Zhoukoudian toolmakers "displayed a cluster of cognitive operations for which no current analogue exists" (p. 243).

Mithen (1996), on the other hand, adopted the modified modularity of Karmiloff-Smith (1992) in his recent overview of human cognitive evolution. Karmiloff-Smith integrated the ideas of Piaget and Fodor by proposing that modules are constructed developmentally and ultimately integrated through a process of "representative-redescription". Mithen argues that human evolution recapitulated these stages, progressing from domain-general cognition to the possession of cognitively isolated "intelligences" and ultimately to "cognitive fluidity" between these intelligences.

With respect to technology specifically, Mithen contends that "technical intelligence" first began to develop with the Oldowan, but was limited to "a few micro-domains". This technical intelligence blossomed in the Acheulean and Middle Paleolithic, producing impressive stone-craftsmanship, but it was still limited by isolation from other domains. In particular, isolation from "natural history intelligence" prevented flexible utilization of alternative raw materials and the manufacture of special purpose or multi-component tools. Only with modern humans was full cognitive fluidity achieved, as reflected in diverse and specialized tool-kits.

Ecological Psychology

In 1979, J.J. Gibson published his classic book, *The Ecological Approach to Visual Perception*, in which he argued that perception arises directly from experience of the environment rather than indirectly via the construction of an internal mental representation. This ecological paradigm blurs or eliminates traditional boundaries between perception, action, subject and object, by defining visual perception as a dynamic activity of the unitary "organism-plus-environment" system. In this view perception is to be understood, not in terms of the information processing and representative capacities of the organism, but rather in terms of the possible relationships or "affordances" encompassed by the organism/environment system.

Together with the dynamic biomechanics of Bernstein (1967), Gibson's ecological perception theory forms the foundation of what is now known as *ecological psychology*, a theoretical approach defined by its focus on systems and dynamics rather than on organisms and structures. Although application of this perspective to the generally static and isolated evidence available to archaeologists and paleontologists can be difficult, ecological psychology has figured prominently in a number of theoretical and empirical contributions to the archaeology of human origins.

One example is the work of Roux *et al.* (1995) with modern stone-bead knappers in Khambhat, India. This "experimental field" research employed an ecological approach in order to explore the foundations of knapping skill, highlighting the importance of coordination and accuracy in the elementary percussive movement. As argued by the authors (p. 83): "This analysis suggests that the action plan depends to a large extent on the control of elementary movement. The less the control, the more difficult to organize an adequate succession of action." This view represents an interesting departure from more traditional approaches to stone tools and cognition, which tend to emphasize mental representation and planning rather than skill and execution.

Other influences of ecological psychology on human evolutionary studies have been more broadly theoretical in nature, having to do with the nature of mind and its relationship to tools, intelligence and language. Noble and Davidson (1996), for example, outline what they call the "social construct" story of the mind. In Noble and Davidson's own words (p. 105) "The 'social construct' story is that 'mental life' is an ongoing interpersonal activity. Far from 'mind' as a personal possession, it is better characterized as *socially distributed*" (emphasis original). Although these authors (p. 96-105) explicitly ground their concept of socially constructed cognition in the philosophical work of Wittgenstein, Ryle and Coulter, it nevertheless has much in common with the concept of "distributed" or "situated" cognition that has developed in cognitive and developmental psychology out of the work of Vygotsky (1976) and others (e.g. Hutchins, 1995; Lave & Wegner, 1988; Rogoff, 1984; Poon *et al.*, 1993).

The distributed cognition paradigm seeks to "move the boundary of the unit of cognitive analysis out beyond the skin of the individual" (Hutchins, 1995:287) in order to encompass the informational content and dynamics of the entire organism-plus-environment system. This includes other individuals, social contexts and artifacts. Although the philosophical stance adopted by Noble and Davidson (1996) leads them to equate "mindedness" exclusively with language (symbol use), and largely to dismiss Paleolithic stone tools as evidence of pre-modern cognition, tools figure prominently in more inclusive considerations of distributed and socially constructed cognition.

Tools may be viewed, not only as physical and perceptual extensions of the body (J.J. Gibson, 1979), but also as cognitive extensions of the mind. Hutchins (1995), for example, discusses the role of tools such as checklists and instrumentation in the distributed cognition that takes place on the bridge of a merchant-marine vessel during navigation. More mundane examples are supplied by Gatewood (1985), who describes the "spatial mnemonic" afforded fishermen by the physical organization of a salmon fishing boat, and by Graves (1994), who illustrates his theoretical discussion of tools and language with the observation that much of the information needed to acquire bike-riding skill is inherent in the design of the bicycle itself.

It is this more inclusive vision of distributed and socially constructed cognition that underlies Tomasello's (1999) theory of human cognitive origins. In contrast to Nobel & Davidson (1996), Tomasello stresses the importance of material as well as symbolic culture in establishing what he calls a "ratchet effect" in cognitive origins. Tomasello (p. 7) argues that a single "uniquely human social-cognitive adaptation" for understanding others as intentional agents forms the biological foundation for cumulative cultural evolution. It is the historical progress of this cultural evolution, rather than additional neurobiological evolution, which has produced the myriad other cognitive capacities commonly considered hallmarks of humanity. As Tomasello (p. 202) argues, "Developing children are…growing up in the midst of the very best tools and symbols their forbearers have invented…as children internalize these tools and symbols…they create in the process some powerful new forms of cognitive representation". Succeeding generations thus construct the cognitive niche in which their children develop, a niche that is saturated with distributed information and structure.

Tools and Language

In stark contrast to the ecological "social construct" views adopted by Noble & Davidson (1996) and Tomasello (1999) is the work of researchers using modular or neo-Piagetian theory to propose direct neurological and evolutionary links between tools and language (e.g. Greenfield, 1991; Calvin, 1993; Wilkins and Wakefield, 1995). As is evident from the work of Chomsky (1972), Pinker (1994) and others, the human language capacity is the single strongest candidate for characterization as an innate, biologically specified and content-rich mental module of the kind stipulated by evolutionary psychologists (e.g. Cosmides & Tooby, 1994). In addition to the universal "deep structure" of language hypothesized by Chomsky and the surprising ease with which children acquire language, the apparent localization of language function in the brain is considered to be a major piece of evidence in support of such a language module. According to the classic view, linguistic processing is localized in two areas: Broca's "motor speech" area and Wernicke's "grammatical comprehension" area. This view has engendered a number of hypotheses proposing that tool behavior spurred language evolution by contributing to the elaboration of these classic cortical language areas.

Greenfield (1991), for example, contends that a cortical region roughly equivalent to Broca's area underlies the hierarchical organization of both speech and object manipulation prior to modularization of these two capacities later in development. This leads her to "posit an evolutionary reconstruction in which tool use and manual protolanguage evolved together" (p. 547). Wilkins and Wakefield (1995) present the closely allied argument that the evolutionary emergence of Broca's area and the parieto-occipito-temporal junction corresponding to Wernicke's area occurred in order to support "motor programs dedicated to manual manipulation and throwing behavior" (p.172). Calvin (1993) on the other hand sees specialized "neural sequencing" regions in left prefrontal and premotor cortex as the common neural foundation for a range of behaviors including toolmaking, language, planning and aimed throwing.

In contrast to these modular (or neo-Piagetian in the case of Greenfield) hypotheses of language evolution is the Baldwinian argument of Deacon (1997). Citing the evolutionary theories of Mark Baldwin, Deacon proposes that the evolution of the human language capacity has actually been a process of co-evolution between language itself and the brains of those who use it. This co-evolutionary relationship arises because "learning and behavioral flexibility can play a role in amplifying and biasing natural selection because these abilities enable individuals to modify the context of natural selection that affects their future kin" (Deacon, 1997: 322). Although concerned with biological evolution rather than cultural-historical development, Deacon's "Baldwinian evolution" is very similar to the niche-constructing "ratchet effect" described by Tomasello (1999). In fact, Deacon's portrayal of language and brain evolution is complementary to that of Tomasello in many ways.

Deacon argues that the key adaptation supporting language is a generalized symbolic capacity supported by prefrontal cortex. In this view, language universals and the rapidity of language acquisition in children are thought to have more to do with the adaptation of languages to suit young minds rather than vice versa. Broca's and Wernicke's areas, instead of being anatomically distinct language centers, are simply motor and auditory association areas that present bottlenecks in information flow during language processing. These areas may have experienced evolutionary changes relating to language use and other behaviors, but they do not represent a separately evolved "language organ" in the brain. Just as Tomasello (1999) credits a generalized capacity for intersubjectivity for many of the more specific cognitive achievements of modern humans, Deacon sees an increased mnemonic and attentional capacity for learning as the fundamental adaptation supporting the specific and varied structures of human languages. This leads Deacon to view the potential evolutionary relationships between tools and language as being relatively indirect, and certainly not as involving the kind of exaptation or correlated evolution envisioned in modular and neo-Piagetian models.

Comment: Stone Tools and the Brain

Even in this brief review of literature dealing with stone tools and cognition, the diversity of everything from basic theoretical orientations to particular interpretations of empirical evidence is striking. Particular tool types might be impressive examples of planning (e.g. Gowlett, 1984) or accidental by-products (Noble & Davidson, 1996). Technical ability might be based on a terminal extension in the evolutionary recapitulation of cognitive development (Parker & McKinney, 1999), or a separately evolved mental module (Mithen, 1996) with discrete neural foundations (Robson Brown, 1993). Tools and language may be dissimilar in many ways (Chase, 1991; Wynn 1995, but might also be alternate expressions of fundamentally similar neural processes (Calvin 1993; Greenfield 1991). In the big picture, stone technology might be a mode of cultural-historic cognitive elaboration (Tomasello, 1999), a primary cause of biologically based cognitive evolution (Parker & Gibson, 1979), merely an indicator of such evolution (Wynn, 1989) or even basically irrelevant to the whole issue (Noble & Davidson, 1996).

Obviously no single research initiative will be able to resolve all of these wide ranging and deep-rooted controversies. It might, however, be argued that part of the reason for this pervasive disagreement is a lack of hard evidence regarding the relationship between tool-behavior and brain function. Exactly what demands does the manufacture of stone tools actually place on the central nervous system? Are these demands really similar to those of language processing? Is there evi-

dence for substantial recruitment of structures associated with spatial cognition and motor imagery, as might be expected by Wynn (1989) and Robson Brown (1993)? What about planning and executive centers underlying the putative "mental templates" of Gowlett (1984) and Clark (1996), or the "neural sequencers" of Calvin (1993)? Is tool manufacture supported by an anatomically discrete module as suggested by the work of Robson Brown (1993) and Greenfield (1991) or by generalized "information processing" in neocortical association areas (Gibson, 1993)? Empirically supported answers to these questions might go a long way toward resolving some of the more long-standing and contentious issues in the study of human cognitive evolution. Used appropriately, PET offers a valuable new opportunity to pursue such answers.

THE ROLE OF POSITRON EMMISION TOMOGRAPHY IN HUMAN EVOLUTIONARY STUDIES

Positron Emission Tomography (PET) is a radiological technique that may be used to produce images of physiological activity in the brains of living subjects (Posner & Raichle, 1994). Pioneered in the mid-70's (Ter-Pogossian *et al.*, 1975; Phelps *et al.*, 1975), PET uses the annihilation radiation produced by positron absorption to provide an image of the *in vivo* spatial distribution of a blood-born radionuclide tracer (Raichle, 1994). It is the decay of the radionuclide that makes imaging possible (Figure 1). Isotopes incorporated into the tracer molecules decay by emitting antimatter particles known as *positrons*. Each positron passes through the surrounding tissue until it encounters an electron, generally within 0.2 - 7 mm (Roland, 1993). The two particles annihilate each other, thereby generating two gamma rays traveling in exactly opposite directions. These rays are detected by a circular array of crystals surrounding the head of the subject. Data regarding the coincident arrival of gamma rays on opposite sides of this ring are used to reconstruct points of origin within the brain. Ultimately an image of the distribution of annihilation events (and thus of the tracer) within the brain is produced.

A variety of radionuclide tracers have been developed for use in PET and it is the kinetics of the particular tracer used that ultimately determine what is being imaged. One example is the radioisotope ^{15}O, which is injected in the form of $H_2^{15}O$ (water). During the relatively short half-life of ^{15}O, this water remains within the subject's blood vessels, and images collected reflect concentrations of blood within the brain. Because local blood flow is sensitive to the physiological action of surrounding neurons (Roland, 1993), such images may be used to identify patterns of functional activation.

In contrast, the tracer FDG (^{18}flouro-2-deoxyglucose) is a radioactively "tagged" glucose analog with a relatively long half-life. This tracer is actually absorbed

Figure 1

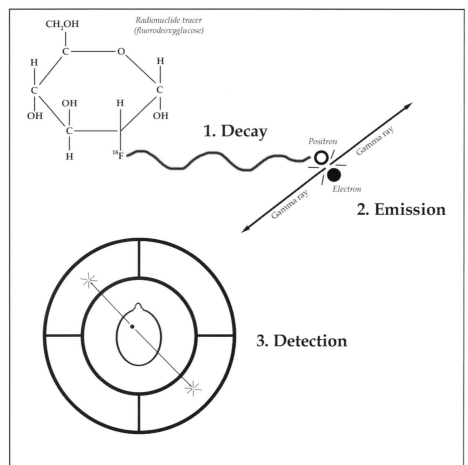

1. The mechanics of Positron Emission Tomography. 1) A radionuclide tracer injected into the bloodstream decays, producing an antimatter particle called a positron. 2) The positron collides with a nearby electron, annihilating both particles and producing two gamma rays that travel outward in opposite directions. 3) Gamma ray pairs are detected by a circular array of sensors surrounding the subject's head, and used to reconstruct points of origin within the brain.

by neurons, producing concentrations that reflect rates of glucose metabolism. Other tracers (*ligands*) have also been developed that have an affinity for specific receptor sites on neuronal membranes, allowing for the *in vivo* investigation of the distribution and activity of ion channels and neurotransmitter receptors (Roland, 1993).

The Meaning of PET Data

Although it is commonly described as a "functional brain imaging" technique, PET does not really produce images of brain function. As detailed above, what PET images actually show is the distribution and concentration of a tracer within the brain. The *interpretation* of brain function on the basis of this information requires consideration of specific tracer kinetics, their relation to neuronal activation, and the further relation of such activation to brain function.

Currently available radionuclide tracers track neuronal metabolism in impressively direct and sensitive ways. The previously mentioned glucose analog, FDG, behaves in the brain in the same way as glucose. Because glucose metabolism is the only major source of energy in the brain, and because neuronal activity regulates energy metabolism (Roland, 1993), FDG concentrations provide an indication of such activity. More specifically, FDG distribution traces the activity of the Na^+ pumps in neuron membranes. Na^+ pumps function to maintain the charge differential across the neural membrane (the *resting membrane potential*) that makes action potentials possible. Each time a neuron fires, restoration of the resting membrane potential creates work for the Na^+ pumps, requiring the metabolism of intracellular ATP that can only be replaced using blood-born glucose. For this reason, FDG accumulates preferentially in brain regions where more neuronal action potentials are occurring.

Another popular tracer is $H_2^{15}O$, which is carried along with the blood moving through the vessels of the brain. The distribution of $H_2^{15}O$ in the brain thus reflects the distribution of blood. Somewhat surprisingly, this gross measure turns out to be a sensitive indicator of neuronal metabolism (and thus activity). As reviewed by Roland (1993), there is a perfect correlation between regional Cerebral Blood Flow (rCBF) and the regional Cerebral Metabolic Rate of glucose (rCMRgl) in all cerebral structures in awake rats. This coupling is sensitive enough to detect variations in rCMRgl at the microscopic level of individual functional columns of neurons. The tight coupling between rCBF and rCMRgl means that the distribution of blood revealed by $H_2^{15}O$ PET images indicates variation in the frequency of neuronal actions potentials across brain regions.

Although FDG and $H_2^{15}O$ PET may be counted on to reveal regional variation in action potential frequency, there is more than one potential cause for such variation. A major confounding factor for researchers interested in task-related variation is the influence of anatomy, and particularly of neuron density. The more neurons that are packed into a region the greater the number of synapses between them and the higher the baseline frequency of action potentials. A particularly relevant example is the primary (striate) visual cortex of the occipital lobe, which is known to have a relatively high neuron density (Blinkov & Glezer, 1968). As expected, the resting rCMR of striate cortex is higher than that observed in surrounding areas (Roland, 1993) with lesser neuronal densities.

Because regional differences in neuroanatomy can lead to differences in baseline metabolism and blood flow, functional interpretations of PET images must always be made with reference to baseline conditions. In practice, images of tracer distribution resulting from the activity of interest are compared with images from a control condition to produce a "subtraction" image. It is the statistically significant differences between control and experimental conditions revealed by this subtraction that are interpreted. This technique effectively eliminates the constant background influences of regional anatomy and reveals neuronal activation that is specific to the experimental task.

From Neuronal Activation to Brain Function

Using the subtraction method, it is possible to identify changes in neuron activation (firing) that are specifically task-related. It should be stressed that PET image subtractions do not show the absolute demands of a particular task but rather show how it differs relative to a control task. In a typical experimental design, the control task replicates the experimental task as closely as possible, excepting only those narrowly defined aspects of behavior that are under investigation. In this way, maximum experimental control and resolution is achieved.

A major problem with this approach, however, is that it ignores the possible functional importance of the resting baseline state. In a recent review of functional imaging research, Gusnard & Raichle (2001) conclude that resting brains exhibit a stable physiological baseline state that corresponds to the continual or "tonic" exercise of a particular set of mental operations. Tonically active regions are located in posterior parietal and medial prefrontal cortex, and are thought by Gusnard and Raichle to support visuospatial, emotional and cognitive aspects of a unified and continuous self-concept. Although activity in these regions is commonly attenuated during absorbing, goal-oriented tasks, it is not yet possible to say what role baseline activity may play in supporting normal human performance. It is important to remember that when we subtract task-related control activity we may also be subtracting task-independent baseline activity that is more than just random background noise.

Even accepting this caveat, there is still the question of what neuronal activation actually means in terms of brain (and mental) function. This is a difficult question from both theoretical and philosophical standpoints. Although a majority of researchers adhere to a computational model of brain function in which neuron firings represent information in much the same way as bits in the binary circuits of a digital computer, others have proposed alternative computational mechanisms (Wallace, 1995) or even suggested that the entire computational paradigm is fundamentally flawed (Penrose, 1994). Similarly, "dynamic systems" theorists have criticized conventional concepts of functional localization and representation in the brain by characterizing the brain "as a dynamic collective…[that] works in a holistic, plastic, self-organizing fashion, with structural boundaries that are less fixed than previously thought" (Thelen & Smith, 1994: 131).

Leaving aside such deep issues as the nature of consciousness and its relationship to the brain, the fact remains that systematic relationships exist between particular behaviors and coincident patterns of brain activation. These relationships are often most easily described in functional terms. For example, the fact that tasks involving motor performance invariably activate the precentral gyrus (the motor strip) to a greater degree than those not involving motor performance is most easily explained by positing that this region plays an important role in motor praxis. This is not to say that such activation necessarily indicates the presence of a "motor program" or some other such static representation, but it does help to reveal the neural systems that act as a medium for motor behavior. This kind of information about the neural substrates of specific behaviors can be of great value to researchers interested in the evolution of the human brain.

Paleopsychological and Evolutionary Interpretations of PET

There are two major ways in which PET data may be applied to human evolutionary studies: (1) as a source of insight into the psychology of archaeologically visible behaviors like stone knapping, and (2) as evidence regarding the selective pressures imposed on the evolving brain by such behaviors. Like all actualistic research, these applications are based on analogy and must be justified using uniformitarian arguments (Foley, 1992).

Paleopsychological Interpretation

The paleopsychological interpretation of PET is based on analogy between the mental demands of an experimental task and those of a similar prehistoric behavior. The basis of this analogy is the argument that similar behaviors imply similar mental processes even though the size and organization of the neural substrate may vary. At the macroscopic level revealed by PET activation images, the "basic" functional organization of the human brain appears to be highly conserved, differing little from that of chimpanzees and macaques (Passingham, 1998). However, it is still possible that evolutionary changes in microscopic organization and/or overall functional capacity have somewhat altered the composition of the neural circuits underlying knapping behavior. Whether or not this is the case, the basic informational and mental demands of the knapping task itself should remain roughly comparable in kind and magnitude.

The critical question is how accurately our interpretation of modern human data characterizes these basic mental requirements. In conventional PET methodologies, researchers use well-understood tasks to explore the functional anatomy of the brain. The application of PET to paleopsychology inverts this methodology by using existing knowledge of functional neuroanatomy to explore the mental demands of poorly understood tasks. For example, if experimental stone knapping tasks were found to activate prefrontal cortex (PFC), we might conclude that knapping is relatively demanding of the planning and problem-solving behaviors associated with PFC. In this way, brain activation patterns observed in modern humans may be loosely interpreted as reflecting task-related mental behaviors like visual perception, motor coordination or strategic planning.

As we have seen, PET subtraction methods allow researchers to identify the unique demands of an experimental task as compared with a control condition. In the current context, this method may be expected to reveal any exceptional demands of Oldowan knapping in modern humans. The two major errors that might arise in applying these results to pre-modern hominins are (1) overestimating the minimal requirements of knapping due to "extraneous" activation in modern humans, and (2) underestimating minimal requirements by failing to account for modern human baseline brain activity.

The first of these two potential errors is the more easily discounted. As long as a reasonable control condition is used, the subtraction method should eliminate extraneous or background activation not related to the experimental task. The only way in which "extraneous" activation would survive the subtraction process would be if modern human subjects uniformly employed a particular inefficient or over-elaborate strategy during knapping but not during a closely related control task.

Ironically, the effectiveness of the subtraction method in preventing the overestimation of mental demands inevitably increases the possibility of underestimation. The use of subtraction to delete "extraneous" background activity equally well removes any baseline activity that is functionally significant. Gusnard & Raichle (2001) have argued that ongoing baseline activity in the resting human brain functions to maintain a stable and unified self-concept. This may include tonic activity in dorsal medial prefrontal cortex relating to "a

continuous 'simulation of behavior', 'inner rehearsal' and 'an optimization of cognitive and behavioral serial programs'" (p. 692). Ongoing rehearsal and optimization, although not specific to any one behavior, might nevertheless provide an important functional foundation for modern human performance, including stone knapping. Subtracting this baseline brain activity could lead to underestimation of the mental demands of stone knapping in modern humans and, by extension, in early hominin toolmakers.

In strict terms, this potential for underestimation means that PET evidence should always be treated as providing only a minimum indication of mental demands. It is nevertheless worth noting that severe underestimation is relatively unlikely. As reported by Gusnard & Raichle (2001), tonic baseline activity tends to be attenuated during goal-oriented activity. This suggests that, while baseline "mental continuity" may provide an important foundation for everyday activity, it is actually subordinated to more immediately relevant processes during focused activity. It is these immediately relevant processes that are revealed by PET activations. PET may not completely characterize a subject's mentality during task performance, but it can identify the most salient demands of a particular activity.

Evolutionary Interpretation

The second way to apply PET evidence to human evolutionary studies is by using it to identify the probable selective influences of particular prehistoric behaviors. Those areas of the brain most heavily recruited by modern humans performing a particular task are the ones most likely to have been subject to selection relating to that behavior during human evolution. Structures not recruited by modern humans may be considered much less likely to have been the direct focus of such selection. This logic is similar to that used by Marzke *et al.* (1998) in their attempt to identify skeletal indicators of habitual Oldowan toolmaking: those areas that experience the greatest physiological stress are the most likely to show adaptation.

A major challenge confronting the evolutionary interpretation of PET evidence is the fact that the basic processes through which brain evolution occurs remain poorly understood. Much research on this subject has been based upon the assumption that brain evolution occurs primarily through macrostructural changes in the absolute or relative size of the brain and its major anatomical components. More recently, however, evidence of adaptation at the microstructural level has been accumulating (Nimchinsky *et al.*, 1999; Preuss *et al.*, 1999; Buxhoeveden *et al.*, 2001). Despite this, the prevalence and importance of microstuctural adaptations in brain evolution remain unknown, as does their relationship with changes in regional brain volumes and/or overall brain size. Because it is not clearly understood how genetic inheritance, environment and developmental processes interact to produce adult brain structure on any of these levels, it is difficult to make a strong argument about the level(s) at which selection is most likely to act.

One important source of insight into macroscopic brain evolution is the study of modern brain size variation. The comparative methods of *evolutionary biology* (Harvey & Pagel, 1991) make it possible to interpret modern variation as evidence of evolutionary history. Unfortunately, students of mammalian brain evolution have been unable to agree even on the nature of modern variation, let alone on evolutionary explanations for it. In the past two years alone, four different groups of researchers have published four different characterizations of modern mammalian neuroanatomical variation, all using the same published set of brain volume data (from Stephan *et al.*, 1981).

At one extreme, Finlay *et al.* (2001) argue that the volume of major brain structures is rigidly covariant across species, showing little evidence of adaptation in individual structures or functional systems. They contend that the volumes of individual brain regions are constrained by a highly conserved order of neurogenesis, with late growing structures expanding disproportionately (but predictably) as the entire brain expands. As a result, they consider "coordinated enlargement of the entire non-olfactory brain" to be the most likely response to selection for "any behavioral ability" (Finlay & Darlington, 1995: 1578).

Barton & Harvey (2000), in contrast, see a mosaic of variation in modern mammalian brain structure, implying a similarly piecemeal history of evolutionary adaptation. These authors find evidence both of "grade shifts" in neocortex size (across insectivores, strepsirhine primates and haplorhine primates), and of independently correlated growth in specific functional systems. In direct opposition to Finlay & Darlington (1995) and Finlay *et al.* (2001), Barton & Harvey conclude (p. 1057) that "the cognitive and ecological significance of species differences in brain size should be evaluated by examining which neural systems in particular have been the target of selection."

Clark *et al.* (2001) adopt an intermediate position between these extremes, arguing that distinct, adaptive "cerebrotypes" are evident within mammals, but that these cerebrotypes are themselves conserved and "scalable" across 100-fold variations in absolute brain size. Although Clark *et al.* (2001) propose that their findings suggest a "reconciliation" between developmental (Finlay & Darlington, 1995) and adaptationist (Barton & Harvey, 2000) models, Barton (2002) has been critical of their conclusions.

Finally, there is the work of Winter & Oxnard (2001), which employs a "hypothesis-free multivariate morphometric approach" to conclude that, while mosaic brain adaptations are evident within mammalian orders, variation between orders "suggests an interplay of selection and constraints" (p. 710). Like Clark *et al.* (2001) and Barton & Harvey (2000), Winter & Oxnard

(2001) thus find evidence of some degree of mosaic evolution in brain organization. However, the evidence and patterns of mosaic evolution described by each of these publications are different.

It is beyond the scope of this chapter to attempt a resolution to the ongoing debate about patterns and processes in mammalian brain evolution. A central problem is that our incomplete understanding of brain genetics, development and function provides no *a priori* reason for preferring one statistical approach or scaling method over another. Without knowing exactly how knapping-related selective pressure on particular brain regions is likely to have affected brain structure (e.g. local expansion, correlated growth, microstructural adaptation), it remains for the current investigation to clarify the nature of that selection. For the time being, results may be interpreted in terms of the multiple possible evolutionary implications suggested by competing models of mammalian brain evolution.

THE PET RESEARCH

In February of 1997 a single-subject pilot study was performed in order to more concretely assess the utility of PET to human evolutionary studies. Results from this study (Stout *et al.*, 2000) not only confirmed the practicality and value of the technique, but also suggested specific hypotheses and methodological improvements for further research. These were incorporated into a six-subject follow-up study that is now in the data analysis stage.

In the pilot study an $H_2^{15}O$ water tracer was used in order to examine patterns of brain activation during simple, Mode I, flake production. The subject of the study, Nicholas Toth, is an experienced Paleolithic archaeologist and experimental flintknapper with over 20 years knapping experience. Results of the study should be viewed in light of the subject's prior experience and may not reflect the brain activation that would occur in a less experienced subject performing the same tasks.

Due to the relatively short half-life of ^{15}O, all experimental tasks were performed in the PET scanner, with emission data being collected during 2 ½ minute long trials. The resulting activation images represent time-averaged emission data collected over three minutes beginning with the initiation of task performance. Images were collected during three trials for each of three task conditions in the single experimental subject.

The three task conditions employed in the pilot study were (1) a control condition, (2) "mental imagery" and (3) "knapping". The purpose and rationale of control tasks in functional PET experiments has been described above; in this experiment the control task consisted of the subject holding and viewing a spherical cobble without any attempt to imagine or carry out knapping. The "mental imagery" task consisted of the subject holding a partially reduced core with both hands while visualizing the removal of flakes from

Table 1

Location number	Centroid location	Functional attribution	Side	Talairach Coordinates (x, y, z)	Volume (mm³)	Mean Z value
1	Superior parietal (Brodmann Area 7)	Dorsal "where" visual pathway	left	21, -49, 56	6,948	5.75
2			right	-30, -53, 61	1,948	5.16
3	Central sulcus (Brodmann Areas 1 and 4)	Primary motor and somatosensory processing	left	33, -26, 52	8,042	5.24
4	Postcentral gyrus (Brodmann Area 1)	Primary somatosensory processing	right	-39, -26, 56	5,889	5.10
5	Cerebellum (hemisphere)	Motor planning and initiation	left	10, -37, -18	1,002	5.22
6			right	-37, -51, -25	604	4.82
7	Cerebellum (vermis)	Motor Coordination	right	-3, -53, -9	1,082	5.07
8	Fusiform gyrus (Brodmann Area 37)	Ventral "what" visual pathway	right	-24, -53, -9	1,287	5.05

Table 1: Regions of differential activation observed during flake production in the pilot study (after Stout et al., 2000)

it. The "knapping" condition consisted of the subject actually striking the core with a hammerstone and removing flakes. For convenience, the subtraction data reported in Stout *et al.* (2000) from the comparison of the two experimental tasks with the control condition are re-presented here in Tables 1 and 2 and in Figures 2, 3, 5 and 6.

It should be stressed at the outset that the pilot study was "an heuristic, initial exploration" (Stout *et al.*, 2000) and involved only a single subject. The primary goal was to investigate the research potential of PET in human origins studies. This potential was firmly demonstrated by the collection of robust and clear-cut activation data. A secondary goal was to use the pilot study to further develop and refine research methods and hypotheses. This is best accomplished through interpretation of the results, even though definitive conclusions would be premature from a single-subject study.

As outlined in the preceding discussion, the proper interpretation of PET subtraction data requires careful consideration of the control task. In this pilot study, the control task was designed to control for visual stimulation by having the subject inspect a target very similar

to the cores that would be used in the experimental conditions. The objective was to identify brain activation associated with thinking about or acting on a cobble in a purposeful, technological fashion that was absent in unmotivated perception. Numerous studies cited by Gusnard & Raichle (2001) indicate that, outside the visual cortices, passive visual inspection is associated with typical "resting" or baseline activation patterns. In addition to visual stimulation, the control condition also involved a low level of bimanual motor activity involved in holding the cobble with two hands. Motor related activity in the subtraction images may thus be interpreted as indicating demands beyond those involved in the use of the two arms as a static support system for the target.

Knapping Activations

As may be seen in Table 1 and in Figures 2 & 3, the knapping task produced activation throughout broad volumes of the cerebral hemispheres and cerebellum. The most notable of these volumes is centered in the superior parietal lobule of the left cerebral hemisphere (Figure 2) and extends from the posterior parietal to the motor regions of the posterior frontal lobe, including the

Figure 2

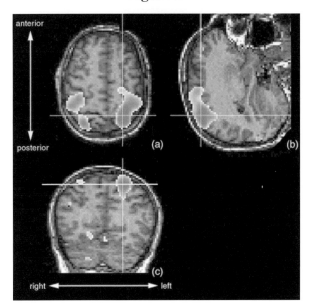

Figure 3

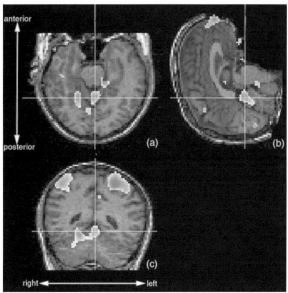

2. ***Parietal activation during Mode 1 stone knapping*** *(after Stout et al., 2000). Crosshairs indicate activation in the superior parietal lobule of the left hemisphere, as seen in transverse (a), saggital (b) and coronal (c) section. This region is commonly associated with spatial cognition, and is part of the dorsal "position and motion" visual pathway. Contiguous activation extends into the more anterior somatosensory and motor areas surrounding the central sulcus. A small volume of activation in the fusiform gyrus of the right inferior temporal lobe is also visible in (c).*

3. ***Cerebellar activation during Mode 1 knapping*** *(after Stout et al., 2000). Crosshairs indicate activation of the cerebellar vermis, as seen in transverse (a), saggital (b) and coronal (c) section. This structure is associated with muscle tone and modulation of movement execution. More lateral activation of the right cerebellar hemisphere, involved in the planning and initiation of movement, is also visible in (c).*

intervening somatosensory areas of the postcentral gyrus. A similar activation pattern is visible in the right hemisphere but it is of a lesser intensity and may be distinguished into two separate volumes in the images presented. Other regions of activation include the cerebellar hemispheres and vermis, as well as the fusiform gyrus (Brodmann's area 37) of the right inferior temporal lobe.

Posterior Frontal Lobe

The posterior frontal lobe consists of three "agranular" motor areas, the premotor cortex (PM), supplementary motor area (SMA), and primary motor cortex (M1), which are anatomically distinguished from more anterior association areas by their lack of an internal granular layer (layer IV) of cortex (Roland, 1993). Without attempting to make overly fine distinctions, the activation seen in this general area (especially in the left hemisphere) may be taken to reflect the greater motor information content of active knapping as compared with simply holding a cobble.

Superior Parietal Lobe

Perhaps the most interesting result of the pilot study is the strong bilateral activation of the superior parietal lobe seen during knapping. As described in Stout *et al.* (2000: 1220) "The superior parietal lobe consists of what is referred to as 'multi-modal association cortex' and is involved in the internal construction of a cohesive model of external space from diverse visual, tactile and proprioceptive input." In particular, the superior parietal lobe is known to contain diverse functional fields involved in primary and secondary somatosensory perception, "remote" somatosensory association and higher-order visual processing (Roland, 1993).

As shown in Figure 2, knapping produced elevated activation bilaterally in the primary somatosensory cortex of the postcentral gyrus and in the classic visual association cortex of the superior parietal lobule. In the left hemisphere, increased activation of the secondary somatosensory and association areas located between these regions was also apparent.

The somatosensory areas activated during knapping have previously been shown to undergo activation during vibration of the fingers (Fox & Applegate, 1988), tactile shape discrimination (Roland, 1985) and movements in extra personal space (Roland *et al.*, 1980). As with the previously discussed motor activity, the observed somatosensory activation is unsurprising in

Table 2

Location number	Centroid location	Functional attribution	Side	Talairach Coordinates (x, y, z)	Volume (mm³)	Mean Z value
1	Superior parietal (Brodmann Area 7)	Dorsal "where" visual pathway	left	21, -53, 56	1,766	5.61
2			right	-30, -55, 58	433	4.84
3	Inferior parietal (Brodmann Area 40)	Visualization, motor imagery	left	42, -35, 43	1,572	5.67
4			right	-51, -33, 50	570	5.15
5	Precentral gyrus (Brodmann Area 4)	Primary motor processing	left	33, -19, 54	1,834	5.19
6	Occipital lobe (Brodmann Area 19)	Secondary visual processing	left	28, -78, 2	558	5.25
7			right	-30, -78, 16	1,037	5.45
8	Fusiform gyrus (Brodmann Area 37)	Ventral "what" visual pathway	right	-28, -51, -9	421	5.22
9	Cerebellum (hemisphere)	Motor planning and initiation	right	-37, -44, -20	649	4.62

Table 2: Regions of differential activation observed during "mental imagery" in the pilot study(after Stout et al., 2000).

light of the increased physical activity associated with the knapping task.

Results from the pilot study thus suggest that knapping is relatively demanding of somatosensory and motor processing, particularly with in the left hemisphere/right hand system that was involved in percussive movements by the right-handed subject of this study. It would appear that the right hemisphere/left hand system responsible for supporting and orienting the core was less heavily recruited.

Visual components appear to be very important in the perceptual-motor dynamics underlying simple stone knapping. Both hemispheres show substantially increased activation of the posterior superior parietal lobule, a visual association area belonging to what is commonly thought of as the *dorsal stream* of visual processing. The existence of two streams of visual processing (dorsal and ventral) in the primate cerebral cortex was first proposed by Ungerleider & Mishkin (1982), who localized these streams to the posterior parietal and inferior temporal cortex respectively (Figure 4). According to the classic model, the dorsal "where"

stream is involved in the perception of location and motion while the ventral "what" stream is implicated in the perception of object characteristics like form and color. More recently, Milner & Goodale (1995) have suggested that the distinction between these streams is more accurately described as being between dorsal visuomotor control and ventral perceptual representation. In either case, the strong observed activation of the posterior parietal may be interpreted as reflecting the greater visual demands of active knapping as compared with passive inspection of a stationary target.

Inferior Temporal Lobe

A small but intriguing volume of activation is also visible in the fusiform gyrus on the medial aspect of posterior inferotemporal cortex (Figure 2). As noted above, the cortex of the posterior inferior temporal lobe is conventionally associated with the ventral stream of visual processing (Figure 4). Although the full extent of the inferotemporal visual association areas in humans is unknown (Roland, 1993), visual activation of the medial bank of the inferior temporal sulcus has been observed (Roland *et al.*, 1990). As reviewed by Bradshaw & Mattingley (1995), more anterior inferotemporal cortex has been linked with face recognition (Tovee & Cohen-Tovee, 1993) and with the integration of visual perception and memory in general (Miyashita, 1993).

As noted in Stout *et al.* (2000), care should be taken not to over-interpret a small activation volume observed in a single subject. Nevertheless, the activation does provide a tantalizing suggestion of ventral visual processing to complement the more robustly evident activation of the dorsal stream. Following the models of Ungerleider & Mishkin (1982) and Milner & Goodale (1995), such processing would be expected to involve the perception and/or representation of object characteristics. Increased demands for such perception during knapping might arise from the visual complexity of the partially reduced core combined with greater attention to technologically relevant morphology (e.g. edges, angles, ridges, depressions). It is unclear why fusiform activation is apparent only in the right hemisphere. This may simply be a reflection of marginal significance lev-

els in this single subject sample, or could possibly indicate preferential attention to the left visual field (the core was held in the left hand).

Cerebellum

A final region of activation observed in the knapping-minus-control subtraction is the cerebellum (Figure 3). This activation is unsurprising as the cerebellum has long been viewed as a center for the control of voluntary movement (Rolando, 1809; cited in Schmahmann, 1997b). In the current study, significant activation may be seen in both cerebellar hemispheres as well as in the medial cerebellar vermis.

The cerebellar vermis is a phylogenetically ancient structure that, together with the most medial parts of the cerebellar hemispheres, makes up the *spinocerebellum*. As summarized by Kandel *et al.* (1991), the spinocerebellum uses somatosensory, auditory, visual and vestibular feedback to control muscle tone and to help in movement execution by compensating for small variations or deviations. The observed activation of the spinocerebellum may be taken as an indication of the

Figure 4

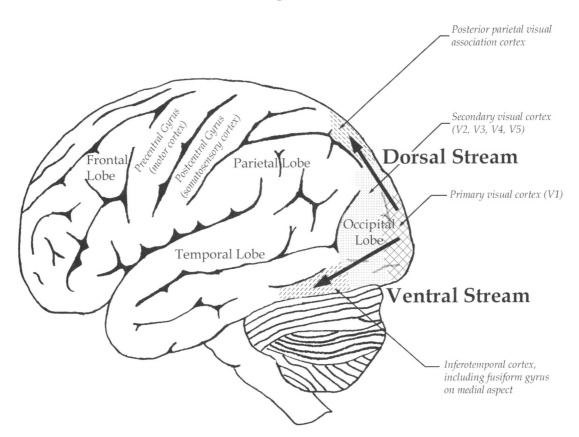

4. ***The two streams of cortical visual processing.*** *In the conventional view, cortical visual processing may be distinguished into two different functional streams: a dorsal "position and motion" stream and a ventral "form and color" stream. A more recent re-appraisal (Milner & Goodale, 1995) has suggested that these streams are better characterized as a dorsal "visuomotor control" system and a ventral "perceptual representation" system. The PET evidence suggests that the dorsal stream in particular is heavily activated during Oldowan-style knapping.*

demands for such control imposed by the knapping task.

The more lateral cerebellar hemispheres or *cerebrocerebellum*, are generally considered to participate in the planning and initiation of movement. The cerebrocerebellum is reciprocally connected with large areas of cortex, including premotor, motor, somatosensory and posterior parietal regions, and is thought to play an important role in the precise timing of complex multi-joint movements (Kandel *et al.*, 1991). Recent perspectives on the cerebrocerebellum have also tended to emphasize its role in cognitive and sensory functions (Leiner *et al.*, 1986; papers in Schmahmann, 1997a), leading to a more general characterization of the cerebellum as a "multipurpose learning machine which assists all kinds of neural control, autonomic, motor or mental" (Ito, 1993: 449). Cerebrocerebellar activation during knapping might reflect any of these functions, but is probably most closely tied to the control of knapping movements.

Mental Imagery Activations

The mental imagery task was incorporated in the pilot study for two reasons: (1) to provide a clear example of what intrinsic brain activity (i.e. abstract thinking) related to knapping might look like, and (2) to explore the possibility of obtaining useful imaging data while avoiding the methodological difficulties posed by vigorous physical tasks. The latter objective was inspired by research (review in Kosslyn *et al.*, 2001) indicating that mental imagery tasks recruit many of the same brain regions as do conventional perceptual and motor tasks. To the extent that this is true for stone knapping, imagery tasks might be used to explore knapping behaviors that are difficult to perform within the constraints of the scanning situation. Although the use of more slowly decaying tracers in research subsequent to the pilot study (see below) has largely obviated the need for this kind of methodological contortion, the results from the imagery-minus-control subtraction remain of theoretical interest.

The concept of internal mental representation has been a particularly important one in archaeological theorizing about the cognitive implications of stone tools. Specific examples include the mental templates of Clark (1996) and Gowlett (1984), the mental rotation of Robson Brown (1993), and the abstracted spatial cognition of Wynn (1989). The mental imagery task was designed as an artificially exaggerated case of such internal representation in a simple toolmaking task.

As expected, activation in the imagery-minus-control subtraction (Table 2, Figures 5 & 6) is quite similar to that seen in the knapping-minus-control subtraction (Table 1, Figures 2 & 3), although somewhat less intense/extensive. All of the major regions recruited during knapping were significantly activated in at least one hemisphere during the imagery condition, including superior parietal, pericentral (precentral gyrus), and inferotemporal (fusiform gyrus) cortex as well as the cerebellum. This is consistent with growing evidence that visuo-motor imagery relies, at least in part, on the same neural substrates as does visuo-motor action. Turning this around, it appears that actual knapping shares many of the mechanisms involved in generating an internal representation of the knapping task.

There is, however, some activation visible in the imagery-minus-control subtraction that is not evident in the knapping-minus-control subtraction. This consists of the bilateral activation of the anterior inferior parietal lobule and secondary visual processing areas in Brodmann's area 19 of the occipital lobe. Activation of these areas indicates processing demands/mechanisms of visuo-motor imagery that are not evident in actual knapping.

As reported in Stout *et al.* (2000), the anterior portion of the inferior parietal lobule has been described as a bi-modal visual and somatosensory association area (Roland, 1993). Siegel & Reed (1998) further describe the role of this region in integrating visual and occulo-motor information to provide a "head-centered representation of space". Activation of the inferior parietal in the mental imagery condition most likely reflects a greater salience of spatial information in supporting visuo-motor imagery as compared both with visual perception in the control condition and with visuo-motor performance in the knapping condition.

Activation of secondary visual areas in the occipital lobe probably reflects a similar emphasis on visual mechanisms in supporting mental imagery. Brodmann's area 19 is made up of the visual areas V3, V4 and V5 (Kandel *et al.*, 1991). The particular volume of area 19 that is activated in this study is located near the conjunction of occipital, parietal and temporal lobes, and may correspond to V5 (also known as MT). V5/MT is an area known for its response to the direction and speed of moving visual stimuli (Maunsell & Newsome, 1987), and is part of the dorsal stream of visual processing. Activation of V5/MT in the imagery condition suggests that visualization of knapping activity not only relies upon mental processes similar to those involved in actual performance, but can actually be more demanding of these processes.

Interpreting the Pilot Study

Although results from this single-subject pilot study require further corroboration, they do provide preliminary evidence regarding the system of brain structures that are recruited during Oldowan-style toolmaking by an experienced modern human knapper. The actual dynamic behavior of this distributed neural system may not visible in the static PET images collected (c.f. Segalowitz, 2000), but the time-averaged activation patterns nevertheless reveal key anatomical substrates. Together with environmental and somatic factors, it is the organization and functionality of these neural substrates that ultimately affords knapping activity.

Figure 5

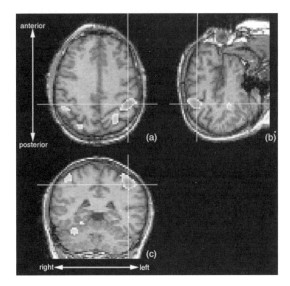

5. ***Inferior Parietal activation during Mental Imagery*** *(after Stout et al., 2000). Crosshairs indicate activation of the inferior parietal lobule, as seen in transverse (a), saggital (b) and coronal (c) section. This region is particularly important in creating internal representations of space. Its activation here reflects the greater representational demands of visualizing the knapping task as compared with its actual visuo-motor execution.*

Figure 6

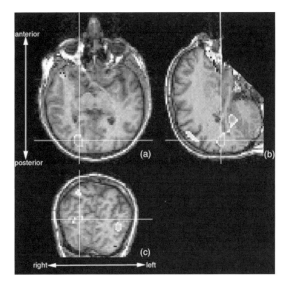

6. ***Occipital activation during Mental Imagery*** *(after Stout et al., 2000). Crosshairs indicate activation of secondary visual cortex in the right occipital lobe, as seen in transverse (a), saggital (b) and coronal (c) section. Corresponding activation of the left occipital is visible in (c). Secondary visual activation during mental imagery reflects the important role that even relatively low-level visual cortices can play in generating internal representations.*

As detailed above, the pericentral, posterior parietal, inferotemporal and cerebellar regions recruited during knapping are known to participate in the processing of motor, somatosensory and visual information. The co-activation of these regions during knapping supports the characterization of Oldowan toolmaking as a relatively demanding perceptual-motor activity that is heavily reliant upon visual guidance. The pilot study does not, however, provide any evidence of the recruitment of "higher order" prefrontal association cortex or classic language processing regions. Baseline activity is of course ongoing throughout the brain, but there is not yet any evidence that Oldowan knapping is particularly demanding of the cognitive processes supported by these regions. By conventional standards, Oldowan-style toolmaking does not appear to be a particularly "intellectual" pursuit.

On the other hand, it is important that the mental demands of skilled perceptual-motor performance not be underestimated. As argued by Reed & Bril (1996: 434), "the ability to construct, coordinate, and modulate movements regardless of the functional context of the organism is itself one of the most sophisticated achievements of human action systems". Artificial Intelligence (A.I.) researchers have similarly discovered the complexity of real-world perception and action. Chess-playing programs capable of competing at the highest levels have been around for decades, but A.I. researchers are still striving to produce computer programs capable of basic perceptual-motor behaviors in simulated "block worlds" consisting of nothing more than colored blocks on a table (e.g. Finney *et al.*, 2001). In the human brain itself, large regions of occipital, parietal, frontal and temporal cortex are involved in perceptual-motor activity, as are numerous subcortical structures (Figure 7). Although many of these regions are not involved exclusively in perceptual-motor processing, they nevertheless account for a large portion of total brain volume.

Recognizing that skilled perceptual-motor performance is itself a sophisticated mental achievement, the question remains as to whether the specific demands of Oldowan knapping are exceptional. Specifically, is Oldowan knapping more demanding than everyday perceptual-motor activities human ancestors may have engaged in prior to 2.5 Ma.? PET subtraction methods are ideally suited to answer this question though direct comparison of the brain activations associated with different task conditions.

In the pilot research presented here, Oldowan-style knapping was compared with a control condition consisting of the static support and visual inspection of a stone cobble. Knapping was found to produce activations in cerebellum, primary motor and somatosensory cortex and posterior regions of superior parietal and inferotemporal cortex. This indicates that knapping exerts greater perceptual-motor demands than the control condition. Particularly interesting is the activation of posterior parietal and inferotemporal cortex. These regions, although closely tied with visual processing, are dominated by intrinsic cortico-cortical connections rather than by the extrinsic input/output connections seen in primary sensorimotor cortex. Their recruitment suggests that simple Oldowan flake removal can involve relatively high-level, intrinsic perceptual-motor processing.

It is, however, important to remember that the control condition used in this pilot study was designed to produce little beyond resting baseline activity. It remains for the follow-up research to indicate whether the demands of Oldowan knapping on visual association cortices really are exceptional when compared with other perceptual-motor activities. Even within the pilot study, it is interesting to note that the mental imagery task produced additional activation of inferior parietal and occipital cortex not seen in the knapping condition. These regions, and particularly the multi-modal association cortex of the anterior part of the inferior parietal, are important substrates for visuo-motor imagery. The fact that they are not significantly activated during actual knapping suggests that mental imagery may not be particularly important part of Oldowan toolmaking. In future research, it will be interesting to see if the neural substrates of visuo-motor imagery play a greater role in more advanced lithic technologies.

WORK IN PROGRESS

By demonstrating the utility of PET in human evolutionary research, the pilot study of Stout *et al.* (2000) set the stage for further research. At the time of this writing, data collection for two follow-up studies has been completed and analysis is ongoing. These studies include a six-subject follow-up investigation of Mode 1 toolmaking and a single-subject, exploratory study of Mode 2 handaxe manufacture (Stout *et al.*, this volume). The experimental design and hypotheses from both studies have been heavily influenced by lessons learned in the pilot study.

Lessons Learned and Improvements Made

There are several important lessons to be learned from the pilot study. Most important is that it is possible to obtain clear and significant evidence of brain activation during stone tool manufacture using PET. Among other things, the strength of the results clears the way for the use of more refined control conditions.

Figure 7

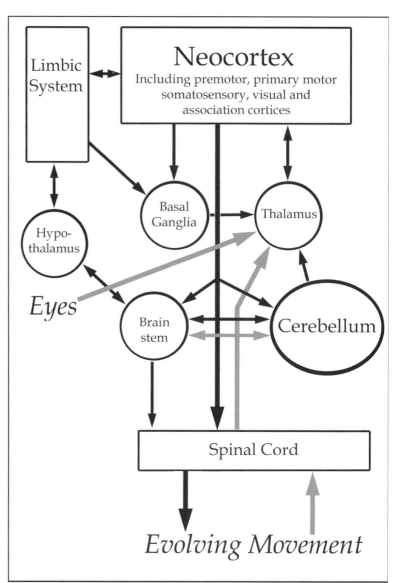

7. ***Neural structures and connections involved in visuomotor performance*** *(after Brooks, 1986). Note that nearly every part of the brain may become involved in motor performance in one way or another: from limbic motivational inputs, to retino-geniculo-cortical visual pathways and cortico-subcortical motor circuits including basal ganglia, thalamus and cerebellum. The question confronting human origins researchers is whether specifically knapping related demands on any of these structures are exceptional.*

Control Task

In the pilot study a relatively simple control task (holding and inspecting a spherical cobble) was employed. This was done in order to maximize the likelihood of obtaining significant results by maximizing the contrast between control and experimental conditions. The drawback of this approach is that it sacrifices specificity. As has been repeatedly stressed, PET subtraction data indicate relative rather than absolute neuronal activation. Results from the pilot study thus indicate that knapping is relatively more demanding than passive visual inspection (i.e. a resting baseline state) but do not provided evidence that it is more demanding than other everyday motor tasks such as reaching for or picking up an object.

In order to address this question, follow-up research employed a more active control task. This task, which consisted of striking two cobbles together without attempting to produce flakes, was designed to isolate any demands on motor accuracy, target perception, visuo-motor imagery or strategic planning that are exceptional to knapping as compared with more generic prehensile and percussive activities that human ancestors may have engaged in.

Tracer and Task Performance

Another major lesson of the pilot study was the difficulty of collecting activation data for a vigorous physical task like stone knapping. As outlined above, the relatively short half-life of ^{15}O requires that PET emission data be collected during actual task performance. This means that tasks must be performed lying prone on the scanner bed, an artificial condition that might conceivably affect numerous sensory and motor aspects of the task. In addition, the subject's head must remain motionless during imaging, which is an obvious problem when it comes to tasks like knapping that require vigorous movement. The movement problem was addressed in the pilot study by having the subject knapp using "approximately one-half normal force" (Stout *et al.*, 2000). Although flakes were nevertheless produced, this half-strength knapping task may not fully reflect the neural demands of the combined speed and accuracy employed in more naturalistic knapping.

In subsequent research (Stout, in prep.), this problem was resolved by switching to a different radionuclide tracer (^{18}flouro-2-deoxyglucose or "FDG") with a longer uptake and decay period. FDG is a glucose analog that is taken up from the blood stream by metabolically active cells (including neurons) for about 30 - 40 minutes following its injection. Tracer concentrations built up during this uptake period are subsequently stable, and may be detected using conventional PET imaging techniques. In practical terms, this means that experimental tasks may be conducted in naturalistic conditions outside the scanner. After 40 minutes of task performance, the subject is simply escorted to the scanner and images are collected. The drawbacks of using FDG are that only a single trial may be performed on a given day and that activation data reflect the average of activity over an even longer (40 minute) period than is the case with ^{15}O.

The use of FDG allowed several major improvements in experimental design. To begin with, subjects were able to perform tasks in a more natural posture: seated in a chair. This eliminates concerns about potentially anomalous visuo-motor and postural demands associated with knapping in a prone position. Even more importantly, subjects were able to engage in full force knapping, in some cases producing multiple generations of flake removals. This task condition not only captured the forceful and accurate motions of knapping, but also allowed for the unfolding of reduction sequences and any associated mental processes. As an additional benefit, artifacts produced during the knapping task were available for analysis as an independent measure of knapping skill. A final improvement allowed by the use of FDG was inclusion of raw material selection in the task (and control) condition. A cart supporting a collection of cobbles of various sizes, shapes and material compositions was placed next to the chair where subjects were seated. During the 40 minutes of task performance, subjects selected cores and hammerstones from this cart. In this way, it was possible to incorporate an additional aspect of knapping knowledge and decision-making.

Experience and Learning

The experimental subject in the pilot study was an experienced stone knapper. As outlined by Stout *et al.* (2000): "Patterns of brain activity observed in the subject were considered to represent those operative in one who has already learned the necessary skills for stone tool-making, and to provide a valuable baseline for future comparisons with less skilled or novice tool-makers." Multi-subject follow-up research has now provided the opportunity for such comparison.

Skill acquisition is thought to proceed through the streamlining or canalization of mental processing and the elimination of extraneous brain activation (Haier *et al.*, 1992). For this reason, it is reasonable to hypothesize that inexperienced knappers might display more intense and/or extensive brain activation than do experts. Such activation might be expected to reflect both the acquisition of task-specific knowledge and selection from multiple potential strategies. In fact, much of what is cognitively interesting about stone knapping is likely to occur in the process of learning rather than during expert performance.

In order to investigate learning, follow-up research was designed as a longitudinal investigation of stone knapping skill acquisition. Six subjects with no prior stone knapping experience were recruited and imaged during a novice trial in which they were asked to produce sharp stone flakes that would be "useful for cutting". Following the novice trial each subject participated in four hour-long uninstructed individual practice sessions. During these sessions, subjects were provided with a range of raw materials as well as pieces of vinyl and wood on which to test the efficacy of the flakes they produced. After completion of the practice regimen, subjects were brought back for imaging of an "experienced" trial. Although the relevant data are still being analyzed, it is hoped that this longitudinal design will help to reveal both the demands and dynamics of stone knapping skill acquisition.

Preliminary Results

Data analysis for the follow-up research described above is still ongoing, but it is nevertheless possible to present some preliminary results and interpretations. These are based on data from three subjects for the "experienced knapping"-minus-control subtraction. It is expected that more detailed analysis with the entire data set will reveal additional areas of activation, particularly in the novice condition. For the time being, only the most robust activations visible in the preliminary images will be considered.

Significant activations are visible in a number of brain regions, but at this stage may be most confidently identified in occipital and peri-central cortex and in the cerebellum. Occipital activation encompasses virtually the entire occipital lobe, certainly including the primary (striate) visual cortex (V1) surrounding the calcharine fissure and most likely extending into the secondary visual cortices (V2, V3, V4 and V5) in Brodmann's area 19. Pericentral activation is somewhat less extensive, but clearly encompasses primary motor cortex in the precentral gyrus and primary somatosensory cortex in the postcentral gyrus. Further analysis will be necessary to determine the extent to which more anterior secondary motor and more posterior sensory association cortices are activated. Observed cerebellar activation once again includes both vermis and hemispheres. It is most striking that, despite major differences in methodology (i.e. tracers, tasks, and subjects), both pilot and follow-up research implicate similar cortical and sub-cortical elements of the visuo-motor system.

The greatest apparent divergence between pilot and follow-up results is that the latter clearly show significant activation of primary visual cortex. The apparent demands of stone knapping on V1, which is generally considered to be a "low-level" visual input/output structure, are somewhat surprising given the similarity of basic visual environments in control and task conditions. Most likely this activation reflects the increased visual attention and acuity needed to guide forceful and accurate percussion. Although it is expected that further analysis of data from the follow-up study will reveal some additional activations outside primary motor and visual cortex, currently available results suggest that the most exceptional demands of Oldowan knapping are concentrated in more peripheral sensorimotor regions rather than in intrinsic association cortices.

SYNTHESIS

At the current stage of analysis, it remains most accurate to characterize Oldowan-style toolmaking as a perceptual-motor task that is particularly demanding of visual perception and guidance. No evidence has yet accrued for the exceptional involvement of the complex planning and associative capacities of prefrontal cortex, or of significant activation of Broca's and Wernicke's classic language processing regions. Nevertheless, it

does appear that simple Oldowan flake production by modern humans generates increased activation in sizeable volumes of cortex, even when compared to a fairly sophisticated bimanual visuo-motor control task. This observation has important implications for both paleopsychological and evolutionary interpretations of the Oldowan Industry.

Paleopsychological Interpretation

The past twenty years have produced a growing consensus that the manufacture of Oldowan artifacts is not a cognitively demanding process. Although the various "core tools" described by Leakey (1971) were originally thought to be the intended products of early toolmakers, Toth (1982; 1985) demonstrated that the vast majority could be explained as arising from least effort flake production. Toth (1982: 328) further characterized this technology as being "quite simple to replicate once bifacial flaking is mastered". The cognitive demotion of the Oldowan Industry has been carried to an extreme by Wynn & McGrew (1989), who argue that Oldowan technology is no more demanding than the tool-use of modern chimpanzees.

Many disagree with Wynn & McGrew, particularly when such issues as raw material selectivity and transport (Schick & Toth, 1993; Gowlett, 1996; Stiles, 1998) or the probable use of stone tools to make other tools (Mithen, 1996) are taken into account. When it comes to the narrowly defined cognitive demands of Oldowan knapping, however, even the most generous commentators attribute little beyond a basic "concept of form" (Gowlett, 1996) that allows the knapper to maintain the viability of a core during sequential flake removals (Mithen, 1996; Roche *et al.*, 1999). On the other hand, these authors and others (e.g. Semaw, 2000; Ambrose, 2001; Ludwig & Harris, 1998) emphasize the motor skills needed to reliably detach useful flakes. PET research with modern humans will not reveal whether or not Oldowan toolmaking is within the capabilities of modern apes (a question better addressed in the work of Schick *et al.*, 1999 and Toth *et al.*, chapter in this volume), but it can refine our understanding of the basic mental processes involved.

To date, PET research has supported the prevailing archaeological assessment of Oldowan toolmaking: that it is a demanding visuo-motor skill but does not call upon sophisticated internal representation, planning or problem solving. The real contribution of the PET data has been to provide concrete empirical support for this intuitive assessment. The experiments described here demonstrate that effective flaking is more neurally demanding than everyday tasks like grasping or striking objects, and that these exceptional demands are concentrated in the cortical visuo-motor regions and cerebellum.

The Representational Perspective

Having identified the functional neuroanatomy underlying Oldowan knapping, it remains to provide a (paleo)psychological interpretation. As we have seen, archaeologists interested in early hominin intelligence commonly subscribe to a representational view of mind. This paradigm defines mentality as an abstract internal construction to which sensation and action are little more than peripheral input/output channels.

Spatial Cognition

In the archaeological literature, the representational paradigm is particularly well expressed in the Piagetian work of Wynn (1989). Wynn's basic assertion that "We construct the space we live in…from the coordination of internalized schemes of action" (p. 81) leads him to the ultimate conclusion that "Early artifacts are crude because early homini[n]s had not yet structured space in the way we so casually understand it…the internalized action schemes required for the manufacture of Oldowan tools were not very complex." (p. 83). Thus the sophistication of Oldowan technology is evaluated in terms of the internal mental representations required and found to be unimpressive.

Within the bounds of this interpretive paradigm, the PET evidence is broadly supportive of Wynn's conclusions. Although the pilot study did reveal increased activation of higher order visuo-motor and spatial association areas during knapping, this was in comparison with a control task involving nothing more than passive visual inspection. Preliminary results from follow-up research involving a more active and demanding control task have yet to demonstrate such activation, suggesting that spatial-cognitive "representation" may not be one of the more exceptional demands of Oldowan knapping. It is also interesting that the mental imagery task used in the pilot study evoked activation of inferior parietal spatial-cognitive association cortex that was not seen during actual knapping.

The PET research presented here, like Wynn's Piagetian analysis, fails to provide evidence of sophisticated internal representation in Oldowan knapping. However, there is some equivocal evidence suggesting involvement of the mental rotation and oblique perception capacities stressed by Robson Brown (1993) in her evaluation of Mode I artifacts from Zhoukoudian, China. Neuroimaging studies conducted over the past decade (refs. in Kosslyn *et al.*, 2001) have shown mental rotation to be associated with activation of superior parietal and right frontal lobes. Superior parietal activation was in fact observed during the pilot study, but preliminary results have yet to reveal such activation in the follow-up research. Mental rotation may not have been a particularly important operation in the simple Oldowan-style knapping observed during these experiments.

PET also provides mixed evidence regarding what Robson Brown (1993) refers to as the "significance of the oblique". It is well known (e.g. Appelle, 1972; Furmanski & Engel, 2000; Gentaz & Ballaz, 2000) that humans perceive horizontal and vertical orientations more accurately than they do oblique orientations. This also applies in the perception of motion (Loffler & Orbach, 2001). Robson Brown, following Rudel (1982), maintains that the perception and construction of oblique lines is cognitively demanding because "To differentiate opposite obliques demands holding one dual coordinate in 'mind' while comparing it with another" (p. 239). In so far as she sees evidence that obliques were "both perceived and manipulated" during the production of the Zhoukoudian artifacts, Robson Brown contends that these artifacts provide evidence of cognitive abilities "far more sophisticated than previous philosophical or psychological studies have assumed" (p. 240).

This conclusion may be evaluated on general archaeological and psychological grounds, as well as in light of the specific PET evidence presented here. To begin with, Robson Brown implicitly assumes that core forms from Zhoukoudian were the intentional end products of knapping plans rather than byproducts of flake production. This is quite possible, but in light of the experiments of Toth (1982, 1985), must be demonstrated rather than assumed. Robson Brown further assumes that the production of these core forms required that a complete internal representation be "held" in the maker's mind. Ecological psychologists would be quick to point out that much of the information needed to arrive at finished artifact forms is present in the stone being worked, and need not necessarily be represented in the mind of the maker. Altering the existing shape of a cobble through the removal of flakes is not conceptually or practically equivalent to the *de novo* construction of oblique lines on a blank sheet of paper.

In order to apply the experimental PET results presented here to the question of the "significance of the oblique", we must consider what is known about oblique perception in the brain. In a recent review, Gentaz & Ballaz (2000) concluded that the "visual oblique effect" or VOE (i.e. impaired perception of oblique orientations) is a multi-component phenomenon that occurs at different levels of processing according to the specific task at hand. At the lower end of the scale, it has been shown (Furmanski & Engel, 2000) that human primary visual cortex (V1) is more responsive to gratings with horizontal or vertical orientations rather than oblique orientations. This suggests that the VOE may result in part from V1 having a relatively smaller population of neurons tuned to the detection of oblique stimuli. The situation is complicated, however, by findings of other researchers that the VOE follows a gravitational rather than retinal reference frame (Buchanan-Smith & Heeley, 1993). In other words, the "definition" of oblique depends on an individual's perception of up and down rather than the actual orientation of their retinas. This would suggest involvement of higher-level spatial processing of the kind that occurs in the associative cortex of the parietal lobe.

It appears that, depending on the context, oblique perception may place unique demands on primary visual cortex and/or parietal association cortex. The PET research presented here provides some evidence for the involvement of both regions in Oldowan knapping. In the pilot study superior parietal and inferotemporal visual association cortex was activated, a phenomenon that might, among other things, reflect task-specific demands for the perception of oblique objects and motion. Interestingly, the pilot study was conducted with the subject prone on the scanning bed, possibly invoking higher-order processes needed to "re-define" the subject's visual frame of reference. On the other hand, follow-up research with subjects in a more natural orientation has so far only revealed primary (V1) visual activation. For the time being, the available PET evidence suggests that (naturalistic) knapping involves relatively demanding visual perceptive processes, perhaps including oblique perception, but that these processes have more to do with structuring extrinsic visual sensations than with generating sophisticated intrinsic representations.

Mental Templates

At the current stage of research, the PET data fail to provide compelling evidence of knapping-related procedural templates or instruction sets (Gowlett, 1996) more elaborate than those employed in everyday motor behaviors. Although the terms "procedural template" and "instruction set" are not commonly employed in experimental neuroscience, imaging studies have revealed that strategic planning tasks are most typically associated with prefrontal activation. One example is the *Tower of London* task (TOL) as studied by Dagher *et al.* (1999). The TOL is "a test of motor planning in which subjects must move colored balls on a computer screen to match a specified arrangement in the minimum number of moves possible." (p. 1973). Dagher *et al.* (1999) found that, although visuo-motor areas were routinely activated during TOL, activation in dorsolateral prefrontal, lateral premotor, anterior cingulate and caudate (basal ganglia) regions was correlated with the complexity of the TOL problem presented. The authors concluded that these latter regions constitute a "network for the planning of movement".

Attempts to assess the sophistication of knapping-related procedural templates and instruction sets might reasonably focus on activity in this network, and particularly in prefrontal regions thought to be involved in handling "sequential contingencies" and criterion-based pattern analysis (Roland, 1993: 344-345). In point of fact, the PET research presented here does not indicate knapping-specific activations of Dagher *et al.*'s planning network, with the possible exception of premotor

cortex, although this dearth of evidence may change with further analysis and experimentation.

In evaluating these observations, it is important to consider modern human baseline brain activity, and especially the tonic activity known to exist in dorsal medial prefrontal cortex. This modern baseline activity might be sufficient to support simple, Oldowan-style knapping plans, but it is quite likely that relatively small-brained early hominin species displayed a lower level of baseline functionality. For such species, the formulation of knapping instruction sets may have required significant mental effort beyond the baseline condition. Concrete data are needed to inform such speculation, but are beyond the resolution of the experimental results presented here. Pursuit of such data should be a priority for future research.

Information Processing Capacity

Gibson (1993) has proposed that human sophistication in the superficially diverse realms of tool-use, language and social behavior shares a common foundation in the generalized information processing capacity of neocortical association areas (prefrontal, parietal and temporal). This premise leads her to conclude, among other things, that hominin intelligence has evolved in a gradual rather than punctuated fashion, that overall brain size provides a good indication of evolved intelligence, and that it would be "misleading to judge the intelligence of fossil [hominins] by the form of their tools alone" (p. 263). Instead, Gibson suggests that such judgments should be based on quantitative estimations of the "degree of information processing capacity necessary to support given sociotechnological systems" (p. 264).

The PET research presented here is focused on stone tool manufacture, and does not address broader social and environmental contexts. In this it fails to meet Gibson's criteria. Despite this, it does provide a valuable step toward the kind of quantitative evaluation that Gibson also calls for. In combination with research into the social foundations of stone toolmaking (Stout, in press), the PET evidence can enhance our ability to estimate the information processing demands of at least one archaeologically visible component of prehistoric sociotechnological systems.

Of particular importance to Gibson's hypothesis is the degree of activation observed in association cortex during toolmaking. Although a substantial volume of superior parietal association cortex was activated in the pilot study, the follow-up research has yet to reveal activation in any of the classic neocortical association cortices. For the time being, it is safest to conclude that Oldowan toolmaking does not rely upon substantial associative information processing. Since expansion of association cortex also accounts for the majority of brain enlargement in human evolution, this is consistent with the fact that the current best candidate for maker of the first stone tools, *Australopithecus garhi*, has an estimated cranial capacity of only 450 cc (Asfaw *et al.*, 1999).

The Perception-Action Perspective

Even within the prevailing representational paradigm, there are numerous different approaches to understanding hominin intelligence, and numerous different ways in which PET evidence might be interpreted. However, the overall interpretation is relatively uniform: that Oldowan toolmaking does not require sophisticated internal representation and therefore is cognitively simple. This evaluation is based largely on the fact that brain activations specific to Oldowan-style knapping currently appear to be limited to primary sensorimotor "input/output" cortex and not to include "higher-order" associative cortex.

This observation is without a doubt both meaningful and important, and the absence of major activations in associative cortex does suggest a lack of emphasis on certain kinds of mental processes. At the same time, a preoccupation with internal mental representation should not lead us to neglect the importance of knapping-specific activations in other parts of the brain. In this respect, an ecological or "perception-action" approach to understanding performance can be helpful.

As previously described, the ecological approach views performance as a dynamic property of the organism-plus-environment system, rather than as the unilateral expression by the organism of a static internal representation. While the representational view tends to privilege internal mental processing as an indication of cognitive sophistication, the ecological perspective emphasizes intelligent action as more broadly embodied in the adaptive combination of environmental *affordances* and organismal *effectivities*. Effectivities (Turvey & Shaw, 1979) are the functional units into which the neuromotor system can potentially be organized (Bongers, 2001), and are defined as much by perceptual capacities and bodily parameters as by abstract planning or representative abilities. The most valuable contribution of the ecological perspective is the realization that perception and action matter, and cannot be viewed as merely peripheral to some ideal, Platonic realm of pure cognition within the brain. As argued by Thelen & Smith (1994: 164, emphasis original): "Perception can be outside the study of concepts and categories only if mind is viewed as representing reality instead of contacting it, if knowledge exists outside of performance, and if the dynamic of knowledge acquisition is divorced from the processes of its storage and use."

In the case of stone knapping at least, knowledge does seem to be inextricably linked with performance. As observed by Roux *et al.* (1995), mastery of the forces needed for individual flake removals is the essential prerequisite for development of an effective knapping plan. Such plans are not rigid templates imposed from above, but arise flexibly from a practical understanding (*savoir-faire*) of knapping processes and potentials (Pelegrin, 1990). In the language of ecological psychology, a knapper's understanding comes not from abstract Euclidean representations but from direct experiential knowledge of flaking dynamics. These dynamics include the effectivities of the neuromotor system and the perceived affordances of the knapping materials.

The concept of an *embodied cognition* (Johnson, 1987; Varela *et al.*, 1991), although controversial in its broader application (Dennett, 1993; Kirsch, 1991; Vera & Simon, 1993), is useful in appreciating the importance of the primary visual and motor cortex activations observed during Oldowan-style knapping. To the extent that understanding is embodied in experience and performance, the fine-grained perception and manipulation supported by these sensorimotor regions is as important to "intelligence" as are more internally directed associative processes. Although Mode 1 knapping is conceptually quite simple, the PET data show that it is also a relatively demanding perceptual-motor interaction with the physical environment. The appearance of Oldowan artifacts in the archaeological record thus provides evidence for a level of behavioral sophistication beyond that evident in the everyday manipulative and percussive behaviors humans share with other primates. This ability to interact with the physical environment in increasingly complex and effective ways is as much a hallmark of hominization as are increasing social complexity and symbolic capacities.

Evolutionary Interpretation

Oldowan toolmaking, though supported by pre-existing somatic (e.g. Marzke *et al.*, 1998) and neural traits, was itself a behavioral innovation. As pointed out by Deacon (1997) behavioral innovations must logically precede the biological adaptations that they foster. It is only when a useful behavior spreads through a population and begins to affect reproductive fitness that it actually leads to the changes in gene frequency that constitute biological evolution. Thus there are two questions we can ask about the evolutionary implications of Oldowan toolmaking: (1) what essential preconditions (minimal capacities) does its initial appearance imply, and (2) what selective pressures might its subsequent spread through hominin populations have created?

With respect to the first question, the PET evidence presented in this chapter is only indirectly applicable. Observed brain activations provide information only about the relative neural demands of stone knapping, not about absolute minimum requirements. It is simply impossible to equate the relative levels of activation seen in healthy, adult modern human brains to some minimum neural mass, neuron number or other such measure that would have been required of the earliest toolmakers.

In functional terms, however, the PET evidence does show Oldowan-style knapping to be a relatively demanding perceptual-motor activity. Oldowan artifacts may thus be taken as evidence of relatively greater behavioral sophistication than might otherwise be assumed. However, the absolute level of sophistication implied remains to be more concretely specified. Research with modern primates (e.g. Schick *et al.* 1999 and Toth *et al.*, chapter in this volume) and human children (e.g. Lockman, 2000; Piaget & Garcia, 1991) may ultimately prove more revealing in this respect. Functional brain imaging with tool making primates would provide a particularly interesting point of comparison if the practical difficulties attending such research could be overcome.

The activation evidence from modern humans is more directly applicable in identifying the selective pressures that Oldowan knapping might have placed on the early hominin brain. Even on this point, the evidence helps to define evolutionary possibilities and probabilities rather than certainties. For Oldowan knapping (as opposed to related aspects of tool production, transport and use) to have exerted any direct selective influence, there would have had to be variations in knapping ability that affected survivorship and reproductive fitness. We do not know if this was the case.

For one thing, we do not know how important Oldowan tools actually were to early hominin lifeways. Long tradition has of course viewed toolmaking as a defining attribute of humanity (e.g. Benjamin Franklin in Boswell, 1887; Darwin, 1871), and the developing field of paleoanthropology rather naturally came to see stone toolmaking as a kind of "prime mover" (Potts, 1993) in human evolution (e.g. Oakley, 1959; Washburn, 1960; Leakey *et al.*, 1964). This stance has more recently been bolstered by concrete evidence of the role of Oldowan tools in facilitating meat procurement (reviews in Isaac, 1984; Schick & Toth, 2001). On the other hand, hard evidence of the actual frequency with which such tool-assisted meat procurement occurred, and of its ultimate adaptive significance, is still lacking. In fact, the spatial and temporal limitations of early archaeological evidence have precluded any secure estimation of the frequency of Oldowan toolmaking and use, its prevalence within hominin groups, or its distribution across populations. We are left with the intuitively compelling yet circumstantial argument that the *potential* utility of Oldowan tools (e.g. Schick & Toth, 1993) implies an *actual* adaptive significance.

If we accept that that the use of Oldowan tools did in fact provide a significant adaptive advantage for early hominins, then there is still the question of whether meaningful variation existed in the ability to manufacture those tools. Growing appreciation of the simplicity of Oldowan tools and of the tool-using capacities of modern apes has led some to conclude that such variation did not exist. If effective Oldowan knapping was within the pre-existing capacities of the average adult, then clearly "toolmaking *per se* cannot have constituted the main 'adaptive wedge' driving the evolution of

hands, brains and behavior in early *Homo*" (Potts 1993: 62).

Although this conclusion may ultimately turn out to be correct, it neglects the issues of skill learning and efficiency. In addition to the simple presence or absence of a behavioral capacity, the ease and reliability with which it is acquired should also be considered. For example, although modern chimpanzees are clearly capable of using stone hammers to crack open nuts, it nevertheless takes them years of learning to acquire proficiency (Boesch, 1993; Matsuzawa, 1996). Similarly, the stone flaking abilities of the bonobo Kanzi, though impressive, have developed slowly over more than a decade of experimentation (Toth *et al.*, chapter in this volume). In contrast, inexperienced modern humans are almost immediately and effortlessly able to produce near-replicas of early stone artifacts (Stout & Semaw, chapter in this volume). Although modern humans are clearly "over-qualified" to acquire Oldowan toolmaking skills, our smaller-brained ancestors quite probably found this leaning process to be more challenging. Even if nearly every healthy adult eventually acquired comparable knapping abilities, variations in learning speed and efficiency could still have provided raw material for selection.

To the extent that Oldowan knapping ability actually was important to survival, neural adaptations that facilitated its rapid and reliable acquisition would have been favored. Such adaptive facilitation would presumably have occurred through increases in the functional capacity of those neural structures most stressed during learning and performance. PET research offers a unique opportunity to identify these structures, although subsequent evolutionary interpretations are somewhat complicated by our limited understanding of the processes by which adaptive increases in functional capacity are achieved.

Mosaic Adaptation

One such process, consistent with the work of Barton & Harvey (2000) and Winter & Oxnard (2001), would be targeted increases in the size of functionally relevant brain regions. Although the relationship between size and function in brain structures is not well understood, it is commonly assumed that increases in size roughly equate to increases in neuron number and associated processing capacity. As shown by the currently available PET activation data, Oldowan-style knapping in modern humans is exceptionally demanding of neuronal activity in primary visual and motor cortices as well as in the cerebellum. Although further analysis may reveal additional areas of activation, these regions are the appropriate focus for the current discussion.

Within modern humans, performance on visual (Demb *et al.*, 1997) and motor (Grafton *et al.*, 1992; Pascual-Leone *et al.*, 1994; Karni *et al.*, 1995) tasks is positively correlated with the intensity and extent of activation in primary visual and motor cortices. Cerebellar volume is similarly correlated with the ability to learn a simple motor response (eyeblink conditioning: Woodruff-Pak *et al.*, 2001). Evolutionary increases in the size of these structures could have been one way in which the human brain became an over-qualified or "fail-safe" (Deacon, 1997) medium for the acquisition of knapping skills.

1) Primary Visual Cortex (V1)

One problem with this hypothesis is the fact that primary sensorimotor regions are actually among the least evolutionarily expanded portions of human cerebral cortex. Although concrete data regarding the size of major cortical subdivisions in humans and other primates are surprisingly hard to come by, Stephan *et al.* (1981) do report primary visual (striate) cortex volumes for 41 primate species. As a result, it is well known that the volume of human striate cortex is less than expected for a primate of its brain size (Holloway, 1979). Of course, human striate cortex is still absolutely larger than that of any other primate, including apes whose body sizes and (presumably) peripheral visual systems are quite similar to those of humans (Table 3). Thus, Deacon (1997: 216) has argued that "The human brain does not have a reduced visual cortex, but the appropriate amount of visual cortex for its retina". It may even be that human primary visual cortex is expanded relative to its retinal inputs, although any such expansion is certainly dwarfed by the much greater expansion of association cortex in the neighboring posterior parietal and elsewhere (Passingham, 1975; Holloway, 1983).

Paleoneurological evidence regarding to the evolving size of hominin striate cortex has been remarkably controversial (e.g. Holloway, 1981b; Falk, 1983), most fundamentally with respect to matters of timing. At the heart of disagreement is whether striate cortex experienced an independent reduction in *absolute* size prior to major allometric increases in hominin brain size (Holloway, 1995), or whether it merely decreased in *relative* size due to those increases (Armstrong *et al.*, 1991). In the former case, striate reduction (proposed to have occurred between 3 and 4 Ma [Holloway, 1995]) would obviously have predated any possible selective influence of Oldowan knapping, which first appears at 2.5 Ma (Semaw *et al.*, 1997). However, such reduction could have had implications for the pre-existing visual capacities of the first toolmakers.

2) Primary Motor Cortex (M1)

Much less is actually known about the comparative size and evolution of primary motor cortex (M1). Deacon (1997: 217) contends that this region is only 35% as large as expected for a primate brain of human size, but acknowledges that data used to make this and other estimates are "incomplete and insufficient for statistical tests". The only published data of which the author is aware are the surface area estimates of Glezer (1958; reprinted in Blinkov & Glezer, 1968), which show human M1 to be both relatively and absolutely smaller than that of chimps and orangutans (Table 4). These data indicate that, while the precentral region as a whole is of roughly the same relative size in humans and apes, a dramatic expansion of human premotor cortex at the expense of M1 has occurred within the region. This "zero-sum" relationship is highly suggestive of cortical reorganization independent of overall expansion.

It should be noted, however, that the data of Blinkov & Glezer (1968) have been questioned with respect to the small sample sizes employed and potential problems with postmortem shrinkage (Semendeferi & Damasio, 2000). Blinkov & Glezer themselves (1968: 5-10) identify numerous methodological problems in measuring the surface area of brains, including slicing deformation that may produce a 4% to 20% reduction in linear dimensions, shrinkage during preservation by up to 40% of surface area, and errors of up to 39% generated by calculating the area of a continuously curved surface from serial sections. Although generalized mathematical corrections have been developed for all of these problems, the errors produced are inevitably variable and particularly sensitive to differences in brain size and shape. This is problematic when comparing small samples across species that display dramatic differences in brain size.

In addition to the surface area work of Glezer (1958), attempts have also been made to compare the motor maps of M1 in humans and macaques. For example, Washburn (1959) concluded that the M1 hand area was relatively enlarged in humans, perhaps as an adaptation for tool use. Passingham (1973), however, used a different macaque motor map and concluded that this apparent difference was actually due to the smaller size of the foot area in humans. Unfortunately, there are no published paleoneurological observations regarding evolutionary changes in the size or morphology of hominin M1 (i.e. the precentral gyrus).

Table 3

Species	Striate Cortex Volume (mm³)	Body Weight (kg)
Pan troglodytes	14,691	46
Gorilla gorilla	15,185	105
Homo sapiens	22,866	65

Table 3: Primary visual (striate) cortex volume and body weight in humans and apes. Although decreased as a proportion of total cerebral volume, human striate cortex is still absolutely larger than that of any other primate.
(data from Stephan et al., 1981)

Thanks to the work of Heffner & Masterson (1975), we do know that the number of direct cortico-motoneuronal (CM) projections from motor cortex to the spinal cord is increased in humans. Interspecific variation in the number of these direct CM connections has also been found to correlate with an index of dexterity (Heffner & Masterson, 1975; Kuypers, 1981; Lemon, 1993). It is apparent from this work that human motor cortices in general have assumed a more direct and important role in controlling dexterous movements of the hands. What is less clear is how the connections and relative contributions of M1 and secondary motor cortices may have altered during this process. In humans, roughly fifty percent of CM projections arise from M1, while most of the rest originate in the more anterior secondary motor areas (Heffner & Masterson 1991). Similar data have not been reported for other species.

There is currently too little evidence to say exactly how the size of human M1 relates to that of other primates. It is at least safe to say that M1 has undergone nowhere near the degree of expansion seen in secondary motor and association cortices. In fact, M1 may even have experienced a real reduction in absolute size during human evolution. It thus seems unlikely that the apparent demands of Mode 1 knapping on M1 were a particularly influential factor in human brain-size evo-

lution. The comparative surface area evidence instead calls attention to secondary motor cortex as a major locus of evolutionary change. At the same time, comparative CM projection data confirm a human evolutionary shift toward increasingly direct cortical control of movement. Further analysis of the PET data will provide a better indication of the degree to which secondary motor cortices are recruited during Mode 1 knapping, and help to clarify potential relationships between stone knapping and the evolution of human motor cortex.

3) Cerebellum

By this point, it should not be surprising that the role of the cerebellum in human brain evolution is also controversial. Both Deacon (1997) and Finlay & Darlington (1995) identify the cerebellum as being one the more preferentially expanded structures in the modern human brain. Clark *et al.* (2001), on the other hand, argue that the cerebellum actually constitutes an invariant fraction of total brain volume across mammals (including humans). Sultan (2002) has questioned the import of Clark *et al.*'s observation by noting that, while cerebellar white matter volume is relatively invariant, cerebellar and cerebral surface area do co-vary. Meanwhile, Barton (2002) is engaged in a disagreement

Table 4

Species	Precentral Region (Brodmann Areas 4 & 6)		Primary Motor Cortex (Area 4)		Premotor Cortex (Area 6)	
	cm^2	% of total hemisphere	cm^2	% of precentral region	cm^2	% of precentral region
Homo sapiens	62.50	8.4	7.34	12	55.1	88
Pan troglodytes	30.60	7.6	8.94	29.8	21.7	70.2
Pongo pygmaeus	41.97	7.6	13.57	33	28.4	67
Hylobates sp.	6.20	7.5	3.04	49	3.18	51
Papio sp.	8.45	6.8	4.83	58	3.63	42
Cercopithecus sp.	6.71	8.3	4.64	69	2.10	31
Callithrix sp.	1.01	5.5	0.80	79	0.21	21

Table 4: Surface areas of the "precentral region" of humans and other primates as reported by Blinkov & Glazer (1968). Note that human primary motor cortex is actually smaller than that reported for chimpanzees and orangutans, although the precentral region in general is somewhat enlarged due to a dramatic expansion of premotor cortex.

with Clark's group (Wang *et al.*, 2002) over the appropriate statistical treatment of the volume data.

The single most important contribution that can be made toward resolution of these and other controversies regarding primate brain-size evolution is the collection and publication of new comparative data. A number of researchers (Semendeferi *et al.*, 1997; Semendeferi & Damasio, 2000; Rilling & Insel, 1998; Rilling & Insel, 1999; Rilling & Seligman, 2002) are currently making such a contribution through the pioneering use of anatomical MRI to collect *in vivo* primate brain volume data. Unfortunately, even these new data have yet to resolve the many questions surrounding primate cerebellar evolution.

Based on a study of 10 humans and 19 other hominoids, Semendeferi & Damasio (2000) conclude that the cerebellum constitutes a smaller percentage of the brain in humans than in apes. A univariate ANOVA conducted by this author on the data presented by Semendeferi & Damasio confirms that between-species differences in cerebellar proportion do exist (p = 0.005), however a subsequent post hoc (Bonferroni) test reveals that these differences arise only in comparisons involving gorillas (Table 5). It is actually the large size of the gorilla cere-

bellum that accounts for the difference between ape and human means reported by Semendeferi & Damasio.

Semendeferi & Damasio themselves comment on the apparently anomalous size of the cerebellum in gorillas, but caution that "The larger mean value for this species is largely due to the large cerebellum of one of the two individuals examined" (p. 329). This mean value (16.1 %) is greater than that (14.7 %) indicated by the data of Stephan *et al.* (1981), but less than the value (17.0 %) derived from the MRI data of Rilling & Insel (1998). To the extent that the data of Semendeferi & Damasio (2000) indicate any deviation from the hominoid allometric trend, it is on the part of gorillas, not humans (Figure 8).

The more phylogenetically inclusive MRI study of Rilling & Insel (1998) allows for an additional level of analysis. Comparing cerebellar and brain volumes across 44 individuals from 11 haplorhine species, Rilling & Insel observed an apparent grade-shift between cercopithecoid and hominoid primates. If humans are excluded from the regression as a presumptively divergent species, then monkey and ape trends with similar scaling relationships (slopes) but different proportions (y-intercepts) are produced (Figure 9a).

Figure 8

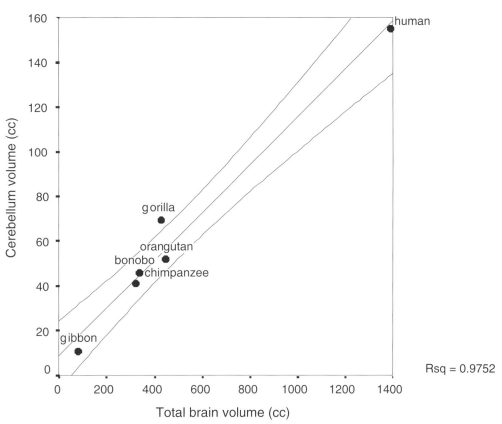

8. ***Allometric plot of species mean values for cerebellum and total brain volume including 95% confidence interval*** *(data from Semendeferi & Damasio, 2000). A regression using species mean values confirms the results from the post-hoc test of individual cerebellar proportions (table 5): gorillas are the only species that falls outside the mean prediction lines (95% confidence interval).*

Table 5

Comparison		Mean Difference (1-2)	Standard Error	Significance
Species 1	Species 2			
human				
	bonobo	-2.1200	.9005	.412
	chimp	-1.4200	.7064	.844
	gorilla	**-4.6200**	**1.0597**	**.003**
	orangutan	-.3450	.8093	1.000
	gibbon	-1.4200	.8093	1.000
bonobo				
	human	2.1200	.9005	.412
	chimp	.7000	.9673	1.000
	gorilla	-2.5000	1.2488	.858
	orangutan	1.7750	1.0448	1.000
	gibbon	.7000	1.0448	1.000
chimp				
	human	1.4200	.7064	.844
	bonobo	-.7000	.9673	1.000
	gorilla	-3.2000	1.1170	.131
	orangutan	1.0750	.8830	1.000
	gibbon	.0000	.8830	1.000
gorilla				
	human	**4.6200**	**1.0597**	**.003**
	bonobo	2.5000	1.2488	.858
	chimp	3.2000	1.1170	.131
	orangutan	**4.2750**	**1.1847**	**.022**
	gibbon	3.2000	1.1847	.191
orangutan				
	human	.3450	.8093	1.000
	bonobo	-1.7750	1.0448	1.000
	chimp	-1.0750	.8830	1.000
	gorilla	**-4.2750**	**1.1847**	**.022**
	gibbon	-1.0750	.9673	1.000
gibbon				
	human	1.4200	.8093	1.000
	bonobo	-.7000	1.0448	1.000
	chimp	.0000	.8830	1.000
	gorilla	-3.2000	1.1847	.191
	orangutan	1.0750	.9673	1.000

Bold indicates a significant difference in cerebellar proportion

Table 5: Post hoc (Bonferroni) test of variation in cerebellar proportion across hominoids (data from Semendeferi & Damasio, 2000). The only significant differences occur in comparisons of gorillas with humans and orangutans. Bold indicates a significant difference in cerebellar proportion

Interestingly, humans appear to be better predicted by the monkey trend. If, however, humans are included in the regression, a different (shallower) hominoid scaling relationship is produced (Figure 9b). Rilling & Insel consider several possible explanations for this pattern, and conclude that "the data can best be explained by a grade shift occurring with the evolution of hominoids, followed by a change in scaling caused by disproportionate cerebral expansion with the evolution of homini[n]s" (p. 313).

An alternative not considered by Rilling & Insel is the possibility that it is actually gorillas that are the divergent hominoid species. This possibility is suggested both by the data of Semendeferi & Damasio (2000) (Table 5) and by regression of Rilling's & Insel's own data (Figure 9b). When gorillas are excluded from the regression, a good allometric fit is observed across the remaining hominoids (Figure 9c). This suggests, not only that something very interesting has occurred in the evolution of the cerebellum in gorillas, but also that Rilling's & Insel's putative "change in scaling caused by disproportionate cerebral expansion" may characterize hominoids in general rather than hominins specifically.

Considering all available evidence, the safest conclusion appears to be that the human cerebellum has indeed undergone considerable evolutionary expansion, even if it has not quite kept pace with the rapidly bal-looning neocortex. In this it appears to have conformed to a primitive hominoid allometric trend. Such proportional human cerebellar expansion is not surprising considering the close functional and anatomical connections between cerebellum and neocortex (Schmahmann & Pandya, 1997), and the involvement of cerebellum in a wide range of perceptual, motor and cognitive behaviors (Leiner *et al.*, 1993; Parsons & Fox, 1997). The co-activation the cerebellum with various cortical regions during knapping is just one example of this pervasive integration. Knapping-related demands on the cerebellum may have been one factor contributing to the rapid expansion of the cerebro-cerebellar system during human evolution.

Correlated Expansion

Another possibility to be considered is that selection on the primary sensorimotor cortices and cerebellum could have led to correlated expansion of the brain as a whole, as suggested by the developmental constraint hypothesis of Finlay & Darlington (1995). There is reason to be cautious in applying this hypothesis to the interpretation of the PET evidence, however. To begin with, the developmental constraint hypothesis predicts the same result (whole brain expansion) from selection on any given brain region or capacity. As Finlay & Darlington (1995: 1583) observe "theories that start from a primary behavioral trait appear to account

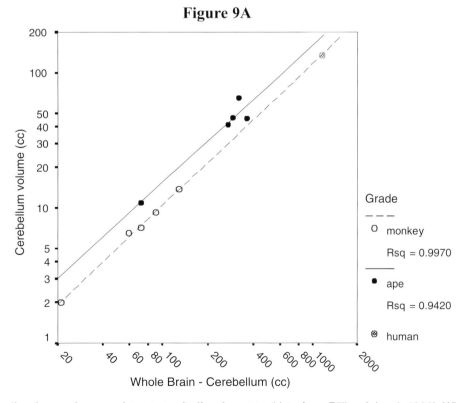

Figure 9A

*9A. **Cerebellar scaling in monkeys and apes, excluding humans** (data from Rilling & Insel, 1998). When humans are excluded from the regression, it appears that apes and monkeys display similar cerebellar scaling relationships but different proportions, a classic grade shift. However, the ape trend depicted is entirely determined by the outlying gibbon value. Within the great apes, there is no significant trend (p = 0.683).*

Figure 9B

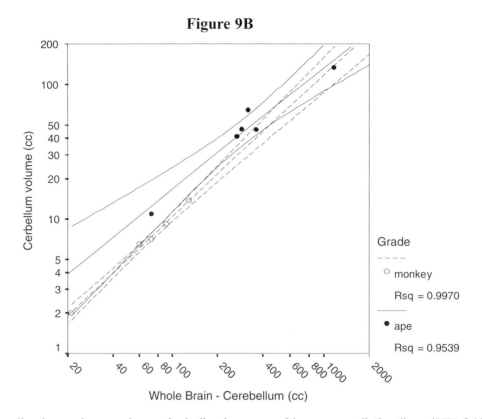

9B. **Cerebellar scaling in monkeys and apes, including humans, with mean prediction lines (95% C.I.).** *(data from Rilling & Insel, 1998). Inclusion of humans in the regression produces a shallower hominoid trend, but once again indicates that the only deviant hominoid value is the exceptionally large cerebellum of gorillas.*

Figure 9C

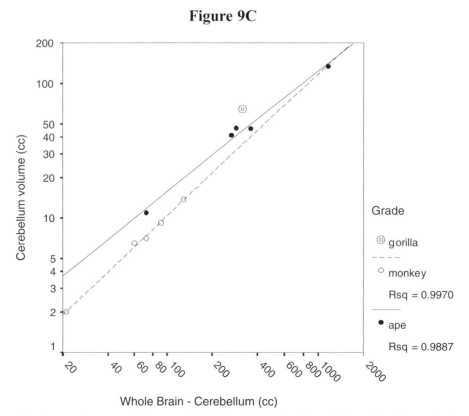

9C. **Cerebellar scaling in monkeys and apes, excluding gorillas** *(data from Rilling & Insel, 1998). Exclusion of the deviant gorilla value produces a very good allometric fit for the remaining hominoid species. This suggests (1) that gorillas have undergone an adaptive specialization in cerebellum size, (2) that hominoid cerebellar scaling relationships differ from those of monkeys, and (3) that humans have a cerebellum roughly the size predicted for a hominoid brain of human size.*

for human [brain] evolution many times over." In theory, knapping-related demands on the cerebellum could have produced the overall brain enlargement seen in early *Homo*, yet the same effect might equally well have been produced by completely different demands on that or other structures. Although the PET evidence of cerebellar activation does confirm the plausibility of knapping-related contributions to overall brain expansion, the inherent equifinality of the constraint hypothesis effectively prevents any more concrete conclusions.

Unfortunately, things become even more complicated when it comes to interpreting knapping-related activations in the neocortex. This is because the constraint hypothesis was developed to explain covariance between major brain structures like neocortex and cerebellum rather than variation in the size of functional areas within neocortex. The possibility of evolution through cortical reorganization greatly complicates any hypothetical relationship between selection on any one cortical region (e.g. motor cortex) and overall neocortical or brain expansion.

As we have seen, the comparative data needed to conduct a rigorous investigation of variation within primate neocortex are simply not available. Nevertheless, it is clear that primary sensorimotor cortices have expanded far less during human evolution than have other cortical regions. The only way in which selection favoring relatively small expansions of sensorimotor cortex could have produced the much larger effects seen in association cortex would have been in the context of extremely rigid allometric constraints on cortical organization. For example, if association cortex necessarily scaled to sensorimotor cortex at an exponent of two, small increases in sensorimotor volume would produce disproportionate expansion of association cortex.

As evidence that such constraint might in fact exist, Finlay & Darlington (1995) cite the work of Nudo & Masterson (1990) indicating that the amount of cortex devoted to forelimb control across species (from "hooves to hands") is highly predictable from total cortex volume. According to Finlay & Darlington (p. 1578), this "suggests that the amount of cortex devoted to forelimb control can increase only as the result of apparently inefficient increase in total cortex volume." In this case, selection on primary sensorimotor cortex could hypothetically have led to more general neocortical (and correlated whole brain) expansion. Important support for this hypothesis would come from the identification of some mechanism or mechanisms accounting for such rigid constraint.

In their developmental constraint hypothesis, Finlay & Darlington (1995) propose that a highly conserved order of neurogenesis is the constraining mechanism underlying the covariation in size of major brain structures. In fact, it does seem that genetic modulation of the overall duration of neurogenesis is the most important mechanism governing the total size of the cortical sheet. There is no evidence, however, that developmental timing similarly influences the formation of functional fields (arealization) within the neocortex. Instead, cortical arealization is thought to be driven by the patterning of incoming thalamocortical projections, guided at a gross level by intrinsic gradients of cortical gene expression (Krubitzer & Huffman, 2000). Comparative evidence of sensory specializations in animals like echolocating bats and blind mole rats illustrates the way in which these mechanisms combine to produce a mammalian pattern of constrained overall cortical topography together with major adaptive variation in the size and number of cortical fields (Krubitzer & Huffman, 2000).

Evidence bearing on the development and evolution of cortical arealization comes almost exclusively from consideration of sensorimotor cortices. On the whole, these regions appear to adapt both readily and independently to changes in peripheral morphology and behavior (Krubitzer & Huffman, 2000). In primates, for example, the size of visual system structures (including striate cortex) has evolved in relative independence of other brain regions (Barton, 1998; Stout, 2001). It seems quite plausible that knapping related selective pressures on primary visual and motor cortices could have resulted in independent expansion of these regions. However, there is little evidence that it actually did.

Expansion of this kind would have been a "zero-sum" game: any expansion of primary sensorimotor cortices through cortical reorganization alone would have entailed commensurate decreases in the size of higher-order sensorimotor and associative cortices. There is no paleoneurological or comparative evidence indicating that such expansion of sensorimotor cortex at the expense of association areas occurred at any point during hominin evolution. In fact, the only proposed paleoneurological example of hominin cortical reorganization involves the exact opposite (Holloway, 1995). If, on the other hand, sensorimotor expansion were accomplished through a combination of overall expansion and adaptive arealization, then some correlated expansion of other cortical regions might possibly occur. One way this could happen would be if the total cortical expansion produced by the stochastic processes of natural selection happened to exceed the volume actually incorporated into sensorimotor cortices. Considering that each additional round of cell division during neurogenesis doubles the number of neurons ultimately produced, such an evolutionary "overshot" is not implausible.

It is also possible that cortical expansion, even if driven by selection on primary sensorimotor cortices, would favor the disproportionate growth of association cortex for functional reasons. Deacon (1997) has pointed out that, since the number of possible connections between neuron increases in geometric proportion to the number of neurons, increasingly large brains will tend to be less thoroughly integrated. As a result, Deacon argues, larger brains will need to devote increasingly more resources to integrative or "managementlike" functions, simply in order to maintain cohesion. Just as growing human organizations seem to require increasing proportions of managerial and administrative employees, cortical expansion might tend to favor disproportionate growth of intrinsic association cortex even if selective pressures were actually operating on extrinsic input/output capacity.

Microstructural Adaptation

Given the current state of knowledge in paleo- and comparative neurology, it is impossible to specify exactly how the neural demands of Oldowan knapping may have influenced hominin brain-size evolution. Even less is known about microstructural specializations in the human brain. In both cases, however, evidence from PET research directs our attention to the specific brain structures that would have been the most likely proximal targets of selection relating to Oldowan knapping.

Even at the current stage of analysis, PET evidence clearly reveals the demands placed on primary visual and motor cortices by Oldowan-style knapping. Results from the pilot study, while in need of further corroboration, also suggest involvement of neuronal fields associated with the dorsal "position and motion" stream of visual processing (Figure 3) in the superior parietal lobe. It should thus be of particular interest to human origins researchers that some of the best evidence of microstructural specialization in the human brain comes from primary visual cortex and appears to focus on the processing stream associated with motion sensitivity.

Working with the carefully sectioned and stained occipital lobes of 29 human and non-human primates, Preuss *et al.* (1999) found that humans display a unique arrangement of neurons and dendrites in layer 4A of primary visual cortex. The neurons involved are part of what is known as the "M-stream" of visual processing, a fast acting system with high contrast sensitivity (Livingstone & Hubel, 1988) that is particularly suited to motion perception. Prior to the work of Preuss and colleagues, the distribution of M-related neurons in V1 was known primarily from studies of macaques, which display a characteristic honeycomb pattern of M-tissue in layer 4A. Preuss *et al.* (1999) have now shown that, although the honeycomb pattern is shared by monkeys and apes, humans display a unique mesh-like architecture in layer 4A that results in a much greater representation of M-tissue.

Preuss and colleagues suggest that this derived characteristic of human visual cortex may represent an augmentation of the M-stream in humans, and furthermore that such augmentation would be consistent with reports indicating that humans are more sensitive to luminance contrasts than are macaques (De Valois *et al.*, 1974; Merrigan, 1980) and have M-related retinal ganglion cells with larger dendritic fields (Dacey & Petersen, 1992). Augmentation of the M-stream in human V1 would be particularly interesting in light of the PET evidence indicating the importance of this region during Oldowan-style knapping. A hominin M-stream specialization that acted to enhance motion perception could quite plausibly have contributed to the initial emergence of stone knapping or been part of an adaptive response to its later spread. If the knapping-related activation of extra-striate visual areas in the superior parietal lobe (i.e. the dorsal visual stream) is confirmed by further analysis, it will provide additional evidence of the critical role played by motion perception in Oldowan-style knapping. As noted by Preuss *et al.* (1999) specializations of primary visual cortex might be expected to have cascading effects on these higher levels of the visual system, which receive most of their input from V1.

It is remarkable that evidence of human microstructural specialization has come first from striate cortex, previously thought to be one of the best understood and most primitive regions of neocortex. This strongly suggests that further research will reveal similar specializations in other parts of the human cerebral cortex. In fact, Buxhoeveden *et al.* (2001) have already reported differences in cortical minicolumn size and morphology in the planum temporale region of human vs. non-human primates. These particular differences are not likely to be related to toolmaking in any direct fashion, but again highlight the need for further comparative research at the microstructural level.

Language

One final point to be considered is the relevance of the PET evidence to various "motor hypotheses" of language evolution (e.g. Greenfield, 1991; Calvin, 1983; Calvin, 1993; Wilkins & Wakefield, 1995). Although differing in particulars, these hypotheses generally posit that some form of neural overlap between tool behavior and language led to co-evolution of the two capacities. In the specific case of Oldowan-style knapping, however, the PET data have yet to provide convincing evidence of such overlap. More detailed consideration of the individual hypotheses offers some insight into what might be expected from further analysis and research.

The neo-Piagetian hypothesis of Greenfield (1991) focuses on the putative role of left ventral-lateral frontal cortex (i.e. Broca's area) in distributed neural circuits underlying both language and "hierarchical object combination". Greenfield posits that a Broca's area homologue present in a common ancestor of apes and humans was elaborated and differentiated during hominin evolution, producing two adjacent but functionally distinct sub-areas. It is an implication of the hypothesis that manual behaviors contributing to or enabled by this evolutionary differentiation should produce activation in the superior "manual object combination" sub-area.

Ideally, PET tests of Greenfield's hypothesis would seek to demonstrate the presence or absence of activation in this particular sub-area during various manual activities. This is made somewhat more difficult by the relatively low resolution of PET and by the fact that Greenfield does not specify the anatomical boundaries of the sub-areas she proposes. In pragmatic terms, almost any observed activation of Broca's area should probably be considered consistent with Greenfield's hypothesis. Currently available PET evidence fails to indicate such activation during Oldowan-style knapping. If these results are born out by further analysis, they will strongly suggest that Oldowan knapping did not play a major role (as cause or consequence) in the elaboration of distinct manual and language circuits within Broca's area. If additional PET research with a wider range of evolutionarily relevant tool behaviors similarly fails to produce evidence of Broca's area activation, then the hypothesis as a whole will be cast into doubt.

The "neurolinguistic preconditions" hypothesis of Wilkins & Wakefield (1995) does not lead to such concrete predictions about the recruitment of Broca's area during stone toolmaking. Unlike Greenfield's hypothesis, which envisions the evolutionary elaboration of a pre-existing Broca's area homologue, Wilkins & Wakefield propose that Broca's and Wernicke's language areas were "re-appropriated" from motor and somatosensory association cortices initially expanded in response to "selectional pressures for…the manufacture and/or use of stone tools (including throwing)" (p. 173). According to this scenario, modern patterns of activation in these re-appropriated cortices may no longer reflect their earlier evolutionary history.

Wilkins' and Wakefield's hypothesis is a specific application of the more general concept of correlated brain expansion, in which additional structure is thought to precede enhanced function (Finlay et al., 2001: 277). In contrast to the developmental constraint hypothesis, however, Wilkins and Wakefield propose that correlated expansion occurred as a result of functional linkage between brain regions. This functional component makes their argument of evolutionary cause and effect at least somewhat testable using modern activation data. Strongest support for the hypothesis would come from evidence that motor and sensory association areas (especially the lateral premotor cortex bordering Broca's area and the inferior parietal cortex adjacent to Wernicke's area) are in fact activated during stone knapping. Such activation would confirm that these regions were likely targets for tool-related selective pressure, and would be consistent with the hypothesis that such selection yielded over-elaborated neural structures ripe for re-appropriation into evolving language circuits. On the other hand, compelling evidence that these regions are not recruited during stone tool behavior would falsify the proposed link. Although the PET evidence is currently equivocal regarding activation in these areas, further research and analysis should produce more concrete results.

Calvin's (1983, 1993) "neural sequencing" hypothesis of language evolution actually makes somewhat similar predictions about knapping-related activations, although for different reasons. The core of Calvin's argument is the contention that linear increases in the speed and accuracy of ballistic movements require exponential increases in the number of neurons recruited to control them. Since even the huge evolutionary increases in hominin brain size evident from the fossil record could not have kept pace with this geometric progression, the requisite neurons must be gained through "borrowing", with "the experts (probably the premotor cortex and the cerebellum) recruiting some temporary help from other brain regions" (1993: 248). Since Calvin "suspects" that such borrowing is easier in a juvenilized brain, he concludes that selection for neoteny was the driving force behind brain enlargement.

There are some important problems with Calvin's hypothesis as framed. These include the fact that the human brain is actually overdeveloped (peramorphic) rather than juvenilized (Gibson, 1991; McKinney, 2002) and that language is more meaningfully characterized as hierarchical in organization rather than sequential (Poeck & Huber, 1977). On the other hand, the basic point remains that the execution of fast and accurate movements is neurally demanding. Calvin's evolutionary scenario stresses throwing because of what he sees as its uniquely stringent neuromotor demands, yet effective stone knapping requires much the same precision. The PET evidence presented here directly confirms that Mode 1 knapping is unusually demanding of neuronal activation, although mostly within areas that might reasonably be characterized as visual or motor "specialists". There is no evidence of the more widespread neuronal "borrowing" envisioned by Calvin.

At this point, the PET evidence suggests only the most indirect of links between toolmaking and language evolution. Oldowan-style knapping is indeed associated with increased activation in large volumes of neocortex and cerebellum. It is quite possible that knapping-related pressures on these regions could have contributed to brain expansion that ultimately provided raw material for the later evolution of language circuits. This is a difficult hypothesis to test. On the other hand, there is no evidence that modern Mode 1 toolmaking and language processing rely upon similar neural substrates, a fact that argues against the existence of more direct co-evolutionary links between the two behaviors. This apparent lack of neural overlap also suggests that proposed structural similarities between language processing and tool behavior (e.g. Calvin, 1993; Greenfield, 1991; Reynolds, 1976; Wilkins & Wakefield, 1995) may be overstated (cf. Wynn, 1991; Graves, 1994).

SUMMARY AND CONCLUSIONS

Direct evidence regarding the evolution of the human brain and intelligence is notoriously difficult to come by. Brains do not fossilize, and fossil cranial endocasts can provide only limited evidence regarding brain size and macroscopic surface morphology. Archaeological evidence of behavior, though valuable, is subject to the vagaries of deposition, preservation and recovery, and requires careful interpretation. Experimental and comparative research in the modern world is thus essential in order to gain the maximum benefit from the available prehistoric evidence.

The PET evidence presented here, including results from the pilot study of Stout *et al.* (2000) and from ongoing follow-up research, is beginning to reveal the specific neural substrates of Oldowan-style stone knapping. The research was conducted with modern human subjects, but careful interpretation nevertheless yields important insights regarding pre-modern cognition and brain evolution. The ideal situation of imaging research with pre-modern hominin species will obviously remain impossible, although research with modern non-human primates may eventually provide a comparative perspective. For the time being, the modern human data provide at least one concrete reference point for the consideration of evolutionary questions.

There are two major ways in which this information may be applied in human evolutionary studies: (1) as evidence of the mental processes involved in stone knapping, and (2) as evidence of the potential targets of evolutionary selection on toolmaking skill. The former, "paleopsychological", approach is based on the assertion that similar behaviors require similar mental operations regardless of the specific cranial capacity or neural organization of the agent involved.

Paleopsychological Conclusions

Known associations between particular patterns of brain activation and particular kinds of experimental tasks make it possible to "read" activation patterns in terms of the mental task demands they reflect. As long as the issue of baseline brain function is properly considered, PET activation evidence may thus be used to provide a general psychological characterization of Mode 1 toolmaking. This leads to two major conclusions:

1. The PET evidence currently supports the prevailing archaeological view that Oldowan technology was cognitively simple. Activation evidence demonstrates the heavy recruitment of primary visual and motor areas during Mode 1 knapping, but remains equivocal regarding recruitment of secondary sensoimotor and association cortices. Final interpretation must await the completion of further analysis, but for the time being Mode 1 knapping seems best characterized as a relatively demanding visuo-motor skill that is not particularly reliant upon internal representation or strategic planning.

2. The PET evidence does not indicate that Oldowan knapping was mentally trivial for its early practitioners. Ongoing baseline activity of the modern human brain may be concealing low level planning and problem solving requirements that would have been more taxing for smaller-brained Oldowan hominins. More fundamentally, the mental significance of visuo-motor skill itself should not be underestimated. Although mainstream views of cognition tend to privilege abstract internal representation as the hallmark of intelligence, ecological psychologists call our attention to the intelligence embodied in effective action. Fine-grained perception and flexible performance are as integral to human mentality as are more internally directed cognitive behaviors. Oldowan artifacts, when evaluated in light of the PET data, provide the earliest concrete evidence of intelligent behaviors more demanding than those that might be assumed in the common ancestor of humans and African apes.

Evolutionary Implications

The second major application of the PET data is in the generation of evolutionary hypotheses. The visuo-motor demands of Mode 1 toolmaking are embodied in large volumes of knapping-related activation in sensorimotor cortices and the cerebellum, and these activations serve to highlight what would have been the most likely neural targets of selection acting on Oldowan toolmaking skill. Additional archaeological, paleoneurological and comparative evidence will be needed in order to determine whether neural adaptations to facilitate toolmaking skill acquisition actually occurred, but the PET evidence at least tells us where in the brain to look.

More particular evolutionary assessment of the PET evidence depends on which model of mammalian brain evolution is employed. Three such models may be considered:

1. In the case of a predominantly mosaic pattern of evolution, we might expect that those brain regions most heavily taxed by knapping activities would experience preferential expansion. There is little evidence that this has occurred. With the exception of the cerebellum, the volumes of most intense activation appear to be located in primary sensorimotor regions that have undergone relatively little expansion during human evolution. These regions, and particularly the occipital visual cortices, may have undergone some limited enlargement that was outpaced by greater expansion elsewhere, but it does not currently appear that the distributed network of structures associated with Mode 1 knapping was a major focus for mosaic brain enlargement. This preliminary conclusion may change as continuing analysis reveals additional regions of knapping-related activation, but the available evidence makes it clear that any perceived relationship between Oldowan knapping and regional brain expansion will be a relatively complex one.

2. Developmental and functional constraint models of brain evolution do suggest ways in which selection on knapping skill might have led to observed patterns of human brain expansion, but these scenarios are not currently falsifiable. According to the developmental constraint hypothesis of Finlay & Darlington (1995), selection on the size of one brain structure may be expected to produce coordinated enlargement of the whole brain. Alternatively, the information management demands attending localized sensorimotor enlargements might require disproportionate expansion of association cortex simply in order to maintain functional integration (Deacon, 1997). In either case, modest expansion of one or more of the structures supporting knapping behavior could plausibly have contributed to the broader pattern of brain enlargement seen in human evolution. In theory, such correlated expansion could also have produced neural precursors ripe for re-appropriation into evolving language circuits (Wilkins & Wakefield, 1995).

3. Finally, there is the possibility of microstructural adaptation. Although comparative research at the microstructural level is only just beginning, human specialization in primary visual cortex organization had already been documented (Preuss *et al.*, 1999). This is particularly interesting considering that some of the strongest activation observed during Oldowan-style knapping was in the primary and secondary visual cortices of the occipital lobe. The primary visual specialization reported by Preuss and colleagues specifically involves the "M-stream" of visual processing, commonly associated with motion perception and known to have strong outputs to the dorsal "position and motion" pathway in the superior parietal lobe. It will be especially interesting to see if further analysis bears out the preliminary indications of knapping-related superior parietal activation observed so far. As Preuss and colleagues point out, the microstructural reorganization of primary visual cortex is likely to have had cascading effects on such downstream visual processing areas. For the time being, it is reasonable to conjecture that M-stream adaptations in hominin primary visual cortex may have facilitated the initial invention(s) of Oldowan technology and/or been selected for in response to its later spread.

Conclusion

Although still preliminary at this stage, results from PET investigations of Mode 1 stone knapping offer tantalizing insights into the paleopsychological and neuro-evolutionary significance of the earliest stone tools. The experimental methods developed also offer exciting opportunities for the future. Research with non-human primates and investigation of more sophisticated bifacial and prepared-core technologies are two particularly interesting directions to be pursued. PET is a valuable new research tool for human origins studies and promises to add an important empirical dimension to inquiries regarding the evolution of the human brain and mind.

ACKNOWLEDGEMENTS

I would like to thank Nicholas Toth and Kathy Schick for the inspiration for the PET research presented here, as well as for advice and assistance in its execution. This research also would not have been possible without the participation and efforts of the experimental subjects. My thanks go to Julie Stout and David Kareken for assistance in data analysis, Gary Hutchins and Rich Fain for help with experimental design and execution, and PET Technologists Kevin Perry and Susan Geiger. Funding for the pilot research of Stout *et al.* (2000) was provided by the Center for Research into the Anthropological Foundations of Technology (CRAFT) at Indiana University, the office of Research and the University Graduate School at Indiana University, and the Indiana University School of Medicine. Funding for the ongoing follow-up research comes from the National Science Foundation (Award # BCS-0105265), the L.S.B. Leakey Foundation, CRAFT and Friends of CRAFT.

REFERENCES CITED

Ambrose, S. H. (2001). Paleolithic Technology and Human Evolution. *Science* **291**, 1748-1753.

Appelle, S. (1972). Perception and discrimination as a function of stimulus orientation: the oblique effect in man and animals. *Psychological Bulletin* **78**, 266-278.

Armstrong, E., Zilles, K., Curtis, M. & Schleicher, A. (1991). Cortical folding, the lunate sulcus and the evolution of the human brain. *Journal of Human Evolution* **20**, 341-348.

Asfaw, B., White, T., Lovejoy, O., Latimer, B., Simpson, S. & Suwa, G. (1999). A*ustralopithecus garhi*: A new species of early hominid from Ethiopia. *Science* **284**(23 April), 629-635.

Barton, R. A. (1998). Visual specialization and brain evolution in primates. *Procedings of the Royal Society of London* **265**, 1933-1937.

Barton, R. A. (2002). How did brains evolve? *Nature* **415**, 134-135.

Barton, R. A. & Harvey, P. H. (2000). Mosaic evolution of brain structure in mammals. *Nature* **405**, 1055-1058.

Belfer-Cohen, A. & Goren-Inbar, N. (1994). Cognition and communication in the Levantine Lower Palaeolithic. *World Archaeology* **26**(2), 144-157.

Bernstein, N. (1967). *Coordination and Regulation of Movement.* New York: Pergamon Press.

Blinkov, S. M. & Glezer, I. I. (1968). *The Human Brain in Figures and Tables: A Quantitative Handbook.* New York: Basic Books.

Boesch, C. (1993). Aspects of transmission of tool-use in wild chimpanzees. In (K. R. Gibson and T. Ingold, Eds) *Tools, Language and Cognition in Human Evolution.* Cambridge: Cambridge University Press, pp. 171-184.

Bongers, R. (2001) *An action perspective on tool use and its development.* University of Nijmegen: Unpublished Ph.D. Dissertation.

Boswell, J. (1887). *Life of Johnson, Vol. III.* Oxford: Clarendon Press.

Bradshaw, J. L. & Mattingley, J. B. (1995). *Clinical Neuropsychology: Behavioral and Brain Science.* New York: Academic Press.

Brooks, V. B. (1986). *The Neural Basis of Motor Control.* Oxford: Oxford University Press.

Buchanan-Smith, H. M. & Heeley, D. W. (1993). Anisotropic axes in orientation perception are not retinotopically mapped. *Perception* **22**(12), 1389-1402.

Buxhoeveden, D. P., Switalla, A. E., Roy, E., Litaker, M. & Casanova, M. F. (2001). Morphological differences between minicolumns in human and nonhuman Primate cortex. *American Journal of Physical Anthropology* **115**, 361-371.

Calvin, W. (1993). The unitary hypothesis: a common neural circuitry for novel manipulations, language, plan-ahead, and throwing? In (K. Gibson and T. Ingold, Eds) *Tools, Language and Cognition in Human Evolution.* Cambridge: Cambridge University Press, pp. 230-250.

Calvin, W. H. (1983). A stone's throw and its launch window: timing precision and its implications for language and hominid brains. *Journal of Theoretical Biology* **104**, 121-135.

Chase, P. G. (1991). Symbols and paleolithic artifacts: style, standardization and the imposition of arbitrary form. *Journal of Anthropological Archaeology* **10**, 193-214.

Chomsky, N. (1972). *Language and Mind.* New York: Harcourt Brace Jovanovich.

Clark, D. A., Mitra, P. P. & Wang, S. S. H. (2001). Scalable architecture in mammalian brains. *Nature* **411**, 189-193.

Clark, J. D. (1996). Decision-making and variability in the Acheulean. In (G. Pwiti and R. Soper, Eds) *Aspects of African Archaeology: Papers From the 10th Congress of the PanAfrican Association for Prehistory and Related Studies.* Harare: University of Zimbabwe Publications.

Cosmides, L. & Tooby, J. (1994). Origins of domain specificity: the evolution of functional organization. In (L. A. Hirschfeld and S. A. Gelman, Eds) *Mapping the Mind: Domain Specificity in Cognition and Culture.* Cambridge: Cambridge University Press, pp. 85-116.

Dacey, D. M. & Petersen, M. R. (1992). Dendritic field size and morphology of midget and parasol ganglion cells in the human retina. *Procedings of the National Academy of Sciences* **89**, 9666-9670.

Dagher, A., Owen, A. M., Boecker, H. & Brooks, D. J. (1999). Mapping the network for planning: a correlational PET activation study with the Tower of London task. *Brain* **122**, 1973-1987.

Darwin, C. (1871). *The Descent of Man.* London: John Murray.

De Valois, R. L., Morgan, H. & Snodderly, D. M. (1974). Psychophysical studies of monkey vision. 3. Spatial luminance contrast sensitivity tests of macaque and human observers. *Vision Research* **14**(1), 75-81.

Deacon, T. W. (1997). *The Symbolic Species: The Co-evolution of Language and the Brain.* New York: W.W. Norton.

Demb, J. B., Boynton, G. M. & Heeger, D. J. (1997). Brain activity in visual cortex predicts individual differences in reading performance. *Proceedings of the National Academy of Sciences* **94**, 13363-13366.

Dennett, D. (1993). Review of F. Varela, E. Thompson and E. Rosch, 'The Embodied Mind: Cognitive Science and Human Experience'. *American Journal of Psychology* **106**, 121-126.

Falk, D. (1980). Hominid Brain Evolution: The Approach From Paleoneurology. *Yearbook of Physical Anthropology* **23**, 93-107.

Falk, D. (1983). The Taung endocast: A reply to Holloway. *American Journal of Physical Anthropology* **53**, 525-539.

Finlay, B. & Darlington, R. (1995). Linked regularities in the development and evolution of mammalian brains. *Science* **268**, 1578-1584.

Finlay, B. L., Darlington, R. B. & Nicastro, N. (2001). Developmental structure in brain evolution. *Behavioral and Brain Sciences* **24**, 263-308.

Finney, S., Hernandez, N. G., Oates, T. & Kaelbing, L. P. (2001). Learning in worlds with objects. *Working Notes of the AAAI Stanford Spring Symposium on Learning Grounded Representations.* http://www.ai.mit.edu/people/lpk/publications.html

Fodor, J. (1983). *The Modularity of Mind*. Cambridge, MA: MIT Press.

Foley, R. (1992). Studying human evolution by analogy. In (S. Jones, R. Martin, D. Pilbeam, and S. Bunney, Eds) *The Cambridge Encyclopedia of Human Evolution*. Cambridge: Cambridge University Press, pp. 335-340.

Fox, P. T. & Applegate, C. N. (1988). Right-hemispheric dominance for somatosensory processing in humans. *Society of Neuroscience Abstracts* **14** pp. 760.

Furmanski, C. S. & Engel, S. A. (2000). An oblique effect in human primary visual cortex. *Nature Neuroscience* **3**(6), 535-536.

Gannon, P. J., Holloway, R. L., Broadfield, D. C. & Braun, A. R. (1998). Asymmetry of chimpanzee planum temporale: Humanlike pattern of Wernicke's brain language area homolog. *Science* **279**, 220-222.

Gardner, H. (1983). *Frames of Mind: The Theory of Multiple Intelligences*. New York: Basic Books.

Gatewood, J. (1985). Actions speak louder than words. In (J. Dougherty, Ed.) *Directions in Cognitive Anthropology*. Urbana: University of Illinois Press, pp. 199-219.

Gentaz, E. & Ballaz, C. (2000). The visual perception of orientation and the "oblique effect". *Annee Psychologique* **100**(4), 715-744.

Gibson, J. J. (1979). *The Ecological Approach to Visual Perception*. Boston: Houghton-Mifflin.

Gibson, K. R. (1983). Comparative neurobehavioral ontogeny and the constructionist approach to the evolution of the brain, object manipulation, and language. In (E. DeGrolier, Ed.) *Glossogenetics*. New York: Harwood, Academic Press, pp. 37-61.

Gibson, K. R. (1991). Myelination and behavioral development: a comparative perspective on questions of neoteny, altriciality and intelligence. In (K. R. Gibson and A. C. Petersen, Eds) *Brain Maturation and Cognitive Development*. New York: Aldine de Gruyter, pp. 29-64.

Gibson, K. R. (1993). Tool use, language and social behavior in relationship to information processing capacities. In (K. R. Gibson and T. Ingold, Eds) *Tools, Language and Cognition in Human Evolution*. Cambridge: Cambridge University Press, pp. 251-267.

Glezer, I. I. (1958). Area relationships in the precentral region in a comparative-anatomical series of primates. *Arkhiv Anatomii, Gistologii I Embriologii* **2**, 26.

Gowlett, J. A. J. (1984). Mental Abilities of Early Man: A Look at Some Hard Evidence. In (R. Foley, Ed.) *Hominid Evolution and Community Ecology*.

Gowlett, J. A. J. (1992). Early human mental abilities. In (S. Bunney and S. Jones, Eds) *The Cambridge Encyclopedia of Human Evolution*. Cambridge: Cambridge University Press, pp. 341-345.

Gowlett, J. A. J. (1996). Mental abilities of early *Homo*: Elements of constraint and choice in rule systems. In (P. Mellars and K. Gibson, Eds) *Modeling the Early Human Mind*. Cambridge: McDonald Institute for Archaeological Research, 191-215.

Grafton, S. T., Mazziotta, J. C., Presty, S., Friston, K. J., Frackowiak, R. S. J. & Phelps, M. E. (1992). Functional anatomy of human procedural learning determined with regional cerebral blood flow and PET. *The Journal of Neuroscience* **12**(7), 2542-2548.

Graves, P. (1994). Flakes and ladders: What the archaeological record cannot tell us about the origins of language. *World Archaeology* **26**(2), 158-171.

Greenfield, P. M. (1991). Language, tools, and brain: The development and evolution of hierarchically organized sequential behavior. *Behavioral and Brain Sciences* **14**, 531-595.

Gusnard, D. A. & Raichle, M. E. (2001). Searching for a baseline: Functional imaging and the resting human brain. *Nature Reviews Neuroscience* **2**(October), 685-694.

Haier, R. J., Seigel, B. V., McLachlan, A., Soderling, E., Lottenberg, S. & Buchsbaum, M. S. (1992). Cortical glucose metabolic changes after learning a complex visuospatial/motor task: A positron emission tomography study. *Brain Research* **570**, 134-143.

Harvey, P. H. & Pagel, M. D. (1991). *The Comparative Method in Evolutionary Biology*. Oxford: Oxford University Press.

Heffner, R. & Masterson, B. (1975). Variation in the form of the pyramidal tract and its relationship to digital dexterity. *Brain, Behavior and Evolution* **12**, 161-200.

Holloway, R. L. (1969). Culture: A human domain. *Current Anthropology* **10**(4), pp. 395-412.

Holloway, R. L. (1979). Brain size, allometry, and reorganization: toward a synthesis. In (M. Hahn, C. Jensen, and B. Dudek, Eds) *Development and Evolution of Brain Size*. New York: Academic Press.

Holloway, R. L. (1981a). Culture, symbols, and human brain evolution: a synthesis. *Dialectical Anthropology* **5**, 287-303.

Holloway, R. L. (1981b). Revisting the S. African Australopithecine endocasts: results of stereoplotting the lunate sulcus. *American Journal of Physical Anthropology* **64**, 285-288.

Holloway, R. L. (1983). Human brain evolution: A search for units, models and synthesis. *Canadian Journal of Anthropology* **3**(2), 215-230.

Holloway, R. L. (1995). Toward a synthetic theory of human brain evolution. In (J.-P. Changeux and J. Chavaillon, Eds) *Origins of the Human Brain*. Oxford: Clarendon Press, 42-54.

Hutchins, E. (1995). *Cognition in the Wild*. Cambridge, MA: MIT Press.

Isaac, G. L. (1981) Stone age visiting cards: approaches to the study of early land use patterns. In (I. Hodder, G. Isaac and N. Hammond, Eds) *Patterns of the Past*. Cambridge: Cambridge University Press.

Isaac, G. L. (1984). The archaeology of human origins: Studies of the Lower Pleistocene in East Africa 1971-1981. *Advances in World Archaeology* **3**, 1-87.

Isaac, G. L. (1986). Foundation stones: early artefacts as indicators of activities and abilities. In (G. N. Bailey and P. Callow, Eds) *Stone Age Prehistory: Studies in Honor of Charles McBurney*. London: Cambridge University Press, pp. 221-241.

Ito, M. (1993). Movement and thought: identical control mechanisms by the cerebellum. *Trends in Neurosciences* **16**(11), 448-450.

Johnson, M. (1987). *The Body in the Mind: The Bodily Basis of Meaning, Imagination and Reason.* Chicago: University of Chicago Press.

Kandel, E. R., Schwartz, J. H. & Jessell, T. M. (1991). *Principles of Neural Science.* Norwalk, CT: Appleton & Lange.

Karlin, C. & Julien, M. (1994). Prehistoric technology: a cognitive science? In (C. Renfrew and E. B. W. Zubrow, Eds) *The Ancient Mind: Elements of a Cognitive Archaeology.* Cambridge: Cambridge University Press.

Karmiloff-Smith, A. (1992). *Beyond Modularity: A Developmental Perspective on Cognitive Science.* Cambridge, MA: MIT Press.

Karni, A., Meyer, G., Jezzard, P., Adams, M. M., Turner, R. & Ungerleider, L. G. (1995). Functional MRI evidence for adult motor cortex plasticity during motor skill learning. *Nature* **377**(6545), 155-158.

Kirsch, D. (1991). Today the earwig, tomorrow man? *Artificial Intelligence* **47**, 161-184.

Kosslyn, S. M., Ganis, G. & Thompson, W. L. (2001). Neural foundations of imagery. *Nature Reviews Neuroscience* **2**(9), 635-642.

Krubitzer, L. & Huffman, K. J. (2000). Arealization of the neocortex in Mammals: Genetic and epigenetic contributions to the phenotype. *Brain, Behavior and Evolution* **55**, 322-335.

Kuypers, H. G. J. M. (1981). Anatomy of the descending pathways. In (J. M. Brookhart and V. B. Mountcastle, Eds) *Handbook of Physiology. The Nervous System II.* Washington, D.C.: American Physiological Society.

Lave, J. & Wegner, E. (1988). *Situated Learning: Legitimate Peripheral Participation.* Cambridge: Cambridge University Press.

Leakey, L., Tobias, P. & Napier, J. (1964). A new species of the genus *Homo* from Olduvai Gorge. *Nature* **202**, 7-9.

Leakey, M. D. (1971) *Olduvai Gorge, Volume 3: Excavations in Beds I and II, 1960-1963.* Cambridge: Cambridge University Press.

Leiner, H. C., Leiner, A. L. & Dow, R. S. (1986). Does the cerebellum contribute to mental skills? *Behavioral Neuroscience* **103**, 998-1008.

Leiner, H. C., Leiner, A. L. & Dow, R. S. (1993). Cognitive and language functions of the human cerebellum. *Trends in Neurosciences* **16**(11), 444-454.

Lemon, R. (1993). Control of the monkey's hand by the motor cortex. In (A. Berthelet and J. Chavaillon, Eds) *The Use of Tools by Human and Non-Human Primates.* Oxford: Clarendon Press, pp. 51-65.

Livingstone, M. S. & Hubel, D. H. (1988). Segregation of form, color, movement and depth: anatomy, physiology and perception. *Science* **240**, 740-749.

Lockman, J. J. (2000). A perception-action perspective on tool use development. *Child Development* **71**(1), 137-144.

Loffler, G. & Orbach, H. S. (2001). Ansiotropy in judging the absolute direction of motion. *Vision Research* **41**(27), 3677-3692.

Ludwig, B. V. & Harris, J. W. K. (1998). Towards a technological reassessment of East African plio-pleistocene lithic assemblages. In (M. Petraglia and R. Korisetter, Eds) *Early Human Behavior in the Global Context: The Rise and Diversity of the Lower Paleolithic Period.* New York: Routledge, pp. 84-107.

Marzke, M. W., Toth, N., Schick, K., Reece, S., Steinberg, B., Hunt, K. & Linscheid, R. L. (1998). EMG Study of Hand Muscle Recruitment During Hard Hammer Percussion Manufacture of Oldowan Tools. *American Journal of Physical Anthropology* **105**, 315-332.

Matsuzawa, T. (1996). Chimpanzee intelligence in nature and in captivity: isomorphism of symbol use and tool use. In (W. McGrew, L. Marchant, and T. Nishida, Eds) *Great Ape Societies.* Cambridge: Cambridge University Press, 196-209.

Maunsell, J. H. R. & Newsome, W. T. (1987). Visual processing in monkey extrastriate cortex. *Annual Review of Neuroscience* **10**, 363-401.

McGrew, W. C. (1992). *Chimpanzee Material Culture: Implications for Human Evolution.* Cambridge: Cambridge University Press.

McKinney, M. J. (2002). Brain evolution by stretching the global mitotic clock of development. In (N. Minugh-Purvis and K. J. McNamara, Eds) *Human Evolution Through Developmental Change.* Baltimore: Johns Hopkins Press, pp. 173-188.

Merrigan, W. H. (1980). Temporal modulation sensitivity of macaque monkeys. *Vision Research* **20**(11), 953-959.

Milner, A. D. & Goodale, M. A. (1995). *The Visual Brain in Action.* Oxford Psychology Series Oxford: Oxford University Press.

Mithen, S. (1996). *The Prehistory of the Mind: The Cognitive Origins of Art, Religion and Science.* London: Thames and Hudson Ltd.

Miyashita, Y. (1993). Inferior temporal cortex: Where visual perception meets memory. *Annual Review of Neuroscience* **16**, 245-263.

Nimchinsky, E. A., Gilissen, E., Allman, J. A., Perl, D. P. & Erwin, J. M. (1999). A neuronal morphologic type unique to humans and great apes. *Proceedings of the National Academy of Sciences* **96**, 5268-5273.

Noble, W. & Davidson, I. (1996). *Human Evolution, Language and Mind.* Cambridge: Cambridge University Press.

Nudo, R. J. & Masterson, R. B. (1990). Descending pathways to the spinal cord IV: Some factors related to the amount of cortex devoted to the corticospinal tract. *Journal of Comparative Neurology* **296**, 584-597.

Oakley, K. (1959). *Man the Tool-Maker.* Chicago: University of Chicago Press.

Parker, S. T. & Gibson, K. R. (1979). A developmental model for the evolution of language and intelligence in early hominids. *The Behavioral and Brain Sciences* **2**, 367-408.

Parker, S. T. & Mckinney, M. L. (1999). *Origins of Intelligence: The Evolution of Cognitive Development in Monkeys, Apes and Humans.* Baltimore: Johns Hopkins University Press.

Parker, S. T. & Milbrath, C. (1993). Higher intelligence, propositional language, and culture as adaptations for planning. In (T. Ingold and K. Gibson, Eds) *Tools, Language and Cognition in Human Evolution.* Cambridge: Cambridge University Press, pp. 314-333.

Parsons, L. M. & Fox, P. T. (1997). Sensory and cognitive functions. In (J. D. Schmahmann, Ed.) *The Cerebellum and Cognition.* New York: Academic Press, 255-271.

Pascual-Leone, A., Grafman, J. & Hallett, M. (1994). Modulation of cortical motor output maps during development of implicit and explicit knowledge. *Science* **263**, 1287-1289.

Passingham, R. E. (1973). Anatomical differences between the neocortex of man and other primates. *Brain, Behavior and Evolution* **7**, 337-359.

Passingham, R. E. (1975). Changes in the size and organization of the brain in man and his ancestors. *Brain, Behavior and Evolution* **11**, 73-90.

Passingham, R. E. (1998). The specializations of the human neocortex. In (A. D. Milner, Ed.) *Comparative Neuropsychology.* New York: Oxford University Press, 271-298.

Pelegrin, J. (1990). Prehistoric lithic technology: some aspects of research. *Archaeological Review from Cambridge* **9**(1).

Penrose, R. (1994). *Shadows of the Mind.* Oxford: Oxford University Press.

Phelps, M. E., Hoffman, E. J., Mullani, N. A. & Ter-Pogossian, M. M. (1975). Application of annihilation coincidence detection to transaxial reconstruction tomography. *Journal of Nuclear Medicine* **16**, 210-224.

Piaget, J. & Garcia, R. (1991). *Toward a Logic of Meaning.* New York: Basic Books.

Piaget, J. & Inhelder, B. (1969). *The Psychology of the Child.* New York: Harper.

Pigeot, N. (1990). Technical and social actors: flintknapping specialists and apprentices at Magdelinian Etoilles. *Archaeological Review from Cambridge* **9**, 127-141.

Pinker, S. (1994). *The Language Instinct: How the Mind Creates Language.* New York: William Morrow and Company.

Poeck, K. & Huber, W. (1977). To what extent is language a sequential activity? *Neuropsychologia* **15**, 359-363.

Poon, L. W., Welke, D. J. & Dudley, W. N. (1993). What is everyday cognition? In (J. Puckett and H. Reese, Eds) *Mechanisms of Everyday Cognition.* Hillsdale, NJ: Lawrence Erlbaum Associates.

Posner, M. I. & Raichle, M. E. (1994). *Images of Mind.* New York: Scientific American Library.

Potts, R. (1993). Archaeological interpretations of early hominid behavior and ecology. In (D. Tab Rasmussen, Ed.) *The Origin and Evolution of Humans and Humanness.* Boston: Jones and Bartlett Publisher, 49-74.

Preston, B. (1998). Cognition and tool use. *Mind and Language.* **13**(4), 513-547.

Preuss, T. M., Huixin Qi & Kaas, J. H. (1999). Distinctive compartmental organization of human primary visual cortex. *Proceedings of the National Academy of Sciences* **96**(20), 11601-11606.

Raichle, M. E. (1994). Images of the mind: studies with modern imaging techniques. *Annual Review of Psychology* **45**, 333-356.

Reed, E. S. & Bril, B. (1996). The primacy of action in development. In (M. L. Turvey and M. T. Latash, Eds) *Dexterity and Its Development.* Mahwah, New Jersey: Lawrence Erlbaum and Associates, Publishers, 431-452.

Reynolds, P. C. (1976). Language and skilled activity. In (Steklis, H. B., Harnad, S. R. & Lancaster, J., Eds) *Origins and Evolution of Language and Speech.* New York: New York Academy of Sciences.

Rilling, J. K. & Insel, T. R. (1998). Evolution of the cerebellum in primates: Differences in relative volume among monkeys, apes and humans. *Brain, Behavior and Evolution* **52**, 308-314.

Rilling, J. K. & Insel, T. R. (1999). The primate neocortex in comparative perspective using magnetic resonance imaging. *Journal of Human Evolution* **37**, 191-223.

Rilling, J. K. & Seligman, R. (2002). A quantitative morphometric comparative analysis of the primate temporal lobe. *Journal of Human Evolution* doi:10.1006/jhev.2001.0537 pp. 29.

Robson Brown, K. (1993). An alternative approach to cognition in the Lower Paleolithic: the modular view. *Cambridge Archaeological Journal* **3**(2), 231-245.

Roche, H., Delagnes, A., Brugal, J.-P., Feibel, C., Kibunjia, M., Mourre, V. & Texier, P.-J. (1999). Early hominid stone tool production and technical skill 2.34 Myr ago in West Turkana, Kenya. *Nature* **399**(May 6), 57-60.

Rogoff, B. (1984). Introduction: Thinking and Learning in Social Context. In (B. Rogoff and J. Lave, Eds) *Everyday Cognition: It's Development in Social Context.* Cambridge, MA: Harvard University Press, pp. 1-8.

Roland, P. E. (1985). Somatosensory detection in man. *Experimental Brain Research, Supplement* **10**, 93-110.

Roland, P. E. (1993). *Brain Activation.* New York: Wiley-Liss.

Roland, P. E., Gulyas, B., Seitz, R. J., Bohm, C. & Stone-Elander, S. (1990). Functional anatomy of storage recall, and recognition of a visual pattern in man. *NeuroReport* **1**, 53-56.

Roland, P. E., Larsen , B., Lassen, N. A. & Skinhöj, E. (1980). Supplementary motor area and other cortical areas in the organization of voluntary movements in man. *Journal of Neurophysiology* **43**, 118-136.

Rolando, L. (1809). *Saggio sopra la vera struttura del cerbello dell'uome e degli animali e sopra le funzoini del sistema nervosa.* Stampeia da S.S.R.M. Privilegiata, Sassari.

Roux, V., Bril, B. & Dietrich, G. (1995). Skills and learning difficulties involved in stone knapping . *World Archaeology* **27**(1), 63-87.

Rudel, R. G. (1982). The oblique mystery. In (M. Potegal, Ed.) *Spatial Abilities.* New York: Academic Press, 129-145.

Rumbaugh, D. M., Savage-Rumbaugh, E. S. & Washburn, D. A. (1996). Toward a new outlook on primate learning and behavior: complex learning and emergent processes in comparative perspective. *Japanese Psychological Research* **38**(3), 113-125.

Savage-Rumbaugh, E. S. & Lewin, R. (1994). *Kanzi: The Ape at the Brink of the Human Mind.* New York: John Wiley & Sons.

Schick, K. & Toth, N. (2001). Paleoanthropology at the Millennium. In (Feinman and Price, Eds) *Archaeology at the Millenium: A Sourcebook.* New York: Kluwer Academic/Plenum Publishers, pp. 39-108.

Schick, K. D. & Toth, N. (1993). *Making Silent Stones Speak: Human Evolution and the Dawn of Technology.* New York: Simon & Schuster.

Schick, K. D., Toth, N., Garufi, G., Savage-Rumbaugh, E. S., Rumbaugh, D. & Sevcik, R. (1999). Continuing Investigations into the Stone Tool-making and Tool-using Capabilities of a Bonobo (*Pan paniscus*). *Journal of Archaeological Science* **26**, 821-832.

Schlanger, N. (1994). Mindful technology: unleashing the *chaine operatoire* for an archaeology of mind. In (C. Renfrew and E. B. W. Zubrow, Eds) *The Ancient Mind: Elements of a Cognitive Archaeology.* New York: Cambridge University Press, 143-151.

Schmahmann, J. D. (1997a). *The Cerebellum and Cognition.* New York: Academic Press.

Schmahmann, J. D. (1997b). Redisovery of an early concept. In (J. D. Schmahmann, Ed.) *The Cerebellum and Cognition.* New York: Academic Press, pp. 3-27.

Schmahmann, J. D. & Pandya, D. N. (1997). The cerebrocerebellar system. In (J. D. Schmahmann, Ed.) *The Cerebellum and Cognition.* New York: Academic Press, pp. 31-59.

Segalowitz, S. (2000). Dynamics and variability of brain activation: searching for neural correlates of skill acquisition. *Brain and Cognition* **42**, 163-165.

Semaw, S. (2000). The world's oldest stone artefacts from Gona, Ethiopia: Their implications for understanding stone technology and patterns of human evolution between 2.6-1.5 Million Years Ago. *Journal of Archaeological Science* **27**, 1197-1214.

Semaw, S., Renne, P., Harris, J. W. K., Felbel, C. S., Bernor, R. L., Fesseha, N. & Mowbray, K. (1997). 2.5-million - year-old stone tools from Gona, Ethiopia. *Nature* **385**(Jan 23), 333-336.

Semendeferi, K., Damasio, H., Frank, R. & Hoesen, G. (1997). The evolution of the frontal lobes: a volumetric analysis based on three-dimensional reconstructions of magnetic resonance scans of human and ape brains. *Journal of Human Evolution* **32**, 375-388.

Semendeferi, K. & Damasio, H. (2000). The brain and its main anatomical subdivisions in living hominioids using magnetic resonance imaging. *Journal of Human Evolution* **38** pp. 317-332.

Siegel, R. M. & Reed, H. L. (1998). Construction and representation of space in the inferior parietal lobule. In (K. S. Rockland, J. H. Kaas, and A. Peters, Eds) *Extrastriate Cortex in Primates.* Kluwer Academic Publishers.

Stephan, H., Frahm, H. D. & Baron, G. (1981). New and revised data on volumes of brain structures in insectivores and primates. *Folia Primatologica* **35**, 1-29.

Stiles, D. (1998). Raw material as evidence for human behaviour in the Lower Pleistocene: the Olduvai case. In *Early Human Behavior in the Global Context: The Rise and Diversity of the Lower Paleolithic Period.* New York: Routledge, pp. 133-150.

Stout, D. (in prep.). *Stone tools and the evolution of human thinking: Cultural, biological and archaeological elements in an Anthropology of human origins.* Ph.D. dissertation, Indiana University, Bloomington.

Stout, D. (in press). Skill and cognition in stone tool production: An ethnographic case study from Irian Jaya. *Current Anthropology* **45**(3).

Stout, D. (2001). Constraint and adaptation in primate brain evolution. *Behavioral and Brain Sciences* **24**(2), pp. 295-296.

Stout, D., Toth, N., Schick, K., Stout, J. & Hutchins, G. (2000). Stone Tool-Making and Brain Activation: Positron Emission Tomography (PET) Studies. *Journal of Archaeological Science* **27** pp. 1215-1223.

Sultan, F. (2002). Analysis of mammalian brain architecture. *Nature* **415**, 133-134.

Ter-Pogossian, M. M., Phelps, M. E., Hoffman, E. J. & Mullani, N. A. (1975). A positron-emission transaxial tomograph for nuclear imaging (PETT). *Radiology* **114**, 89-98.

Thelen, E. & Smith, L. (1994). *A Dynamic Systems Approach to the Development of Cognition and Action.* Cambridge, MA: MIT Press/Bradford Books.

Tobias, P. V. (1979). Men, minds and hands: Cultural awakenings over two million years of humanity. *South African Archaeological Bulletin* **34**, 85-92.

Tobias, P. V. (1991). *Olduvai Gorge, Vols. 4A and 4B. The Skulls Endocasts and Teeth of Homo Habilis.* New York: Cambridge University Press.

Tomasello, M. (1999). *The Cultural Origins of Human Cognition.* Cambridge, MA: Harvard University Press.

Tomasello, M., & Call, J. (1997). *Primate Cogniton.* New York: Oxford University Press.

Toth, N. (1982). *The stone technologies of early hominids at Koobi Fora, Kenya: an experimental approach.* University of California, Berkeley: Unpublished Ph.D. Dissertation.

Toth, N. (1985). The Oldowan Reassessed: A Close Look at Early Stone Artifacts. *Journal of Archaeological Science* **12** pp. 101-120.

Toth, N. & Schick, K. D. (1993). Early stone industries and inferences regarding language and cognition. In (K. R. Gibson & T. Ingold, Eds) *Tools, Language and Cognition in Human Evolution.* Cambridge: Cambridge University Press, pp. 346-362.

Tovee, M. J. & Cohen-Tovee, E. M. (1993). The neural substrates of face processing models: A review. *Cognitive Neuropsychology* **10**, 505-528.

Turvey, M. T. & Shaw, R. E. (1979). The primacy of perceiving: An ecological reformulation of perception for understanding memory. In (L. G. Nillson, Ed.) *Perspectives in Memory Research: Essays in Honor of Uppsala University's 500th Anniversary.* Hillsdale, NJ: Lawrence Erlbaum Associates.

Ungerleider, L. G. & Mishkin, M. (1982). Two cortical visual systems. In (D. J. Ingle, M. A. Goodale, and R. J. W. Mansfield, Eds) *Analysis of Visual Behavior.* Cambridge, MA: MIT Press, pp. 549-586.

Varela, F., Thompson, E. & Rosch, E. (1991). *The Embodied Mind.* Cambridge, MA: MIT Press.

Vera, A. H. & Simon, H. A. (1993). Situated action: a symbolic interpretation. *Cognitive Science* **17**, 7-48.

Vygotsky, L. S. (1976). *Mind in Society: The Development of Higher Psychological Processes.* Cambridge, MA: Harvard University Press.

Wallace, R. (1995). Microscopic computation in human brain evolution. *Behavioral Science* **40**(2), 133-158.

Wang, S. S. H., Mitra, P. P. & Clark, D. (2002). Reply to Sultan and Barton. *Nature* **415**, 135.

Washburn, S. L. (1959). Speculations on the interrelations of the history of tools and biological evolution. In (Spuhler, Ed.) *The Evolution of Man's Capacity for Culture.* Detroit: Wayne State University Press.

Washburn, S. L. (1960). Tools and Human Evolution. *Scientific American* **203**(9), 63-75.

Wilkins, W. & Wakefield, J. (1995). Brain evolution and neurolinguistic preconditions. *Behavioral and Brain Sciences* **18**, 161-226.

Winter, W. & Oxnard, C. E. (2001). Evolutionary radiations and convergences in the structural organization of mammalian brains. *Nature* **409**, 710-714.

Woodruff-Pak, D. S., Vogel, R. W., Ewers, M., Coffey, J., Boyko, O. B. & Lemieux, S. K. (2001). MRI-assesses volume of cerebellum correlates with associative learning. *Neurobiology of Learning and Memory* **76**, 342-357.

Wynn, T. (1989). *The Evolution of Spatial Competence.* Illinois Studies in Anthropology 17. University of Illinois Press.

Wynn, T. (1991). The comparative simplicity of tool-use and its implications for human evolution. *Behavioral and Brain Sciences* **14**, 576-577.

Wynn, T. (1995). Handaxe enigmas. *World Archaeology* **27**(1), pp. 10-24.

Wynn, T. & McGrew, W. (1989). An ape's view of the Oldowan. *Man* **24** pp. 383-398.

CHAPTER 10

KNAPPING SKILL OF THE EARLIEST STONE TOOLMAKERS: INSIGHTS FROM THE STUDY OF MODERN HUMAN NOVICES

BY DIETRICH STOUT AND SILESHI SEMAW

ABSTRACT

Defining the knapping skill level and associated mental capabilities of the earliest stone toolmakers is a major objective for the archaeology of human origins. This requires that technological indicators of hominin knapping skill be identified and differentiated from raw material influences. To this end, products from two experiments with novice stone knappers were compared to a sample of artifacts from the site of EG10 in the Gona study area, Ethiopia (Semaw, 1997, 2000; Semaw et al. 1997). In the first experiment, three individuals were given cobbles of a highly distinctive variety of trachyte identical to that used by the Pliocene toolmakers at EG10. The continued availability of this material in the Gona study area provides an ideal opportunity for the experimental control of raw material influences. Data from a second experiment conducted as part of an unrelated research project (Stout, in prep., this volume) was also considered in order to provide an additional dimension of comparison. In this experiment, a single novice subject was given a variety of raw materials from a local (Martinsville, Indiana) quarry, including quartz, quartzite and limestone. Finally, an archaeological sample consisting of all surface and *in situ* EG10 artifacts manufactured in the distinctive "Gona trachyte" was taken. Comparison of these three samples revealed important raw material influences as well as material-independent differences in knapping techniques and flake attributes. The EG10 toolmakers were found to have employed a more uniform knapping technique and to have produced thicker flakes than modern human novices. These differences indicate a relatively high skill level among the EG10 hominins, and have important implications for the mental capabilities of the earliest stone toolmakers.

INTRODUCTION

Flaked stone artifacts dating from as much as 2.55 Ma. (Semaw *et al.*, 1997) provide some of our earliest and most fine-grained evidence of evolving hominid behavior and intelligence. Understanding this ancient evidence requires diverse actualistic studies in the modern world, including ethnographic (Stout, in press), replicative (Bordes, 1947; Crabtree, 1966; Callahan, 1979; Toth, 1985), mechanical (Speth, 1974; Dibble & Pelcin, 1995), biomechanical (Marzke *et al.*, 1998), and neurological (Stout *et al.*, 2000; Stout, this volume) research.

This chapter describes an application of the replicative approach to the question of Oldowan/Mode I (Clarke, 1969) knapping skill. Previous work (Toth, 1985) has established that the simple goal of flake production is sufficient to account for the vast majority of artifact types known from the earliest Stone Age. The question remains, however, as to the degree of skill and sophistication that was deployed by Plio-Pleistocene hominids in pursuit of this goal.

Isaac (1984) advocated a stepwise approach to understanding variation in early artifact assemblages, arguing that "among the known very early assemblages, raw-material and least-effort factors account for most differences, leaving very little if any residual variance on which to base either activity facies or culture-historic kinds of interpretations" (p. 161). The exact meaning of "least-effort" flaking was not, however, defined by Isaac, who instead called for the experimental investigation of flaking strategies "as responses to particular forms of raw material."

The residual approach has been very useful for Early Stone Age archaeologists, but tends to favor artificially exaggerated distinctions between social, functional and mechanical aspects of stone knapping. Stone knapping, like other primate tool-using behaviors (Boesch & Boesch, 1990; McGrew, 1992; van Schaik *et al.* 1999), is a goal-oriented activity that is learned and performed in a physical, social, and (in humans) cultural environment (Stout, in press). No knapping strategy, however simple, exists apart from the social environment in which it is learned, the raw materials in which it is executed and the functional purposes that it serves. One way in which to achieve a more integrated perspective is to consider stone knapping as a form of skilled action.

Isaac (1984) did not explicitly consider the issue of knapping skill but did emphasize the psychological significance of understanding the "design concept and motor patterns actually involved in making [stone tools]" (p. 161). Despite the simplicity of the goals and products of Mode I flaking, more or less complicated knapping processes and degrees of skill may be employed (Schick & Toth, 1993:133-134). With this in mind, researchers have begun to comment on the perceptual-motor sophistication apparent from early stone tools (e.g. Semaw, 2000; Ambrose 2001). Along these lines, Roche *et al.* (1999) describe a pattern of "unidirectional or multidirectional removals [from] a single debitage surface" at the Pliocene (ca. 2.34 Ma.) site of Lokalelei 2C in Kenya, which they argue is reflective of relatively advanced strategic and motor skill on the part of the toolmakers. In contrast, the high incidence of step fractures on artifacts from the similarly aged (ca. 2.35 Ma.) site of Lokalalei GaJh 5 has been seen as providing evidence of poorly understood and executed flaking techniques (Roche, 1989; Kibunja, 1994).

Technological case studies like these are essential in order to document the nature and diversity of early hominid knapping skills, but need to be interpreted in light of evidence from experimental replication. Quantitative and replicable experimental observations provide a consistent, external standard of comparison that is not available in the archaeological record itself. For example, Kibunja (1994) asserts that step fractures at Lokalalei GaJh 5 are indicative of unskilled knapping because his "surface observation" of raw materials and inspection of cores did not reveal any material flaws. Ludwig (1999; Ludwig & Harris, 1998), however, reports that his inspection of the GaJh 5 materials suggests that "subtle material flaws" did in fact play "a major role in the formation of these step fractures." This basic difference in qualitative evaluation leads to major differences in theoretical interpretation, but could be relatively easily and conclusively resolved through quantitative experimentation with authentic raw materials and knappers of known skill level.

Such an approach is adopted here with respect to the Pliocene site of EG10 from the Gona study area in the Afar region of Ethiopia (Semaw, 2000; Semaw 1997; Semaw *et al.* 1997). The site is well dated to between 2.52 and 2.6 Ma on the basis of [40]Ar/[39]Ar and paleomagnetic evidence, making it one of several sites in the Gona study area that are the oldest known in the world. As such, EG10 is of special significance to researchers interested in the origins and evolution of stone knapping.

The Experiments

Rationale

Raw material variation is a major confounding factor in attempts to assess the skill-level of pre-historic Mode I (Oldowan) stone knappers. In the current study, it was possible to control for raw material influences by conducting experiments with raw materials identical to those from a selected archaeological sample. A further experiment with a single knapper and a variety of raw materials provided complementary insights into the effects of raw material variation with skill level held constant. By carefully differentiating between raw material and skill related influences, these experiments provided an empirical baseline for assessing the knapping skills evident at EG10.

Typically, evaluation of the skill-level of prehistoric knappers has been based upon qualitative assessments by experienced experimental knappers. Systematic and quantitative studies of skill-related variation are quite rare (e.g. Ludwig, 1999), although Schick & Toth (1993: 133-4) have made the more general observation that "the efficient [Mode I] flaking of stone...is a skill that can take a number of hours to master, even for a modern human." While it is impossible to directly study knapping skill acquisition in extinct hominin species, studies of modern humans (and other extant primates) offer an essential reference point. The current study evaluated the knapping products of novice modern humans in order to complement previous work with experienced human knappers and non-human primates (Schick *et al.*, 1999; Toth, 1985, 1997; Toth *et al.*, 2002, this volume; Wright, 1972).

Archaeological Context

The research presented here consists of the comparison of two experimental knapping samples with a subset of the artifacts recovered from EG10 by Semaw and colleagues (Semaw *et al.*, 1997). EG10 comprises a high-density scatter of lithic artifacts eroding out of fine-grained sediments on the east side of the Kada Gona drainage, and represents an excellent opportunity for experimental study due both to its sedimentary and stratigraphic context and to the nature of the raw materials used by the toolmakers. The fine-grained sedimentary context, the freshness of the artifacts, and the lack of vertical dispersion or apparent size sorting all suggest that the assemblage is relatively undisturbed. The site sits in floodplain silts stratigraphically above sands and a cobble conglomerate (the *Conglomerat Intermediaire* or "Intermediate Conglomerate" of Roche & Tiercelin [1977,1980]) all laid down by the same meandering channel. The cobble conglomerate below the EG10 site represents the remains of what, on the paleolandscape, would have been exposed cobble bars in this channel. Evidence of a channel cut-and-fill located between the sites of EG12 and EG13 (Semaw, 2000) further indicates that cobbles in the Intermediate Conglomerate were exposed in smaller feeder-channels on the floodplain itself. The EG10 artifacts, clearly made from river cobbles, almost certainly originated from either the axial channel or its tributaries. In fact, proximity to raw materials may have been an important factor leading to hominin activity at this site.

The Intermediate Conglomerate provides an exceptional record of the raw materials available to the EG10 hominids. One particularly distinctive raw material found both at EG10 and in the conglomerate is a variety of trachyte. The Gona trachyte is fine-grained, light brown or gray in color often with phenocrysts and dark brown cortex. It is a high-quality raw material with excellent fracture properties, and was clearly preferred by the EG10 hominids as shown by its disproportionate representation in the artifact assemblage compared with the cobble conglomerate (Semaw 2000). The distinctiveness of the Gona trachyte and its continued availability in the cobble conglomerates within the study area provide an ideal situation for the experimental control of raw material variation.

METHODS

Two experiments in the Mode I production of flakes by modern human novices were carried out. The first took place during the 2000 field season in the Gona research area, Afar, Ethiopia. Cobbles (~ 80 - 160 mm in maximum dimension) of the distinctive "Gona trachyte" were selected by one of the researchers (D.S.) from conglomerates in the study area and made available to three novice experimental knappers. These individuals were familiar with the general appearance of the archaeological materials found in the area, but had never previously attempted to replicate them and had no prior instruction, either practical or theoretical, in stone knapping or artifact typology. They were asked to produce stone flakes like those they were familiar with from the study area. Knapping occurred in 2-3 approximately 15-minute long sessions per individual, amounting to a total of no more than one hour's experience for any one individual. All products were collected and analyzed, and are currently stored at the National Museum

in Addis Ababa. This experiment allowed for exploration of variation in the flaking of a single raw material type between modern novices and the EG10 toolmakers.

A second experiment took place in Indianapolis, Indiana in August, 2001, as part of an unrelated research project being conducted by one of the authors (D.S.). This project, involving the use of Positron Emission Tomography to explore the neural substrates of Mode I knapping skill (Stout, this volume), is still underway, but a small sample of the experimental artifacts thus far produced have been analyzed and are presented here. These artifacts were produced during a single 40-minute session by one novice stone knapper with no prior experience. In an important difference from the Gona experiment, raw materials made available to the knapper came from a local (Martinsville, Indiana) quarry. These included an assortment of limestone, quartz, and quartzite cobbles of approximately 100 - 200 mm in maximum dimension. This experiment allowed for exploration of raw material influences on flaking by a novice knapper, and comparison with the flaking of Gona trachyte by both modern novices and prehistoric hominins.

The experimental samples from Gona and Indianapolis were compared both with each other and with a sample taken from the EG10 archaeological collection stored at the National Museum in Addis Ababa. The archaeological sample included all *in situ* and surface artifacts > 20 mm in maximum dimension made in the distinctive "Gona trachyte" described above. Because raw material identification from small hand specimens is not always definitive, questionable cases were excluded from the sample.

A Student's t-test indicated that, of the metric attributes analyzed for this study, the only significant difference between *in situ* and surface flakes was the slightly more obtuse interior platform angle of the *in situ* flakes ($110°$ vs. $103°$, $p = 0.093$). Surface flakes also included a small percentage (17.2 %) with non-cortical platforms, which were completely absent from the *in situ* flakes. Neither of these differences would affect the conclusions and interpretations presented in this chapter.

For each sample, all artifacts > 20 mm in maximum dimension were analyzed. The resulting numbers and types of artifacts in each sample are summarized in Table 1, with relative frequencies represented in Figure 1. Data from the cores, core fragments and whole flakes will be dealt with here. Cores and core fragments were described in typological and qualitative terms, as well as by quantitative attributes. Quantitative core attributes recorded included length (defined as maximum dimension), breadth, thickness, # of flake scars (> 20 mm), # steps and hinges and % cortex remaining. The total number of detached pieces per core was recorded, as was the number of whole flakes per core. Whole flake

Figure 1: Frequency of Artifact Types

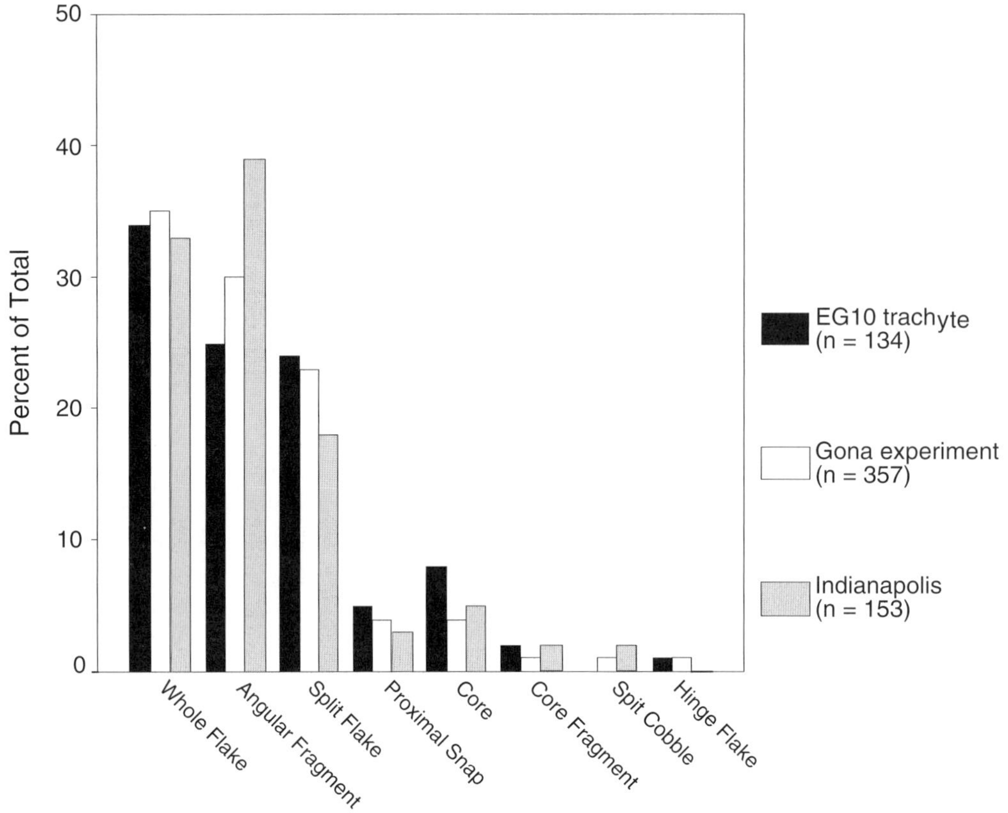

1. *Percentage representation of major artifact types in the three samples. Percentages are remarkably similar across samples with the exception of the greater representation of cores at EG10 and the higher proportion of angular fragments in the Indianapolis experiment. The latter is likely a reflection of differences in raw material - both in terms of fracture properties and expression of distinctive flake morphology.*

Table 1: The Experimental and Archaeological Samples

Sample	Cores	Core Fragments	Split Cobbles	Whole Flakes	Split Flakes	Proximal Snaps	Angular Fragments	Hinge Flakes	Total
EG10	11 (8%)	3 (2%)	0 (0%)	45 (34%)	32 (24%)	7 (5%)	34 (25%)	2 (1%)	134 (100%)
Gona Experiment	15 (4%)	5 (1%)	3 (1%)	126 (35%)	81 (23%)	16 (4%)	107 (30%)	4 (1%)	357 (100%)
Indianapolis Experiment	7 (5%)	3 (2%)	3 (2%)	50 (33%)	27 (18%)	4 (3%)	59 (39%)	0 (0%)	153 (100%)

attributes recorded included length (perpendicular to the striking platform), maximum breadth, maximum thickness, maximum dimension, platform thickness, platform breadth, external core angle, internal bulb angle, # dorsal scars, # steps and hinges, # platform scars, % dorsal cortex, % platform cortex and Toth's (1985) flake type. Artifacts from the Indianapolis experiment were also described in terms of raw material.

Unless otherwise noted, comparisons between samples were made using two-tailed Student's t-tests. Because many samples were not normally distributed, non-parametric Kolmogorov-Smirnov and Mann-Whitney tests were also employed. Results of these tests were consistent with those obtained using the parametric test, and did not suggest alternative interpretations.

RESULTS AND INTERPRETATION
Archaeological and Experimental Samples from Gona
Knapping Strategy

Comparison of the modern experimental sample from Gona with the archaeological materials from EG10 reveals important technological differences. To begin with, there is a difference in basic flaking strategy. The EG10 trachyte artifacts are dominated by the products of unifacial flaking, identifiable directly in terms of core morphology and indirectly in terms of flake attributes. Only one out of eleven EG10 trachyte cores examined for this study displayed flake scars indicative of bifacial reduction. In contrast, four out of fifteen experimental cores provide definitive evidence of bifacial reduction.

Also strongly indicative of unifacial flaking is the observed distribution of cortex on the whole flakes in the archaeological sample (Figure 2). In order to describe this patterning, the flake classification system of Toth (1985) is useful. This system recognizes six different flake types based on the presence or absence of cortex on the striking platform and dorsal surface: Types I, II and III have cortical platforms with completely cortical (I), partially cortical (II) or completely non-cortical (III) dorsal surfaces, while Types IV, V and VI have non-cortical platforms with completely cortical (IV), partially cortical (V) or completely non-cortical (VI) dorsal surfaces.

The EG10 trachyte whole flake sample is clearly dominated (86%) by Types II and III. The prevalence of cortical platforms demonstrates that the toolmakers were not exploiting scars from previous flake removals as platforms for subsequent removals. The unifacial reduction of four quartzite cobbles by one of the authors

Figure 2: Distribution of Flake Types in the Three Samples

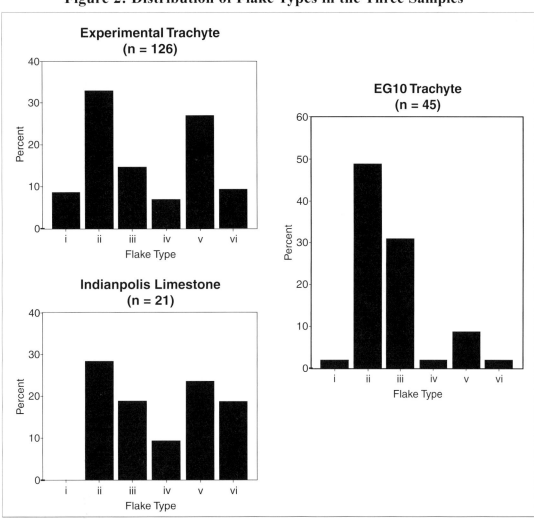

2. *The relative frequency of Toth's (1985) flake types in the three samples under discussion. While both experimental samples show a similar distribution, the archaeological sample is exceptional in its near total domination by flake types II &III (flakes with cortical platforms and a partial [II] or total [III] absence of dorsal cortex). This reflects unifacial flaking of moderate to heavy intensity in the EG10 trachyte. The absence of whole flakes of type I from the limestone samples indicates that the relatively small number of cortical flake removals that did occur shattered to produce fragments rather than whole flakes.*

(D.S.) confirms this relationship between flake types and reduction strategy: of 26 whole flakes with intact platforms, none displayed non-cortical platforms. Instead, the sample consisted entirely of Types I (27%), II (50%) and III (23%). The relatively high representation of Type I flakes in this experiment reflects reduction much less exhaustive than that seen in the EG10 trachyte.

In the EG10 trachyte, the predominance of flakes with a partial (Types II & V, 61% of flakes) or total (Types III & VI, 33% of flakes) absence of dorsal cortex indicates moderate to heavy reduction intensity. Although some initial flaking may have occurred off-site (Schick, 1987), flakes types represented at EG10 clearly indicate substantial subsequent reduction. The dorsal morphology of EG10 Type II & V flakes, which display an average of 3 flake scars and only 27% cortex, is also indicative of relatively intense reduction. These trends, coupled with inspection of the EG10 trachyte cores, reveal a unifacial knapping strategy in which successive flakes were removed from the same area through reiterated strikes to the cortical surface of the core. This uniform strategy produces cores which are easily classified using Mary Leakey's (1971) typology, and which are dominated by unifacial side and end choppers (7 out of 11 or 64%). This strategy has the further effect of producing relatively spherical cores though preferential reduction of the long axis. The result is a ratio of mean core length:width:thickness that is approximately (1.6):(1.3):(1.0). In contrast, a random sample of 20 cobbles from one of the conglomerates exposed in the Gona study area produced a ratio of approximately (2.1):(1.4):(1.0).

The pattern of reduction evident from the Gona experimental cores and flakes is much different. Both bifacial and unifacial reduction are evident from inspection of cores, and knapping does not tend to be concentrated in any one region of the core. The experimental cores in general are much more difficult to classify according to Mary Leakey's (1971) typology, and many are best described as either "irregular polyfaceted" or "casual" cores using Isaac's (1977: 176-7) typology. Because there is no preferential reduction of the long axis, experimental cores tend to be more elongated, with a length:width:thickness ratio of (2.0):(1.5):(1.0) that is very similar to that found in unmodified cobbles (above). The distribution of flake types (Figure 2) also shows a different pattern, with a much greater representation of flakes with non-cortical platforms (43.5% compared with 14% from EG10). A Chi-Square test confirms that the distribution of flake types in the two samples is significantly different (p < 0.001).

This difference in the pattern of flake type representation indicates that experimental subjects were much more likely to exploit scars from previous flake removals in selecting a striking platform. A majority (60%) of experimental whole flakes displayed dorsal surfaces that were partially cortical. On average, these flakes possessed cortex on 45% of their dorsal surface

and 1.6 flake scars. Thus, although flakes with partially cortical dorsal surfaces dominated both experimental and archaeological samples, these partially cortical flakes displayed a much higher percentage of dorsal cortical surface area (45% vs. 27%, p < 0.001) in the experimental sample, as well as fewer flake scars (1.6 vs. 2.8, p = 0.023). These differences reflect a combination of less intense and less uniformly localized flaking by experimental subjects, with dorsal flake surfaces being less heavily modified by previous flake removals from the same part of the core.

Although it should be remembered that the stone artifacts recovered from EG10 include only those that "fell out" (Isaac 1984: 150) of the system of hominin tool transport and modification at this particular location, and which further survived subsequent site formation processes to be recovered by archaeologists, it would be difficult to construct a plausible scenario of transport and/or winnowing to account for the multifaceted and consistent pattern observed. Observed patterns in core morphology, flake type representation, percentages of dorsal cortex and numbers of dorsal flake scars all combine to produce a picture of reduction in the EG10 trachyte sample that is much different from that evident in the experimental sample. Whereas the EG10 toolmakers pursued a uniform strategy for the localized, unifacial reduction of trachyte cobbles, experimental subjects were more "opportunistic" in their approach, exploiting potential striking platforms wherever they occurred.

Flake Metrics

Despite these major differences in basic knapping strategy, whole flakes from the Gona experimental and archaeological samples are actually quite similar metrically (Table 2). In fact, the only significant differences are in the greater mean thickness of flakes and flake platforms in the archaeological sample. The fact that the two samples differ in thickness but not other measures suggests that the archaeological flakes are relatively as well as absolutely thicker. This is supported by the observation that the ratio of thickness to breadth is significantly greater in the archaeological sample and the ratio of maximum dimension to thickness is significantly less.

The Experimental Sample from Indianapolis

The small experimental sample from Indianapolis displays interesting commonalities and contrasts with the experimental sample from Gona, as well as with the archaeological materials from EG10. To begin with, it should be noted that the raw materials used in this experiment, predominantly limestone, quartz and quartzite, were quite different both from the fine-grained Gona trachyte and from each other. Raw material variation clearly exerted a major influence on the technological patterns observed in the Indianapolis sample.

Table 2: Flake Attributes

Attribute	Statistic	EG10 Trachyte	Gona Experiment	Indianapolis Limestone
Length (mm)	mean	**42.5**	**40.2**	**42.6**
	s.d.	*14.6*	*16.7*	*17.7*
	(range)	(18.2 – 73.8)	(9.8 – 97.0)	(9.6 – 90.4)
Breadth (mm)	mean	**40.6**	**39.8**	**39.5**
	s.d.	*14.3*	*16.9*	*11.0*
	(range)	(18.0 – 82.8)	(9.8 – 87.5)	(19.0 – 64.2)
Thickness (mm)	mean	**16.0**	**12.8****	**13.7**
	s.d.	*6.8*	*7.7*	*7.5*
	(range)	(5.7 – 36.0)	(1.7 – 42.5)	(4.1 – 32.3)
Maximum Dimension (mm)	mean	**51.8**	**52.7**	**53.1**
	s.d.	*16.1*	*19.4*	*17.5*
	(range)	(26.3 – 86.9)	(20.5 – 109.0)	(25.5 – 94.6)
Platform Breadth (mm)	mean	**27.1**	**27.5**	**20.6*†**
	s.d.	*12.0*	*14.1*	*8.4*
	(range)	(4.0 – 62.0)	(4.0 – 70.5)	(8.7 – 38.8)
Platform Thickness (mm)	mean	**12.5**	**10.1***	**9.0**
	s.d.	*6.4*	*7.1*	*6.4*
	(range)	(0.1 – 34.5)	(1.4 – 40.6)	(4.0 – 22.6)
External Angle (°)	mean	**77**	**74**	**87*††**
	s.d.	*15*	*16*	*17*
	(range)	(52 – 115)	(40 – 121)	(65 – 120)
External Angle (°)	mean	**105**	**109**	**99*††**
	s.d.	*17*	*15*	*18*
	(range)	(62 – 135)	(55 – 170)	(68 – 120)
Dorsal Scars (#)	mean	**2.8**	**1.6****	**2.9††**
	s.d.	*1.8*	*1.2*	*1.8*
	(range)	(0 – 10)	(0 – 6)	(0 – 7)
Breadth/Length	mean	**0.99**	**1.1**	**1.2**
	s.d.	*0.3*	*0.4*	*1.0*
	(range)	(0.6 – 1.8)	(0.2 – 2.9)	(0.4 – 5.5)
Thickness/Breadth	mean	**0.4**	**0.3****	**0.3**
	s.d.	*0.1*	*0.2*	*0.2*
	(range)	(0.2 – 0.8)	(0.1 – 1.7)	(0.2 – 0.9)
Max. Dim./ Thickness	mean	**3.4**	**5.3****	**4.5**
	s.d.	*0.7*	*2.7*	*1.7*
	(range)	(2.0 – 5.7)	(1.8 – 18.0)	(1.9 – 8.8)

*   *significantly different from EG10 trachyte at p < 0.1,*
**   *significantly different from EG10 trachyte at p < 0.05,*
†   *significantly different from Gona experiment at p < 0.1,*
††   *significantly different from Gona experiment at p < 0.05*

Table 2. Comparison reveals that flakes from the Gona experiment are significantly thinner than EG10 flakes, but of similar size and shape in plan view. Limestone flakes from Indianapolis differ from EG10 in the same way, but not significantly so. The Indianapolis sample differs from both of the other samples in platform breadth and platform angles, and from the Gona experiment in number of dorsal scars.

The degree of reduction seen on individual cobbles varied dramatically according to raw material. Thus, three cobbles of relatively soft and easily flaked limestone yielded 75 flakes and fragments > 20 mm in maximum dimension, while three quartzite cobbles yielded only 38. Two quartz cobbles produced 15 detached pieces. These differences in reduction intensity were also reflected in core morphology. The three limestone cobbles produced three relatively heavily reduced cores that were easily assigned to "typical" typological categories, including 2 unifacial choppers and one polyhedron. In contrast, three quartzite cobbles yielded 1 unifacial core, 1 "casual" core showing only a single flake scar, 3 core fragments and one split cobble. In an archaeological context, the artifactual nature of at least two of these quartzite pieces would be open to question.

Looking to flake characteristics, we find further evidence of material-related differences in reduction intensity. Limestone whole flakes (n = 21) have a much higher proportion of non-cortical platforms (types IV, V & VI = 52.4%), and a much lower proportion of completely cortical dorsal surfaces (types I & IV = 9.5%) when compared with quartzite flakes (n =19; types IV, V & VI = 26.3%; types I & IV = 47.3%). These differences clearly reflect the more intense reduction of the relatively easily flaked limestone cobbles. The intuitively puzzling absence of any type I flakes in the limestone whole flake sample results from a combination of reduction strategy/intensity and the fact that the small number of initial cortical removals that were executed on these three cores produced fragments rather than whole flakes.

Raw material differences also affected flake metrics, although not at a high level of significance for this small sample. Limestone flakes are larger on average than quartzite flakes in all measured dimensions (length, breadth, thickness, maximum dimension), but only at around an 85% confidence level or less. The greater mean weight of limestone flakes is, however, significant at the 90% level (p = 0.086). Other significant differences include the greater number of dorsal flake scars on limestone flakes (p < 0.001) and the sharper edge angles of limestone flakes (p = 0.031).

Giving due consideration to the importance of raw materials, the Indianapolis experiment may be compared with the experimental and archaeological samples from Gona. In fact, the Indianapolis limestone sample is quite similar to the Gona experimental sample, and the two differ in the same ways from the EG10 trachyte. On the other hand, the Indianapolis quartzite differs markedly from all other samples. Like the Gona experimental sample, the Indianapolis limestone displays evidence of relatively variable, "opportunistic" flaking, including some bifacial reduction. Although two of three limestone cores are classified as unifacial choppers, some bifacial reduction is evident on the third, a polyhedron. Furthermore, limestone flakes display a high percentage of non-cortical platforms (52.4%) sim-

ilar to that observed in the Gona experimental flakes (43.5%) and much higher than that found in the EG10 trachyte (14.0%).

Qualitative inspection of limestone cores reveals that two of three (a unifacial side chopper and a polyhedron) display an opportunistic pattern of flake removals around the entire surface of the core, as seen in the Gona experimental cores, while the third (a unifacial end chopper) shows removals concentrated in a single area, as is characteristic of the EG10 trachyte cores. The Indianapolis limestone sample represents a mixture of reduction strategies that, considered as a whole, is more similar to the variable and opportunistic flaking seen in the experimental sample from Gona than to the uniform and localized flaking in the EG10 trachyte sample.

Steps and Hinges

It has recently (Ludwig, 1999) been argued that the frequency of steps and hinges on cores and dorsal flake surfaces is a good indicator of Mode I knapping skill. This measure produces interesting results in the current study, in which the number of steps or hinges per flakes scar observed on EG10 cores is significantly less than on the Indianapolis limestone (p = 0.034) or Gona trachyte (p = 0.077) experimental samples. These experimental samples, despite differences in raw materials, are not statistically distinguishable (p = 0.429). Following Ludwig (1999), these results would suggest greater knapping skill on the part of the Pliocene toolmakers as compared with modern human novices.

DISCUSSION

Three major conclusions may be drawn from the results presented above. First, raw materials impose important constraints on knapping strategies and products. This is especially true for knappers of a low skill level. Second, when raw materials allow, technologically naïve modern human knappers tend to follow more variable and less patterned reduction strategies than those seen in the EG10 trachyte artifacts. Third, technologically naïve modern human knappers tend to produce flakes thinner than those known from EG10, even when working in the same raw material.

The Importance of Raw Materials

The importance of considering raw material variation in evaluating the skill or sophistication of stone knappers cannot be overemphasized. Many researchers have made this point in the past (e.g. Jones, 1979; Clark, 1980; Isaac, 1984; Ludwig & Harris, 1998) and it is particularly applicable in the case of Mode I technologies, where skill is defined by little more than the ability to efficiently detach flakes. In the current study, a single novice individual working over a 40-minute period produced artifacts of completely different character depending on the raw materials used. In contrast, novice knappers working only with Gona trachyte produced a set of flakes that differed from the archaeological sample only in thickness. The immediate success of experimental subjects at Gona reinforces the point that the Gona trachyte is a relatively easy material to work with, and may not have been particularly taxing for early toolmakers. Comparison of limestone and quartzite artifacts from the Indianapolis experiment further suggest that easily flaked materials do not provide as sensitive an index of knapping skill as do more challenging raw materials.

The selection of high-quality material in the first place, however, is itself revealing of the mental capabilities of the toolmakers. Schick & Toth (1993) and Stiles (1998) have previously described evidence of raw material selectivity in Lower Pleistocene Mode I artifact assemblages from Koobi Fora and Olduvai Gorge. Ongoing research at Gona is similarly revealing a high degree of raw material selectivity among late Pliocene stone toolmakers. Such selectivity has been taken as evidence of "relatively complex thought" (Stiles 1998: 133) or, more conservatively, as reflecting "some expertise" in distinguishing the quality of materials (Schick & Toth 1993: 125-6). In terms of mental capacity, the ability of the earliest stone toolmakers to understand and evaluate the suitability of materials for flaking and to locate, identify and preferentially collect higher quality materials is perhaps as significant as their ability to master the controlled fracture of stone.

Knapping Strategies

The particular knapping strategy employed by the EG10 toolmakers provides tantalizing evidence of perceptual-motor skill and mental sophistication. The trachyte artifacts from EG10 reveal a uniform reduction strategy of localized, unifacial flake removals. This differs from the more opportunistic and variable strategy adopted by experimental knappers both at Gona and in Indianapolis. The Gona experiment in particular, conducted with raw materials identical to those in the archaeological sample, indicates that the uniform patterning observed in the EG10 trachyte is underdetermined by material constraints and naïve approaches to least-effort flake production. Why and how did the EG10 hominins adopt this particular strategy?

The "residual approach" to early stone tools would suggest that, having accounted for raw material and least effort factors, explanations in terms of function and/or tradition should be considered. For example, localized, unifacial flaking might be a strategy used to produce certain useful artifact characteristics (e.g. thicker flakes), or to maximize the efficiency of flake production. Alternatively, it might represent an idiosyncratic "culture-historical" knapping tradition. From an evolutionary and psychological perspective, the knapping strategy might also be explained in terms of cognitive differences between the EG10 toolmakers and modern humans.

These potential explanations are not, however, mutually exclusive. To begin with, all stone knapping is cultural behavior, at least in the sense commonly used by animal behaviorists and Early Stone Age archaeologists (i.e. it is socially learned). The identification of a functional advantage for the mode of flaking employed with the EG10 trachyte certainly would not make this practice any less "cultural" in nature. Explanation of EG10 knapping practices in terms of function would also have important implications for the inferred mental capacities of the toolmakers. Selection of a particular, advantageous strategy implies technical intent and understanding beyond that often attributed to early Oldowan toolmakers (e.g. Wynn & McGrew, 1989), and it would be hard to dismiss the converse argument that the repetitive, unifacial strategy employed on the EG10 trachyte actually resulted from a lack of such understanding.

The argument from functional intent would, of course, be strengthened by evidence that the EG10 strategy is actually advantageous. For example, Roche *et al.* (1999) have argued that the unifacial reduction strategy pursued by Pliocene hominins at the site of Lokalelei 2C in Kenya was a particularly efficient one, producing up to 50 flakes per core (Roche & Delanges, 2001). It is impossible to know the exact number of flakes produced per trachyte cobble by the EG10 toolmakers, although the opportunistic strategy employed by Gona experimental subjects produced an average of 22.5 detached pieces per cobble. Further experimentation is needed to determine if a localized, unifacial strategy favors the production of more flakes per cobble. It must be remembered, however, that the reduction intensity observed at archaeological sites may be influenced by other factors, including raw material availability and the duration/intensity of occupation.

Instead of attempting to distinguish between overlapping functional, cultural and cognitive explanations, it makes a lot of sense to evaluate the EG10 artifacts in terms of acquired knapping skills. Such skills constitute demonstrated abilities, rather than residual explanations inferred through a process of elimination. Knapping skill acquisition (Roux *et al.*, 1995; Stout, in press) occurs through the discovery of dynamically stable behavioral solutions (cf. Bernstein, 1996; Thelen & Smith, 1994) to the inherently variable problems presented by lithic raw materials. Whereas skilled performance is embodied in the stable articulation of perception and action, skill learning is characterized by highly variable experimentation and exploration.

This is exemplified in the performance of the novice experimental knappers described in this chapter, who generally pursued "opportunistic" or exploratory knapping strategies and produced diverse, difficult-to-classify cores. In contrast, the EG10 toolmakers displayed a treatment of trachyte that was much more uniform and stable. A strategy of localized, unifacial flaking dominates the sample, cores are more easily classified into a small number of specific categories, and flake dimensions are less highly variable (Table 2). This uniform pattern is indicative of experienced performance, and implies that knapping was a habitual, skilled activity among the EG10 hominins. Future experimentation with novice and experienced knappers will be valuable in order to further test and refine this hypothesis.

Also of interest is comparative evidence of knapping skill acquisition in bonobos. Schick *et al.* (1999) report results from stone toolmaking experiments conducted with the bonobo Kanzi at a time when he had had 3 years (roughly 120 hours) of toolmaking experience. Although encouraged to practice hand-held, direct percussion, Kanzi independently adopted throwing as his preferred flaking technique. Schick *et al.* suggest that this was due to Kanzi's difficulties in producing sufficient percussive force though hand-held flaking. Using the relatively uncontrolled throwing technique, Kanzi produced 12 cores that were classified by Schick *et al.* as consisting of 8 (66.7%) casual or "minimally modified" cores, two bifacial end choppers, one two-edged bifacial end chopper and one heavy duty scraper. This extreme of variation in both technique and products is beyond that seen in novice human knappers (at least under experimental conditions), but nevertheless reflects a similar tendency toward behavioral exploration and variability during skill learning.

With further practice and encouragement, both Kanzi and his sister Panbanisha were able to develop more stable and controlled hand-held knapping skills (Toth *et al.* 2002, this volume). This generated a much more uniform pattern of core types, including an overwhelming majority (78.8%) of one-edged (63.6%) and two-edged (15.2%) unifacial cores, and 21.3% mixed bifacial/unifacial (15.2%) or exclusively bifacial (6.1%) cores. This shift to more uniform flaking techniques reflects the acquisition and stabilization of knapping skills by bonobos, and provides a remarkable parallel to observed differences between modern human novices and the EG10 toolmakers. Bonobos take longer to acquire knapping skill, and may never achieve parity with experienced humans, but the broad pattern of skill acquisition in both species appears quite similar. For both humans and bonobos, stable performance and uniform products are achieved through effortful practice. This would also have been the case among the Pliocene toolmakers of EG10.

Flake Metrics

One of the most fundamental aspects of stone knapping skill is the ability to combine force and accuracy during percussion (Stout, in press). This was a major stumbling block for Kanzi, leading him to adopt the much less controlled throwing technique. Because the amount of force (mass * acceleration) required in detaching a given flake is positively correlated with its mass (Dibble & Pelcin, 1995), flake metrics provide an indication of knapping forces employed. With raw material held constant, the production of larger flakes requires blows that are more powerful. Unless this increased force is accompanied by a compensatory decrease in accuracy (throwing being an extreme example), it constitutes an increase in perceptual-motor task difficulty (Fitts, 1954). Direct evidence of striking accuracy is a little more difficult to derive from artifacts, but platform dimensions and degree of battering can provide some indication.

Raw Material Influences

Out of the samples presented here, the experimental quartzite flakes from Indianapolis are the least massive, and weigh significantly (p = 0.086) less than limestone flakes produced by the same individual in the same knapping session. This difference in flake size clearly reflects the constraints of raw material, in that greater amounts of force are required to produce flakes of a given size in the fracture-resistant quartzite. Only blows that were relatively close to core edges resulted in successful fracture, producing small, thin flakes.

Another pattern attributable to raw material influences is the significantly steeper (closer to 90°) platform angles seen in limestone flakes compared with the experimental and archaeological Gona trachyte samples (Table 2). Sahnouni *et al.* (1997: 710) have previously noted the "special flaking qualities of limestone", and specifically its tendency to yield unusual platform angles. The pattern seen in this study is not identical to that reported by Sahnouni *et al.*, possibly due to differences in blank morphology and reduction intensity, but the idiosyncratic fracture mechanics of limestone are evident in both cases.

Flake Thickness

There is a significant difference in flake thickness between the archaeological and experimental trachyte samples (Table 2) that cannot be attributed to raw material. A similar but non-significant difference in thickness also exists between EG10 trachyte flakes and experimental limestone flakes. Although archaeological and experimental flakes are not appreciably different in length, breadth, maximum dimension, or platform angles, EG10 flakes are thicker and have thicker platforms. Flake weights for these samples are not available, but the observed differences in thickness nevertheless indicate that the EG10 flakes are more massive (cf. Dibble & Pelcin, 1995; Stout, in press).

There are several potential explanations for the difference in flake thickness between the archaeological and experimental samples, but a technological difference in flaking behavior is the most likely. Alternative explanations would involve some kind of preferential winnowing of relatively thin flakes from the archaeological sample, either through site formation processes or as a result of selective transport on the part of hominins. As previously discussed, artifact condition, sediment composition, artifact size distribution and the lack of preferential artifact orientation all indicate that EG10 is an undisturbed site. The preponderance of artifacts in the 15-25 mm size range at EG10 (Semaw, 1997: 106) argues against the probability that some flakes in the 20-90 mm range analyzed for this study were preferentially carried off due to minor differences in weight or shape. An explanation invoking the selective removal of thin flakes by hominins is more difficult to discount. The idea is plausible, but not very testable. Winnowing by hominin activity stipulates a form of behavior for which there is no other evidence and which presupposes undemonstrated mental sophistication and discriminative abilities. Nevertheless, it cannot be ruled out.

If the greater mean thickness of EG10 flakes is accepted as technological in origin, it indicates both that the EG10 toolmakers tended to strike further from core edges, and that they employed sufficient percussive force to detach the thicker flakes that resulted. At the same time, there is no evidence for a compensatory decrease in striking accuracy. Flaking is well controlled, producing the uniform unifacial pattern described above, and battering is, if anything, less pronounced in the archaeological sample than in the experimental materials. Platform thickness is also less variable in the archaeological sample (Table 2), further reflecting accuracy and consistency of percussion (i.e. striking at a consistent distance from core edges). The consistent production of thick flakes at EG10 provides further evidence of well-developed knapping skills, and particularly of the ability to combine force and accuracy during percussion.

This is not to say that Mode I knapping skill is necessarily defined by the production of thick flakes. In fact, highly skilled modern knappers working with Gona trachyte did not produce flakes as thick as those from EG10 (Toth *et al.* 2002, this volume). Technological habit and intent also become important factors here. What is significant is that the EG10 toolmakers imposed a bias on flake morphology (toward thicker flakes) that is not simply a reflection of raw materials and least-effort flaking. Experiments with bonobos (mean flake thickness = 10.4 [Toth *et al.*]), human novices (mean flake thickness = 12.8 mm [this study]), and human experts (mean flake thickness = 13.8 mm [Toth *et al.*]) using Gona trachyte all produced flakes thinner than those found at EG10 (mean thickness = 16.0 mm [this study]). Flake production at EG10 clearly did not follow an inevitable path of least resistance, and provides evidence of acquired knapping skill.

This leaves the question of why the EG10 toolmakers favored thicker flakes in the first place. Anecdotal evidence (Toth, pers. comm.) suggests that thicker flakes may be easier to hold during butchery and have edges that are more durable, but additional experimental work is needed in order to test these and other ideas. Although it is well known that early stone tools were used at least some of the time for carcass processing

(Isaac, 1984; de Heinzelin *et al.*, 1999; Schick & Toth, 2001), and microwear evidence (Keeley & Toth, 1981) further indicates use in woodworking and cutting of soft plant material, many details regarding the function of Oldowan artifacts remain unknown. Tool-use experiments can at least provide insight into the potential relations between artifact characteristics and utility in various tasks.

Knapping Skill and Hominin Mental Capabilities

Comparison of experimental and archaeological samples indicates a relatively high level of knapping skill among the Pliocene toolmakers of EG10. Although the evidence does not reveal the specific combination of social, functional and cognitive factors that allowed for the development of this skill, it does suggest that knapping was an habitual activity associated with more than just a few hours of skill learning. Even modern humans require practice in order to achieve the kind of uniform and controlled knapping seen at EG10, and bonobos can take hundreds of hours to develop flaking skills that are even broadly comparable. It is most likely that the EG10 hominins fell somewhere between these extremes.

In humans and other apes, skill acquisition is a social process (Boesch & Boesch 1990; Inoue-Nakamura & Matsuzawa, 1997; van Schaik *et al.* 1999; Stout, in press). The mental demands of social interaction are widely appreciated (e.g. Humphrey, 1976; Byrne & Whiten, 1988; Dunbar 1992), and the acquisition of increasingly sophisticated knapping skills in prehistory provides an important indication of evolving social cognitive capabilities. The level of skill evident at EG10, though fairly low by human standards, suggests the presence of social contexts and mechanisms for skill acquisition beyond those seen in modern apes.

Research presented here and elsewhere (Stout *et al.*, 2000; Stout, this volume) is also beginning to reveal the perceptual-motor demands of Mode I knapping. Derived neural substrates for the effective coordination of perception and action may have been important in the development of the knapping skills seen at EG10. In any case, it is clear that some combination of social facilitation and/or perceptual-motor sophistication is necessary to explain the knapping skill of the EG10 toolmakers. Similar factors may also have underlain the raw material selectivity evident at EG10.

CONCLUSION

It would be inappropriate to attempt a general assessment of Oldowan hominin knapping skill and mental capabilities on the basis of artifacts from a single site or research area. To begin with, stone artifacts provide an indication only of minimum required competencies (Wynn 1989), and are unlikely to reflect the full extent of the makers' mental capabilities. This would be true even if evidence from all known sites were synthesized; it is certainly the case for this more limited investigation. Plio-Pliestocene hominin technological behavior is demonstrably variable over time and space. The relative importance of environmental, traditional, functional, population and even species level differences in explaining this variation remains to be evaluated, but it is clear that no one site can provide a full picture.

With this said, the narrowly focused investigation presented here does provide important information about the technological practices of at least one group of early toolmakers. What is sacrificed in breadth is gained in experimental control, increasing the utility of results and the robusticity of interpretations. Particularly important is the control achieved over raw material variation. Experimentation with Gona trachyte confirms its excellent flaking properties, and reveals that novice modern humans using this material are easily and immediately able to produce flakes comparable to those found at EG10. In contrast, artifacts produced by a modern novice on difficult-to-flake quartzite were quite distinct from the EG10 trachyte materials. Due to its high quality, the EG10 trachyte does not provide evidence of knapping skills that might have been evident (or demonstrably absent) in more difficult raw materials. On the other hand, evidence of the identification, selection and procurement of high-quality materials at EG10 and other Gona sites has much to reveal about the mental capabilities of the earliest stone toolmakers. It is even possible that the behavioral innovation leading to the first blossoming of knapped stone technology at Gona 2.5 million years ago lay as much in the selection of suitable raw materials as in the mastery of stone fracture mechanics.

Nevertheless, the trachyte artifacts from EG10 do provide important evidence of skilled flaking. The uniform reduction strategy of localized, unifacial flake removals employed at EG10 differs from the more opportunistic approach of novice knappers, and is indicative of experienced performance. This is also true of the production bias toward thicker flakes seen at EG10, which reflects consistent knapping practices that are underdetermined by raw materials and naïve approaches to least-effort flake production. The size and thickness of flakes from EG10 provide evidence of the skillful combination of percussive force and accuracy by the EG10 toolmakers.

The skill evident at EG10 is indicative of habitual knapping behavior associated with a relatively extended learning period. It is impossible to say exactly how long this learning period might typically have lasted, but it was most likely somewhere between the several hours required by modern humans and the hundreds of hours required by bonobos. This suggests the presence of relatively elaborate social contexts and mechanisms for skill learning, and has important implications for hominin social cognitive capabilities. It is striking that such skilled performance is evident from the very first appearance of stone knapping in the archaeological record.

ACKNOWLEDGEMENTS

We would like to thank the experimental knappers whose efforts made this research possible. We would also like to thank Professors Kathy Schick and Nick Toth (Co-directors of CRAFT) for inviting us to contribute to this volume, providing helpful comments on a draft version of this chapter, and supplying the bonobo toolmaking data discussed herein, as well as for the overall support provided by CRAFT and the Friends of CRAFT. Research at Gona is made possible by field permits from the Authority for Research and Conservation of Cultural Heritage (ARCCH) of the Ministry of Sports and Culture of Ethiopia, the comradeship and field support of the Afar people of Eloha and the administration at Asayta, and funding from the L.S.B. Leakey Foundation, the National Science Foundation, the Wenner Gren Foundation, the National Geographic Society and the Boise fund. Funding for research in Indianapolis was provided by the L.S.B. Leakey Foundation, the National Science Foundation, CRAFT and Friends of CRAFT. The site of EG10 was excavated in collaboration with Professor J. W. K. Harris of Rutgers University.

References Cited

Ambrose, S.H. (2001). Paleolithic Technology and Human Evolution. *Science* **291**, 1748-1753.

Bernstein, N. (1996). On Dexterity and its Development. In (M.L. Latash and M.T. Turvey, Eds) *Dexterity and its Development*. New York: Pergamon Press, pp. 3-244.

Boesch, C., & Boesch, H. (1990). Tool use and tool making in wild chimpanzees. *Folia Primatologica* **54**, 86-99.

Bordes, F. (1947). Etude comparative des differentes techniques de talle du silex et des roches dures. *L'Anthropologie* **51**, 1-29.

Byrne, R. & Whiten, A. (Eds) (1988) *Machiavellian Intelligence*. Oxford: Oxford University Press.

Callahan, E. (1979). The basics of biface knapping in the Eastern Fluted Point Tradition: A manual for flintknappers and lithic analysts. *Archaeology of Eastern North America* **7(1)**, 1-172.

Clark, J.D. (1980). Raw material and African lithic technology. *Man and Environment* **4**, 44-55.

Clarke, G. (1969). *World Prehistory*. Cambridge: Cambridge University Press.

Crabtree, D. (1966). A stoneworker's approach to analyzing and replicating the Lindenmeier Folsom. *Tebiwa* **9(1)**, 3-39.

de Heinzelin, J., Clark, J.D., White, T., Hart, W., Renne, P. WoldeGabriel, G., Beyene, Y. Vrba, E. (1999). Environment and behavior of 2.5-million-year-old Bouri hominids. *Science* **284**, 625-629.

Dibble, H.L. & Pelcin, A. (1995). The effects of hammer mass and velocity on flake mass. *Journal of Archaeological Science* **22**, 429-439.

Dunbar, R.I.M. (1992). Neocortex size as a constraint on group size in primates. *Journal of Human Evolution* **20**, 469-493.

Fitts, P.M. (1954). The information capacity of the human motor system in controlling the amplitude of movement. *Journal of Experimental Psychology* **47**, 381-391.

Humphrey, N.K. (1976). The social function of intellect. In (P. Bateson and R. Hinde, Eds.) *Growing Points in Ethology*. Cambridge: Cambridge University Press, pp. 303-317.

Inoue-Nakamura, N., & Matsuzawa, T. (1997). Development of stone tool use by wild chimpanzees (*Pan troglodytes*). *Journal of Comparative Psychology* **111(2)**, 159-173.

Isaac, G. (1977). *Olorgesaiile: Archaeological Studies of a Middle Pleistocene Lake Basin in Kenya*. Chicago: University of Chicago Press.

Isaac, G. (1984). "The archaeology of human origins: studies of the Lower Pleistocene in East Africa 1971-1981". In (F. Wendorf & A. Close, Eds) *Advances in Old World Archaeology*, vol. 3. New York: Academic Press, pp. 1-87.

Jones, P. (1979). Effects of raw material on biface manufacture. *Science* **204**, 835-836.

Keeley, L. & Toth, N. (1981). Microwear polishes on early stone tools from Koobi Fora, Kenya. *Nature* **293**, 464-465.

Kibunja, M. (1994). Pliocene archaeological occurrences in the Lake Turkana basin. *Journal of Human Evolution* **27** 159-171.

Leakey, M.D. (1971). *Olduvai Gorge, Volume 3: Excavations in Beds I and II, 1960-1963*. Cambridge: Cambridge University Press.

Ludwig, B.V. (1999). *A Technological Reassessment of East African Plio-Pleistocene Lithic Artifact Assemblages*. Unpublished Ph.D. Dissertation. New Brunswick, NJ: Rutgers University.

Ludwig, B.V. & Harris J.W.K. (1998). Towards a technological reassessment of East African Plio-Pleistocene lithic assemblages. In (M. Petraglia and R. Korisetter, Eds) *Early Human Behavior in the Global Context: The Rise and Diversity of the Lower Paleolithic Period*. New York: Routledge, pp. 84-107.

Marzke, M., Toth, N., Schick, K., Reece, S., Steinberg, B., Hunt, K., Linsheid, R.L., & An, K-N. (1998). EMG study of hand muscle recruitment during hard hammer percussion manufacture of Oldowan tools. *American Journal of Physical Anthropology* **105**, 315-332.

McGrew, W.C. (1992). *Chimpanzee Material Culture: Implications for Human Evolution*. Cambridge: Cambridge University Press.

Roche, H. (1989). Technological evolution in the early hominids. *OSSA, International Journal of Skeletal Research* **14**, 97-98.

Roche, H. & Delagnes, A. (2001). Evidence of controlled and reasoned stone knapping at 2.3 Myr, West Turkana (Kenya). Poster presented at "Knapping Stone: A Uniquely Hominid Behavior?" workshop held 21-24 November, 2001, Abbaye des Premontres, Pont-au-Mousson, France.

Roche, H. Delagnes, A., Brugal, J.P., Feibel, C., Kibunjia, M., Mourre, V. & Texier, P.J. (1999). Early hominid stone tool production and technical skill 2.34 Myr ago in West Turkana, Kenya. *Nature* **399 (6731)**, 57-60.

Roche, H., Tiercelin, J.J. (1977). Découverte d'une industrie lithique ancienne *in situ* dans la formation d'Hadar, Afar central, Éthiopie. *C.R. Acad. Sci. Paris D* **284**, 187-174.

Roche, H., Tiercelin, J.J. (1980). Industries lithiques de la formation Plio-Pléistocène d'Hadar: campagne 1976. In (R.E.F. Leakey and B.A. Ogot, Eds.) *Proceedings, VIIIth PanAfrican Congress of Prehistory and Quaternary Studies*. Nairobi, pp. 194-199.

Roux, V., Bril, B. & Dietrich, G. (1995). Skills and learning difficulties involved in stone knapping. *World Archaeology* **27(1)**, 63-87.

Sahnouni, M., Schick, K.D., Toth, N. (1997). An experimental investigation into the nature of faceted limestone "spheroids" in the early Paleolithic. *Journal of Archaeological Science* **24**, 701-713.

Schick, K.D. (1987). Modeling the formation of Early Stone Age artifact concentrations. *Journal of Human Evolution* **16**, 789-807.

Schick, K.D. & Toth, N. (1993). *Making Silent Stone Speak: Human Evolution and the Dawn of Technology*. New York: Simon and Schuster.

Schick, K.D. & Toth, N. (2001). Paleanthropology at the Millennium. In (Feinman and Price, Eds) *Archaeology at the Millennium: A Sourcebook.* New York: Kluwer Academic/Plenum Publishers, pp. 39-108.

Schick, K.D., Toth, N., Garufi, G., Savage-Rumbaugh, E.S., Rumbaugh, D. & Sevcik, R. (1999). Continuing investigation into the stone tool-making and tool-using capabilities of bonobo (*Pan paniscus*). *Journal of Archaeological Science* **26**, 821-832.

Semaw, S. (1997). Late Pliocene archaeology of the Gona River deposits, Afar, Ethiopia. Unpublished Ph.D. dissertation. New Bruswick, NJ: Rutgers University.

Semaw, S. (2000). The world's oldest stone artifacts from Gona, Ethiopia: Their implications for understanding stone technology and patterns of human evolution between 2.6-1.5 million years ago. *Journal of Archaeological Science* **27**, 1197-1214.

Semaw, S., Renne, P., Harris, J.W.K., Feibel, C.S., Bernor, R.L., Fesseha, N., & Mowbray, K. (1997). 2.5-million-year-old stone tools from Gona, Ethiopia. *Nature* **385**, 333-336.

Speth, J.D. (1974). Experimental investigations of hard-hammer percussion flaking. *Tebiwa* **17(1)**, 7-36.

Stiles, D. (1998). Raw material as evidence for human behavior in the Lower Pleistocene: the Olduvai case. In *Early Human Behavior in the Global Context: The Rise and Diversity of the Lower Paleolithic Period.* New York: Routledge, pp. 133-150.

Stout, D. (in prep.). Stone tools and the evolution of human thinking: Cultural, biological and archaeological elements in an Anthropology of human origins. Ph.D. dissertation, Indiana University, Bloomington.

Stout, D. (in press). Skill and cognition in stone tool production: An ethnographic case study from Irian Jaya. *Current Anthropology* **45**(3).

Stout, D. Toth, N., Schick, K., Stout, J. & Hutchins, G. (2000). Stone tool-making and brain activation: Positron Emission Tomography (PET) studies. *Journal of Archaeological Science* **27**, 1215-1223.

Thelen, E. & Smith, L. (1994). *A Dynamic Systems Approach to the Development of Cognition and Action.* Cambridge, MA: MIT Press/Bradford Books.

Toth, N. (1985). The Oldowan reassessed: a close look at early stone artifacts. *Journal of Archaeological Science* **12**, 101-120.

Toth, N. (1997). The stone artifact assemblages: A comparitive study In (G.L. Isaac & B. Isaac, Eds) *Koobi Fora Research Project,* Volume 5: *Plio-Pleistocene Archaeology.* Oxford: Oxford University Press, pp. 262-299.

Toth, N., Schick, K., Semaw, S. (2002). A technological comparison of the stone toolmaking capabilities of *Australopithecus*/early *Homo, Pan Paniscus,* and *Homo sapiens,* and possible evolutionary implications. Abstract from the Paleoanthropology Society meetings, 2002. *Journal of Human Evolution* **42**(3): A36-A37.

van Schaik, C., Deaner, R.O., Merrill, M.Y. (1999). The conditions for tool use in primates: implications for the evolution of material culture. *Journal of Human Evolution* **36**: 719-741.

Wright, R.V. (1972). Imitative learning of a flaked tool technology: the case of an orangutan. *Mankind* **8**, 296-284.

Wynn, T. (1989). *The Evolution of Spatial Competence.* Urbana: University of Illinois Press.

Wynn, T. & McGrew, W.C. (1989). An ape's view of the Oldowan. *Man* **24**, 383-398.

CHAPTER 11

COMPARING THE NEURAL FOUNDATIONS OF OLDOWAN AND ACHEULEAN TOOLMAKING: A PILOT STUDY USING POSITRON EMISSION TOMOGRAPHY (PET)

BY DIETRICH STOUT, NICHOLAS TOTH, AND KATHY SCHICK

ABSTRACT

Functional brain imaging technologies provide human origins researchers with the unique opportunity to examine the actual neural substrates of evolutionarily significant behaviors. This pilot study extends previous brain imaging research on stone toolmaking (Stout et al., 2000; Stout, this volume) by using Positron Emission Tomography (PET) to compare Mode II, Acheulean biface production with Mode I, Oldowan flake and core production. Results from this single-subject pilot study are not sufficient for statistical analysis, but do confirm the applicability of PET research methods to Mode II and later technologies as well as providing some indication of what may be expected from future research.

INTRODUCTION

Recent work using Positron Emission Tomography (PET) to examine the brain activation associated with Mode I, Oldowan-style toolmaking (Stout et al., 2000; Stout, this volume) has begun to shed light on the psychological and evolutionary implications of the earliest stone tools. Applying these experimental methods to the study of more recent stone technologies will be an important next step for this research program. By identifying the actual neural foundations of the stone technologies associated with various periods of human evolution, functional brain imaging research will facilitate the psychological interpretation of archaeological evidence and potentially help to chart the evolutionary emergence of the human brain and intelligence.

As a step in this direction, the authors conducted a preliminary experiment in the use PET to examine Mode II, Acheulian-style biface manufacture. This experiment was intended primarily as a feasibility study, and confirmed that methods previously used to investigate Oldowan-style knapping (Stout, this volume) were also applicable to handaxe-making. Results obtained from this single-subject experiment are not sufficient for statistical analysis, but do provide a suggestion as to what may be expected from future research.

The emergence of Mode II technology, dated to at least 1.5 Ma (Isaac & Curtis, 1974; Asfaw et al., 1992), has long been regarded as a milestone in hominin cognitive evolution. Compared with the simple cores and flakes of the preceding Oldowan, early Acheulean bifaces clearly reveal the appearance of more regularly patterned and technically demanding toolmaking activities. However, it is also important to appreciate the variation, both temporal and spatial, that exists within the broadly defined Acheulean Industrial Complex (Clark, 2001). Differences between early and later Acheulean artifacts are especially striking, and may reflect further important developments in hominin cognitive evolution (Wynn, 1989). It should thus be noted that the handaxe production undertaken for this pilot experiment is representative of later, rather than earlier, Acheulean technology.

The differences between earlier and later Acheulean handaxes reflect the emergence of more meticulous and skill-intensive knapping practices. Later Acheulean handaxes typically display more intense overall reduction, with a greater number of flake scars per unit of surface area and little or no preservation of original blank surfaces. Flake scars are generally shallower, being left by the thin, spreading flakes that are produced by striking close to edges that have been steepened through careful platform preparation. In some cases, soft hammers may have been used. Later Acheulean handaxes also tend to be thinner relative to breadth, with carefully thinned tips, straighter, less scalloped edges, and greater symmetry in both plan form and cross-section. Within assemblages, there is a tendency toward greater uniformity in handaxe size and shape in the later Acheulean.

Although some researchers have commented on the skill required to actually make refined later Acheulean handaxes (Callahan, 1979; Bradley & Sampson, 1986; Schick, 1994; Edwards, 2001; Clark, 2001; Stout, 2002), most psychological interpretations have focused on the degree to which imposed symmetry and "arbitrary form" are evident (or absent) in the finished artifacts (Wynn, 1979; Gowlett, 1984; Isaac, 1986; Wynn & Tierson, 1990; Noble & Davidson, 1996; McPherron, 2000; Noll, 2000). The presence of such imposed form is considered to provide evidence of relatively advanced spatial conceptualization, strategic planning and stylistic (cultural) awareness.

This orthodox, *representational* (Stout, this volume), approach defines the sophistication of prehistoric stone technologies in terms of their reliance upon internally constructed mental images, plans and concepts. Applied to PET research, this perspective calls our attention to those parts of the brain that are characteristically associated with representational and introspective activities, and especially to the classic "planning and problem solving" areas of the prefrontal cortex (e.g. Brodmann's Areas 9, 10, 45 and 46). Such regions are not significantly activated during Mode I stone knapping (Stout *et al.*, 2000; Stout, this volume), and their activation during Mode II biface production would provide support for the conventional view that Acheulean handaxes reveal "higher conceptual and cognitive abilities" than do Oldowan cores and flakes (Ambrose, 2001: 1750).

In order to enlarge upon this traditional perspective, Stout (this volume) proposes an additional, *perception-action* approach to understanding the brain activation associated with stone toolmaking. This approach emphasizes the importance of dynamic knapping skill, rather than static mental representation, in supporting stone toolmaking activities. Knapping skills are embodied in effective actions in the world and emerge from the purposeful coordination of outwardly directed perception and action (Stout, 2002). The resulting focus on external perception and action as opposed to internal representation draws our attention to different parts of the brain, including the visual and motor cortices of the occipital and frontal lobes and the sensory association cortex of the superior parietal lobe. These regions do appear to be recruited during Oldowan-style knapping (Stout *et al.*, 2000; Stout, this volume), but the greater technical demands of handaxe-making might be expected to produce relatively more intense and/or extensive

activation. Of particular interest would be the level of activation observed in the premotor areas of the posterior frontal lobe (Brodmann's Area 6) and the polymodal association cortex of the superior parietal (Brodmann's Area 7), regions that provide essential neural substrates for the dynamic coupling of complex patterns of perception and action.

THE EXPERIMENT

Brain imaging methods used in this pilot study very closely follow those previously employed to examine Mode I flake production, which are discussed in detail elsewhere (Stout, this volume). As in previous experiments, the slowly decaying tracer FDG ([18]flouro-2-deoxyglucose) was used so that knapping could occur in a relatively naturalistic setting outside the scanner. A single subject (Nick Toth, an experimental stone knapper with over 25 years experience) was imaged during one trial for each of three task conditions. This constituted the maximum acceptable research-related radiation exposure for a one-year period. The three conditions were: (1) a control condition that consisted of striking two cobbles together without attempting to produce flakes, (2) Mode I flake production, and (3) Mode II handaxe production. All experimental activities were videotaped and all products were collected for analysis (Figure 1).

Activation data from each knapping condition was compared with the control condition in order to reveal any regions of increased neuronal activity. Where present, such increases reflect neural demands of stone knapping in excess of those associated with the simple bimanual control task (Stout, this volume). Unfortunately, small sample sizes (n=1) in the current pilot study do not allow the statistical significance of observed differences to be assessed (see below).

OLDOWAN CORE AND FLAKE PRODUCTION

As in the Oldowan experiment reported by Stout (this volume), the subject in this pilot study was presented with an assortment of water-rounded cobbles of a variety of raw materials and asked to produce sharp, useable flakes through hard-hammer percussion. Both hammerstones and cores were selected from this assortment during the 45-minute duration of the experiment. Cores were reduced until they were exhausted, usually because the edge angles became so steep that further reduction was difficult. The resulting cores (in Mary Leakey's typological system) included nine choppers, four polyhedrons, two heavy-duty scrapers and one casual core (modified cobble). The flakes produced were also typical of the Oldowan, with thick striking platforms and cortex on the dorsal surfaces of most of the flakes.

Technical Acts

The videotape of the flake and core production was reviewed in order to quantify number and rate of technological acts employed. In Mode I knapping, technological action was limited to hard hammer direct percussion. Over the 45-minute period there were 1165 percussive blows, or roughly one blow every 2.3 seconds. With 16 cores produced, this equates to an average of 68.5 blows per core.

LATE ACHEULEAN HANDAXE MANUFACTURE

Acheulean handaxe production also took place during a 45-minute experimental period. As in the Oldowan condition, no clocks or timepieces were visible to the subject, who also made a deliberate attempt not to mentally "verbalize" the operation or count sequential technological acts (percussion blows, grinding). The materials used in this handaxe replication included the large obsidian flake blank for handaxe manufacture; a larger, denser sandstone spherical hammerstone for the roughing out of the biface; a smaller, less dense limestone disc-shaped hammerstone for striking platform preparation, shaping of the plan form, and abrasion of the striking platform; a soft hammer of elk antler to remove thinning flakes from the handaxe and for final shaping and straightening of sinuous edges; and a gazelle skin to protect the subject's leg, which supported the obsidian biface.

The blank used in this handaxe manufacture was a large obsidian flake that had been previously struck from a discoidal boulder-core with a very large hammerstone. The quarrying of such large flake blanks for handaxe and cleaver manufacture is a recurrent technological strategy seen in the Acheulean of much of Africa (see Toth, 2001) as well as sites in the Near East, Iberia, and the Indian subcontinent. The flake blank used in this experiment was a corner-struck, sub-rectangular thick flake with approximately 50% of continuous cortex on the dorsal face. The blank weighed 5,596 gm and measured 30 cm x 25 cm x 12 cm with a large, thick multi-faceted striking platform measuring 22 cm x 11 cm. Both the proximal and distal ends of the flake were quite thick, with a thick striking platform and prominent hinged termination at the distal end.

Figure 1

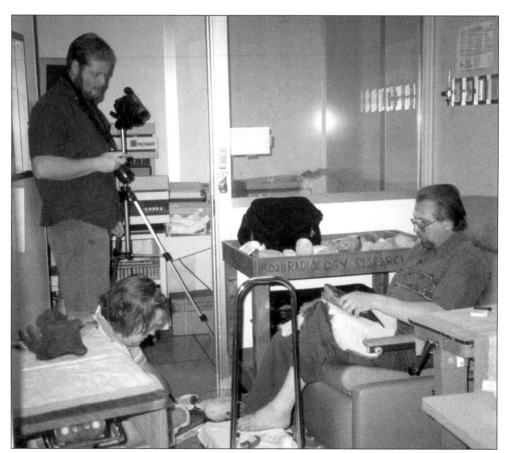

1. *Setting up the handaxe manufacture experiment: The slowly-decaying radioactive tracer (FDG) is being injected into the subject's foot, the video recorder is being set in place, and the subject is seated with the obsidian flake blank in his lap and the stone and antler hammers within easy reach on his right. (Photo by Kathy Schick).*

Although it is true that stone tool manufacture can be quite fluid rather than rigidly divided into sequential stages of reduction, nonetheless four major stages of manufacture were envisioned and could be identified in this experimental replication. These were:

1. Examination of the flake blank (3.5 minutes)

2. Roughing-out of the biface (8 minutes)

3. Primary bifacial thinning and shaping of the handaxe (24 minutes)

4. Secondary thinning and shaping of the handaxe (9.5 minutes)

Each of these stages of reduction will be discussed, with consideration of the mental operations and the technological acts that were employed.

Figure 2

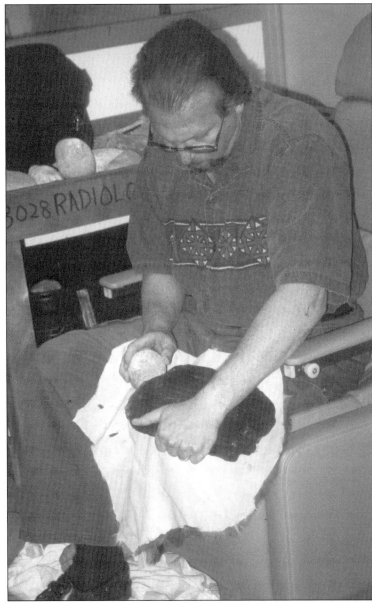

2. *The subject begins roughing-out the biface from the obsidian flake blank using a sandstone hammer. (Photo by Kathy Schick).*

1. Examining the flake blank (3.5 minutes)

After the radioactive tracer was injected into the subject's foot and entered his bloodstream (Figure 1), the 45 minute experimental period began. The subject first inspected the flake blank, examining the overall morphology, looking for potential flaws or inclusions in the raw material, and testing the obsidian blank by tapping with the sandstone hammer to listen to its acoustic properties (a good obsidian flake will have a clear, glassy ring when tapped, while a flake with a serious flaw will often have a dull, muted sound).

2. Roughing-Out the Biface (8 minutes)

This first stage of lithic reduction was carried out with the larger sandstone hammer, and attempted to create a continuous, sharp edge around the perimeter of the biface and producing a continuous acute edge that was generally centered between the two faces of the biface (Figure 2). Reduction was conducted to make the blank more symmetrical and to generate a well-centered edge. Lighter hammerstone blows were used to remove overhangs and spurs from edges; more forceful hammerstone blows were used to drive off larger, longer flakes in this first stage of reduction.

At first it was unclear exactly where the long axis of the biface would be, but as reduction continued, the long axis began to emerge in the rough-out: the right and left sides of the flake delineated the long axis, while the thicker proximal and distal ends of the blank became the sides of the biface. The proximal and distal ends of the blank were relatively thick due to presence of the original striking platform and bulb on the one hand and a prominent hinge release surface on the other. These surfaces provided the platforms for bifacial thinning. The thinner right side of the flake became the tip of the handaxe, and the somewhat thicker right side became the butt.

The flakes produced tended to be thick, with prominent bulbs of percussion and usually one or two scars on the thick striking platform; normally there were only a few bold scars on the dorsal surfaces of the flakes. Lithic analysts would generally classify these flakes as "hard-hammer percussion" flakes.

3. Primary Thinning and Shaping (24 minutes)

During this phase of reduction, the overall shape of the final handaxe could be envisioned in the still irregular, relatively asymmetrical and thick biface. Large thinning flakes were removed from the biface with a smaller, less dense limestone hammer to decrease the overall thickness. Striking platforms were first carefully prepared by intensive, light flaking performed along the edge from which the thinning flake was to be removed but directed toward the opposite face. This edge preparation was done to steepen and strengthen the edge to receive the forceful blows of an antler soft hammer. Edges were also abraded with the limestone hammer, creating roughened areas that provided greater purchase for the antler hammer. During this intensive platform preparation it was possible to control the shape of the plan form of the handaxe, making it bilaterally symmetrical and beginning to shape the pointed tip end and the steeper, wider butt end.

The flakes produced in this process tended to be thin and slightly curved in side view, with a diffuse bulb of percussion, a thin or punctiform striking platform, a slight lipping on the ventral surface near the point of percussion, numerous scars (facetting) on the striking platform, a steep exterior platform angle, and occasional evidence of hammerstone abrasion on the platforms; often there were numerous shallow scars on the dorsal surfaces of the flakes as well. Lithic analysts would generally classify these flakes as "soft-hammer percussion flakes", although these flakes can also be produced with a hard hammer by employing careful platform preparation and marginal flaking near the edge of the biface.

4. Secondary Thinning and Shaping (9.5 minutes)

A new round of bifacial thinning and shaping occurred in the last 9.5 minutes of reduction. Platforms were prepared by robust light flaking and abrasion with the limestone hammer to produce regular, strong edges to support the robust blows from the antler soft hammer and remove invasive thinning flakes. During this flaking all of the original cortex, and almost the entire original blank surface, was removed, the pointed tip and steepened butt were carefully shaped, and any sinuous edges straightened. Much of the final flaking was carried out with light blows from the antler baton. The flakes produced tended to be morphologically similar to those produced in the primary thinning and shaping stage, but smaller in overall size.

The Finished Piece

After 45 minutes, the final form of the biface was a large, elongate cordate handaxe characteristic of the late Acheulean (Figures 3 & 4). Retouch was extensive, shallow, and invasive, with all of the cortex (and all of

Figure 3

3. The finished product of the handaxe-making experiment: A large, elongate cordate handaxe. Handaxes with such a high degree of symmetry and extensive retouch, with removal of many invasive, shallow flakes, are characteristic of the later Acheulean. (Photo by Kathy Schick).

Figure 4

4. The finished handaxe along with the flakes and fragments produced in the 45 minutes of fashioning the tool. The antler soft hammer is in the top center. (Photo by Kathy Schick).

the original dorsal flake surface) having been removed from the dorsal face, and only one small area on the ventral face (3.4 by 2.4 cm) showing the original release surface. The final form of the handaxe had the following attributes:

Weight: 1,960 gm. (30.5% of the original flake blank weight)

Length: 25.0 cm (83.3% of the original blank length)

Breadth: 14.0 cm (56.0% of the original blank breadth)

Thickness: 5.6 cm (46.7% of the original blank thickness)

Flake scars (one cm or greater) on dorsal face: 48

Flake scars on ventral face: 58

Total flake scars: 106

Maximum dimension of largest flake scar: 9.8 cm

It should be noted that the final form of the large handaxe could still have been resharpened and thinned a number of times if there had been more time. Nonetheless, the 45 minutes of biface production was typical of the all of the technological operations and cognitive decisions that were required to make a late Acheulean handaxe.

Technical Acts

The videotape of the handaxe manufacture was reviewed a number of times in order to quantify number and rates of different technological acts employed to modify the stone and produce the handaxe. These technological acts did not include shifting from one knapping tool to another, turning the biface over from one face to another, or brushing off detached flakes from the animal skin, but rather only acts of physical force such as percussion and grinding on the obsidian artifact itself.

Figure 5

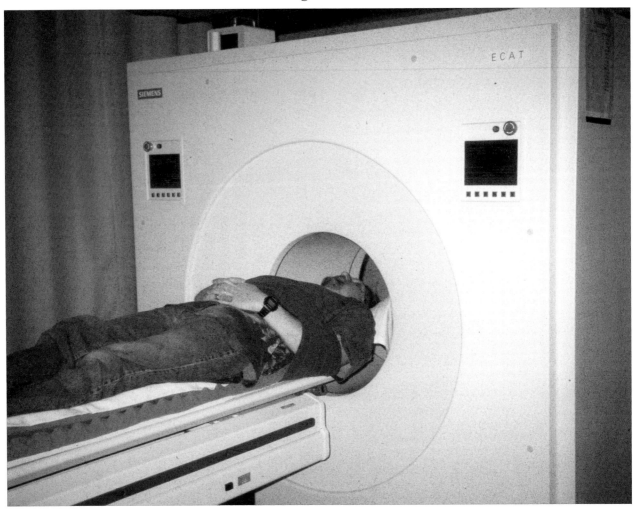

5.   *The subject being scanned immediately after the 45-minute tool-making session. (Photo by Kathy Schick).*

Roughing-out stage:

> Light (preparation) blows with larger sandstone hammer: 80
>
> Strong blows with larger sandstone hammer: 57
>
> Rate: One technological act every 3.5 seconds

Primary & secondary thinning and shaping

> Striking platform preparation blows with smaller limestone hammer: 1640
>
> Grinding striking platforms with limestone hammer: 270
>
> Strong antler hammer thinning blows: 76
>
> Light antler hammer shaping blows: 286
>
> Rate: One technological act every 0.89 seconds

Interestingly, at the end of 45 minutes of intensive late Acheulean flaking, the subject felt more "mentally fatigued" than after 45 minutes of Oldowan flaking. The Acheulean flaking required much more concentration and attention to detail, more complex attention to three-dimensional space, and continuous imagining of the final handaxe shape inside the stone as reduction proceeded. The subject repeatedly would examine the underside of the biface (where flakes would be detached) before hammerstone or antler hammer blows were struck, and often platforms would be re-prepared as knapping proceeded.

At the end of the knapping, the subject immediately went into the PET scanner and was immobile for the next 45 minutes of scanning (Figure 5).

PET RESULTS

The activation data collected during this experiment were analyzed using the *Statistical Parametric Mapping (SPM99)* software package developed by the Wellcome Department of Cognitive Neurology, Institute of Neurology, at the University College London. This software conducts statistical comparisons (t-tests) between the individual *voxels* (essentially three-dimensional pixels) in control and experimental data sets in order to generate an image showing significant differences. This process requires multiple scans in each condition in order to provide the data necessary for significance testing. However, in the pilot experiment presented here each condition is represented by only a single scan. Statistical analysis is thus impossible at this point.

In order to obtain images for use in preliminary evaluation and hypothesis generation in this pilot project, data from each condition were entered three times, as if representing three separate trials. The resulting images reveal differences in activation between experimental and control conditions, but do not indicate the statistical significance of these differences. Any interpretations must therefore be regarded as highly provisional in nature. Nevertheless, it is encouraging that the regions of greatest difference between knapping (both Acheulean and Oldowan) and control conditions observed in this pilot experiment very closely approximate regions of significant activation observed in a more systematic six-subject study of Oldowan-style knapping (Stout, this volume). More specifically, these regions comprise an arc extending from the cerebellum through the occipital and parietal lobes and into the posterior frontal lobe (Figures 6 & 7).

Figure 6

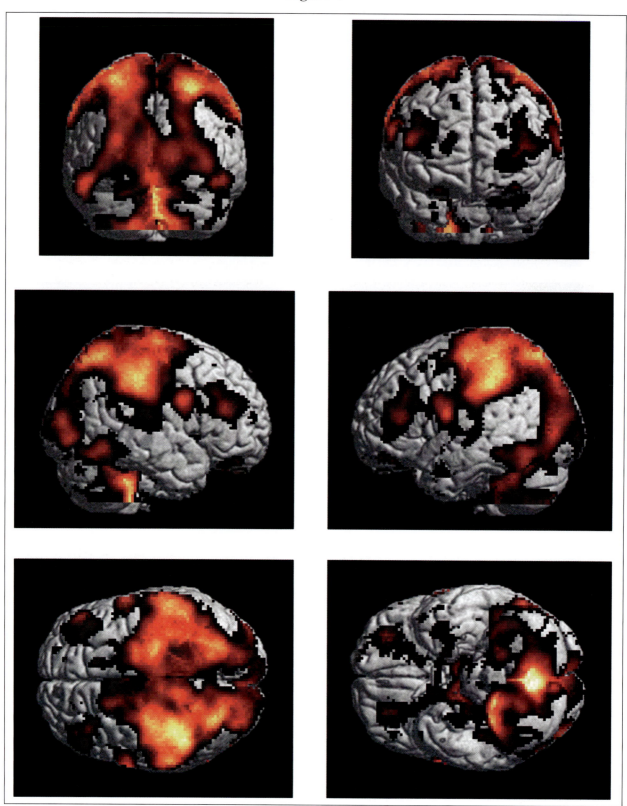

6. *Brain activation during Acheulean handaxe production: Six views (posterior, anterior, right hemisphere, left hemisphere, superior, and inferior) of brain activation during Acheulean-style handaxe production. Activation is extensive and bilateral, occurring in a broad arc from the cerebellum through the occipital and parietal lobes and into the posterior frontal. Regions involved are those commonly associated with visuomotor action and spatial cognition.*

Figure 7

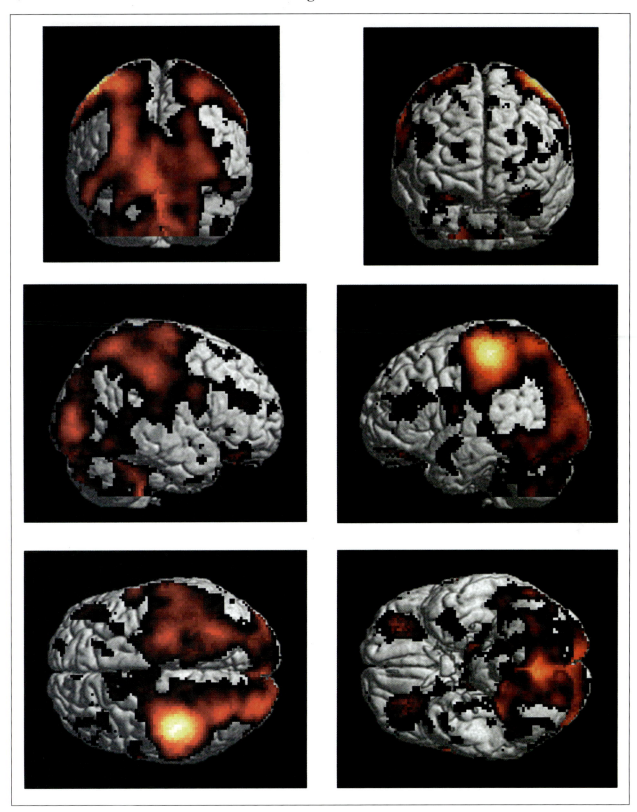

7. *Brain activation during Oldowan flake production: Six views (posterior, anterior, right hemisphere, left hemisphere, superior, and inferior) of brain activation during Oldowan-style flake production. Once again, the characteristic "stone knapping pattern" of activation in cerebellar, occipital, parietal and frontal regions is visible, however activation is less intense/extensive and more clearly lateralized when compared with Acheulean handaxe production (Figure 6). In particular, activation of the primary motor and somatosensory cortex surrounding the central sulcus appears to be much stronger in the left hemisphere (corresponding to the right hand) than in the right hemisphere (left hand).*

DISCUSSION

The appearance of this characteristic "stone knapping pattern" in images contrived from a single-trial pilot study strongly suggests that FDG PET will be an effective means for investigating the brain activation associated with Acheulean-style biface production and later prehistoric technologies. It also provides some suggestion that differences in the neural foundations of Mode I and Mode II knapping will be more on the order of variations on a theme, with some areas activated in Mode I knapping being more intensely activated in Mode II knapping, rather than of drastic differences in overall organization. Unraveling these differences, and their import, will be a relatively subtle matter of identifying quantitative differences in activation intensity and extent.

For example, images produced from this pilot study seem to show a much more bilateral pattern of activity in Acheulean-style biface production (Figure 6) as compared with Mode I knapping (figure 7). Each activity produces activation in both hemispheres, but activation of the primary somatosensory and motor areas of the right hemisphere (corresponding to the left arm) appears to be less robust during Mode I knapping. Statistical comparison in multi-subject studies will be necessary in order to determine if this is actually the case. If so, it might possibly reflect greater demands on the left or "postural" hand in carefully positioning the core during handaxe production, as compared with more unilateral right-hand dominated percussion during Oldowan-style knapping.

CONCLUSION

This pilot study confirms the feasibility of using FDG PET to investigate the neural foundations of Acheulean-style handaxe production and of comparing these with the substrates of Mode I flake production. The results of the pilot study do not support detailed analysis or interpretation at this point, but do suggest that differences in activation between Mode I and Mode II knapping will relate more to quantitative differences in intensity and extent than to qualitative differences in pattern. Future applications of the methods developed here will test this and other hypotheses and begin to clarify the psychological and evolutionary implications of the major technological changes that accompanied human evolution.

ACKNOWLEDGEMENTS

We would like to thank David Kareken for assistance in PET data analysis, Gary Hutchins and Rich Fain for help with experimental design and execution, and PET Technologists Kevin Perry and Susan Geiger. Funding for this research was provided by CRAFT and the Friends of CRAFT.

REFERENCES CITED

Asfaw, B., Beyene, Y., Suwa, G., Walter, R. C., White, T. D., WoldeGabriel, G. & Yemane, T. (1992). The earliest Acheulean from Konso-Gardula. *Nature* **360** pp. 732-735.

Bradley, B. & Sampson, C. G. (1986). Analysis by replication of two Acheulian artefact assemblages. In (G. N. Bailey and P. Callow, Eds) *Stone Age Prehistory.* Cambridge: Cambridge University Press.

Callahan, E. (1979). The basics of biface knapping in the Eastern Fluted Point Tradition: A manual for flintknappers and lithic analysts. *Archaeology of Eastern North America* **7**(1), pp. 1-172.

Clark, J. D. (2001). Variability in primary and secondary technologies of the Later Acheulian in Africa. In (S. Miliken and J. Cook, Eds) *A Very Remote Period Indeed: Papers on the Palaeolithic Presented to Derek Roe.* Oakville, CT: Oxbow Books, pp. 1-18.

Edwards, S. W. (2001). A modern knapper's assessment of the technical skills of the Late Acheulean biface workers at Kalambo Falls. In (J. D. Clark, Ed.) *Kalambo Falls Prehistoric Site, Volume III.* Cambridge: Cambridge University Press, pp. 605-611.

Gowlett, J. A. J. (1984). Mental abilities of early man: A look at some hard evidence. In (R. Foley, Ed.) *Hominid Evolution and Community Ecology.* New York: Academic Press, pp. 167-192.

Isaac, G. L. (1986). Foundation stones: early artefacts as indicators of activities and abilities. In (G. N. Bailey and P. Callow, Eds) *Stone Age Prehistory: Studies in Honor of Charles McBurney.* London: Cambridge University Press, pp. 221-241.

Isaac, G. Ll. & Curtis, G. H. (1974). Age of Early Acheulian industries from the Peninj Group, Tanzania. *Nature* **249** (June 14), pp. 624-627.

McPherron, S. P. (2000). Handaxes as a measure of the mental capabilities of early hominids. *Journal of Archaeological Science* **27** pp. 655-663.

Noble, W. & Davidson, I. (1996). *Human Evolution, Language and Mind.* Cambridge: Cambridge University Press.

Noll, M. P. (2000). *Components of Acheulean lithic assemblage variability at Olorgesailie, Kenya.* Ph. D. Dissertation: University of Illinois at Urbana-Champaign.

Schick, K. D. (1994). The Movius Line reconsidered: Perspectives on the Earlier Paleolithic of Eastern Asia. In (R. S. Corruccini and R. L. Ciochon, Eds) *Integrative Paths to the Past: Paleoanthropological Advances in Honor of F. Clark Howell.* Englewood Cliffs, NJ: Prentice Hall, pp. 569-596.

Stout, D. (2002). Skill and cognition in stone tool production: An ethnographic case study from Irian Jaya. *Current Anthropology* **45**(3), pp. 693-722.

Stout, D., Toth, N., Schick, K., Stout, J. & Hutchins, G. (2000). Stone tool-making and brain activation: Positron Emission Tomography (PET) Studies. *Journal of Archaeological Science* **27** pp. 1215-1223.

Toth, N. (2001). Experiments in quarrying large flake blanks at Kalambo Falls. In (J. D. Clark, Ed.) *Kalambo Falls Prehistoric Site, Volume III.* Cambridge: Cambridge University Press, pp. 600-604.

Wynn, T. (1979). The intelligence of later Acheulean hominids. *Man* **14** pp. 371-391.

Wynn, T. (1989). *The Evolution of Spatial Competence.* Illinois Studies in Anthropology 17. University of Illinois Press.

Wynn, T. & Tierson, F. (1990). Regional comparison of the shapes of later Acheulian handaxes. *American Anthropologist* **92** pp. 73-84.

CHAPTER 12

THE BIOMECHANICS OF THE ARM SWING IN OLDOWAN STONE FLAKING

BY JESÚS DAPENA, WILLIAM J. ANDERST, AND NICHOLAS P. TOTH

ABSTRACT

The biomechanics of the arm swing in Oldowan stone flaking was analyzed using three-dimensional motion analysis methodology. The analysis calculated the joint torques (and therefore the dominant muscular actions) at the three main joints of the swinging arm.

The flexor, external rotator and abductor muscles of the shoulder and the flexor muscles of the elbow were activated by the subject after each impact to complete the braking of the downward motion of the hammerstone, and subsequently to speed up its upward motion. The extensor, internal rotator and adductor muscles of the shoulder and the extensor muscles of the elbow were then activated to slow down the upward motion of the hammerstone, and subsequently to accelerate its downward motion toward the core for the next impact.

The hammerstone traveled through a downward distance of 0.45-0.48 m, and reached a final velocity of 18.6-20.1 mph, which implied a kinetic energy equivalent to that of a baseball thrown at 39-42 mph. Since the archeological evidence indicates that Oldowan hominins were able to flake basalt cobbles very efficiently, it is probable that they achieved speed and kinetic energy values similar to those found in this study.

INTRODUCTION

A cursory examination of the motion of the swinging arm in stone flaking suggests that it should be classified as an "overarm" motion. This is a class of motions that includes a wide variety of human activities, such as hammering, baseball pitching, javelin throwing, tennis serving, water polo throwing, and football quarterback passing.

Sometimes the muscular actions that drive human motions seem evident. However, fast dynamic movements can be deceptive. For instance, in baseball pitching the elbow of the throwing arm extends at a very fast rate, but the elbow extensor musculature is not very active (Feltner & Dapena, 1986). The extension of the elbow is produced mainly by inertia through a rather complex flail-like mechanism driven by the shoulder musculature. This implies that a baseball pitcher needs great strength in the shoulder musculature but only moderate strength in the elbow extensor musculature. To find out the muscular actions that drive a human motion, it is necessary to use kinetic chain analysis.

In kinetic chain analysis, precise measurements of the motions of body segments are combined with information on the inertial parameters of those body segments (such as masses and moments of inertia) to calculate the force and the torque exerted by a body segment on its immediate distal neighbor. This force and this torque are exerted through the joint that connects the two segments, and therefore they are called the joint force and the joint torque. The joint force is the sum of all the forces exerted by muscles, bones, ligaments and other structures. It is generally not a very informative parameter. The joint torque is the sum of all the torques exerted about the center of the joint. Since these torques are often exerted exclusively by muscles, the joint torque reflects the predominant muscular effort at the joint, and therefore it is a very informative parameter. The purpose of this project was to use kinetic chain analysis to determine the joint torques (and therefore the muscular actions) at the three main joints of the swinging arm during stone flaking.

METHODS

Calculation of Locations, Velocities and Accelerations

A skilled right-handed male (standing height = 1.75 m; mass = 75 kg) was filmed with two motion-picture cameras while he flaked lava cobbles. Both cameras were set at nominal frame rates of 200 frames/second. The cameras were placed to the right and in front of the subject, respectively.

Two typical trials (subsequently named Trial 1 and Trial 2) were selected for analysis. The film images were projected onto a digitizing tablet. The locations of 21 anatomical body landmarks (vertex, gonion, suprasternale, and right and left shoulders, elbows, wrists, knuckles, hips, knees, ankles, heels and toes) and of the approximate centers of the hammer and of the core were measured manually with the digitizing tablet in each film frame between an instant immediately after a stone impact and an instant immediately before the following stone impact. To minimize known problems in data acquisition in impact situations (see below), the trials were digitized only up to the last film frame prior to impact. The digitized locations taken from both cameras were stored in an Apple PowerBook G4 computer, which was also used for all subsequent calculations. (See Levanon & Dapena (1998) for further details on the methodology.)

The Direct Linear Transformation (DLT) method, developed by Abdel-Aziz & Karara (1971) and described in detail by Walton (1981), was used to compute the three-dimensional (3D) coordinates of the 23 landmarks from the digitized data. The 3D coordinates of the landmarks were expressed in terms of a right-handed orthogonal reference frame R1. The X1 and Y1 axes of R1 were horizontal, and perpendicular to each other; the Z1 axis was vertical, and pointed upward.

Coordinate data based on landmark locations obtained through manual digitization contain random errors that become magnified in the subsequent calculation of velocities and accelerations. To reduce this problem, the 3D location data were smoothed with quintic spline functions (Wood & Jennings, 1979) fitted to the time-dependent X_1, Y_1 and Z_1 coordinates of each landmark. An appropriate degree of smoothing requires a compromise between two conflicting goals: the reduction of high frequency noise resulting from errors inherent in manual digitization, and the preservation of the true (lower frequency) patterns of the activity. For the

stone flaking trials, the best compromise was reached with a smoothing factor of $N \cdot 10 \cdot 10^{-6}$ m^2 for each landmark and direction, where N was the number of frames in the trial. (The smoothing factor determines the sum of squares of the differences between the smoothed coordinates and the raw coordinates; larger smoothing factors produce a greater amount of smoothing.) The first and second derivatives of the quintic spline functions yielded smoothed landmark velocity and acceleration values, respectively, in the X_1, Y_1 and Z_1 directions.

Computation of Joint Torques

The swinging arm was modeled as a four-link kinetic chain composed of upper arm, forearm, hand, and hammerstone. The mass and the location of the center of mass (c.m.) of each arm segment in relation to its two endpoint landmarks were taken from Dempster's cadaver data (Dempster, 1955). Moment of inertia values were taken from Whitsett (1963), and were personalized for the subject using a procedure described by Dapena (1978). The mass of the hammerstone was determined with a scale (m = 0.625 kg); the moment of inertia of the hammerstone about its own c.m. was assumed to be zero.

The hammerstone was assumed to be subjected to two forces: weight, acting at the c.m. of the stone, and a proximal force exerted by the hand through the c.m. of the stone. The hand was assumed to be subjected to the force of its own weight, the reaction to the force exerted by the hand on the stone, and a proximal joint force and a proximal joint torque exerted by the forearm on the hand at the wrist. The forearm and upper arm segments were each assumed to be subjected to the force of their own weight, acting through the c.m., plus a joint force and a joint torque at both the proximal and distal joints.

The instantaneous c.m. location and local angular momentum of each segment about its own c.m. were computed following procedures described by Dapena (1978), modified to use instantaneous landmark velocities. The net force exerted on each segment was calculated from the mass of the segment and the second derivative of its c.m. location. The net torque about the segment c.m. was computed as the first derivative of its angular momentum about its own c.m. A procedure described by Andrews (1974, 1982) was then used to calculate the force and torque exerted by the proximal segment on the distal segment at the shoulder, elbow and wrist joints.

Reference Frames for the Expression of Joint Torques

To aid in the interpretation of the joint torques, non-inertial orthogonal reference frames were defined for the shoulder, elbow and wrist joints. (See Figure 1.) The origin of reference frame R_S was located at the shoulder joint. Axis S_1 was perpendicular to the plane formed by the longitudinal axes of the upper arm and forearm; S_2

Figure 1

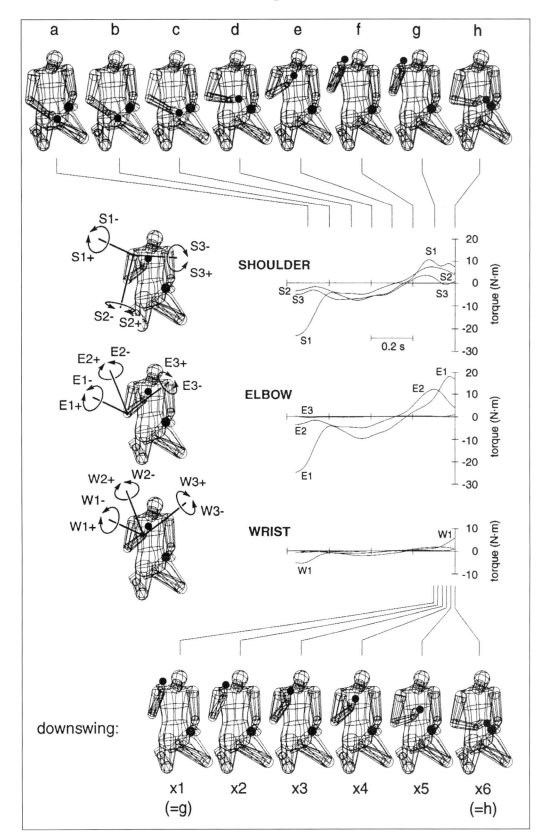

1. *Torques at the joints of the swinging arm in Trial 1. The computer graphics wireframe sequence at the top shows the entire stone-flaking cycle; the sequence at the bottom shows the downswing in greater detail. The wireframe drawings in the mid-left part of the Figure show the possible directions of the three torque components at each joint, and their signs. The plots show the values of the joint torques versus time. The wireframe sequences and patterns for trial 2 were similar to those of Trial 1.*

was aligned with the longitudinal axis of the upper arm; S_3 was perpendicular to S_1 and S_2. The origin of reference frame R_E was located at the elbow joint. Axis E_1 was perpendicular to the plane formed by the longitudinal axes of the upper arm and forearm (and was therefore parallel to S_1); E_2 was perpendicular to E_1 and to the longitudinal axis of the forearm; E_3 was aligned with the longitudinal axis of the forearm. The origin of reference frame RW was located at the wrist joint. Its three axes, W_1, W_2 and W_3, were parallel to the corresponding axes of the R_E reference frame.

Problems with Data Smoothing in Activities Involving Impacts

The impact of the hammer against the core produces a sudden deceleration of the right arm. If pre-impact and post-impact data are included in the input to a smoothing program, the program will not be able to distinguish between the true acceleration (deceleration) produced by the impact and spurious accelerations due to noise in the data. Consequently, the data will be over-smoothed, and the deceleration produced by the impact will seem to start before the impact itself. This will add false "braking" torques at the shoulder, elbow and wrist joints prior to impact. To prevent such systematic errors, no post-impact data were input to the smoothing program. The locations, velocities and accelerations corresponding to the instant of impact were estimated by extrapolation from the quintic spline coefficients of the last time interval prior to impact.

RESULTS AND DISCUSSION

Stone Movements

The downward distance of travel of the hammerstone from its highest position to impact was 0.45-0.48 m in the two trials. The maximum speed of the hammerstone was 8.3-9.0 m/s (18.6-20.1 mph), and it occurred immediately before impact. This implied a kinetic energy of 21.5-25.3 Joules (J), equal to the kinetic energy of a baseball thrown at 39-42 mph. (For comparison, the kinetic energy of a baseball thrown by a major league pitcher at 90 mph is 117 J.) In both trials, the subject brought the core upward to meet the hammerstone at 0.6-1.3 m/s (1.3-2.9 mph). Although the motion of the core was not exactly opposite to the motion of the hammerstone, it contributed to increase the combined impact speed to 8.8-10.1 m/s (19.7-22.6 mph). Since the archeological evidence indicates that Oldowan hominins were able to flake basalt cobbles very efficiently, it is probable that they achieved speed and kinetic energy values similar to those found in this study.

Sequences

The wireframe sequence at the top of Figure 1 (images a-h) shows the complete cycle of the stone flaking action in Trial 1, from an instant shortly after the impact of the previous swing until the impact of Trial 1; the sequence at the bottom of the Figure (images x_1-x_6) shows the downswing in greater detail. The sequences of Trial 2 were similar to those of Trial 1. The computer graphics sequences, as well as computer animations, showed that the motion of the swinging arm was not a simple planar flexion and extension, but a clear three-dimensional overarm motion.

Torques

The plots in the center-right section of Figure 1 show the torques at the three main joints of the right arm (shoulder, elbow and wrist). Joint torques give an indication of the muscular activity of the subject, and they occur in three different directions at each joint, as indicated by the images to the left of the plots. For instance, the negative E_1 torque in the early part of the cycle indicates that the elbow flexor muscles (such as the biceps) were dominant at that time, i.e., that the torque produced by the elbow flexor muscles about the center of the elbow joint was larger than any torque produced by the elbow extensor muscles (triceps); the positive E_1 torque in the late part of the cycle indicates that the elbow extensor muscles were dominant at that time. Two torques (E_2 at the elbow and W_3 at the wrist) are exceptions in that they are not produced by muscles; instead, they are passive torques exerted by the proximal segment on the distal segment through bony and ligamentous structures of the joint.

The torque plots indicate that in the early stages of the cycle (a-e) the shoulder muscles were active in the directions of flexion, external rotation and abduction (negative S_1, S_2 and S_3 torques, respectively), while the elbow muscles were active in the direction of flexion (negative E_1 torque). These muscular actions served to stop the downward motion of the arm after the impact of the previous cycle (images a-b), and later to lift the upper arm and rotate it outward, and to flex the elbow (images b-e). About 0.25 seconds prior to impact, the torques reversed direction: The shoulder muscles became active in the directions of extension, internal rotation and adduction (positive S_1, S_2 and S_3 torques, respectively), and the elbow muscles became active in the direction of extension (positive E_1 torque). These new muscular actions first stopped the upward and outward motion of the upper arm and the flexion of the elbow (images e-g), and then produced downward and inward rotation of the upper arm, and extension of the elbow (images g-h and x_1-x_6). The muscles that cross the wrist joint did not play a major role in the generation of the arm swing. In contrast with baseball pitching, the torques of the flaking action were in good agreement with what might have been expected prior to the analysis: The muscles that produce upward motion of the hammerstone were activated after the impact to complete the braking of the downward motion of the hammerstone, and subsequently to speed up its upward motion; the muscles that produce downward motion of the hammerstone were activated to slow down the upward motion of the hammerstone, and subsequently to accelerate its downward motion toward the core.

In comparison to the joint torques used in sports activities, the joint torques exerted by the shoulder and elbow muscles in stone flaking were relatively small. In a typical collegiate level baseball pitch, torques S_1, S_2 and S_3 all reach maximum values of around 70 N · m (Feltner & Dapena, 1986), while the maximum shoulder torque values during the last 0.20 seconds prior to impact in the two flaking trials were: $S_1 = 11\text{-}14$ N · m;

$S_2 = 7\text{-}9$ N · m; $S_3 = 3\text{-}4$ N · m. The maximum elbow extension torque in the two flaking trials was $E_1 = 18\text{-}21$ N · m. This was not much smaller than the corresponding values in a typical baseball pitch (20-30 N · m), but it is important to bear in mind that the elbow extension torque is small in baseball pitching. For comparison purposes, it is useful to consider the fact that 20 N · m would be the approximate joint torque that the elbow extensor muscles would need to exert in order to hold up in the air a 5 kg load with the forearm in a horizontal position (Figure 2). This would be quite easy for most people. It is not surprising that stone flaking does not require a large amount of strength, since in prehistoric times this activity probably needed to be accessible to a large number of individuals.

Figure 2

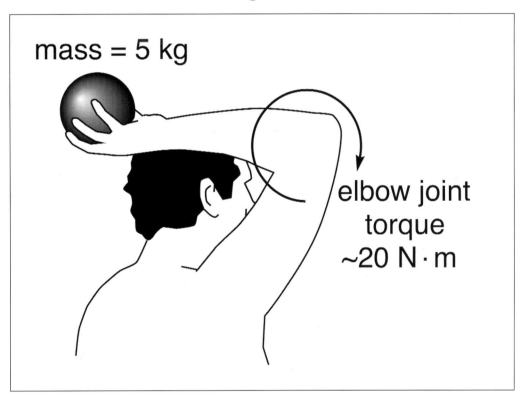

2.   *Elbow joint torque necessary to hold a 5 kg mass up in the air.*

REFERENCES CITED

Abdel-Aziz, Y.I. and H.M. Karara. Direct linear transformation from comparator coordinates into object space coordinates in close-range photogrammetry. In: *Proceedings of ASP/UI Symposium on Close Range Photogrammetry*, Falls Church, VA: American Society of Photogrammetry, pp. 1-18, 1971.

Andrews, J.G. Biomechanical aspects of human motion. In: *Kinesiology IV*, Washington, D.C.: American Association for Health, Physical Education and Recreation, pp. 32-42, 1974.

Andrews, J.G. On the relationship between resultant joint torques and muscular activity. *Med. Sci. Sports Exerc.* 14:361-367, 1982.

Dapena, J. A method to determine the angular momentum of a human body about three orthogonal axes passing through its center of gravity. *J. Biomechanics* 11:251-256, 1978.

Dempster, W.T. Space requirements of the seated operator. *WADC Technical Report 55-159*, Dayton, OH: Wright-Patterson Air Force Base, 1955.

Feltner, M.E. and J. Dapena. Dynamics of the shoulder and elbow joints of the throwing arm during a baseball pitch. *Int. J. Sport Biomechanics* 2:235-259, 1986.

Levanon, J. and J. Dapena. Comparison of the kinematics of the full-instep and pass kicks in soccer. *Med. Sci. Sports Exerc.* 30:917-927, 1998.

Walton, J.S. Close range cine photogrammetry: a generalized technique for quantifying human motion. Ph.D. Dissertation, Pennsylvania State University, University Park, PA, 1981.

Whitsett, C.E. Some dynamic response characteristics of weightless man. *AMRL Technical Report 18-63*, Dayton, OH: Wright-Patterson Air Force Base, 1963.

Wood, G.A. and L.S. Jennings. On the use of spline functions for data smoothing. *J. Biomechanics* 12:477-479, 1979